NORTH WEST WETLANDS SURVEY **6**

THE LOWLAND WETLANDS OF

Cumbria

D Hodgkinson
E Huckerby
R Middleton
C E Wells

With contributions by

P Cundill
P Hughes
I Miller
R M Newman
M Ryder
S Stallibrass
T Wilmott

LANCASTER IMPRINTS

2000

Published by
Lancaster University Archaeological Unit
Storey Institute
Meeting House Lane
Lancaster
LA1 1TF
(*Phone:* 01524 848666; *Fax:* 01524 848606)
(*E-Mail:* archunit@lancaster.ac.uk)
(*Internet* —www. http://www.lancaster.ac.uk/)
Further details of the North West Wetlands Survey can be obtained from the LUAU Home Page at the above address

Distributed by
Oxbow Books
Park End Place
Oxford
OX1 1HN
(*Phone:* 01865-241249; *Fax:* 01865-794449)

Printed by
Kent Valley Colour Printers, Kendal, Cumbria

ISBN 1 86220 082 3
ISSN 1343-5205

NWWS Series editor
Rachel Newman
Copy editor
Martin Lister
Indexer
Peter Gunn
Design
Robert Middleton
Layout and formatting
Andrea Scott

Front Cover: Flooded peat cuttings on Scaleby Moss

LANCASTER IMPRINTS

Lancaster Imprints is the publication series of Lancaster University Archaeological Unit. The series covers work on major excavations and surveys of all periods undertaken by the Unit and associated organisations.

Contents

List of illustrations

Figures

A key for figures 33–6 is located on a flap on the inside back cover

Tables

Plates

Abbreviations

AMS	Accelerator Mass Spectrometry
BAP	Biodiversity Action Plan
CLAU	Cumbria and Lancashire Archaeological Unit (now LUAU)
EH	English Heritage
GIS	Geographic Information System
LUAU	Lancaster University Archaeological Unit
MAFF	Ministry of Agriculture, Fisheries, and Food
MBARS	The Morecambe Bay Archaeological Research Society
NNR	National Nature Reserve
NWWS	North West Wetlands Survey
OD	Ordnance Datum
OS	Ordnance Survey
PPG16	Planning Policy Guidance 16: Archaeology and Planning
PRO	Public Record Office, London
RCHME	Royal Commission for Historic Monuments (England) (now part of English Heritage)
SAC	Special Areas of Conservation
SM	Scheduled Monument Number
SMR	Cumbria Sites and Monuments Record
SPA	Special Protection Areas
SSSI	Site of Special Scientific Interest
SURRC	Scottish Universities' Research andReactor Centre at East Kilbride

Contributors

Peter Cundill
School of Geography and Geosciences, Purdie Building, North Haugh, St Andrews, Fife, KY16 9ST

David Hodgkinson
Wardell Armstrong, 2 Hassell Street, Newcastle under Lyme, Staffordshire, ST5 1AG

Elizabeth Huckerby
NWWS, Lancaster University Archaeological Unit, Storey Institute, Meeting House Lane, Lancaster, LA1 1TF

Paul Hughes
Quaternary Environmental Change Research Group, Dept of Geography, University of Southampton, SO17 1BJ

Robert Middleton
Farming and Rural Conservation Agency, 5 Hanover Court, Manaton Close, Matford Business Park, Exeter, EX2 8QJ

Ian Miller
Lancaster University Archaeological Unit, Storey Institute, Meeting House Lane, Lancaster, LA1 1TF

Rachel Newman
NWWS, Lancaster University Archaeological Unit, Storey Institute, Meeting House Lane, Lancaster, LA1 1TF

Michael Ryder
4 Osprey Close, Southampton, SO16 8EX

Sue Stallibrass
School of Archaeology, Classics and Oriental Studies, William Hartley Building, University of Liverpool, Brownlow Street, Liverpool, L69 3BX

Colin Wells
Scottish Natural Heritage, 2 Anderson Place, Edinburgh, EH6 5NP

Tony Wilmott
Centre for Archaeology, English Heritage, Fort Cumberland, Fort Cumberland Road, Eastney, Portsmouth, PO4 9LD

Frontispiece: peatcutting tools from the Lyth Valley

Acknowledgements

The archaeological and palaeoecological survey has involved a large number of people, taking place as it did over the entire life of the NWWS project. In particular, we would like to thank the following for assisting with fieldwork: Ken Browell, Imogen Grundon, Phil Mills, Tove Oliver, Ruth Parkin, Mike Peace, Richard Short, Dr Karen Stanley, and Dr Jon Westaway. We would also like to thank Ruth Parkin, Andrea Scott and Mark Tidmarsh for the illustrations in this volume. The maps have been output via AutoCAD 14; the raw data from which they were taken have been manipulated in ArcInfo; this work was initiated by Sid-Ali Cherchali, with processing being undertaken by Andrea Scott and Richard Short.

The authors would like to express their gratitude to those landowners, tenants, and managers who have allowed access to land during the course of the survey, in particular: Mrs Metcalf of Crook Hall Farm; Chris Palmer of Skelsmergh; the staff of L & P Peat Limited at Solway Moss; Mr Alex Laidlaw (Boothby and Penicuik Peat Company Limited, Bolton Fell Moss); Mr Jonathan Cox (Fisons Horticulture Ltd) and Mr Donald Manley (Levingtons) at Wedholme Flow; and English Nature for access to SSSIs.

During the excavations at Bardsea in 1992, Ruth Parkin ably assisted with the fieldwork. Members of Morecambe Bay Archaeological Research Society — Ian McNicol, Stella Clarkson, Cath Baxter, Tammy Murray, and Stan Wieclawska, led by Doug Stables — gave generously of their time. South Lakeland District Council and English Nature both permitted the work to be undertaken.

Ian Richardson, Ian Maxwell, and the staff of the L & P Peat Limited, Solway Moss, alerted us to the animal remains in 1991 and provided substantial help during both the routine survey work and the excavations. The radiocarbon dating was facilitated by David Jordan and Alex Bayliss of English Heritage, zoological analysis was by Sue Stallibrass (then of the University of Durham), and the remains were conserved by Vanessa Fell, University of Manchester. The principal site excavators were Ruth Parkin and Deirdre Winstanley. Anne Ross is to be thanked for providing much valuable information on the context of such a find.

We are extremely grateful to the contributors to the volume, several of whom provided unpublished information. We would particularly like to thank the following: the staff of all the branches of the Cumbria Record Office (Barrow, Carlisle, Kendal and Whitehaven), particularly Richard Hall at Kendal, and Susan Dench at Carlisle; English Nature, especially Frank Mawby (SSSIs); the Countryside Agency (Access and Countryside Character); the Environment Agency (Water Level Management Plans); Doug Stables (Morecambe Bay Archaeological Research Society); Caroline Earwood (Ehenside Tarn); Paul Hughes (Solway Mosses); Jacqui Huntley (Hadrian's Wall); Roger Leech (Meathop Moss); and Margaret Cox (Glasson Moss). In addition, Geoffrey Halliday kindly allowed us to manipulate his map of the solid geology of Cumbria and also freely allowed information from *The Flora of Cumbria* (1997) to be used. John Todd of St Bees was unendingly helpful during fieldwork in the area and has offered support during the compilation of the volume, as did Angus Winchester of Lancaster University. We also thank Gordon Cook of SURRC for facilitating the bulk of our radiocarbon dating. Bette Hopkins of Cumbria County Council (CCC) deserves special thanks for her unending patience and advice during the compilation of the SMR data. Jon Robinson (CCC) and John Clayson of the Lake District National Park Authority are also thanked for their help in transferring the data digitally. Guy Wimble's unpublished PhD thesis was consulted extensively during compilation of this volume, although he was not prepared to see the results published in full at this stage. In addition, Caroline Skinner and Jim Innes are to be thanked for providing much information on recent research on a wider area than could be covered by the survey and Richard Newman gave freely of his knowledge of the South Cumbrian iron industry.

The following people have greatly assisted the authors by supplying photographs and information: Bob Bewley of English Heritage (aerial photographs from Solway Plain), Margaret Cox of Southampton University (Glasson Moss), Chris Cox (Glasson Moss), Tim Padley of Tullie House, Carlisle Museum and Art Gallery (Salta Moss rapier), Richard Hall of Cumbria Record Office (Kendal) (maps of the Lyth Valley and many queries in connection with the documentary research) and Susan Dench of Cumbria Record Office (Carlisle) (Black Moss Enclosure map and early maps of the Solway Plain). In particular, Anne-Marie Colmann, Bill Miller and Cordelia Ehrhardt are thanked for the speed and erudition with which they produced the foreign language summaries.

Fisons Horticulture Ltd (now Levingtons) kindly provided the project with the long-term use of a Land Rover.

The North West Wetlands Survey was wholly funded by English Heritage and has been overseen by an Advisory Panel, chaired by Prof Bryony Coles, and the Committee for the North West Wetlands. For English Heritage, the project was overseen by Gerry Friell and monitored by Tony Wilmott. The Ancient Monuments Laboratory was represented at various times by Alex Bayliss, Wendy Carruthers, Dominique des Moulins, David Jordan, and David Weir. Rachel Newman acted as Project Manager for Lancaster University Archaeological Unit.

Finally, we would like to thank the external readers, Dr Peter Moore and Dr Charly French, for their patience and enormous help and encouragement over the years, and also all the English Heritage staff who commented on the text, in particular, Tony Wilmott, Peter Marshall, and James Wells.

Foreword

Cumbria is to many people, and certainly to most visitors, a land of lakes and mountains, synonymous with 'The Lake District'. There is, however, a large part of the county beyond the central mountain massif: in the east, the Lune and Eden valleys form a corridor flanked by the Pennines, separating Cumbria from Yorkshire, County Durham, and Northumberland; in the south and west, narrow coastal plains divide the mountains from Morecambe Bay and the Irish Sea; and in the north, a wider plain looks towards the Solway and Scotland. It is a county with an enormous wetland resource, with numerous tarns, both in the central massif and the coastal plains, lake sediments, large raised mires in the north, and many kettle holes in the south.

The remit of the North West Wetlands Survey has been to examine the lowland wetlands within the seven counties of Shropshire, Staffordshire, Cheshire, Merseyside, Greater Manchester, Lancashire and Cumbria. In the other counties, a spatial division between upland and lowland clearly defined the areas of lowland wetland, but Cumbria is unique in the diversity of its topography, with elements of lowland contained within the central massif. The main emphasis of this volume, however, has been placed on the coastal lowlands, in part because these can be characterised as finite land units containing major areas of wetland, but also to complement the weight of research, particularly palaeoecological, although also archaeological, already undertaken on the upland areas of the central massif.

The North West Wetlands Survey has provided an overview of wetland development from the last Ice Age to the large-scale reclamations of the eighteenth and nineteenth centuries, and it has assessed the condition of the extant wetland resource, and the threats to it. Such a task could only be achieved by the integration of archaeological and palaeoecological techniques to create an holistic approach, and by the collaboration of many bodies, both national and local. It is evident from this volume that such an approach can produce important results; in particular, the recovery of the remains of early medieval cattle from a commercial peat extraction site on Solway Moss demonstrates the value of liaison and understanding between all those involved with wetlands. The way forward is therefore clear: there must be dialogue and genuine understanding between those individuals and bodies involved with the landscape in its widest sense, to take into account all interests, including those of local communities and visitors, and reconcile these in objective-led and sustainable conservation management strategies. In all this, however, the human dimension of the wetland landscape, and the part man has played in its development, must not be forgotten. It is the largely hidden archaeological and palaeoecological potential of the wetlands that is vital in providing this historical perspective and it is hoped that this volume makes a valuable contribution to the ongoing debate by assessing this potential for Cumbria.

Rachel Newman
Deputy Director
Lancaster University Archaeological Unit

Summary

This volume is the sixth of seven County-based books resulting from the North West Wetlands Survey, a long-term project design to identify the archaeological and palaeoecological resource of the lowland wetlands of the North West of England. The title is explicit in excluding the upland wetlands which are very extensive in the County, and which include the Lake District. The survey focused on lowland wetlands on the coastal plain in the south, west and north of the County, with rapid palaeoecological assessments of small basin mires on the upland fringe, to supplement the large body of extant literature.

The three largest areas of wetland are in south Cumbria, on the fringes of Morecambe Bay, the West Cumbrian coast, and the area on the south coast of the Solway on the North Cumbrian Plain. Many wetlands survive well, though most are in decline due to a range of factors such as commercial peat-cutting and forestry. The predominantly pastoral land-use has meant that archaeological remains are difficult to detect from either surface or aerial survey, and this may mean that ancient sites and landscapes remain well preserved. Most of the evidence for the use of the wetlands throughout the study area derives from the excellent palaeoecological archive, which records climate change, clearance episodes and marine transgressions.

The earliest human use of the mosslands in all areas was probably during the late Palaeolithic period, when there is artefactual evidence from deposits in caves. Mesolithic activity is both sealed by barrows in the south and has been identified on the raised beaches of former coasts in the west, where it was possible to take advantage of the biodiversity of these environments. A concentration of sites exist in the Esk estuary, and the archaeological potential for the Mesolithic is also shown by sites revealed by survey on the northern side of Solway.

In West Cumbria, the type site for the early Neolithic is the famous habitation site of Ehenside Tarn, but occupation of this date appears more widespread than during the Mesolithic. Later Neolithic use may be reflected by the deliberate deposition of polished stone axes in mossland sites, and there are visible ritual monuments in the west. Some Beaker-period burials have been identified in the south, but the distribution of Bronze Age material broadly results from chance finds. Palaeoecological evidence suggests more intensive land-use, coupled to an increased material assemblage as the late Bronze Age is reached, with settlement expanding onto marginal land, where numerous sites survive. Unlike earlier periods there is no real evidence for the preferential use of wetlands at this time, though ritual deposition of metalwork seems to have occurred.

The late Bronze Age clearances were clearly not maintained though the early Iron Age, as vegetation regenerated, but a sustained and extensive clearance for cereal production and grazing took place in the late Iron Age. Clearance in the Hadrian's Wall area had clearly taken place before the Wall was built, and was not predominantly a product of its construction. Despite the palaeoecological evidence for this phenomenon, there is little associated archaeological material attesting Iron Age and Romano-British settlement. Settlement sites, detected on the North Cumbrian Plain through aerial photography, may in many cases be of this period, though one example at Plasketlands on excavation proved to be Neolithic in date; such sites may represent a long period of landscape development. It seems that settlement had moved away from wetlands in the late prehistoric period though their continued use for ritual is shown by the bog bodies from Seascale and Scaleby Mosses.

For the early medieval period there is evidence of ritual activity in the shape of the Solway cattle, and there is also some evidence for hemp retting on Glasson Moss. Later in the middle ages, large-scale management of mossland exploitation was undertaken by the great monastic houses, such as Furness in the south and Holm Cultram in the north. The survey undertook substantial documentary work to trace the medieval and post-medieval uses of the mosslands. The principal means of exploitation was demonstrated to be the cutting of peat for both domestic and industrial fuel supply in the absence of timber, though this was carefully regulated. Enclosure and reclamation has only taken place on any scale since the eighteenth century.

It is clear that the mosslands retain a huge archaeological potential, particularly in the palaeoecological archive contained. There are many threats to the survival of the archaeological archive of the wetlands, including commercial peat extraction, drainage, forestry and, ironically, wetland restoration schemes. These can, however, be minimised by sensitive management.

Résumé

Ce volume est le sixième de sept livres régionaux qui résultent de l'étude menée dans les Wetlands (tourbières) du nord-ouest, un project à long terme qui devait identifier les ressources archéologiques et paléo-écologiques des tourbières de la plaine du nord-ouest de l'Angleterre. Le titre exclut explicitement les tourbières de la région montagneuse qui sont nombreux dans la région et dont le Lake District fait partie. L'étude s'est concentrée sur les tourbières de la plaine côtière dans le sud, l'ouest et le nord de la région, avec de rapides estimations paléo-ecologiques de petits bassins de boue en bordure de la région montagneuse, pour ajouter un supplément à la volumineuse litérature qui existe déjà.

Les trois plus grandes tourbières sont en Cumbria du sud, à la limite de la Baie de Morecambe, la côte ouest de Cumbria et la région sur la rive sud du Solway dans la plaine du nord de Cumbria. Quelques tourbières survivent bien, bien que la plupart d'entre eux sont en déclin à cause de différents facteurs tels que la coupe commerciale de tourbe et la sylviculture. Parce que les terres sont principalement utilisées en tant que pâturages, les vestiges archéologiques sont difficiles à détecter à partir d'études au sol ou d'études aériennes et ceci pourrait supposer que les sites et les paysages anciens ont été bien préservés. La plupart de l'évidence de l'utilisation des tourbières dans la région étudiée dérive de l'excellente archive paléo-écologique où les changements de climat, les périodes plus clémentes et les transgressions marines sont bien enregistrés.

La première utilisation par l'homme des tourbières dans toutes les régions remonte probablement à la fin de la période Paléolithique, pour laquelle des objects faconnés ont été trouvés dans les caves. L'activité Mésolithique est à la fois enfouie sous les tumulus dans le sud et a été identifiée sur les anciennes plages des côtes d'autrefois dans l'ouest, où il était possible de profiter de la diversité biologique de ces environements. Il existe une concentration de sites dans l'estuaire d'Esk et le potentiel archéologique pour la période Mésolithique est aussi démontré par des sites révélés par des études faites sur la rive nord du Solway.

En Cumbria de l'ouest, le site type pour le début de la période Néolithique est le fameux site d'habitation d'EhensideTarn, mais l'occupation durant cette période semble être plus importante que pendant la période Mésolithique. La présence d'axes en pierre polie, déposés délibérément dans les tourbières indique l'utilisation ultérieure de ces habitations en période Néolithique, ainsi que l'existence de monuments pour la pratique de rituels qui sont reconnaissables dans l'ouest. Certaines tombes qui datent de la période Beaker ont été identifiées dans le sud, mais la distribution de matériel qui date de l'Âge de Bronze résulte dans la plupart des cas de trouvaille chanceuse. L'évidence paléo-écologique suggère une utilisation plus intensive de la terre, liée avec un assemblage de matériel en progression vers la fin de l'Âge de Bronze, avec une colonisation qui s'étend jusqu'aux terres marginales où de nombreux sites ont survécus. A l'opposition des périodes antérieures, il n'y a aucune évidence réelle pour l'utilisation préférentielle des tourbières pendant cette période, bien que des dépôts d'objects en métal semblent avoir eu lieu pendant les rituels.

Les zones défrichées à la fin de l'Âge de Bronze n'étaient de toute évidence pas maintenues pendant le début de l'Âge de Fer, où la végétation s'est regénérée, mais un défrichement extensif et soutenu pour la production de céréales et les pâturages a eu lieu à la fin de l'Âge de Fer. Le défrichement du site du mur de Hadrian a de toute évidence eu lieu avant que le mur ait été construit, et n'était pas priorotairement dû à sa construction. Malgré l'évidence paléo-écologique pour ce phénomène, il y a peu de matériel archéologique associé qui pourrait attester la colonisation pendant l'Âge de Fer et la colonisation romano-britannique. Des sites colonisés, détectés dans la plaine dans le nord de Cumbria à l'aide de nombreuses photographies aériennes, pourraient dans beaucoup de cas être de cette période, bien qu'un exemple à Plasketlands a prouvé être de période Néolithique après des fouilles; de tels sites pourraient représenter une longue période de dévelopement des paysages. Il semblerait que la colonisation se soit éloignée des tourbières à la fin de la période préhistorique bien que leur utilisation ininterrompue pour les rituels est démontrée par les corps humains trouvés dans les tourbières de Seascale et Scaleby.

Pour le début de la période médiévale, il y a évidence d'activités rituelles sous forme d'os de bétail de Solway, et il y a également évidence de la fabrication de chanvre dans la tourbière de Glasson. Plus tard, au Moyen-Âge, les tourbières ont été exploitées à grande échelle par les grandes maisons monacales, comme les Furness dans le sud et les Holm Cultram dans le nord. L'étude a été apronfondie par de sérieuses recherches documentaires dans le but de tracer toute utilisation médiévale et post-médiévale des tourbières. Il a été démontré que l'exploitation principale des tourbières était la coupe de tourbe pour l'approvisionnement domestique et industriel d'énergie en l'absence de bois, bien que ceci fut attentivement contrôlé. Le clôturage et le défrichement n'ont eu lieu qu'à partir du dix-huitième siècle.

Il est clair que les tourbières conservent un potentiel archéologique énorme, particulièrement au niveau paléo-écologique. La survie de l'archive archéologique dans les tourbières est menacée par de nombreux facteurs, comme l'extraction industrielle de tourbe, le drainage, la sylviculture et, ironiquement, les projects de restauration des ces zones. Ces menaces peuvent être néanmoins minimisées par une gestion sensibilisée.

Traduction: *Anne-Marie Colmann et Bill Miller*

Zusammenfassung

Dieser Band ist das siebente von sechs auf Grafschaft basierenden Büchern der North West Wetlands Survey, einem Langzeitprojekt zur Identifikation der archäologischen und paläo-ökologischen Ressourcen des Moorgebietes im Tiefland von Nordwestengland. Die Überschrift schliesst explizit das Hochland des Grafschaft aus, das einen Grossteil des Gebietes umfasst und zu dem auch der Lake District gehört. Die Untersuchung konzentriert sich auf die Moorgebiete des Tieflandes der Küstenebene im Süden, Westen und Norden des Grafschaft. Eine kurze paläo-ökologische Beurteilung der kleinen Morastbecken am Rande des Hochlandes ergänzt die Fülle der vorhandenen Literatur.

Die drei grössten Moorgebiete befinden sich am Rande der Bucht von Morecambe im Süden Kumbriens, an der Westküste Kumbriens und im Gebiet der Südküste von Solvay in der Nordebene Kumbriens. Die meisten Moorgebiete sind gut erhalten, gehen aber zurück durch Faktoren, wie kommerzielles Torfstechen und Forstwirtschaft. Die vorherrschende Landnutzung hat zur Folge, dass archäologische Überreste auf Oberflächen- oder Luftaufnahmen schwer zu erkennen sind. Der beste Beweis für die Nutzung des Moorlandes im Studienzeitraums stammt aus dem exzellenten paläo-ökologischen Archiv, dass Klimaveränderungen, Zeiten der Rodung und Überflutungen registriert hat.

Die erste menschliche Nutzung des Moorlandes in allen Gebieten geht wahrscheinlich zurück auf das späte Paläolithikum. Aus dieser Zeit stammen menschliche Zeugnisse in Höhlen nachgewiesen werden. Aktivitäten aus dem Mesolithikum sind in den Hügelgräbern im Süden verborgen und konnten an den heute erhobenen Stränden der früheren Küsten im Westen identifiziert werden, wo die Möglichkeit zur Nutzung der Artenvielfalt der Umgebung bestand. Eine Ansammlung von Besiedlungsstätten befindet sich in der Mündung des Esk. Das archäologische Potential des Mesolithikums beweisen auch Untersuchungen des Nordens von Solway.

Typisch für das frühe Neolithikum im Westen Kumbriens ist die berühmte Wohnstätte Ehenside Tarn. Allerdings scheint die Bewohnung in dieser Zeit ausgedehnter gewesen zu sein als im Mesolithikum. Eine Nutzung im späteren Neolithikum zeigt sich in der absichtlichen Plazierung polierter Steinäxte in den Stätten es Moorlandes und den sichtbaren Ritualmonumenten im Westen. Im Süden konnten einige Grabstätten der Beaker-Periode identifiziert werden. Aber die Verteilung der Bronzezeitfunde beruht hauptsächlich auf dem Zufall.

Die paläo-ökologischen Zeugnisse lassen auf eine intensivere Landnutzung schliessen, verbunden mit einer erhöhten Ansammlung von Funden mit Erreichen der späten Bronzezeit. Die Besiedlung dehnt sich bis auf die Randgebiete aus, in denen sich zahlreiche Stätten erhalten haben. Es gibt keinen richtigen Beweis für die bevorzugte Nutzung des Moorlandes zu dieser Zeit, wie es in den früheren Perioden der Fall war, aber offensichtlich fanden rituelle Hinterlegungen von bearbeitetem Metall statt.

Die Rodungen der Bronzezeit wurden offensichtlich während der frühen Eisenzeit nicht aufrechterhalten, aber nachhaltige und weite Rodungen für den Anbau von Getreide und als Weideland fanden in der späten Eisenzeit statt. Die Rodung im Gebiet des Hadrian Walls fanden erkennbar vor dem Bau des Walls statt und waren keine Folge des Baus. Trotz des paläo-ökologischen Beweises für dieses Phänomen existieren wenig zusätzliche archäologische Funde, die die Besiedlung in der Eisenzeit und der Römisch-Britischen Zeit bezeugen.

Siedlungsstätten, die auf den Luftaufnahmen der Nordkumbrischen Ebene entdeckt wurden, stammen in den meisten Fällen vermutlich aus dieser Zeit. Das Beispiel Plasketlands stellte sich jedoch während der Ausgrabung als neolithisch heraus. Solche Stätten können einen langen Zeitraum landschaftlicher Entwicklung verkörpern. Es scheint, dass sich die Besiedlung in der späten prähistorischen Periode vom Moor weg bewegt hat, aber die Moorleichen der Moore von Seascale und Scaleby beweisen die weiterführende Nutzung für rituelle Zwecke. Für rituelle Aktivitäten im frühen Mittelalter gibt es Beweise durch Funde von Knochen des Solway Rindes. Ausserdem konnte verrotteter Hanf im Moor von Glasson nachgewiesen werden. Im späteren Mittelalter wurde das Moor im grossen Stil von den grossen Herrschaftshäusern, wie Furness im Süden und Holm Cultram im Norden ausgebeutet. Dieser Untersuchung liegen intensive Nachforschungen in den Dokumentationen zugrunde, um die mittelalterliche und nach-mittelalterliche Nutzung des Moores zurückzuverfolgen. Die grundsätzliche Nutzung war das Torfstechen für die häusliche und industrielle Brennstoffgewinnung, da kein Nutzholz zur Verfügung stand. Allerdings wurde das Torfstechen sorgfältig reguliert. Einzäumungen des Gebietes und Rekultivierung in grossem Massstab gab es erst seit dem 18. Jahrhundert.

Es steht fest, dass das Moorland grosses archäologisches Potential hat und das besonders durch sein paläo-ökologischen Archiv. Es gibt viele Bedrohungen für die Erhaltung dieses paläo-ökologischen Archivs. Zu diesen gehören kommerzielles Torfstechen, Drainage, Forstwirtschaft und, ironischerweise, Rekultivierungspläne für das Moorgebiet. Diese können jedoch durch sensibles Management minimiert werden.

***Übersetzung**: Cordelia Ehrhardt*

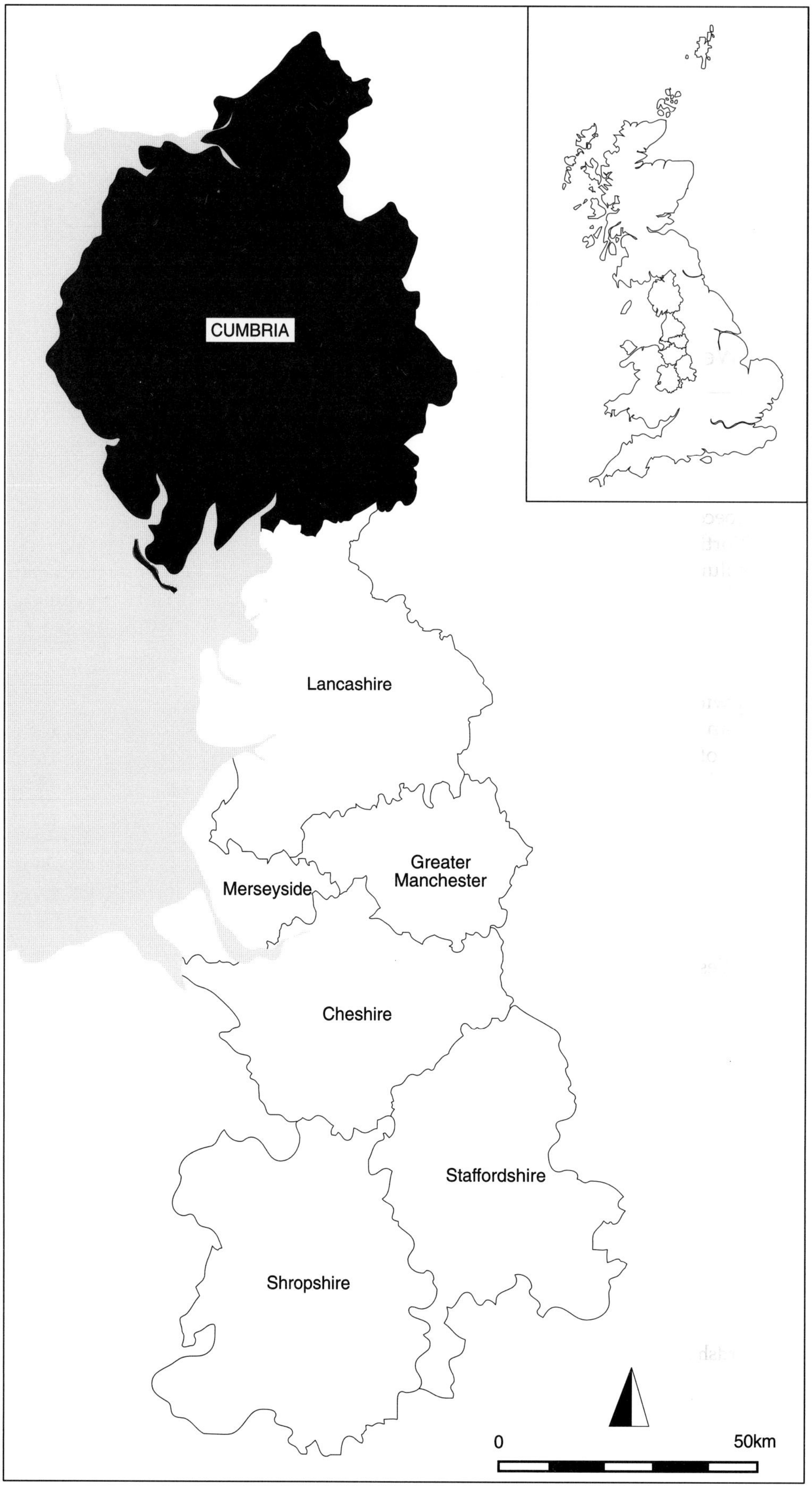

Fig 1: The location of Cumbria in the North West Wetlands Survey area

1

INTRODUCTION

The North West Wetlands Survey

This volume is the sixth in the seven volume series of North West Wetlands Survey monographs, each of which outlines the results of detailed archaeological, historical, and palaeoecological investigations in areas of lowland peat in North West England. The structure and format of the volumes are similar but reflect the diversity of wetlands, available evidence, and scope of fieldwork in the different counties. This volume describes a programme of archaeological survey undertaken in lowland Cumbria (Fig 1) between 1991 and 1997, along with a description of the previous evidence for human use of the wetlands. The southernmost area of Cumbria, which includes Arnside Moss, has already been described in *The wetlands of North Lancashire* (Middleton *et al* 1995).

The study presented here forms one component of the North West Wetlands Survey (NWWS), a long-term project which aimed to identify the archaeological and palaeoecological resource in the lowland wetlands of North West England. The origins of the work lie in the two previous major wetland surveys — the Somerset Levels Project (Coles and Coles 1986) and the Fenland Project (Hall and Coles 1994) — although the project reflected the regional variation in the nature, extent, and visibility of the archaeological evidence.

The project was initiated in 1987 when English Heritage commissioned a desk-based assessment of existing sites and finds, complemented by a gazetteer which summarised previous palaeoecological work (Howard-Davis *et al* 1988). This established the need for a long-term programme of systematic archaeological survey complemented by a comprehensive examination of the region's palaeoecology based on detailed and broad-scale analyses of peatland landscapes. An eight-year programme of work was commissioned in 1990. The interim results of the work have been published in a series of Annual Reports (Middleton 1990; 1991a; 1992a; 1993a; Middleton and Newman 1994). The programme of fieldwork and final publication for the counties covered by the NWWS is presented in Table 1 and Figure 2.

County	Surveyed	Publication
Merseyside	1990–91	Cowell and Innes 1994
Lancashire	1990–96	Middleton *et al* 1995, Middleton *et al* forthcoming
Greater Manchester	1992–93	Hall *et al* 1995
Cheshire	1993–95	Leah *et al* 1997
Shropshire / Staffordshire	1994–96	Leah *et al* 1998
Cumbria	1990–97	This volume

Table 1 Publications in the North West Wetlands Survey series

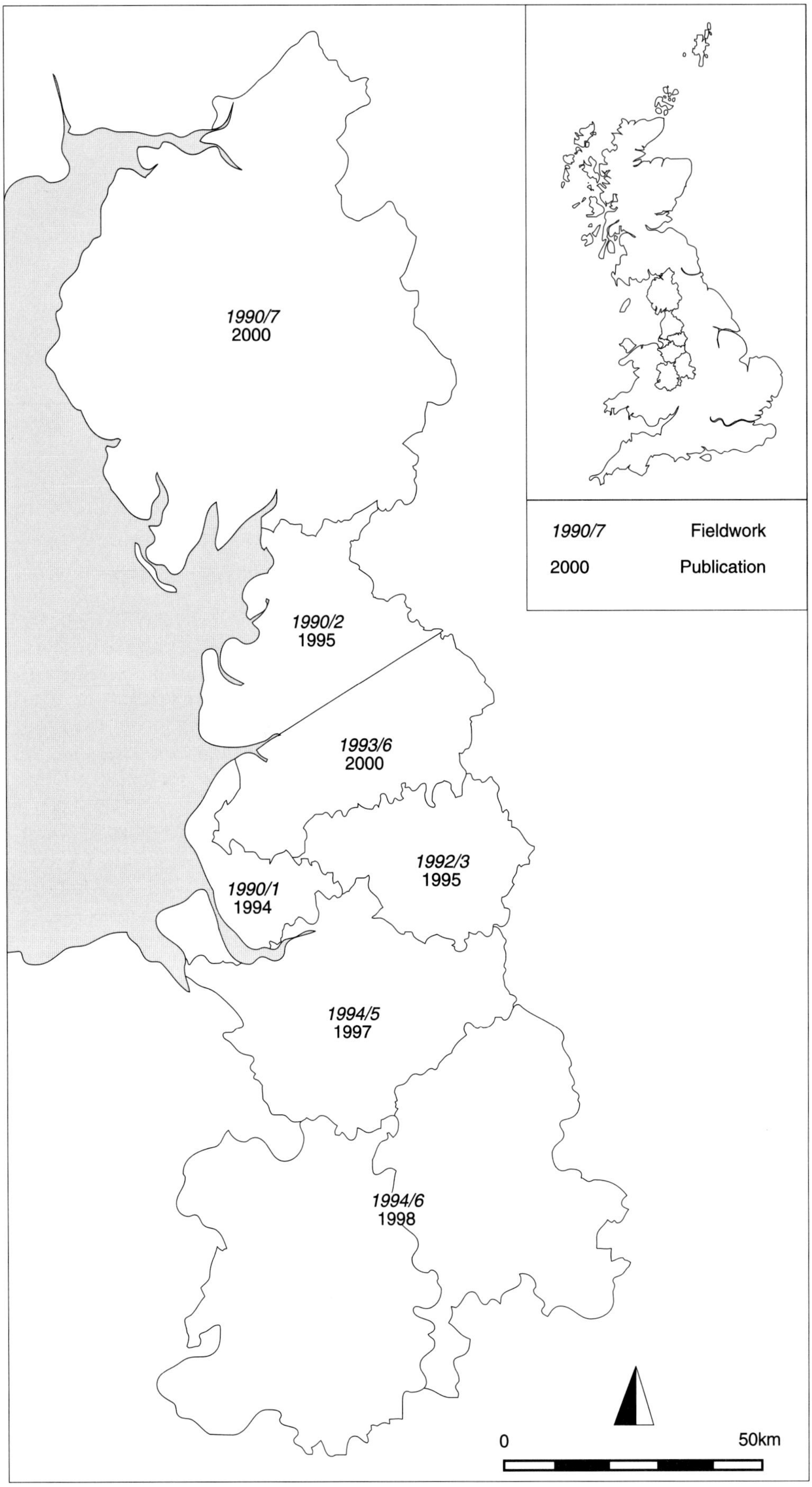

Fig 2 Location of North West Wetlands Survey areas showing fieldwork and volume publication dates

It is estimated that there are 36,689 ha of peat in the lowlands of the North West (Burton and Hodgson 1987, appendix). The term *lowland* is here defined on hydromorphological criteria and includes a diverse variety of mire types. These include raised, basin, floodplain, and valley mires and some intermediate types (Hall *et al* 1995, 32-3), which are spread unequally across the region; large raised mires are concentrated in the Mersey Basin, the Lancashire Coastal Plain, and the Solway Plain, whilst basin peats and valley mires predominate in South Cumbria and on the Shropshire / Cheshire Plain.

The term 'wetland' can embrace coastal marshes, upland peats, meres, river valleys, and lowland peats. The decision to exclude all but lowland peat from the project was taken for three reasons:

1. North West England contains large areas of wetlands which could not be examined within reasonable time-scales;
2. lowland peat forms a discrete landscape type;
3. lowland peat is one of the most threatened landscapes in the country.

Description of the survey area

This volume is concerned with the archaeology and palaeoecology of the lowland peat areas in the modern county of Cumbria. In numerous ways, particularly in terms of topography and geology, Cumbria is an extremely diverse, and in many respects unique, county, since it contains both the highest mountains in England and extensive deposits of lowland peat, concentrated on the extended coastline, which stretches around three sides of the county. Given the brief of the NWWS, the emphasis of the survey was inevitably focused upon the lowland areas around the periphery of the county. These encompass the Lyth, Winster, and Duddon valleys in the south, the coastal plain between Ravenglass and St Bees in the west, and the Solway Plain in the north of the county. The eastern boundary of Cumbria is broadly defined by the Pennines to the east of the Eden Valley in the northern part of the county, and is marked by the Howgill Fells in the south.

The upland and lowland areas are not so spatially separated in Cumbria, as they are in other counties in England, and narrow pockets of low-lying land exist within the upland region. Nevertheless, the defined lowland component of Cumbria correlates with the general distribution of the county's major wetlands and the regional distinction provided a useful framework for the fieldwork.

Relief

Cumbria is dominated by the mountains of the Lake District massif, which lie along the axis of an ancient upfolded mountain chain formed during the Caledonian orogeny. Successive folding and uplifting of a great central dome, followed by extensive denudation of younger, superficial rocks and deep erosion along a developing radial drainage pattern, gradually formed the Lake District. The radial drainage pattern has persisted throughout the history of the area and is manifested by a dearth of major rivers in the central massif (Halliday 1997, 8).

The south of the county is largely dominated by undulating low fells and ridges, from which a pastoral landscape with substantial woodlands has developed. The southern edge of the area is defined by the broad expanse of Morecambe Bay and the surrounding limestone lowlands, which are penetrated by the valleys and estuaries of the Rivers Duddon, Leven, and Kent, all of which support wetland environments in their lower reaches.

The West Cumbrian Plain, a narrow strip of land stretching from Haverigg on the Duddon Estuary to St Bees Head, does not extend further than 7 km inland from the coast and for the most part is low-lying, although impressive sea-cliffs exist at St Bees. The North Cumbrian Plain is defined as all land below 200 m OD, and is situated to the north and west of the Lake District massif. The landscape is dominated by a series of low ridges forming a gently undulating terrain, typical of a boulder clay landscape, giving way to a variety of coastal forms cumulatively forming the Solway Firth, comprising sand dunes, intertidal mudflats, estuarine and lagoonal complexes, and saltwater marshes. Inland from these are situated considerable expanses of wetland.

Climate

In general terms Cumbria has a cool oceanic climate but with diverse meteorological conditions which are exacerbated by the relatively extreme topography. A steep gradient of increasing rain, snowfall, and atmospheric humidity exists between the coastal lowlands and the high fells of the central massif. Moreover, there is a sharp altitudinal decrease of temperature between the valleys and the mountain summits (Halliday 1997, 19–26).

The driest part of the county is the inner Solway, which experiences a mean annual precipitation of 750 mm. The southern coastal belt, from St Bees towards Morecambe Bay, is slightly wetter with rainfall totalling 1250 mm. The extremes of rainfall are attained, as may be expected, around the Central Fells where the rainfall at Sprinkling Tarn averages

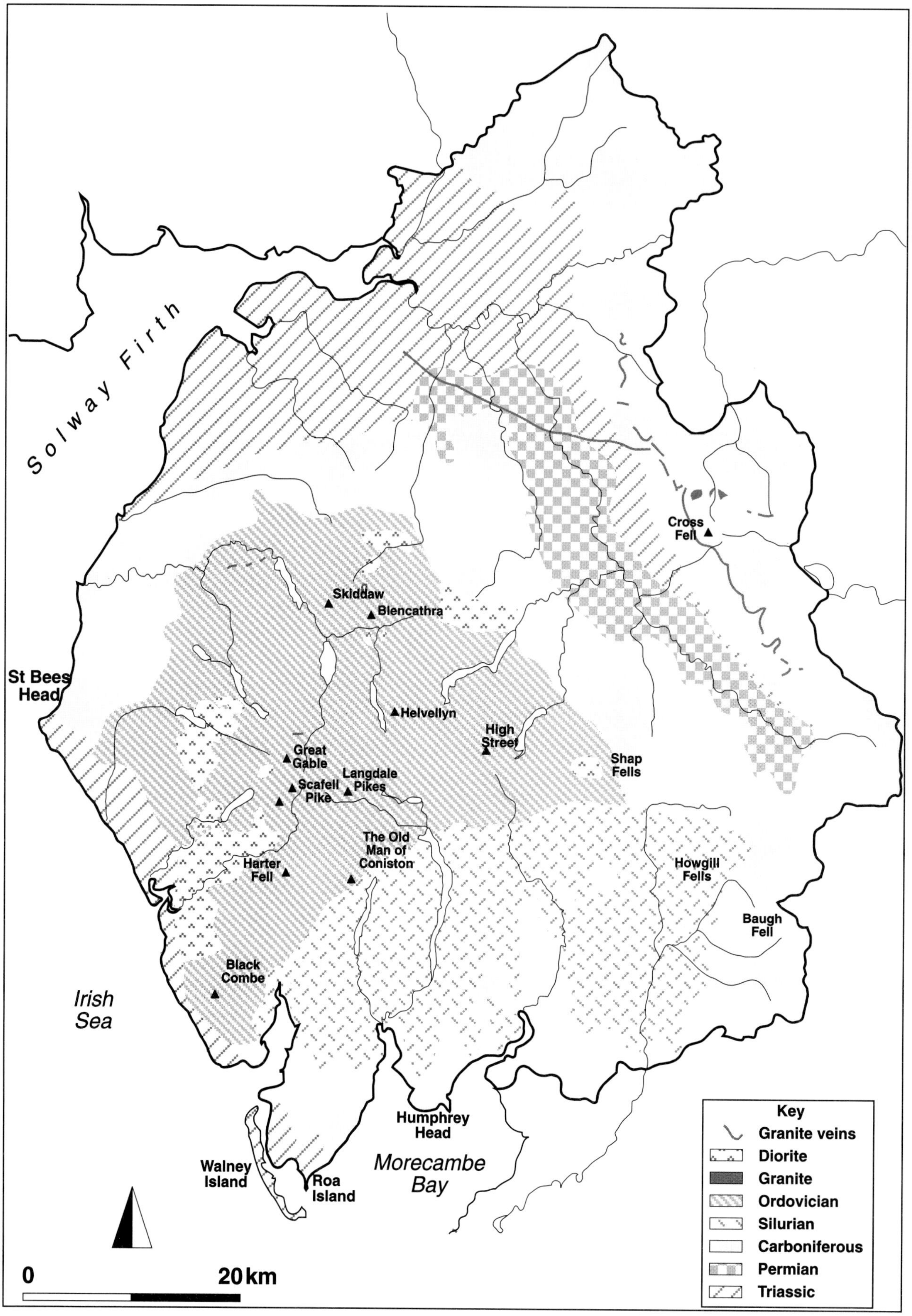

Fig 3 Simplified geological map of Cumbria (after Halliday 1997)

4700 mm *per annum*. The distribution of rainfall, in addition to the total amount, is important to soil development and vegetation. There is a greater humidity from the prevailing low cloud base and an increased incidence of mist along the hills and the valleys close to the west coast than is experienced in the east of the county, despite similar rainfall totals. Similarly, during anticyclonic conditions, quite prolonged droughts may be experienced in central areas (*ibid*). The great drought of the summer of 1976 gave rise to some catastrophic fires, notably at Glasson Moss, on the Solway Coast.

Temperatures are likewise affected by altitude, and the length of the growing season varies greatly. The average maximum temperatures along the coastal sector are 18.4–19 °C, rising to 20.0–20.6 °C east of Shap. January minima vary from 2.0 °C to 2.8 °C along the coastal belt to 0.5–1.0 °C over the Pennines and Borders (*ibid*). Thus, in general terms, the temperature gradient extends from the western coastal belt, which has a relatively narrow temperature range between winter and summer, to an eastern inland zone with colder winters but warmer summers. This has implications for frost, with the south–west coastal strip experiencing the lowest incidence and the least severe winter temperatures.

These climatic characteristics to some extent dictate the vegetational features of Cumbria. On the coastal plains the water balance is sufficient, when coupled with impermeable soils, to have allowed the development of large raised bogs with deep acidic peat. On the steep fells the climate has encouraged the formation of patchy blanket bog, and on the lower slopes, where the abundant rainfall emerges and collects in hollows, there are frequent valley mires (*ibid*).

Geology and soils

Cumbria has a very complex geology (Fig 3), comprising the central mass of older Ordovician and Silurian rocks and the peripheral series of younger rocks of Carboniferous, Permian and Triassic age. It is the younger rocks that form the bulk of the coastal plain areas, with those of Carboniferous age rising eastwards into the massive northernmost part of the Pennine Chain (Halliday 1997, 8–18).

The solid geology of the lowland areas is for the most part completely masked by drift deposits. Two soil associations, namely those of Clifton and Brickfield, dominate the soils of the North Cumbrian Plain. The Clifton series comprises seasonally waterlogged soils developed over tills, and forms the characteristic soil type of the plain to the west and south of Carlisle (Jarvis *et al* 1984, 135–7). Further to the south and west, the predominant soils belong to the Brickfield Association, a group characterised by seasonally waterlogged fine loams and clay-rich surface water gleys that have developed over drift deposits derived from sandstone and shales (*op cit*, 123). The bulk of the coastal plain to the south of St Bees is characterised by soils of the Wick 1 Association. Deep, well-drained coarse loamy brown earths, intermixed with gleyic brown earths and brown sands, are typical, derived from fluvio-glacial deposits and drift of variable stoniness (*op cit*, 302). These provided a striking contrast with the deep, well-drained sandy soils of the Sandwich Association which dominate the dune topography around the Esk Estuary. In the south of the county are calcareous soils that occur in a discontinuous arc following the outcrops of Carboniferous limestones on the rim of Morecambe Bay. This geology produces a variety of brown earth and rendzinas of the Warton Association.

Land-use

The land in lowland Cumbria is predominantly under pastoral agriculture, with arable farming restricted to the driest and most fertile areas (such as the Solway Plain), which see some root crop production. Some peripheral areas include small patches of woodland and scrub. Much of the agricultural land has been improved, which in many cases has involved widespread drainage, with obvious implications for wetland areas.

Additional to agriculture, forestry and nature conservation, several other important industries are also in operation in the county and, whilst some of these are removed from the wetland areas, several of the smaller mosses have been affected. Cumbria, and its south coast in particular, is well documented as an early post-medieval centre of the iron industry and the West Coast was also associated early with the coal and iron industries (Phillips 1977). Whilst deep-shaft mining has been discontinued, open-cast mining is still actively pursued with dramatic implications for the topography. Frequently associated with coal seams are deposits of fireclay, which in Cumbria is obtained from the vicinity of Whitehaven and Harrington on the West Coast. Moreover, boulder clay provides the raw material for most bricks made in Cumbria, and there are several extraction pits in the Carlisle area. Also from the West Coast, gypsum and anhydrite occur as bedded deposits: a rich deposit of anhydrite in the St Bees Evaporites supplies a nearby sulphuric acid and cement works. Peat extraction is, of course, of direct impact on the wetland areas, and there is a wealth of documentary evidence of this industry in Cumbria.

Date	Work undertaken	Publication
October/November 1989	Initial survey of Solway Moss, Bolton Fell, and Wedholme Flow	Middleton 1990; this vol: *ch 4*
April/May 1990	Initial assessment of wetland areas affected along the line of the North West Ethylene Pipeline	Middleton 1990; Lambert 1996
March 1991	Fieldwalking around Tarns on the Abbeytown Ridge	Middleton 1990; this vol: *ch 4*
June/July 1991	Surveys of Solway Moss, Bolton Fell Moss, and Wedholme Flow. Rapid assessments of mosses on North Cumbrian Plain	Middleton 1991a; this vol: *ch 4*
October 1991	Fieldwalking on the Abbeytown Ridge and adjacent areas	Middleton 1991a; this vol: *ch 4*
July 1991	Excavation of the Solway Cow, Solway Moss	Middleton 1992b; Stallibrass 1996; this vol: *ch 4*
August 1991	Watching Brief during construction works on the edge of Wedholme Flow	NWWS Archive
February 1992	Watching Brief during construction works on the edge of Salta Moss	NWWS Archive
June 1992	Excavation of the Solway Sheep, Solway Moss	Middleton 1992b; Huckerby and Wells 1993; this vol: *ch 4*
June 1992	Excavations on Bardsea Beach	Middleton 1992b; this vol: *ch 2*
March 1993	Fieldwalking on the Abbeytown Ridge and adjacent areas	Middleton 1993a; this vol: *ch 4*
August 1993	Watching Brief during nature conservation works at Wedholme Flow, Glasson Moss, and Whitrigg Common	NWWS Archive
December 1993	Watching Brief during nature conservation works at Scaleby Moss	NWWS Archive
January–March 1994	Fieldwalking and survey in Lyth and Winster Valleys and Rusland Pool	This vol: *ch 2*
June/July 1994	Survey of mosses on North Cumbrian Plain. Surveys of Solway Moss, Bolton Fell Moss, and Wedholme Flow	This vol: *ch 4*
July 1995	Examination of restoration works at Sawmill Wood SSSI	NWWS Archive
August 1995	Survey of mosses on West Cumbrian Coastal Plain and Duddon Mosses	This vol: *ch 2;* this vol: *ch 3*
November 1996	Desk-based assessment of Wedholme Flow ahead of revised planning permission	NWWS Archive
February 1997	Reporting find of horse skull from the Winster Valley	NWWS Archive
March/April 1997	Field survey on Abbeytown Ridge	This vol: *ch 4*
Winter 1997/98	West Cumbrian Coastal Plain	This vol: *ch 3*

Table 2 North West Wetlands Survey: archaeological work in Cumbria

History of research in Cumbria

The archaeological study of wetland areas in Cumbria has a long history, and includes some of the most nationally significant work undertaken to date. Chance finds in the nineteenth century, such as the Scaleby and Seascale bodies (Turner 1988; 1989), and the Foulshaw trackway (Munn Rankin 1910), alerted archaeologists to the potential of wetland areas, but it was the excavations at Ehenside Tarn in the early 1870s (Darbishire 1873) that first established the national importance of the wetland resource of Cumbria; indeed, the area became the type site for the Neolithic period in the British Isles for a considerable period.

The later part of the nineteenth century, and the first quarter of the twentieth, was a time of great antiquarian interest in the county, with the Cumberland and Westmorland Antiquarian and Archaeological Society being formed in 1866, with W G Collingwood, Ruskin's secretary, as an early luminary. His interest in the past, and his ability to organise and publish disparate data in his inventories of the counties, was crucial to the establishment of an understanding of the distribution of sites throughout the area, and also to the first attempt at categorising sites by type and period (Collingwood 1923; 1926). Indeed, these attempts to synthesise the data stood alone until the fashion for regional syntheses grew in the 1960s and 1970s, exemplified for Cumbria by the work of Rollinson (1967).

The *Transactions of the Cumberland and Westmorland Antiquarian and Archaeological Society*, however, throughout that time continued to publish reports on both excavations and chance finds, and these volumes are a vital source for any review of the archaeology of the region. In particular, the work of Mary Fair, Clare Fell, and the Cherry family is of importance for a study of the archaeology of lowland Cumbria, and this is reflected in the *Bibliography* of this volume. Mary Fair, working largely in the 1930s and 1940s, was a great recorder of information, and was instrumental in highlighting the archaeological potential of the Drigg and Eskmeals sand dunes (Fair 1936; 1937). This work was taken up by the Cherry family who have revolutionised archaeological knowledge of the west coast of Cumbria. Working for almost 30 years from the early 1960s, the Cherrys systematically examined the coastal zone from Eskmeals to St Bees Head for evidence of prehistoric activity, and amassed an enormous volume of information (*see*, for instance, Cherry 1963; 1965; 1966; 1967; 1969; 1982; Cherry and Cherry 1973; 1983; 1984; 1985). In addition, working in collaboration with Clive Bonsall at Williamson's Moss, they helped to examine a site which potentially rivals Ehenside Tarn in importance for the understanding of the Mesolithic/Neolithic transition (Bonsall *et al* 1991). Perhaps the most significant influence in the development of the study of the archaeology of Cumbria, particularly its prehistory, however, is Clare Fell, who since the Second World War has not only recorded sites and finds from the region, and also discussed their significance (Fell 1940; 1973), but encouraged others, such, indeed, as the Cherry family, to engage in fieldwork.

More recently, the aerial photographic work of Higham and Jones (1975) has greatly increased the number of known archaeological sites throughout the county, but particularly on the North Cumbrian plain, and this work has been consolidated by the work of Bob Bewley (1986; 1992; 1993; 1994). He has not only undertaken further aerial photographic survey, but also excavations on selected sites, which have produced surprising results, leading to a radical questioning of the validity of dating sites by form alone (Bewley 1993). On the North Cumbrian Plain, Barri Jones undertook many years of small-scale excavations, which produced a previously unknown Roman frontier, predating that of Hadrian's Wall (Jones 1989). In the south of the county, the ongoing work of Doug Stables and the Morecambe Bay Archaeological Research Society is demonstrating the importance of repeated fieldwalking, particularly in those largely pastoral areas where ploughing for reseeding is an irregular event.

The lowland archaeological resource of Cumbria cannot, of course, be seen in isolation, but must be placed in the context of the archaeology of the county as a whole. The work on the Langdale axe factories (Claris and Quartermaine 1989; Bradley and Edmonds 1993) has demonstrated that interaction between the upland fells and the lowlands was already occurring in the Neolithic period, whilst the upland areas, especially those marginal lands which are now permanent pasture, contain a wealth of upstanding archaeological remains, particularly of Bronze Age date. These give an indication of the amount of archaeological sites on the lowlands which may have been destroyed or masked by later, more intensive agricultural activity. The Lake District National Park Survey of the 1980s (Quartermaine and Leech forthcoming), and subsequent work on major upland estates (for instance, LUAU 1994; 1995), have mapped many of these extensive landscapes. The recent work by Caroline Skinner (pers comm), building on the work of Webster (1969), has, in the same vein as NWWS, attempted the integration of archaeological and palaeoecological survey in the Pennine uplands, and this is adding to the greater understanding of the archaeology of the county.

Previous palaeoecological research in Cumbria is also extensive, and was initially stimulated by the wetland archaeological finds of the Foulshaw Moss trackway and the settlement at Ehenside Tarn in the nineteenth century (*see p 7*). The accrual of such a wealth of palaeoecological information for Cumbria, which is greater than for any other county in Britain, is largely due to the work of Walker (1966) and of Pennington (1947; 1970; Pearsall and Pennington 1973). Walker studied the infilled lakes and bogs of the north and west of the county, which complemented the work of Smith (1958; 1959) and Oldfield (1960) to the south, around Morecambe Bay, whilst Pennington's work, for the Freshwater Biological Association at Windermere, was more widespread, covering a large number of freshwater and peat deposits in both the lowland coastal fringes and the uplands of the Central Fells. Between these workers, a broad picture of the regional development of the vegetation was built up, for the major part of the county, excepting, perhaps, the area east of a line from Penrith in the north to Milnthorpe in the south.

Their research demonstrated the way in which the vegetation developed in response to climatic amelioration as the ice of the last glaciation began to retreat, its continuing development through to the clear evidence of human clearance of woodland and the first evidence of arable agriculture in the Neolithic and Bronze Age, and onward to the historic period. This early work is, however, largely undated and therefore can provide no firm evidence as to the archaeological period of these changes, but it does provide some indication as to the extent of the manipulation of the vegetation.

This work has provided the basis for much new research in the last 20 years, by a number of workers. In the south of the county, Wimble carried out a detailed palynological study of three mosses, using multiple cores and a suite of radiocarbon dates, which has provided a detailed picture of transient vegetational changes from the late Neolithic to the medieval period (Wimble 1986). Tipping (1994) has analysed pollen from Williamson's Moss on the West Cumbrian Plain, thereby increasing knowledge of the Mesolithic/Neolithic transition first expounded by Walker (1966). In the north, Barber and his co-workers at Southampton have worked extensively

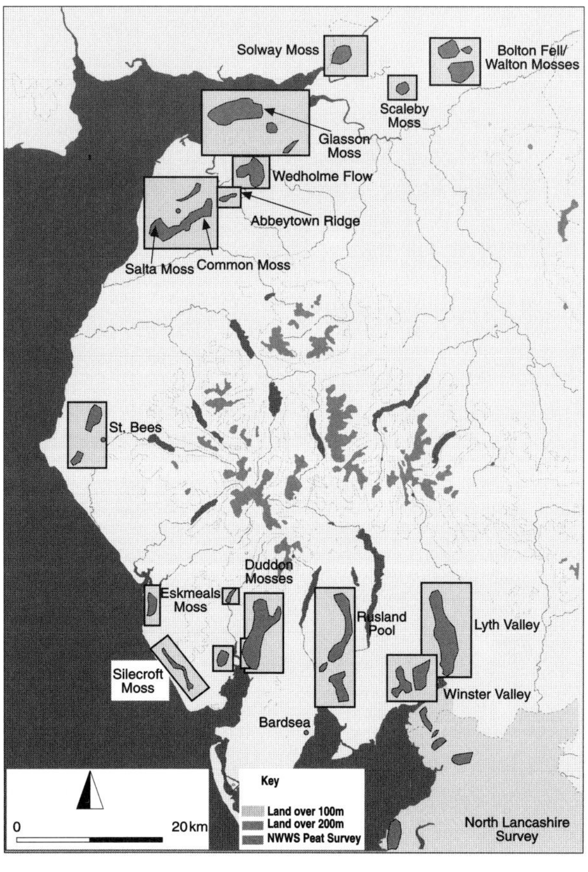

Fig 4 North West Wetlands Survey: archaeological survey areas in Cumbria

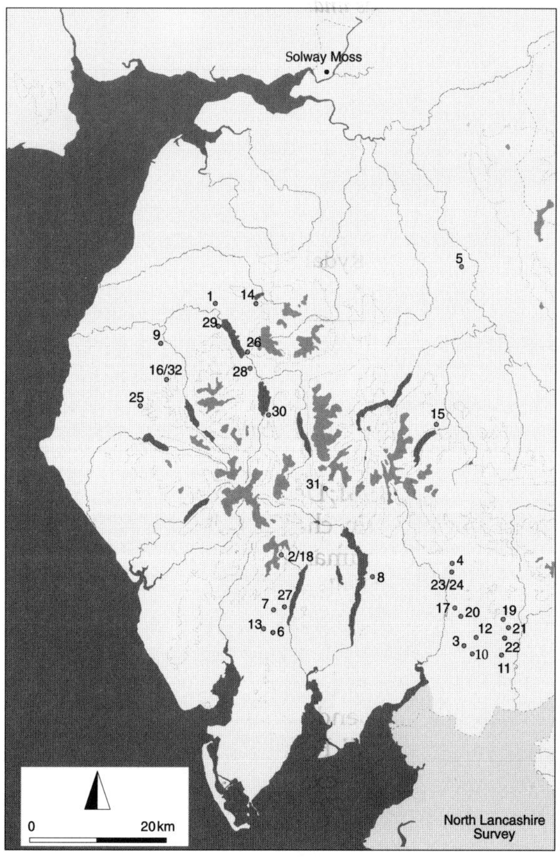

Fig 5 North West Wetlands Survey: palaeoecological survey sites in Cumbria (see *Table 3*)

	Site	NGR	Description	No. of Sample
1	Glebe Farm	NY 185 346	Stratigraphy / Spot pollen samples	(6)
2	Beacon Tarn	SD 273 902	Stratigraphy / Spot pollen samples	(2)
3	Birkrigg Bog	SD 537 866	Stratigraphy / Spot pollen samples	(5)
4	Black Moss Tarn	SD 548 971	Stratigraphy / Spot pollen samples	(3)
5	Brownrigg	NY 520 367	Stratigraphy / Spot pollen samples	(3)
6	Burney	SD 255 859	Stratigraphy / Spot pollen samples	(10)
7	Subberthwaite Common	SD 257 881	Stratigraphy / Spot pollen samples	(9)
8	Crook Hall	SD 448 944	Stratigraphy / Spot pollen samples	(8)
9	Dubbs Moss	NY 103 293	Stratigraphy / Spot pollen samples	(8)
10	Gatebeck	SD 549 859	Stratigraphy / Spot pollen samples	(6)
11	Heights	SD 603 861	Stratigraphy / Spot pollen samples	(8)
12	Hood Tarn	SD 571 876	Stratigraphy / Spot pollen samples	(8)
13	Heathwaite	SD 255 869	Stratigraphy / Spot pollen samples	(3)
14	Little Tarn	NY 249 338	Stratigraphy / Spot pollen samples	(5)
15	Littlewater	NY 509 170	Stratigraphy / Spot pollen samples	(12)
16	Mosedale	NY 109 234	Stratigraphy / Spot pollen samples	(4)
17	Oxenholme	SD 529 891	Stratigraphy / Spot pollen samples	(7)
18	Stable Harvey	SD 278 917	Stratigraphy / Spot pollen samples	(7)
19	Eskrigg	SD 586 889	Stratigraphy / Spot pollen samples	(9)
20	St Sunday's Beck Bog	SD 553 895	Stratigraphy / Spot pollen samples	(7)
21	Three Mile Bog	SD 594 875	Stratigraphy / Spot pollen samples	(10)
22	Wyndhammere	SD 594 852	Stratigraphy	
23	Garnet Farm Moss	SD 537 967	Stratigraphy	
24	Skelsmergh Tarn	SD 534 967	Stratigraphy	(two cores)
25	Kidburngill*	NY 053 211	Stratigraphic transects	
	**This work was undertaken outwith the project by LUAU*			
Lake margins				
26	Bassenthwaite	NY 236 272	Stratigraphy / Spot pollen samples	(6)
27	Coniston	SD 288 905	Stratigraphy / Spot pollen samples	(11)
28	Braithwaite Moss	NY 236 249	Stratigraphy / Spot pollen samples	(2)
29	Castle How	NY 202 308	Stratigraphy / Spot pollen samples	(3)
30	Derwentwater	NY 262 189	Stratigraphy / Spot pollen samples	(7)
31	White Moss, Rydal	NY 251 065	Stratigraphy / Spot pollen samples	(11)
32	Graythwaite	NY 113 230	Stratigraphy / Spot pollen samples	(4)

Table 3 North Wetlands Wetlands Survey: palaeoecological assessment sites in Cumbria

on Bolton Fell, Solway, and Walton Mosses (for instance, Barber 1981; Dumayne 1992; Hughes 1997), examining climatic change as well as providing evidence for the human impact on the area. On the east of the Central Fells and from the Pennine fringes, the picture is more sketchy, but this is being redressed to some extent by Chinn and Innes (1995) in Wet Sleddale, and Innes (pers comm) and Skinner (pers comm) have independently worked on a number of sites in the Howgill Fells and the Pennine fringes. Cundill (1976) has examined peat deposits from Archer Moss at 460 m OD in the Howgill Fells, above the dramatic eroded gully of Carlingill, and also from intercalated peats from the bottom of the gully. This work attempted to relate extensive erosion features to land clearance, as has research by Boardman and his co-workers in the Central Fells (Parker *et al* 1994).

The wetlands of Cumbria

Cumbria contains extensive wetland deposits in a variety of topographic locations, notably the uplands and coasts. The present survey focused on the lowland wetlands on the coastal plain in the south, west, and north of the county, and some of the higher land to the east of the River Eden, also in the north (*see* Fig 4 and Table 2). Rapid palaeoecological assessments were, however, undertaken of small basin mires on the upland fringe (over 200 m OD) and the lake margins of the Central Fells to expand the main ground survey, placing the results in context for the county, and complementing the existing corpus of research in the area (*see* Fig 5 and Table 3).

A significant problem in the county is that many of the smaller mosses and basin mires are not well mapped and so are difficult to target by rapid ground survey. In order to work around this problem, a trial study was undertaken with NWWS data to assess the possibilities of mapping new areas of peat using satellite remote sensing imagery. It had some success, although needed further development to be of immediate use to the project (Cox 1992). It did, however, allow many very small areas of peat to be classified and showed differences in soil moisture not visible on aerial photographs. The next step in the verification of the data generated would be to output through a GIS-compatible computer to create maps to allow field checking. At the time of this trial study the lack of spatial resolution of the remote sensing imagery meant that it was unable to provide more specific archaeological site evidence than competent aerial photograph data can yield (*ibid*).

The regional wetland variation within the North West is reflected in the structure of each county volume and is important in understanding the fieldwork strategies employed in each county. In Cumbria, a number of factors dictated strategy and the presentation of results. The predominantly pastoral land-use meant that fieldwalking in arable areas was always going to be localised to areas such as the Abbeytown Ridge (on the Solway; *Ch 4*; Pl 1), where a variable number of fields was available every year. Because of this, a small area was walked from 1991 onwards as and when they became available; field seasons in North Cumbria were thus restricted to two weeks at most. Later in the project, when resources became available as other county programmes were completed, more systematic survey was undertaken to link the previously fieldwalked areas.

In the south and west of the county, where pastoral land-use was almost universal, archaeological fieldwork was initiated in 1995, comprising a rapid scan of the fields, logging peat extents and depths where available. It was anticipated that the detection of archaeological remains was going to be problematic, and so it proved, but it was recognised that the main value of the work was to identify areas of great archaeological and palaeoecological potential and to collate the available evidence for the past use of

Plate 1 Fieldwalking on the Abbeytown Ridge

Plate 2 Cleaning a ditch during survey of peat cuttings on Solway Moss

wetlands which had previously been scattered and was difficult to assess.

There are three sites in Cumbria on which peat is cut commercially on a large scale: Wedholme Flow, Solway Moss, and Bolton Fell. The importance of these sites was identified at the outset of the project and a programme of annual systematic survey was initiated in 1990. This involved a rapid scan of the ditch faces and worked surfaces to identify archaeological remains, and, at Solway Moss, a more detailed appraisal of the sub-peat surfaces for traces of human activity (*see* Pls 2 and 3).

The survey work also involved discussing possible sites and finds with the staff working at these sites. The interest thus generated provided dividends in terms of both support (Fisons Horticulture Ltd (now Levingtons) provided a Land Rover for the use of the project), and in the identification of archaeological remains as they were uncovered during peat extraction. It is worth noting that both the Solway Cow and the Solway Sheep (*Ch 4*) were discovered through the vigilance of staff at Solway Moss.

Cumbria stands out from the remainder of the NWWS area in the number of projects within it aimed at the management and restoration of mire systems. This reflects the good condition of the mires, especially in the north of the county, as well as the designation of some as Sites of Special Scientific Interest (SSSIs) and, in two cases, as National Nature Reserves (NNRs), managed by English Nature and the Cumbria Wildlife Trust. Much of this work has focused on management works aimed at raising water levels to re-flood former peat workings at, for example, Glasson Moss and Wedholme Flow in the north, and clearing trees to prevent desiccation, such as at Meathop Moss in the south of the county. More active management has involved, for example, the construction of dragonfly ponds on Scaleby Moss. Whilst such works are generally of benefit to the archaeology of the mosses, the survey has permitted the impact of these works to be assessed more fully (*Ch 5*).

The NWWS work in the county has been extended by the more detailed investigation and excavation of three finds of animal remains, a cow and sheep from Solway Moss, and a red deer from Bardsea (*Chs 4* and

Plate 3 Profile through peat from a perimeter dyke on Solway Moss

2 respectively). These interventions were dictated by the circumstances of their identification and were essential adjuncts to the project's survey programme. In conjunction with the excavation of the remains of the cow and sheep from Solway Moss, a detailed stratigraphic survey of the area was carried out to place the finds in context and to provide a picture of the development of the moss and the environment at the time of deposition of the bodies. In the case of the sheep, this was supplemented by more detailed palynological and macrofossil analysis of the peats underlying the remains (*Ch 4*; Pl 4).Table 2 and Figure 6 contain details of the programme of archaeological fieldwork in Cumbria.

Other, palaeoecological, work was restricted in nature, in the light of the large amount of lowland wetland in the county, and was targeted to those parts which had seen little previous study. Thus, in the main areas of wetland in South Cumbria, no work was undertaken since Wimble (1986) had already undertaken a comprehensive pollen analysis of three mires to give a broad picture of vegetational change. A similar situation pertained along the West Coast, where substantial previous pollen analysis by Pennington (1970) and Walker (1966), amongst others, had, in some measure, permitted a regional picture of landscape/vegetation change to be compiled, despite the low numbers of radiocarbon dates obtained during these investigations.

The major part of the palaeoecological work in the county comprised a rapid assessment of the margins of selected lakes in the Central Fells, and of a representative sample of the small basin mires scattered around the fringes of the Lake District and in East Cumbria, to identify their palaeoecological potential. These sites were beyond the areas in which archaeological fieldwork was undertaken, but it was considered that, to understand the wetlands of Cumbria better in the future, some assessment of them was necessary. The collation of all these data is designed to identify those wetlands of significant archaeological value and to aid the monitoring and devising of strategies to manage those which may be threatened. This extension of the palaeoecological survey area was justified further by the fact that much of the earlier palaeoecological work in Cumbria was concentrated in the Central Fells, the major mires north of Morecambe Bay, and the West and North Cumbrian Plains; the NWWS has always tried to highlight areas where there was either no record or it was rudimentary.

In most cases a single core was taken, the stratigraphy noted, and samples taken for spot pollen analysis. The results of these investigations are presented in *Appendix 5*, whilst the sites examined are listed in Table 3 and shown Figure 5. This approach differs greatly from that if full analysis were to be considered, when multiple cores would be required.

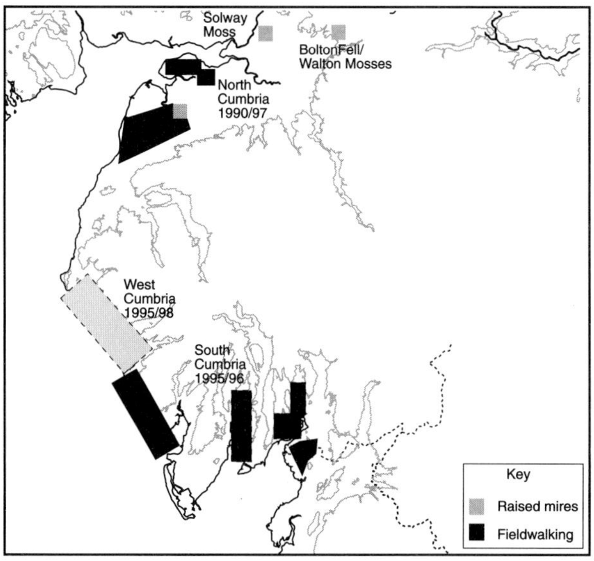

Fig 6 North West Wetlands Survey survey areas and dates of fieldwork

Plate 4 Palaeoecological fieldwork on Solway Moss

Central Lake District

In addition to the palaeoecological survey, more detailed work was undertaken in conjunction with Oxford University on a peat deposit sealed beneath colluvium at Seathwaite (Parker *et al* 1994), in the Central Fells. As this lies outwith the topographic constraints of the project, the detailed results of this work will be reported elsewhere (Wild *et al* forthcoming).

Threats to the archaeological resource

Peat and the past (Howard-Davis *et al* 1988) indicated both the scale of the threat to wetland archaeological remains and the rapidity of their degradation; systematic survey has confirmed the initial findings and revealed the extent of the destructive processes across the region. The main threats to the peat resource in North West England are summarised below.

Agriculture

Since the medieval period, most of the mires of the North West have seen some reclamation activity. In almost all cases this was initially restricted to small peripheral enclosures which followed peat cutting, but from the eighteenth century onwards the movement towards agricultural improvement and increasing prosperity led to more systematic and rapid enclosure of the mosslands. Reclamation was stimulated by the availability of clay tile drains and public money for drainage from the 1840s onwards. Drainage schemes have removed water from the mosses resulting in peat humification over the past century or more. The removal of water affects not only the surface peats, but also those buried underneath clays, silts, and sands: thus it is both the surface and the sub-strata that are de-watered.

Refuse disposal

The unsuitability of most mires for building means that they have often remained open spaces on the

Plate 5 Peat cutting on Solway Moss

fringes of urban areas. Peat is also a saleable commodity which is relatively easy to remove. These two factors makes peatlands ideal for the disposal of refuse. Several mires in the North West are under immediate threat, particularly those in Greater Manchester (*cf* Hall *et al* 1995) and Merseyside (Cowell and Innes 1994); in these cases, mitigating archaeological and palaeoecological action is essential.

Peat extraction

One of the main uses for the mosses of the North West has been as a source of peat for burning. Until the nineteenth century this was carried out on a small scale by local people, although cumulatively many metres of peat were removed over the centuries (Middleton *et al* 1995). In the nineteenth century industrial processes were introduced to extract peat for bedding for both horses and chickens, whilst in the twentieth century the use of this peat has been confined to horticulture. In some areas of the North West, notably on the Solway Plain, this process is still contributing to the erosion of the peatland resource. Uniquely amongst the counties of the North West, commercial peat extraction remains a major threat to the lowland peat resource in Cumbria, for example at Bolton Fell Moss, Solway Moss, and Wedholme Flow (Pl 5).

Urban expansion

After agriculture, perhaps the major long-term threat to the mosslands remains urban expansion. Building, road construction, and other developments by necessity require drainage and the removal of peat. This problem is particularly acute where mosses remain the only area through which roads can be built, such as Ashton Moss in Greater Manchester (Hall *et al* 1995, 63-70).

In Cumbria, the impact of landscape change has not been as rapid as elsewhere in the NWWS survey region, largely owing to its distance from major cities and the unsuitability of much of the landscape for agricultural intensification. In the Central Fells, early adoption of landscape conservation, notably through the National Trust, has prevented large-scale destruction of environmentally important landscape features.

This does not mean, however, that the wetlands of Cumbria are not under threat; aside from the major peat-cutting sites, where the threat is very obvious, wetland drainage, largely on a local scale, and domestic peat-cutting have led to the severe desiccation and truncation of most mires. The nature of the threat is most clearly seen in the case of Ehenside Tarn where the local drainage of a tarn led to many Neolithic and later finds being recovered as the water ebbed away (Darbishire 1873). The subsequent, and continuing, desiccation damaged the adjacent peat deposit (Walker 1966, 85-96), although it still contained waterlogged pollen and wood, and it is likely that the deposits will have degraded further in the subsequent 40 years.

Aims and objectives

The aims and objectives of the North West Wetlands Survey were originally set out in an initial Project Design (Middleton and Wells 1990) and were subsequently amplified and expanded (LUAU 1993). A summary of the aims and objectives of the survey both at regional and county scales is given below.

The main aim of the project was to elucidate the extent and nature of settlement from prehistory to the medieval period in and around the lowland peat of the North West. All evidence from systematic, non-destructive survey for these periods from the mires and their drainage basins was recorded. The very wide remit of the project reflects a real lack of previous work over most of the area; with notable and isolated exceptions (*eg* in West Cumbria and the Abbeytown Ridge), the NWWS work forms the first systematic survey in most of the areas in question.

The aim of the archaeological work has been to survey and record the nature and extent of past human activity, within and adjacent to areas of lowland peat. Where archaeological activity has been detected, the quality of preservation and the nature of any threat has been assessed. Even where archaeological information has not been detected, the state of peat preservation has been assessed and recorded, in recognition of the fact that further peat wastage may expose currently buried sites. Crucially, an attempt has been made to place any archaeology within an environmental context, by means of palaeoecological studies.

As well as acting as an essential adjunct to more conventional archaeological survey, however, the palaeoecological element of the project has also been regarded, in a broad sense, as a survey tool in its own right. In this context, a major aim of the survey was to locate and assess the degree of preservation of environmental archives across the county. The results presented in this volume therefore build upon an existing corpus of information in an attempt to create a clearer understanding of the county's surviving environmental history resource.

The collation of these data, as well as providing a picture of how the county's varied wetlands developed and were utilised at different periods, has also increased the amount of information available with regard to the extent, preservation, and archaeological potential of individual wetland sites. This will allow the development of management strategies to monitor and manage the archaeological and palaeoenvironmental resource.

At the regional scale, the aims of the project as originally formulated are as follows (LUAU 1993, 15-20):

1. to survey and record the character and extent of archaeological remains within, and on the margins of, areas of lowland peat;
2. to place these remains within an environmental context by means of an integrated programme of palaeoecological studies;
3. to identify anthropogenic activities in the palaeoecological record and climatic and other factors which may have affected past environments;
4. to elucidate mire development through broad-scale and detailed palaeoecological studies.

In addition to the regional aims, a Research Agenda for the specific survey area was also compiled, which was based on previous research and the results of NWWS survey to the south, particularly in North Lancashire (Middleton *et al* 1995). The localised aims reflected the particular issues of site visibility, peat survival, and mire management in Cumbria.

1 There was a need to elucidate the nature of activity through time and to determine its visibility in ground survey. Work by others, notably in West Cumbria and the Solway Plain, had identified the particular importance of the coast and wetlands for prehistoric settlement. Allied to this aim was the need to evaluate the important remains from Ehenside Tarn which have influenced expectations of results from wetland investigations since their publication in 1873 (Darbishire 1873).

2 There was a need to assess whether the large variety of wetlands, in terms of both their morphology and hydrology, and survival, was reflected in similar variations in the archaeological record through time.

3 There was a need to evaluate the impact of conservation projects on the archaeological resource, which has rarely been considered as an integral part of nature conservation. Owing to the relatively widespread survival of mosses in the county, and the survival in places where mire vegetation is of international importance, there have been several initiatives concerned with restoring and improving the condition of the mosses.

4 A major aim of the survey was to locate and assess the degree of preservation of environmental archives across the county, since, as well as acting as an essential adjunct to the archaeological survey programme, the palaeoecological side of the project has been employed as an archaeological survey tool in its own right. Within Cumbria, the strategy adopted comprised a palynological assessment of as many of the numerous small basin mires and lake margins which lay outside the main areas of survey as was feasible within the timescale of the project.

In the final chapter each of these aims will be examined individually.

Methodology

Archaeological survey

The main elements of the archaeological survey are common to all of the counties and differ little from the procedure outlined in the Project Designs (Middleton and Wells 1990; LUAU 1993). These in turn were based on the experience of previous wetlands projects, notably the Fenland Project (Hall and Coles 1994), although some modifications to this basic strategy were required to suit conditions local to the North West. The main aim of the methodology was to survey the mosses rapidly and effectively, whilst retrieving maximum information from a scattered and diverse dataset.

The principal methodology was systematic fieldwalking (Pl 1). All available arable fields were walked on a farm-by-farm basis and all existing and former peatland was examined, along with the dryland fringe which varied in width according to local topography and access to fields. The small number of available arable fields was walked at an interval of 10 m (or 12 m tramlines); on deep peat and in other areas where sites were unlikely to be found, the interval was 30 m. The difference in transect spacing reflects the recognition that most evidence for human activity, detectable by fieldwalking, is likely to be found on the solid ground around the wetland margins rather than in deep peat. The act of fieldwalking, however, permitted the fields to be checked for peat depth and to locate any small areas of mineral soil that may have contained finds. The primary aim of this technique was to locate concentrations of artefactual material, plot their spatial extent, collect enough material for dating if feasible, and map the current extent of peat. It should be noted that the transect intervals chosen for the NWWS are much closer than for comparable projects, such as the Fenland Project (*op cit*, 10): this reflects the fact that the artefact densities were known to be much lower in the North West than in other parts of the country.

Once artefacts were located, a 20 x 20 m box around them was walked at a 1 m interval to determine the site's extent, and every item was collected and recorded with the exception of post-medieval finds. All sites and finds were assigned site numbers with a CU– prefix with sites less than 50 m apart assigned to the same number. The site locations were plotted onto 1:10,000 scale Ordnance Survey (OS) maps, but occasionally, more detailed sketch plans were required. There was no intention, at this stage, to carry out any detailed analysis of artefact distribution within individual scatters. All previously-known and aerial photographic sites were visited and all field evidence logged. All pasture areas were also walked with a transect interval depending upon the size of the field and the visibility within it. All earthwork sites were sketch-plotted onto the 1:10,000 field maps, although occasionally more detailed plans were required. *Pro-forma* sheets were completed for each site, which were subsequently added to the NWWS database.

It was recognised that in an area of low artefact densities, the degree of soil weathering would have a crucial effect on artefact recovery rates and, whilst every effort was made to survey fields in optimum conditions, the time-scale of the project meant that inevitably some fields were not seen at their best. All arable fields were therefore coded to describe the visibility of artefacts in the soil:

	Ground conditions
1	Good
2	Moderate
3	Poor
4	No access / not surveyed.

Details about the thickness of the peat and the land-use were also noted on the 1:10,000 maps. These were recorded on the following scales:

	Peat depth
A	Untruncated mire
B1	Peat over 1 m deep
B2	Peat under 1 m deep
C	Organic soil
D	No peat

	Land-use categories
A	Arable
P	Pasture
W	Woodland
SCR	Scrub
RG	Recreational grassland (golf courses etc).

On the maps each field was coded, for example P / B1 or 1 / C. The field data were transcribed onto neat copies of the 1:10,000 maps, which form the basis of the data presented in the maps in this volume (*rear of volume*, Figs 33–6).

It was recognised from the very beginning of the project that, across much of the North West's wetlands, arable was likely to be restricted in extent and limited to the more favourable locations, such as the North Cumbrian Coastal Plain and West Lancashire. Consequently, survey of areas of pasture and, on occasions, woodland, would assume a greater prominence than in a primarily arable area such as the Fens of East Anglia. Obviously, the chances of locating artefactual evidence in such circumstances would be much reduced but, in compensation, the chances of discovering earthworks were correspondingly higher. The methodology adopted was to examine areas of pasture and woodland, within or on the borders of wetlands, on the ground and to plot details of archaeological features, land-use, and peat depth onto 1:10,000 scale maps. On occasions, it proved necessary to record some earthworks at a larger scale, usually 1:2500. Land-use and peat depth were recorded using the codes listed above.

Plate 6 Watching brief, Moss Side Farm, Wedholme Flow

In areas of peat-cutting, a systematic programme of annual survey was initiated early in the project, each site being divided into sections and surveyed where peat-cutting had occurred since the previous survey. At the three sites in Northern Cumbria (*eg* Solway, Pls 2 and 3), all cleared drainage dykes were examined, as were any other areas of active extraction. The purpose of this work was:

1 to identify archaeological remains and evidence for human activity;

2 to assess the nature of the deposits which were being removed at the time of survey, and those revealed in cleaned sections, to identify the potential of the site for archaeological material, for example, the presence or absence of sand islands, which may have provided suitable habitats for settlement;

3 to raise awareness amongst the site staff of archaeological issues to ensure prompt reporting of finds.

In addition to the survey programme, the project undertook a series of watching briefs in conjunction with commercial development (Pls 6 and 7) and three excavations in the county, two on Solway Moss, and one at Bardsea on the south Cumbrian coast. All of these were initiated by casual finds within either mined or naturally eroding wetland contexts and the methodologies adopted reflected the nature of the finds, their contexts, and the circumstances of discovery. The methodology and aims of each excavation have been described in the relevant sections in this volume (*Chs 2, 4).*

Radiocarbon dating

All radiocarbon dating undertaken as a part of the survey was carried out by the Scottish Universities' Research and Reactor Centre at East Kilbride (SURRC), with accelerator dates provided by the University of Arizona AMS facility (through SURRC) and by the Oxford University Radiocarbon Accelerator Unit. Full details of methods and procedures can be obtained from Dr Gordon Cook of SURRC and Dr Paul Pettitt of the University of Oxford.

Where possible, radiocarbon dates have been calibrated using the conventions established at the 1986 Trondheim Radiocarbon Conference (Switsur 1986; Pearson 1987). All dates have been calibrated using the CALIB computer program (University of Washington Quaternary Isotope Laboratory 1987) using a curve based on a 20-year atmospheric record dating to 7210 BC (Pearson and Stuiver 1986; Pearson *et al* 1986; Stuiver and Pearson 1986), and a bi-decadal

Plate 7 Assessment of 'wet' area for Shell (UK) Ltd's North West ethylene pipeline, Drybeck Moss

weighted average of data (Linick *et al* 1985; Kromer *et al* 1986; Stuiver *et al* 1986).

All dates are cited in the following form, for example:

2559–1950 cal BC (3370	±70		GU5167)
calibrated date range	uncal date	standard deviation	lab code

Calibrated radiocarbon dates (cited as cal BC or cal AD) are cited as a range at a two sigma confidence level (96.4 % confidence). Beyond the limits of the 1986 dataset (taken to be 8100 BP), dates have been calibrated using the calibration curve of Stuiver and Reimer (1993). Uncalibrated dates are cited as BP (Before Present).

Aerial photography

No new aerial photography survey work was undertaken by the NWWS in Cumbria. Within selected areas (principally the North Cumbrian Coastal Plain), all existing aerial photographic data known from the county Sites and Monuments Record (SMR) was obtained; the project also drew heavily on the work of Bob Bewley (Bewley 1994).

Palaeoecological survey

Full details of the general rationale behind the palaeoecological survey and research strategies are described in the North West Wetlands Survey Project Design (LUAU 1993). In this section, the methods specific to survey in Cumbria are described and explained. Variations to the general methods described *below* are outlined in the introductions to specific studies where they occur.

The combination of numerous mires and the limited time available necessitated a selective approach to palaeoecological survey in the county. Consequently, it was carefully targeted at a range of sites aimed at covering as wide a geographical spread as possible.

The many basin mires in the south of the county possess a different value from an archaeological perspective to the large raised mires prevalent in the north. In the case of the latter, artefacts have been and will be recorded, thus demonstrating human activity, whereas in most cases the minor basin mires offer opportunities for passive recording of the effects of human disturbance to the environment. Taken individually, such small mires have the capacity to record events mostly of purely local significance but, when viewed as a county-wide network, they possess the potential to provide crucial data to augment Cumbria's archaeological record. A selection of such basin mires was therefore taken for palaeoecological assessment with the aim of testing their capacity to generate suitable information.

While the archaeological fieldwalking programme concentrated on both the mire fringes and any exposed mineral ground within the mire complex, stratigraphic survey sought to describe the deep organic terrain in order to establish the sequence of changing environmental conditions which would have influenced human activity in the vicinity. In addition, the studies helped to determine the degree of preservation of the peat and to provide a contextual framework in which to place any future detailed studies.

A 30 mm bore Eijkelkamp gouge auger was used to obtain cores of peat for rapid field description, whilst a Zeiss optical level was used for determination of relative surface levels of core positions. The exact location, number, and core intervals were varied according to the different conditions and depositional histories encountered in the various types of mire bodies occurring in the county. The more homogeneous large-scale peat deposits contained in many of the extensive 'raised mire' deposits required relatively wide-spaced cores (*ie* >100 m), whereas the more complex seral developments recorded in small basin mires necessitated close sampling (<30 m) in order to achieve a coherent picture.

Field descriptions of stratigraphic elements were made to a standard format and these were frequently complemented by sampling from the cores thus collected. These were examined in the laboratory using a Wild M3 Stereozoom microscope in order to check and add detail to field determinations of material.

A final complete stratigraphic description from each core was synthesised from amendments and additions derived from laboratory examination of peat samples and transcribed to computer database. This information was then used to draw a series of stratigraphic sections through each major mire system surveyed. The sum total of all the data was then used to interpret mire ontogeny. The stratigraphic descriptions throughout and the final diagrams use symbols based on those of Troels-Smith (1955) with modifications (Fig 7).

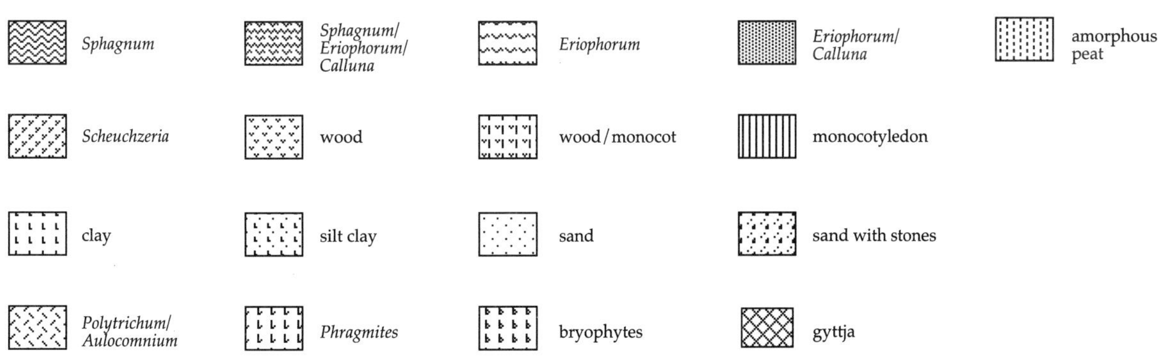

Fig 7 Key to stratigraphic symbols

Palynology

Samples of known volume, measured by volumetric displacement (Bonney 1972), were prepared for pollen analysis using the standard techniques of NaOH, acetolysis, and hot 40 % HF acid treatment if necessary (Faegri *et al* 1989). Tablets of *Lycopodium* spores were added at the start of the preparation so that pollen concentration values could be calculated (Stockmarr 1971); these were omitted in the spot pollen samples. The residues were mounted in silicone oil and examined with an Olympus BH-2 microscope using x 400 magnification routinely and x1000 for critical grains. Counting continued until a sum of at least 500 pollen grains excluding aquatic and mire types had been reached on two or more complete slides. This was done to reduce the possible effects of differential dispersal under the coverslip (Brooks and Thomas 1967). A sum of 250–300 grains was used in the pollen assessment samples. Pollen identification was carried out using the standard keys of Faegri *et al* (1989) and Moore *et al* (1991) and a limited reference collection held at Lancaster University Archaeological Unit (LUAU), which restricted the identification of the more unusual grains. Cereal-type grains were defined using the criteria of Andersen (1979); indeterminate grains were recorded using groups based on those of Birks (1973). Pollen of *Cannabis/Humulus*-type is referred to variously as *Humulus* or *Cannabis*-type in the text; these two taxa are difficult to distinguish one from another (Moore *et al* 1991) and either may be represented, although it is probable that *Humulus*-type occurred during the earlier periods and *Cannabis*-type in more recent times. Charcoal particles >5 µm were also counted following the procedures of Peglar (1993). In the detailed palynological work from Solway Moss, a crude measure of the mineral content of the peat was assessed by loss-on-ignition of the sediments: the samples were dried for 48 hours at 105 °C and then ashed at 550 °C overnight.

Plant macrofossil

Cores were sub-sampled into equal-sized blocks of varying volume and at varying intervals depending on the site investigated. These units are specified in the individual site descriptions. Sample volumes were measured by water displacement and samples were disaggregated by soaking in water for a minimum of 24 hours. Those containing clay or *Phragmites*-clay were dispersed using 10 % potassium hydroxide and dispersed samples were washed through two sieves with mesh sizes of 500 µm and 178 µm. Material was retained from both sieves and was systematically examined for plant macrofossil content using a Wild M3 lower power stereozoom microscope (x 6–40 magnification). Plant material was identified using a reference collection and the keys of Beijerinck (1974), Berggren (1969; 1981); Grosse-Brauckmann (1972), Jessen (1955), and Katz *et al* (1965).

Analysis and Preservation

Analysis and storage of palynological and macrofossil data was accomplished using computer-based tools, particularly the TILIA software package (Grimm 1991), to categorise data and aid its interpretation. Pollen count sheets, microscope slides, and the residue of prepared samples are stored at LUAU. Data were analysed to ascertain possible human influences on the natural environment and to relate this to known and newly-discovered archaeological remains.

Pollen data have been presented as percentage and, where appropriate, concentration pollen diagrams of selected taxa using the computer program TILIA and TILIA*GRAPH (*ibid*). The percentage values are based on a pollen sum of all types excluding aquatics, Cyperaceae, *Sphagnum* spores, Filicales, and indeterminate grains; all types excluded from the pollen sum are expressed as a percentage of the pollen sum. Charcoal (>5 µm) values are expressed as a percentage of the pollen sum plus charcoal counts. The individual curves on the concentration diagram are drawn at various scales depending on their values. The pollen diagrams are generally divided visually into local pollen assemblage zones to aid description; where possible, the boundaries of such zones are dated using calibrated radiocarbon dates (*eg in Appendix 6*).

Macrofossil data are presented as a series of histograms indicating frequencies of vegetative remains per unit volume and seeds from each level sampled. In the diagrams and descriptions the word 'seed' is used to describe both fruits and seeds in order to simplify the presentation. The series of histograms display the frequency value for seeds, leaf fragments etc, found in each sample examined from stratigraphical elements down the core. The results display relative abundance of plant remains assessed on a semi-quantitative five-point scale *viz*:

1 = rare
2 = occasional
3 = frequent
4 = very frequent
5 = abundant

where:

Rare – vegetative material occurring only once or one seed;

Occasional – vegetative material occurring only a few times or two to five seeds;

Frequent – vegetative material occurring regularly or seeds five to 20 seeds;

Very frequent – vegetative material occurring in every portion of the sample examined or seeds 20+;

Abundant – vegetative material occurring in field of view all the time and dominating the sample, or 40+ seeds.

The appearance of a taxon name on the diagrams without additional qualifying information indicates the representation of totals of vegetative/reproductive remains found referable to that particular taxon. Watts and Winter (1966) are followed in the relative certainty of determination afforded to the macrofossils encountered. Individual histograms are arranged such that taxa having similar stratigraphical occurrences are grouped together, as far as is possible.

Documentary studies

The documentary research in Cumbria had two principal aims: firstly to identify and locate as many of the county's former mosses as possible and, secondly, to elucidate the recent history of several of the major lowland mosses, in particular their exploitation and reclamation, as this will have a bearing on their potential for survival in terms both of peat and archaeological material.

The principal source used for the identification of 'moss' and related place-names was the OS First Edition 6" maps of Cumbria held at the County Record Office. These are, however, quite late, the majority having been surveyed in the mid-to-late 1870s and published in the early 1880s, although they do provide the most comprehensive and accurate listing of late nineteenth century place-names, since all the names were verified in the field by the OS surveyors. The maps are also at a scale where the land boundaries can be accurately transferred to modern 1:10,000 maps. The earliest OS maps for Cumbria are the First Edition 1":1 mile series, published in 1842. Although useful in being 40 years earlier, the scale means they are much less comprehensive.

In addition, tithe maps were consulted where available and a general search of mostly published literature was made, supplemented by some primary documents where these were known to be pertinent. In particular, great use was made of the records of Furness Abbey, St Bees Priory, and Holm Cultram Abbey to shed light on the medieval exploitation of the mosses of the south, west and north of the county respectively.

Geographic Information System

A Geographic Information System (GIS) has been used throughout the project to permit the graphic display of data taken from a central database. The GIS has had three major uses within the project: firstly, allowing the production of publication-quality maps directly from the database at a variety of scales and formats; secondly, allowing large amounts of information of many different types, both archaeological, palaeoecological, and topographical, to be stored and manipulated in a much more efficient manner than is otherwise possible; thirdly, and perhaps most importantly, it may permit the construction of models predicting the fate of peatlands, based on all the data recovered during the project. The reconstruction of past landscapes can play a key role where modern threats and developments need to be assessed.

The computer-based system is also the main vehicle for the project archive; it holds, both textually and graphically, the major elements of the archaeological and palaeoecological datasets. The use of this system as the main archiving medium permits flexible output to the county SMRs and National Monuments Record in a variety of formats, to suit local circumstances. The use of a single, centrally-based system also ensures the use of standard data structures throughout the work of the project. Storage of project data in this form will permit the archive to be continuously updated and maintained so that it provides an up-to-date database upon which long-term management strategies can be based.

This volume differs from previous publications in that a decision was taken to restrict the GIS mapping output to the main peat expanses on the North Cumbrian Plain, since the other areas surveyed were small by comparison and few new sites were found on them. In the event, all map data has been manipulated for publication through AutoCad 14, thereby speeding the process.

Archive

The data generated by the survey work in Cumbria has been archived and forms part of the larger North West Wetlands Survey archive, which is held at LUAU. The Cumbria element of the archive, information from which has been copied to Cumbria SMR, contains the following sections:

Archaeology

- Fieldwork data, including note books (day-to-day record of fieldwork), farm record sheets (notes on land-use, field conditions on individual farms), finds record sheets (notes on individual finds and findspots), field copies of 1:10,000 maps (with details of land-use, fields surveyed, and archaeological data), 1:25,000 scale maps (land ownership details), and original copies of sketch field surveys;

- SMR information relating to wetlands survey areas, ordered by 1:10,000 scale OS quarter sheet;

- All correspondence relating to the archaeological survey;

- Original copies of final publication illustrations;

- Catalogue of slides, relating to the Cumbria fieldwork;

- The Solway Cow remains are archived at the Centre for Archaeology, Portsmouth.

Palaeoecology
- Original pollen and macrofossil count sheets stored as raw data;

- Electronic archiving of the same in both numerical and graphical form using the computer software TILIA/TILIA*GRAPH;

- Pollen slides are archived at LUAU along with residues from each preparation;

- Selected peat samples were retained in cold storage at the Institute of Environmental and Biological Sciences, University of Lancaster. At the conclusion of the survey almost all samples were discarded due to the lack of available long-term storage facilities.

Documentary research
- All data gathered during the course of the documentary research, ordered by wetland area;

- A copy of the final report is also included in the archive;

- The finds from the survey, apart from those held by the owners, are in the possession of LUAU, but will be deposited at appropriate museum(s).

Organisation of the volume

The writing of the text has been split between the principal authors: David Hodgkinson (fieldwork, archaeology, and documentary research), Elizabeth Huckerby (palynology and assessment), Bob Middleton (fieldwork, archaeology, synthesis), and Colin Wells (palaeoecological survey and synthesis), although there has been a degree of overlap on all sides. Major contributions to parts of the text have been made by Paul Hughes (Solway Plain Mosses), and Sue Stallibrass and Michael Ryder (the Solway Moss animal finds).

The archaeological survey programme was divided into three major areas, each of which contained the main areas of lowland peat in the county: South Cumbria, the West Cumbrian Plain, and the North Cumbrian Plain. This structure is reflected in the volume, which has five chapters, three of which have the same format: a description of the area and an assessment of the peat resource, followed by descriptions of the known archaeological and palaeoecological information and fieldwork results, and concluded with a discussion of the potential of the mosses in the area concerned. The descriptive text is preceded by an introduction, which discusses the history, aims, and methodology of the survey, as well as a brief introduction to the county, and the volume is completed by a final discussion chapter which summarises the findings of the project and looks to the future of Cumbria's mossland archaeology.

The last part of the volume consists of a series of Appendices containing information on finds and sites discovered by the NWWS, including a brief discussion on the flint material, SMR sites on and around the wetlands studied, sites investigated palaeoecologically, information on wetlands gathered as part of the NWWS and those identified from documentary sources, and a list of radiocarbon dates cited in this volume either obtained by the NWWS or other authors.

2

SOUTH CUMBRIA

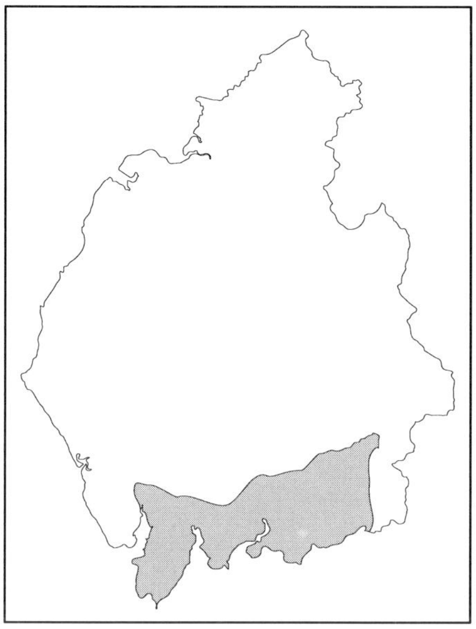

Fig 8 The extent of the area covered by Chapter 2

'I was wandering over the flat-topped ridge of mountain limestone which runs south between the valleys of the Gilpin and the Kent — an area in every way suitable for primaeval man. Here was open ground for pasture when domestic animals came in, as these limestone terraces never carried much timber. On the west was the peat-covered estuary of the Gilpin, affording a rich hunting ground and a safe retreat to man and beast that knew its ways, but more encroached upon by the tides in those days than now. There was the Kent round the southern end of the promontory, with plenty of salmon for food and otters for fur… In such a district we might expect to find traces of primaeval man of almost every age' (McKenny Hughes 1904a, 71).

Introduction

The subject matter of this chapter is an area stretching west from the Kent Estuary to the Duddon incorporating the Furness Peninsula (Fig 8). Its northern boundary is taken as the lower slopes of the Coniston mountains and the upper River Duddon, encompassing the uplands of moderate altitude lying between the estuaries. The area is mainly confined to Countryside Character Area (CCA) 20 *'Morecambe Bay Limestones'*, although it extends into the southern tip of CCA 7 *'West Cumbria Coastal Plain'* (Countryside Commission 1998; Fig 9); the eastern part of CCA 20 has been discussed as part of the North Lancashire survey (Middleton *et al* 1995). The area also forms the southernmost strip of the *'Cumbria Fells and Dales'* Natural Area (English Nature 1997; Fig 10). Adjacent to the lowlands, the rich archaeological record of the upland fringe (Blawith, Subberthwaite, and Torver Commons; LUAU 1994; 1995) necessarily informs any discussion of South Cumbrian landscape history.

As defined, South Cumbria possesses a richly varied topography that incorporates many distinctive landscape elements. Extensive estuaries mark its southern fringe, their saline influence extending well up into the hinterland of rolling knolls smothered in dense woodland, thanks to a deeply indented coastline. The foothills are formed from limestone in the east around the Rivers Kent and Leven, while mudstones and siltstones underlie the broad Lyth Valley with its reclaimed wetlands (Pl 8). To the west, the craggier Duddon Fells comprise ancient Silurian geology, while the Furness Peninsula is largely covered by Pleistocene tills, with Holocene dune and marine deposits on its southern tip. By way of contrast, the northern hinterland of this zone rises to impressive stretches of open moor, leading to the cliffs and ridges of the Coniston Fells (Ratcliffe 1997, 8–17).

Geology and soils

The solid geology of the northern, more elevated (>200 m OD) portion of South Cumbria is dominated by Silurian rocks, mainly Bannisdale Slates and Coniston Grits, with significant outcrops of Brathay Flags occurring in High Furness. The craggy, higher ground which distinguishes this terrain tends to be dominated by soils of the Grizedale Association, a mixture of peaty and groundwater gleys, podzols, and bare rock, while steeper slopes are typically

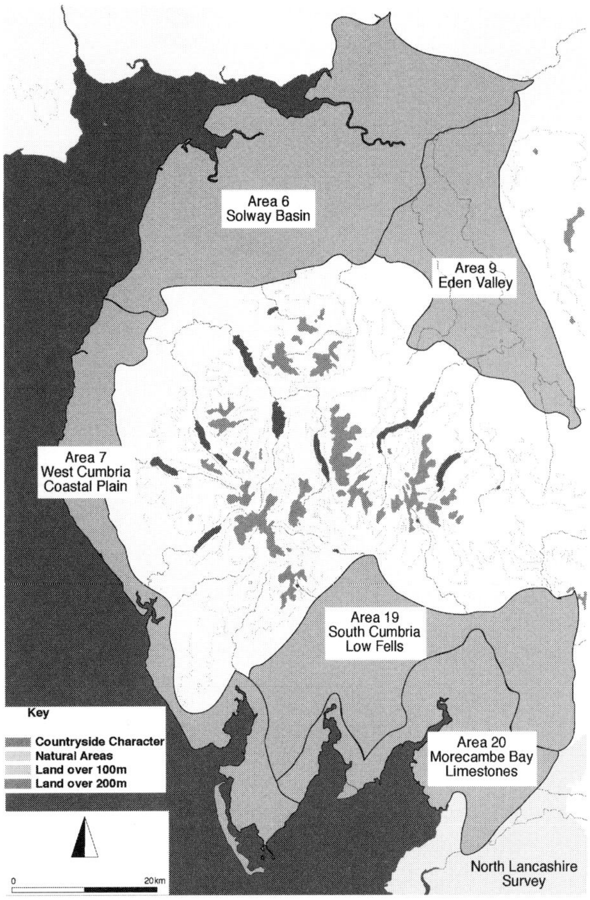

Fig 9 Countryside Character Areas (CCAs) in Cumbria

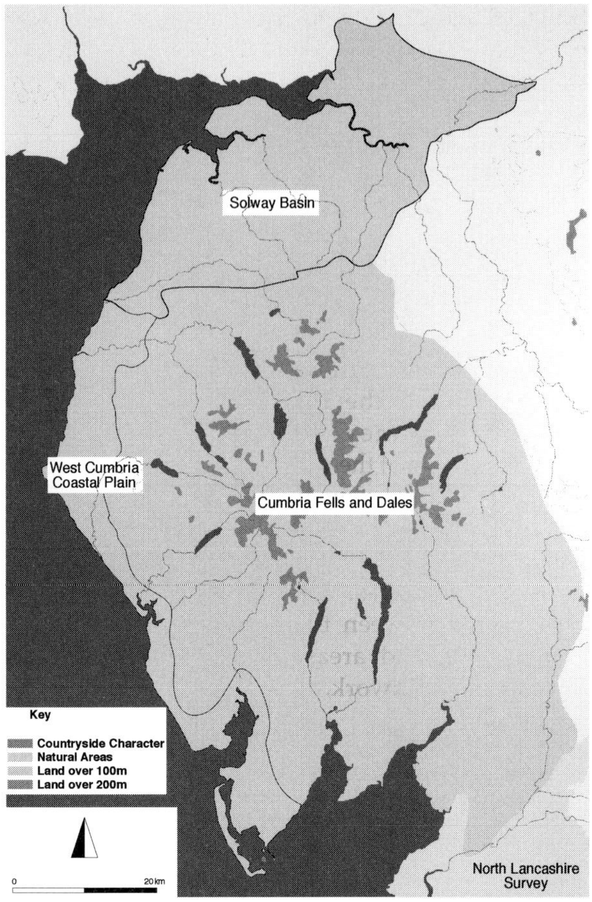

Fig 10 Natural Areas in Cumbria, as defined by English Nature (1997)

characterised by deeper brown earths and rankers of the Brantwood Association (Hall and Folland 1970).

To the south, Carboniferous limestone is the major rock type, although a liberal mantling of Pleistocene tills modifies its surface expression. Much of the lower-lying ground (<200 m OD) is characterised by a mixture of brown earths, gleys, and podzols of the Lowick and Lindal Associations, derived from the tills (*ibid*).

Significant and distinctive components of the area's pedology are calcareous soils which occur in a discontinuous arc following the outcrops of Carboniferous limestones on the rim of Morecambe Bay from Levens to Bardsea. This geology produces a variety of brown earths and rendzinas of the Warton Association, often carrying a semi-natural vegetation, characterised by a mixture of deciduous woodland and calcareous grassland, as well as areas of bare limestone pavement (*ibid*).

The southernmost tip of the Furness Peninsula is formed from Triassic sandstones and mudstones, although most are cloaked in drift deposits and carry similar brown earths and ground water gleys found to the north. Finally, Holocene blown sand deposits form the basis for the most recently developed raw soils of the Ainsdale Association on the coastal dunes of Walney Island, forming the south side of Barrow Harbour (*ibid*).

Archaeological and palaeoecological survey

Archaeology

As with much of lowland Cumbria, the chief determinant on archaeological fieldwork strategy and also of the level and the type of results is the predominantly pastoral land-use. As a result of the limitations of visibility, fieldwork has, in the past, been restricted to the uplands fringing the wetlands where upstanding sites have survived, for example at Levens Park (SD 4985–5085) on the eastern edge of the Lyth Valley (Fig 11). This has meant that there has been no previous systematic fieldwork in the wetland areas, and little recognition of their potential for well-preserved archaeological remains. The Morecambe Bay Archaeological Research Society

(MBARS) has undertaken the only recent fieldwork in Furness, although little of this has been directed specifically at wetlands. Other previous work has tended to be patchy in quality, which is a particular problem with respect to the coastal sites on Walney Island (SD 1768; SD 170680–180720 and SD 1872) where much of the record is made up of casual flint finds from exposures. A more accurate record of such sites may have permitted the coastal archaeology of Morecambe Bay to be characterised more closely.

Whilst the 'dry-land' sites have undoubted importance for the interpretation of wetland archaeology, they do not allow a precise characterisation of the wetland resource, such as the Foulshaw Moss trackway (described further *p 26–7*, Fig 12): such a find does not have 'dry-land' parallels and cannot be predicted from comparisons with non-wetland areas. The understanding of the relationship between the archaeology of wetland and non-wetland areas is important, given the imbalance in fieldwork, and will be discussed further in *Chapter 5*.

The dominance of pasture on the reclaimed wetlands and wetland edge, combined with semi-natural vegetation in the areas of peat survival, along with large areas of forestry, meant that the field survey programme was restricted to a rapid walk-over, in common with other NWWS areas where such conditions prevailed. Peat depths and land-use were noted and detailed fieldwalking and sketch earthwork surveys undertaken where conditions permitted.

The archaeological fieldwork was undertaken in two stages (Fig 13):

> Lyth (Fig 11) and Winster Valleys (Fig 14), Rusland Pool (Fig 15):
> January–May 1994;
>
> Duddon Mosses, Arnaby Moss (Fig 16):
> August 1996.

Survey was restricted to surviving peat, and areas of remnant organic soils where peat was formerly thought to exist, such as the Lyth Valley. There was no attempt to survey all of the coastal silts, such as the large expanse south of Foulshaw Moss, nor were foreshores examined. These omissions are consistent with NWWS survey methodology in other areas (for instance, Merseyside, (Cowell and Innes 1994); and

Plate 8 Lyth Valley, South Cumbria, showing topography

Fig 11 Lyth Valley: predicted peat survival and actual area surveyed (Scale: 1:50,000)

Based upon the 1989 Ordnance Survey 1:50,000 maps with permission of the Controller of Her Majesty's Stationery Office © Crown Copyright

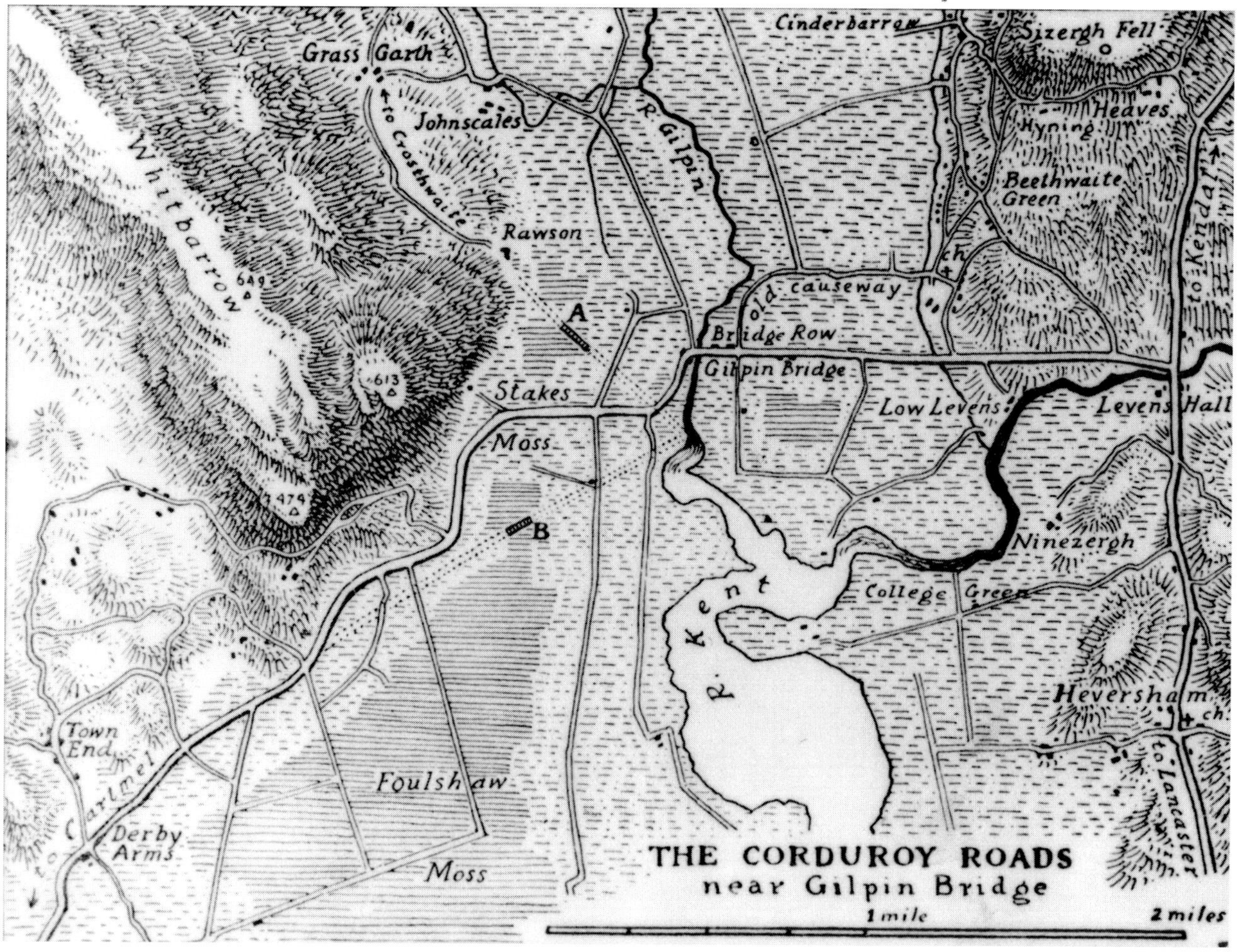

Fig 12 Foulshaw Moss trackway: location plan (from Barnes 1904)

Lancashire, (Middleton *et al* 1995; Middleton *et al* forthcoming)). At the start of the fieldwork, however, one major category of peatland was excluded from the archaeological survey for reasons of time and practicality. There are many hundreds of small basin mires in South Cumbria, particularly in the moraine-covered landscape south and west of Kendal, most of which are unmapped. It has been shown by the NWWS (Middleton *et al* 1995) that these contain valuable palaeoecological deposits which may be particularly important for the identification of small-scale human interference in the vegetation (Leah *et al* 1997, 129-31) and, despite being excluded from the present archaeological survey, future mapping and survey would be desirable for the identification of human activity.

In addition to the landscape survey element, details have been included in this chapter both of a small excavation undertaken in collaboration with MBARS at Bardsea within foreshore peats, and the results of an earlier project, undertaken by LUAU, which examined a pipe trench during construction, in Meathop Moss. In addition, as a result of the vigilance of a member of the public, a horse's head with mandibles was reported during the excavation of a fish pond in the Lyth Valley (SD 4105 9275). The skull was examined at the Department of Archaeology at the University of Durham and was found to be that of Dales mare-type; it was thought because of the state of preservation, to be modern (Gidney and Huntley 1996).

Palaeoecology

Owing to time constraints, no new palaeoecological work was undertaken in South Cumbria as a part of the NWWS field programme, although a limited rapid assessment programme was undertaken on some of the small basin mires east of Kendal (*see Appendix 5*). Recently much detail has, however, been added to the sequence of post-Neolithic environmental change by Guy Wimble, who, in his study (1986), concentrating on the mosses bordering the northern coast of Morecambe Bay and the surrounding limestone uplands, has produced the most detailed and precisely–dated palaeoecological study yet attempted for a segment of Cumbrian landscape. By virtue of their location (Foulshaw,

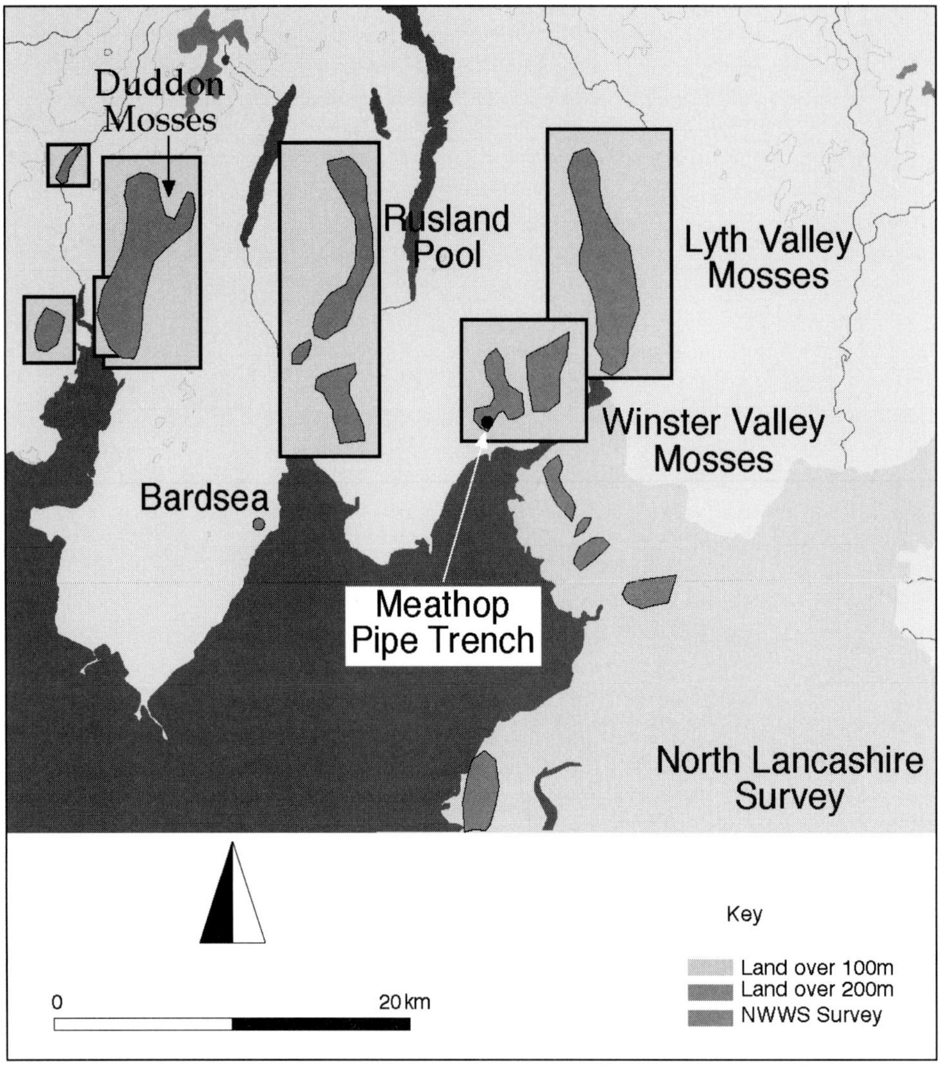

Fig 13 South Cumbria: areas of archaeological survey

Helsington, and White Mosses; Figs 11 and 16), his pollen sampling sites give a comparative vegetational record of a wide area of north Morecambe Bay. The use of multiple radiocarbon dates in all the areas studied has provided Wimble with an opportunity to construct an accurate and relatively-precise chronology of changes ranging from the early Bronze Age to recent times. As well as making a particularly significant contribution to the environmental history of the area, and the relationship of human activity to it (much of the vegetational change recorded in the diagrams was attributed to human intervention), the study's relatively high resolution allows reappraisal of previously unpublished but largely undated pollen diagrams from South Cumbrian mosslands (Smith 1959; Oldfield 1960; Oldfield and Statham 1963). Despite its importance, however, none of Wimble's data has been published, except for a brief summary in the *North West Wetland Survey Annual Report* for 1991 (Wells 1991).

Palaeoecological evidence for early prehistory in South Cumbria is limited because few archives have been identified and so the environmental conditions for this period are difficult to identify. Slightly better data relating to this early period, however, are available from topographically-similar landscapes in North Lancashire. This information is used as additional proxy evidence to provide a broad palaeoenvironmental context for the discussion of the pre-Neolithic archaeology of South Cumbria.

Wetland distribution

Each of the main wetland 'units' was assessed separately during the archaeological field survey. There are many areas, however, where the divisions between the wetlands are largely arbitrary, for example between Foulshaw Moss in the Lyth Valley, and the neighbouring Meathop Moss in the Winster Valley. The distribution of peat is shown in Figures 11 and 15.

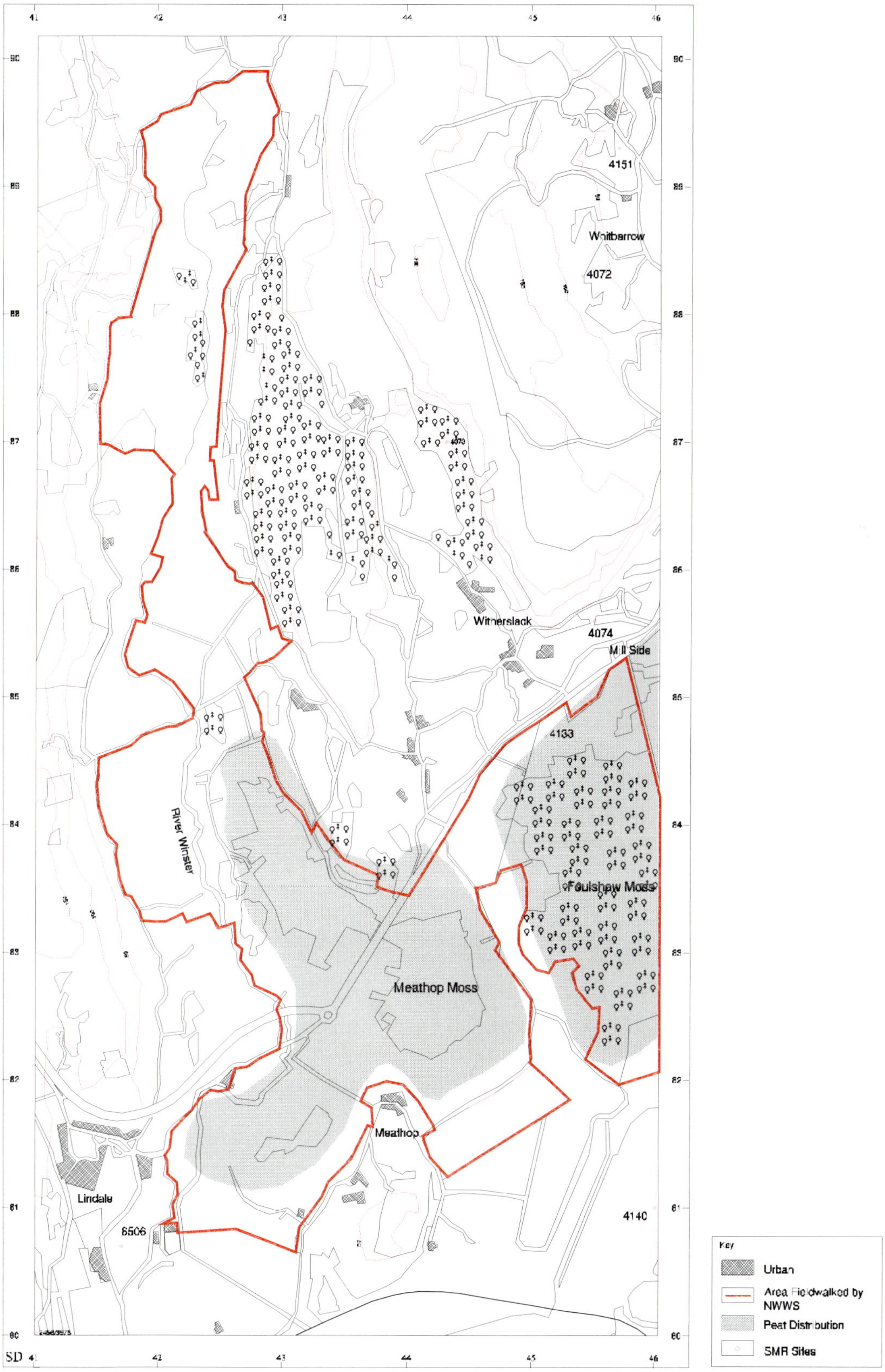

Fig 14 WinsterValley: predicted peat survival and actual area surveyed (Scale: 1:50,000)

Based upon the 1989 Ordnance Survey 1:50,000 maps with permission of the Controller of Her Majesty's Stationery Office © Crown Copyright

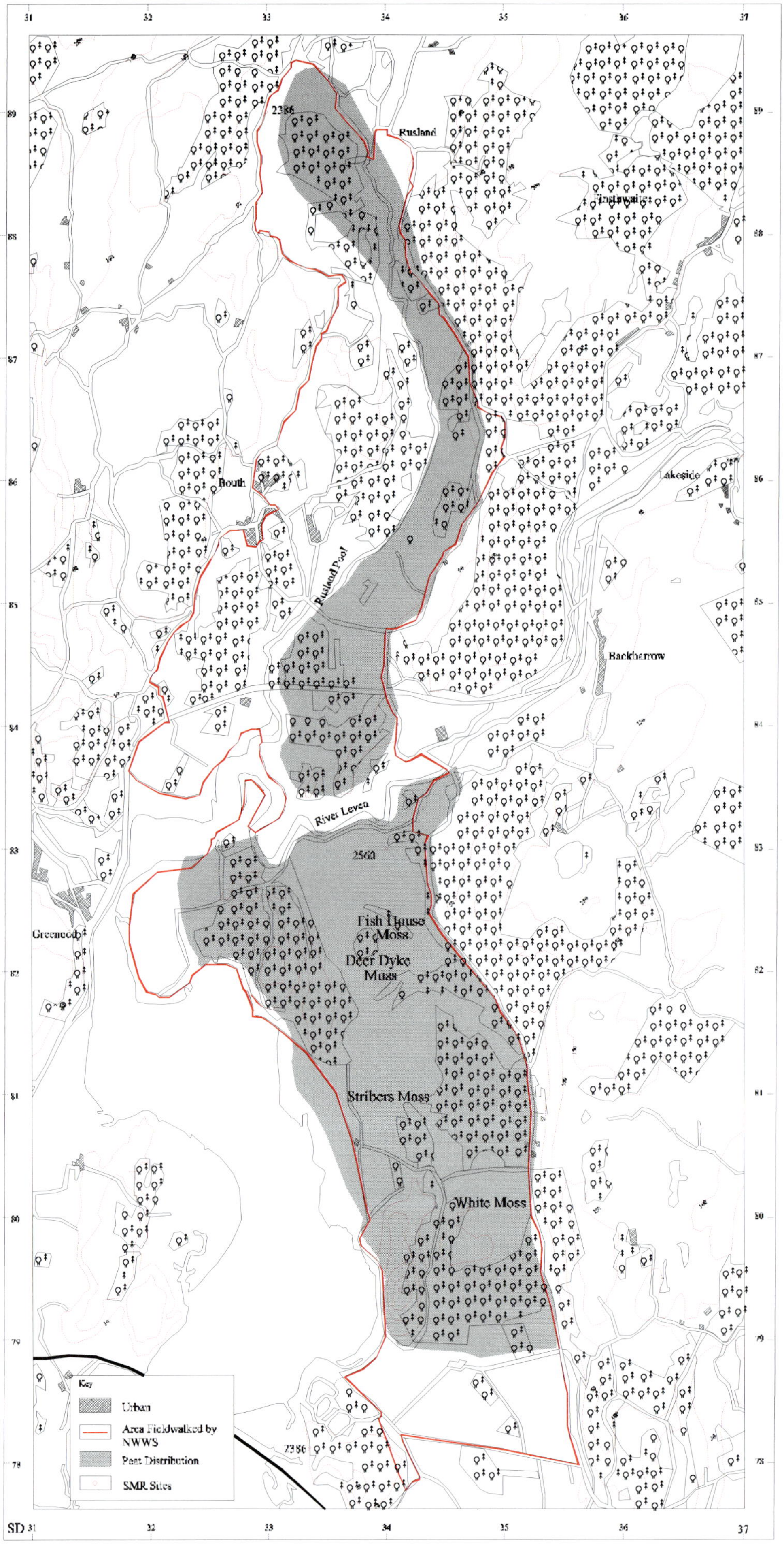

Fig 15 Rusland Pool: predicted peat survival and actual area surveyed (Scale: 1:50,000)

Based upon the 1988 Ordnance Survey 1:50,000 maps with permission of the Controller of Her Majesty's Stationery Office © Crown Copyright

Lyth Valley

The combined agencies of drainage and peat cutting have led to the disappearance of much of the vast raised mire systems which formerly occupied much of the Lyth Valley and low ground west of the Kent Estuary (*see Appendix 4*). Many of the mires are now indistinguishable and have been completely removed by peat cutting. Some of the larger mosses, however, despite having been extensively cut, contain small remnants of deep peat surviving in the former centre of the moss. During fieldwork, each of the mosses was assessed individually in terms of both peat survival and land-use, summaries of which are presented in *Appendix 4*. It should be noted that the moss names are those derived mainly from First-Edition OS maps and may originally have been either hydrologically discrete mire units or parts of mires belonging to particular settlements.

Winster Valley

The smaller, enclosed valley to the west of the Lyth appears in complete contrast with its wide and open valley floor and shelving sides on the west and north. There are, however, several points in common, notably that the valley floor in each was generally moss-covered in the past and that both open out at the southern end to merge with the marine and estuarine silts on the edge of Morecambe Bay.

The Winster Valley stretches from Winster in the north to Lindale in the south (Fig 14), although the mosslands were located where the valley opens a little, from Swallow Mire south to Meathop Moss. To the west, the valley is defined by Newton Fell and Cartmel Fell, and to the east by Whitbarrow, which separates the Winster from the Lyth Valley. The Winster mosses have not suffered the same level of large-scale cutting as those in the Lyth Valley and tend to be better preserved accordingly, although most of the mires have been subject to some hand-cutting.

In *Appendix 4* the major mosses have been identified, although many small mires must have existed in the past, as evidenced from some place-names, notably Swallow Mire and Cowmire House in the north of the valley. There is now no evidence for the wetlands to which these referred, although the presence of many unrecorded basin mires in South Cumbria has been noted *above*.

Rusland Pool

The Rusland Pool mosses are located in the small enclosed valley stretching from Ickenthwaite in the Furness Fells, southwards along the eastern side of the Leven Estuary (Fig 15). To the west, the valley is defined by the steep slopes of Yew Barrow and, south of the River Leven, Stribers Bank. Rusland Pool displays similar characteristics to those seen in the Winster Valley, in that the valley floor was formerly dominated by a wetland environment which opened out at the southern end to meet with marine and estuarine silts on the edge of Morecambe Bay.

The majority of the mosses in this area are low-lying, situated at a height of less than 10 m OD, with the exception of a small group located on the slightly higher ground to the west, in the vicinity of Colton. Whilst drainage and small-scale hand-cutting activity, together with the construction of the A590 trunk road, have caused some damage to the wetland environments, the level of preservation is generally higher than, for instance, in the Lyth Valley.

Duddon Mosses and Arnaby Moss

Encompassing the northern end of the Duddon Estuary, where the narrow Duddon and Lickle Valleys converge, an area of lower-lying land contains numerous mosses (Fig 16). To the west of the Duddon Estuary, Arnaby Moss occupies the valley plain beyond the estuarine salt-marsh, whilst to the east the Angerton Mosses extend in a north-easterly direction, bounded by the steep slopes of the Dunnerdale and Blawith Fells. Small valley mires occupy the Duddon and Lickle Valleys, which open out at the southern end to meet with the marine and estuarine silts on the edge of Morecambe Bay. This area also includes the slightly higher ground of the Furness Peninsula to the south.

Extensive drainage activity across the Angerton Mosses, together with small-scale peat-cutting, has inevitably resulted in the complete loss of some smaller mires and caused desiccation and damage to others. On the western side of the Duddon Estuary, however, Arnaby Moss is relatively well-preserved despite a degree of hand-cutting.

Landscape Evolution

Late Palaeolithic

The late Quaternary origins of the modern South Cumbrian landscape can only be vaguely discerned, given the paucity of environmental archives which stretch back so far. From study of what stratigraphical evidence is available, it seems likely that after the maximum glaciation of the Devensian, perhaps *c* 16,000–14,000 cal BC (*c* 15,000–13,200 BP), deglaciation of the Morecambe Bay area was sufficiently well advanced by *c* 12,000 cal BC (*c* 11,900 BP) to allow the deposition of fluvio-glacial clays, gravels, and sands over a wide area (King 1976). This probably formed the raw material for re-

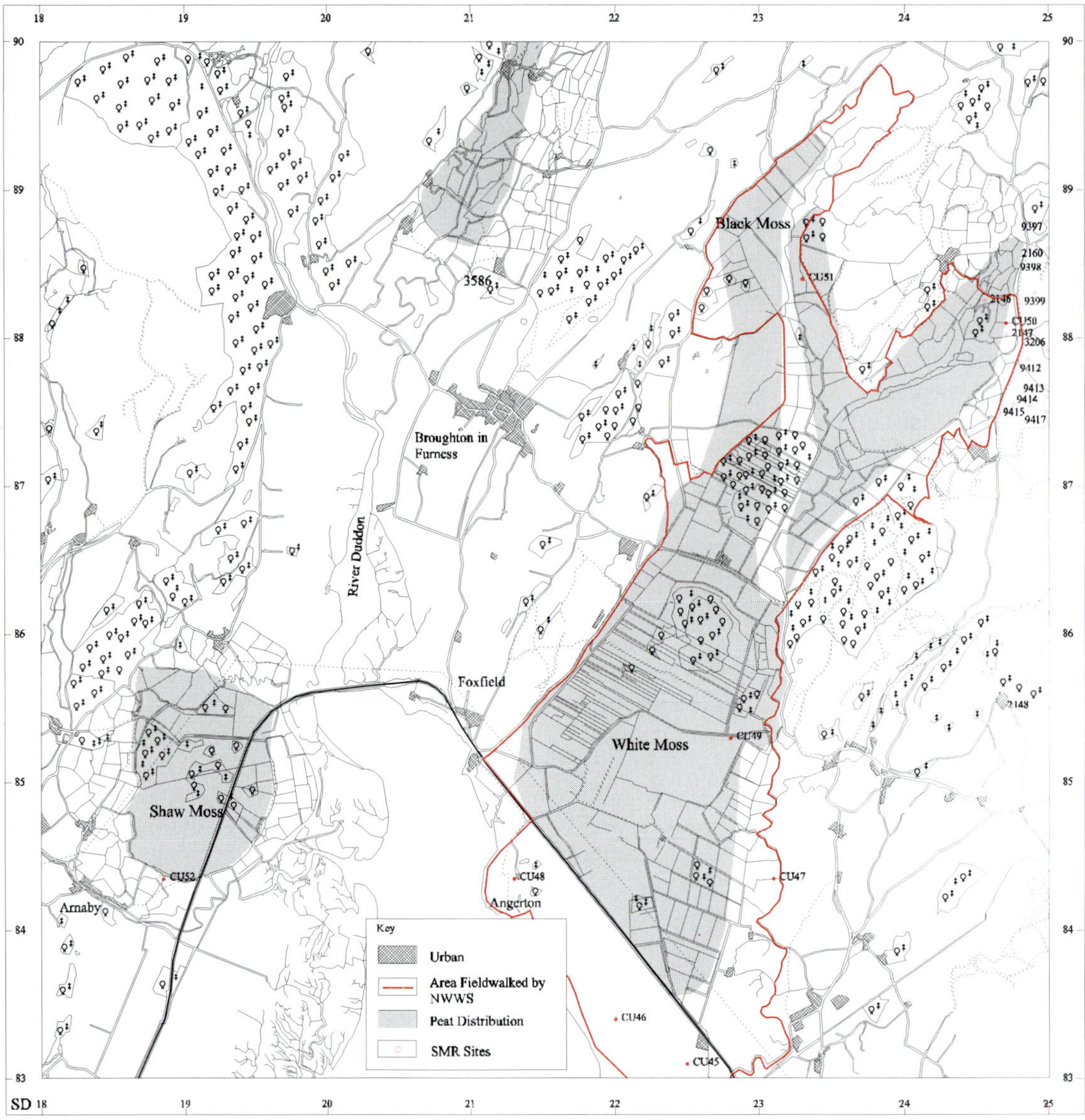

Fig 16 Duddon Mosses: predicted peat survival and actual areas surveyed (Scale 1:50,000)

sorting by subsequent aeolian processes. These deposits formed on the *Sandur* which probably developed in Morecambe Bay and resulted in the mantling of adjacent coasts and hills with loess as the glaciers retreated (Vincent and Lee 1981). Although the fragile soils which developed on such material have long since been removed by overland flow in most areas, modified loess deposits survive to this day on the limestone knolls surrounding the Bay, thanks to the relative permeability of the bedrock causing cementation as a result of chemical reaction (Catt 1977, 226–7). Today, a very visible legacy of these events in South Cumbria is the way in which these hills — Whitbarrow, Farleton Fell, Arnside Knott, Hutton Roof, and Warton Crag — rise above the till-covered ground below, carrying a distinctive *Sesleria* grassland vegetation which, in this area, is mostly restricted to their free-draining loessic soils (Pl 9).

The first evidence of human activity comes from the limestone coasts of southern Cumbria. Kirkhead Cave near Grange (SMR 2415) appears to have been occupied during the Upper Palaeolithic and has produced artefacts defined typologically as belonging to this period, while three Palaeolithic-type blades have been claimed from Lindale Low to the north-east of Kirkhead, near the mouth of the River Kent (Salisbury 1988). Elk (*Alces alces*) antlers, considered by Gale and Hunt (1985) to be associated with the artefacts in Kirkhead Cave, produced a radiocarbon date of *c* 11,075–10169 cal BC (10,700±200 BP, Har-1059) and, on the basis of very limited pollen assemblages preserved in cave sediments, they estimated occupation to date to Late Devensian III (LDe III). The association of the antlers with the artefacts may, however, be suspect due to sediment disturbance by previous attempts at excavation (Salisbury 1986) and Tipping (1986) has suggested that the pollen samples were contaminated. Possible Upper Palaeolithic evidence has also been recovered from ongoing excavations at Bart's Shelter, Scales, in Furness, although this has yet to be established with certainty (D Stables pers comm).

In the absence of an accurate chronology it is difficult to establish the precise date of this presumed late Upper Palaeolithic occupation in South Cumbria, although its importance as the most north–westerly late Upper Palaeolithic human activity recorded in the British Isles (Wymer 1981, 77) is undoubted. There may be an intuitive case for suggesting occupation of these cave sites during the relatively benign Windermere Interstadial (LDe II), rather than during the renewed glaciation of the following Loch Lomond Stadial (LDe III; *cf* Gale and Hunt 1985) and possibly contemporary activity has been found at Poulton-le-Fylde in the North Lancashire wetlands (Hallam *et al* 1973; Middleton *et al* 1995, 86-7). Both this and the South Cumbria material represent scattered fragments of evidence for early hunters exploiting the megafauna of the tundra behind the edge of the retreating ice. The evidence from Poulton may suggest that wetlands would have been exploited for hunting opportunities, with cave sites such as Lindale Low (SMR 6506) and Kirkhead (SMR 2415) offering look-outs across the wide plain of what is now Morecambe Bay.

Mesolithic and early Neolithic

Palaeoecological evidence

The opening of the Holocene was marked by the rapid amelioration of the climate and the establishment of a shrubby mixed woodland on the newly formed soils. Birch (*Betula*) led the invasion, followed by hazel (*Corylus*), which was to become a dominant member of the flora after *c* 9000–8000 cal BC (*c* 9500–9000 BP). The issue of whether the marked hazel expansion at the Flandrian 1b/c boundary was wholly natural, or partly aided by the effects of human activity, in particular fire, remains to be established with certainty (*eg* Huntley 1993). At Little Hawes Water in North Lancashire, however, Taylor *et al* (1994) recorded the presence of charcoal in peat stratigraphy which coincided with the expansion of hazel. This in turn was replaced by pine (*Pinus*), oak (*Quercus*), elm (*Ulmus*), and, subsequently, alder (*Alnus*).

Plate 9 Sesleria *grassland, Hutton Roof (© T N Huckerby)*

There is evidence that at least half a dozen marine transgressions, of varying intensity, affected the coasts around Morecambe Bay during the prehistoric and early Roman periods (Kidson and Tooley 1977, 138–9). The early Holocene transgression phase been recorded starting before *c* 8980-7943 cal BC (9270±200 BP, Birm–141), so that by 7060–6707 cal BC (7995±80 BP, Hv–3362) the lowest reaches of the Kent and Levens Estuaries had been inundated. At the same time the transgression was probably reaching its maximum landward penetration north of the Levens Estuary in the valley of the Rusland Pool. Here, grey-blue clays containing the Foraminifera *Globigerina* sp and fragments of *Radiolara* and sponge spicules have been found inland as far north as Crooks Bridge (*op cit*, 138; Dickinson 1975).

A second transgression, dating to between *c* 5000 cal BC and 4000 cal BC (5865±115 BP, Q–261; 5734±129 BP, Q–256; 5435±105 BP, Hv–3844), has been recorded from Silverdale, Helsington, and Ellerside Mosses, and was probably short-lived, although its influence seems to have been intense, with sea-level rising to *c* +3.72 m to +3.85 m OD at its maximum (Oldfield 1960; Huddart *et al* 1977, 138). A third transgression, reaching nearly +5.0 m OD in places, seems likely to have affected Helsington Moss and the Duddon Estuary in the north of Morecambe Bay, and Arnside Moss to the east, around 4400–3600 cal BC (5277±120 BP, Q–85; 5015±100 BP, Hv–3460; Smith 1958; Kidson and Tooley 1977, 139, 142).

Palaeoecological evidence of Mesolithic activity from South Cumbria is lacking, to some extent due to a paucity of suitable environmental archives, and a lack of general research, although it may reflect a genuine absence of activity in the area. The only published study relevant to the period focused on the Carboniferous limestone belt fringing the South Cumbrian mires at Roudsea (Birks 1982). This work was aimed primarily at elucidating the nature of mid-Flandrian woodland in this area with a view to testing how far the general picture, derived primarily from open-mire sites and thought to represent the regional pollen rain, compared with a pollen record from a small hollow presumed to have been situated actually within the 'wildwood' at the time. Birks' (*ibid*) profile spanned a short period between *c* 6000 cal BC, and 4000 cal BC (*c* 7200–5200 BP), but it nevertheless confirmed that forests of late Mesolithic date displayed a high biodiversity, qualitatively not dissimilar to the present-day assemblage found in the wooded area. The findings tend to support the notion that many of the woodlands presently occupying the limestone headlands round the northern edge of Morecambe Bay are derived from original woodland cover, albeit heavily modified by millennia of subsequent human disturbance. The late Mesolithic woodland assemblage represented in this particular study, however, displayed no evidence of local forest disturbance until the uppermost sediments, in which an Elm Decline is clearly discernible. However, given the very localised nature of the pollen catchment of the sample site (a small hollow set within dense woodland), and the very subtle and scattered nature of the clearance impacts characteristic of the Mesolithic populations, the lack of disturbance indicators is not, perhaps, surprising. It is possible, however, that more extensive investigation of suitable deposits in the area will reveal similar episodes of small-scale clearance to those recorded a few kilometres to the east around the rim of Morecambe Bay, such as at Little Hawes Water. There palynological evidence of possible Mesolithic activity *c* 7531–6646 cal BC (*c* 9200–8200 BP) occurs, marked by a rise in open ground indicators such as the plantains (*Plantago*) and bracken (*Pteridium*), together with a band of charcoal (Taylor *et al* 1994). These signals are amplified by the findings of recent work from Thwaite House Moss, a small basin mire located 8 km to the south–east in the southern part of the Lonsdale limestone country. Here, what appears to be a small-scale clearance, associated with evidence of fires in the vicinity, seems to have occurred towards the close of Flandrian I, perhaps sometime between *c* 6500 cal BC and 5500 cal BC (*c* 7700–6500 BP; Middleton *et al* 1995, 183–9, 193). At this time a dense hazel-dominated woodland occupied the fertile limestone upland bordering the coast of Morecambe Bay. The landscape would have provided a variety of resources attractive to Mesolithic communities and it is possible that the disturbance from the moss may be recording a local example of this.

In support of this, fieldwork at Malham Tarn Moss (Pigott and Pigott 1963; Simmons 1996, 53–8), in the limestone uplands 40 km to the east, has demonstrated that, during much the same period, a series of declines in tree pollen, similar to those described above from Lonsdale and again associated with charcoal, has been ascribed to Mesolithic disturbance. The presence of several substantial Mesolithic flint scatters around the Tarn lend considerable credence to this interpretation (Williams *et al* 1987).

In the light of all this evidence it seems probable that a similar level of Mesolithic activity, encompassing small-scale clearings and the use of fire, extended across the broad band of limestone country stretching across North West England from the Pennines through South Cumbria. Furthermore, it is not inconceivable that this particular landscape was especially attractive, the fertility of its calcareous soils and its attendant biodiversity perhaps holding particular advantages for hunter-gatherer groups, especially when allied to the coastal resources available around Morecambe Bay.

As with the Mesolithic period, palaeoecological information directly relevant to South Cumbria during the early Neolithic is limited, thanks to the scarcity of suitable time-depth segments of stratigraphy, and the neglect of this period by more recent studies. Our incomplete picture of human activity at this time, therefore, is still informed by the relatively low-resolution data available from older research. This hints at small-scale human activity from the coastal fringes of both High Furness (Smith 1958; 1959), and Low Furness (Oldfield and Statham 1963), suggested by the presence of open ground indicators such as ribwort plantain (*Plantago lanceolata*) immediately prior to, and continuing after, the primary Elm Decline (dated regionally to *c* 4034–3700 cal BC (5060±80 BP, Q–913; Hibbert *et al* 1971)). The inference is that minor clearance was being undertaken at this time by a small population which does not appear to have been engaged in cultivation of crops and which, qualitatively, appears to have had a similar impact on the palynological record to that recorded immediately to the south–east during the Mesolithic.

Archaeological evidence

There is a certain amount of evidence that the caves around Morecambe Bay witnessed continued occupation during the Mesolithic period. Continuing work at Bart's Shelter (D Stables pers comm) and Bonfire Scar Cave (SMR 2352) at Scales has revealed evidence for Mesolithic activity including bone and stone artefacts and worked faunal material. A fragment of bone harpoon point has been dated to 6110–5883 cal BC (7160±60 BP, OxA–8069), which may also date the late Mesolithic lithic material from a similar location on the platform outside Bonfire Scar Cave, although the harpoon point may have been re-deposited from within the cave (D Stables pers comm). It is too early to say what the material from Bart's Shelter represents in terms of settlement and subsistence activity in general. The presence of a harpoon point, however, may suggest links with the coast either in terms of the transport of the catch or people bringing back fishing equipment, although it could equally have been used for freshwater fishing.

Numerous excavations within other caves in the area have revealed faunal remains of the period, in some cases human, at Capeshead (SMR 2386), Whitbarrow Bone Cave (SMR 5331), and Baycliff Haggs (SMR 2334), although some is likely to be later in date. The relatively meagre amounts of Mesolithic material currently known from such cave deposits may suggest their use on an occasional basis, or the disturbance of occupation levels during subsequent use of the caves. It is also likely that smaller and finer Mesolithic lithic and faunal material was overlooked during antiquarian excavations. At Kirkhead, J P Morris and M R Bolton were keen to 'develop and explore the hidden mysteries of this primitive abode' and removed some four feet of cave earth in their investigations (Baines 1893, 628). The new work by the Morecambe Bay Archaeological Research Society (MBARS) in Furness will amplify greatly our knowledge of both Mesolithic occupation and the taphonomic processes. Similar work at other sites with surviving stratigraphy may expand considerably our knowledge of Mesolithic cave use in other parts of South Cumbria.

It is already apparent that not all Mesolithic activity was confined to the caves of the limestone crags. Scattered late Mesolithic material has been recovered from fieldwalking in Furness (D Stables pers comm) and evidence for a lakeside settlement has been claimed from excavations at Gleaston Mill (Salisbury and Coupe 1995). Excavations on a Bronze Age ring cairn at Levens (SMR 2650) revealed abundant lithic material with Mesolithic affinities from the buried soil; initial assessment of the assemblage (of *c* 2000 pieces) has suggested late Mesolithic affinities (Turnbull and Walsh 1996, 17). Extensive concentrations of ash and charcoal were also within the buried soil horizon, perhaps suggesting some degree of settlement and subsistence activity. Further to the west, significant concentrations of Mesolithic material have been found, particularly on Walney Island, mainly chance finds from eroding dunes and coast (Wymer 1977, 162), at Trough Head, North End, and at the site of the Sandy Gap School (Barnes 1970). Although the material has not been properly characterised, the sites are of sufficient density to indicate that the island was well frequented, with a preference for the western, seaward coast.

The overall pattern of Mesolithic activity in South Cumbria is clearly more dependent on the survival and recognition of material, rather than a real variability in the level of activity. The Levens Park (SD 5052 8615) material had survived owing to the protection afforded from erosion by the overlying cairn and the creation of Levens Park in 1393 (Farrer 1923–4, II, 126). Similarly, artefactual material from caves may be indicative not only of the use of convenient shelters but also of the good conditions for the survival of archaeological material in such locations. The results of recent fieldwalking in Furness, which has produced a less intensive distribution pattern (D Stables pers comm) compared to that on Walney Island and in Levens Park, may suggest that there was concentrated activity in favourable parts of the landscape. This same pattern was noted in Lancashire (Middleton 1993b; Middleton *et al* 1995; Middleton *et al* forthcoming) and it was suggested there that reliable coastal locations, associated with favourable land, including wetlands, would have provided attractive areas for repeated visits. A similar case has been made for the late Mesolithic of West Cumbria where work at Eskmeals has indicated the long-term and consistent use of favoured estuarine areas (Bonsall *et al* 1991; *see Ch 3*). In South Cumbria such conditions are likely to have existed throughout the Mesolithic in variable configurations as relative sea-levels fluctuated. On Walney Island and in Levens Park we may be seeing the upper altitudinal limits of coastal Mesolithic sites similar to those identified on the eastern and southern sides of Morecambe Bay at Heysham Head (Salisbury and Sheppard 1994) and in Pilling Moss (Middleton *et al* 1995, Ch 2).

Whilst it has proved possible to characterise Mesolithic settlement across the landscape in a rather crude manner, the transition to the Neolithic and the adoption of farming has proved elusive in South Cumbria. The lack of clear differences between assemblages of late Mesolithic and Neolithic date has meant that there is a period in the late fifth and early fourth millennia for which sites are difficult to identify, particularly in the absence of well-dated ceramic and lithic assemblages and monuments.

Owing to these problems, archaeologists have traditionally been dependent on the type artefact for the Neolithic, the polished stone axe, to identify activity across the landscape. It has long been recognised that South Cumbria, and Furness in particular, has produced a significant number of such artefacts (Clough and Cummins 1988, map 6). This is likely to be a function, at least in part, of the proximity of the Groups VI and XI rock sources in the Lake District hills, but the distribution of these artefacts has resulted from a complex of processes relating to both past activity and modern recovery. It should also be borne in mind that axes were in active use for *c* 1500 years during which the role they had in society and manner in which they were used, circulated, and discarded would have changed.

The dependence on the presence of stone axes for the identification of Neolithic activity is due to the paucity of finds of other material, both ceramic and flint, and monuments from the area. There are obvious limitations to interpreting a whole period when the main body of evidence is characterised by a single artefact type, particularly when the bulk of the assemblage comprises chance or imprecisely provenanced finds. The longevity of use of the stone axe also restricts the precision with which accurate dates can be ascribed. However, with these caveats in mind, a brief survey of what represents by far the most predominant artefact type of the region can help to shed some light on human activity in the area during this period.

Of the 67 stone axes recovered from the coastal fringe of South Cumbria, only eight were found east of the River Leven and, of these, four were recovered from a valley bottom context. These include SMR 2417 and SMR 4375 from Winder Moor, near Cartmel, and another from Grange-over-Sands (SMR 3593). The other eastern axes, such as SMR 4106 from Underbarrow and SMR 4072 and SMR 4074 (both flint axes) from the Whitbarrow area, originated from the high, steep-sided limestone scarps, fringing the extensive valley mires of the Lyth Valley and the Leven. A further reworked axe fragment has recently been recorded from this area at Little Strickland Hill (SD 4370 8530; SMR 16720), unusual in that it is a broken end which has been re-flaked to fit into a wood, or antler, sleeve leaving the original cutting edge intact (Fell 1991). An adze (SMR 4070) has also been recovered from Brigsteer. These axes, along with a single leaf arrowhead from Underbarrow (Annable 1987, 385; SD 4880), form the total Neolithic artefact assemblage from the eastern moss fringes.

To the west of the Leven there is a similar pattern with the stray finds restricted to the higher ground on the moss edge. These finds include an axe from the low-lying clay hills adjacent to Newland Moss (SMR 5563), a rough-out axe from Leece (SMR 17930) on the fringes of Stank Moss (Robinson 1985, 36), and polished examples also from Leece (SMR 2610; SMR 3089). A large sandstone polishing stone (*polissoir*), recovered from within the peat of Roose Moss, and associated with bog oak (SMR 2605; *ibid*), has been interpreted as being Neolithic in date, after comparison with similar artefacts from Ehenside Tarn (Darbishire 1873; *Ch 3*). There is also a reputed 'celt' (SMR 2133) from Kirkby-in-Furness, close to Angerton Moss.

Some hints exist that the pattern of intentional deposition of axes in mosses tentatively identified in Lancashire (Middleton 1996, 38) may be reflected in South Cumbria. It has been noted that a number of axes come from the mossland edge, and it is possible that some derived from peat deposits. The axe and polisher from Roose Moss may be a more direct example of intentioned deposition and mirrors the similar association of artefacts with bog oaks from Pilling Moss (Middleton *et al* 1995, 57). It may be significant, however, that none of the five axe hoards from South Cumbria appear to have been associated with wetlands, although the context of these finds in largely unknown (Annable 1987).

There has been much discussion of the Neolithic axe 'trade' and the role on the lowland settlements in the production and dissemination of completed axes

Plate 10 Bardsea 1992: excavation in progress

from the quarry sites in the centre of the Lake District massif (Bradley and Edmonds 1993, 144–53). The association of henges, stone circles, and long mounds with the movement of axes appears established, although the precise mechanisms of the axe movement are imperfectly understood, as is the changing role of the monuments through time (*op cit*, Chs 8, 9). The location of these monuments within Cumbria does, however, appear to suggest a shift in the emphasis of activity from the coastal plain to the edge of the Lake District hills and the Eden Valley. In South Cumbria, the absence of monuments on the coastal plain is notable, perhaps suggestive of a real lack of Neolithic activity in lower-lying land and around wetlands although the largely recent pastoral/agricultural regime may also have contributed to this distribution. As we have seen with the Mesolithic material, however, it is unlikely that the picture is straightforward, particularly when the poor visibility of archaeological material is considered. The distribution of axes does indicate continuing Neolithic activity, and it is likely that coastal and wetland resources were of some importance; the evidence for pre-Elm Decline clearance from Furness may be a reflection of this. Such exploitation could have been seasonal as was the use of the upland pastures in the southern Lake District (*op cit*, 138–43).

Excavations at Bardsea (SD 305 744)

Whilst excavation was outside the remit of the NWWS survey, there were occasional opportunities to investigate certain wetland finds to elucidate matters of stratigraphy and survival. One such opportunity (Pl 10) arose in June 1992 when members of MBARS identified animal leg bones protruding from eroding foreshore peats and silts at Bardsea (Fig 17; Middleton 1992b, 27). A short, three-day excavation established that the bones were located under a very thin band of minerogenic, highly humified peat which formed the upper component of a deposit of dark-brown, organic silt (layer 2), containing numerous organic remains including brushwood, leaves, and hazelnuts. This merged into a thin deposit of grey/brown silt with a lower organic component (layer 4), which formed the upper part of a thick deposit of grey silt with a small organic fraction (layer 6). These deposits were bounded to the west by a very stiff, clay/silt deposit (layer 3) and to the east by a mid-grey, clay/silt loam (layer 5) which merged into the overlying

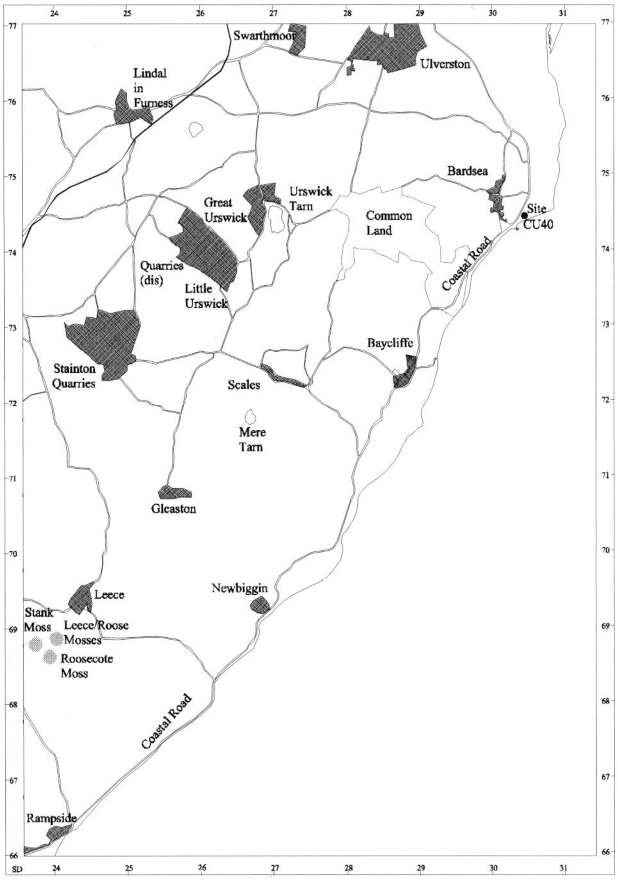

Fig 17 Low Furness: showing location of Bardsea 1992 (Scale: 1:100,000)

Based upon the 1988 Ordnance Survey 1:50,000 maps with permission of the Controller of Her Majesty's Stationery Office © Crown Copyright

*Plate 11 Bardsea 1992: animal bones (*in situ*)*

deposits (Fig 18). The skeletal remains were located at the interface between layers 2 and 4 (Pl 11). Two further bones were recovered from the excavations adjacent to the area of the initial find and at the same level (Pl 12); both of these were disarticulated and could not be related directly to the remainder of the material (Fig 19). All of the bones are of a very large Red Deer (S Stallibrass pers comm).

This sequence suggests that the remains were located in the upper portion of a sequence of clays and silts which filled a channel incised into marine clays (layer 3). Layer 5 may represent the fill of the main channel, the centre of which lay to the east of the excavation. Layers 2, 4, and 6 are successive, and merging, deposits of silt which had a larger organic component higher in the sequence. This may represent the gradual in-filling of a smaller channel which was incised into the initial channel filling (layer 5) but, as this smaller channel silted up, the water flow became less dynamic, permitting the accumulation of organic debris. This included not only wood and leaves from adjacent trees but also the skeletal remains which either washed into position, or were brought to the site. The most likely agency for such a deposition may be a dog, whose teeth marks are evident on some of the bones (S Stallibrass pers comm). This deposit may, therefore, represent a secluded area of stagnant water where a dog brought (presumably) scavenged remains. The date of this episode is not securely established, although pollen from the upper part of layer 2 was post-Elm Decline in date (NWWS assessment). The marine deposit (layer 5) may, therefore, correlate with the Lytham VI transgressive overlap dated between *c* 3680–3350 cal BC (4725±65 BP, Hv–4348), 3970–3525 cal BC (4895±95 BP, Hv–4344), and 3970–3340 cal BC (4830±140 BP, Hv–4347; Tooley 1978, 13). This suggests that the faunal remains were deposited sometime after the sea level had declined sufficient for a channel to be incised into the marine clays.

The value of the work lies not so much in providing a context for the faunal remains, but the insight it gives into the nature of the prehistoric evidence in Cumbria. The remains recovered were ephemeral,

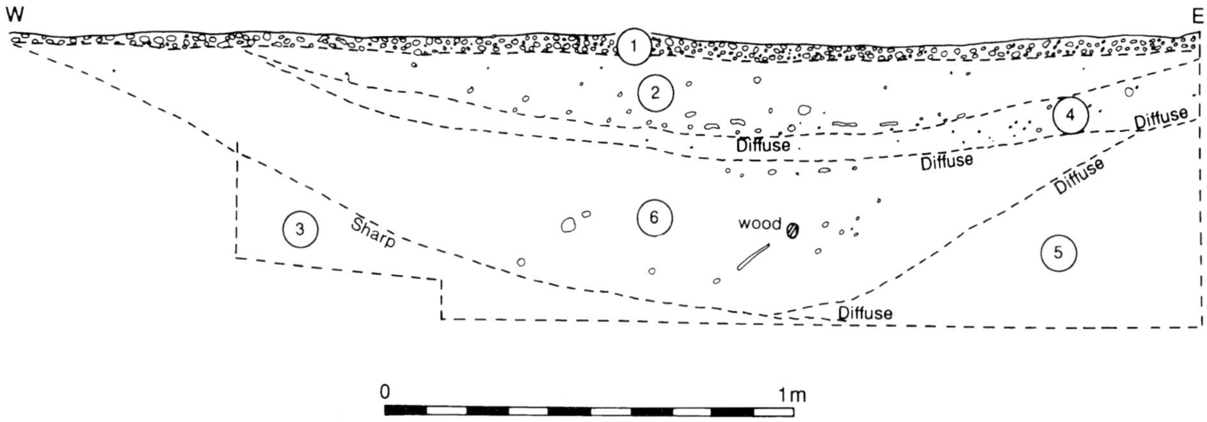

Fig 18 Bardsea 1992: south-facing section

but manage to suggest several small-scale, local processes which would otherwise not be present in the archaeological record. The first is the possible presence of settlement in the vicinity of the site. If the bones were brought to the site by a dog, even a wild one, they may have been dragged from an area where the carcass had lain cooked. The second point is the fact that this was the remains of a wild animal hunted and possibly processed by Neolithic people. This may be a reflection of the continued use of wild resources in the Neolithic, and hunted game would surely have been an important part of the diet, even when crops were being cultivated. The survival of the material is due to its waterlogged preservation

Plate 12 Bardsea 1992: general view of excavation

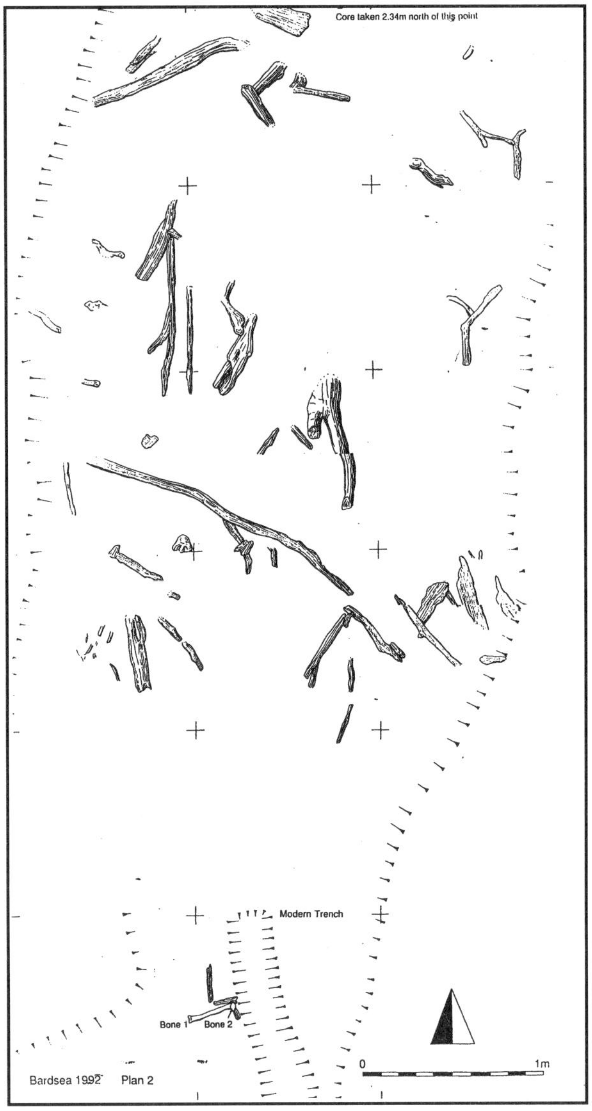

Fig 19 Bardsea 1992: plan of bones excavated

within a palaeochannel and the rapid covering of the remains by silts and peats. Red deer remains have been found widely in 'peat deposits all over Lakeland' and in 'the sands of Morecambe Bay' (MacPherson 1892, li), suggesting that the Bardsea finds may be a fragment of a much larger picture.

Excavations at Meathop Moss, 1982

In Autumn 1982 LUAU (then the Cumbria and Lancashire Archaeological Unit) maintained a watching brief during the construction of a water pipe from Lindale eastwards for 2.8 km towards Meathop Moss (Fig 20; CLAU 1983). The intention of the project was to identify organic remains, perhaps associated with identified trackways (Munn Rankin 1910) and to detect any traces of human activity within surviving peat deposits. Further observations were made to the east where the pipe trench passed to the north of Meathop Moss and Catcrag Moss, although technical difficulties made recording problematic (Pl 13).

The contractor's borehole data suggested that there was little surviving peat in the affected area with a maximum depth of *c* 1 m of peaty topsoil. At one site, however, north of Castle Head Cottage, Lindale (SD 4209 8021), three vertical posts were observed at a depth in excess of 2.3 m within silts beneath vestigial peat deposits. They were aligned north to south and were associated with horizontal planks which seemed to have been modified in antiquity, and extended north and south beyond the trench. Also present was 'worked timber' on the nearby spoilheap, removed by the trench excavation machinery. The wood was recovered for radiocarbon age determination, although no results were obtained in the event.

In the light of the circumstances of the discovery, and the absence of dating, there are real problems in the interpretation of this site. From the topography of the area, it may be assumed that a track was built to link Meathop or Castle Head with the 'mainland' across salt-marsh or similar. It is important as an apparently well-preserved trackway buried deep in silts that may be of pre-Neolithic date, if they can be related to the Lytham VI marine transgression, although this needs to be confirmed by absolute dating. It also highlights the potential archaeological importance of the pre-peat deposits and the excellent conditions for preservation that obtain in the right circumstances. In terms of human activity, the find is significant as an indicator of the potential importance of the coastal areas in early prehistory, prior to the development of wetlands in the valley bottoms.

Late Neolithic and Bronze Age

Palaeoecological evidence

Following *c* 3000 cal BC (*c* 4400 BP), information on human interaction with the South Cumbrian landscape becomes much more detailed, due to extensive studies undertaken across three mire complexes (Foulshaw, Helsington, and White Mosses) by Wimble (1986). His use of multiple pollen sampling sites has allowed a comparative vegetational history of an area stretching from the north of Morecambe Bay through to High Furness and the Duddon Estuary to be constructed, the time period encompassed stretching from the late Neolithic to the post-medieval periods (Wells 1991).

Evidence for human impact on the vegetation in the late Neolithic and early Bronze Age is hardly evident in South Cumbria, only being found at Foulshaw Moss. There, it comprises a small peak of cultural indicators and falls in arboreal pollen, and probably represents a local clearance episode during the late Neolithic, *c* 2570–2140 cal BC (3870±70 BP, CAR–544; Dresser 1985). The trees affected were mainly those characteristic of more fertile soils (Wells 1991). Nevertheless, the limited palynological record

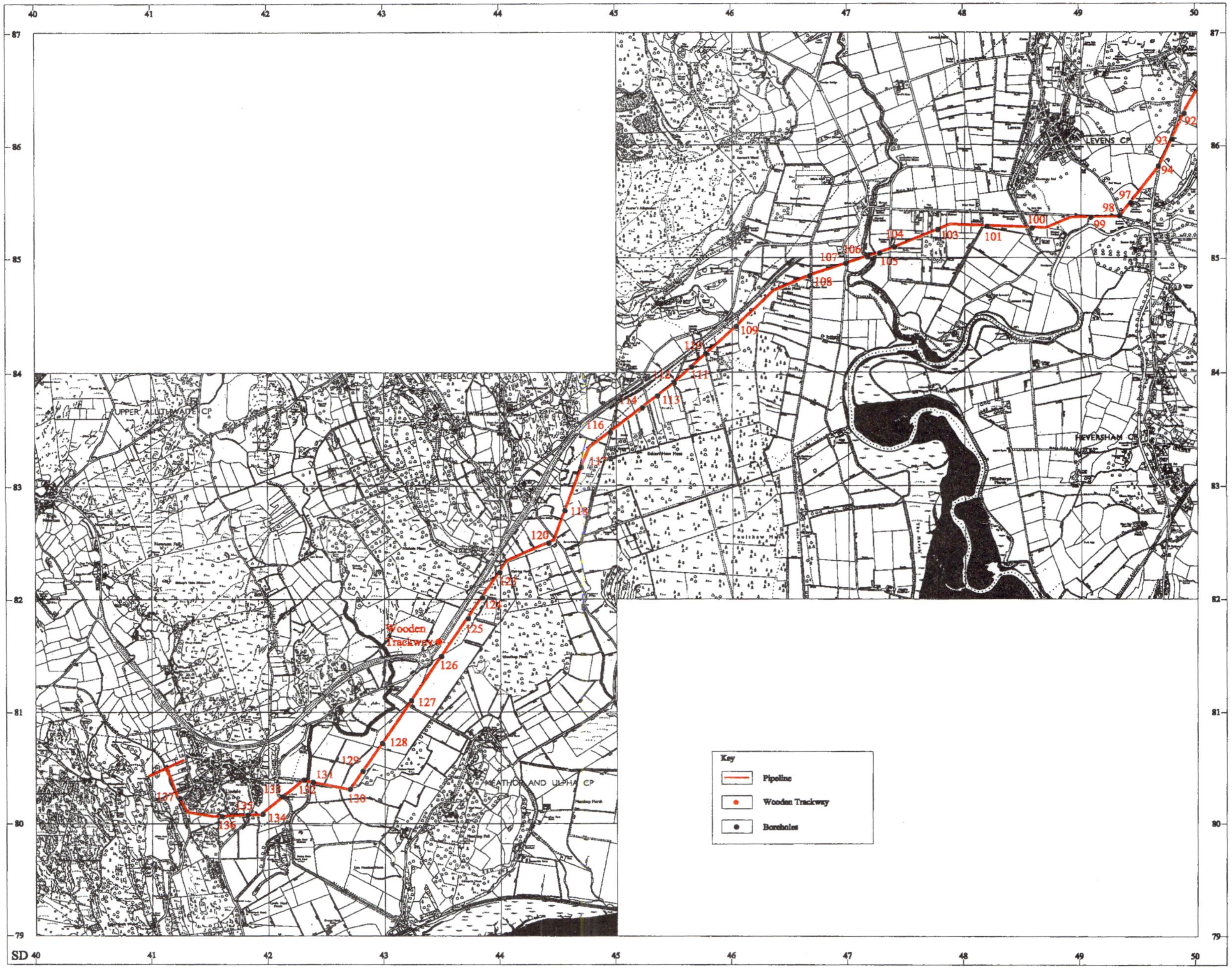

Fig 20 Meathop Moss: pipeline trench location (Scale: 1:50,000)

Based upon the 1989 Ordnance Survey 1:10,000 maps with permission of the Controller of Her Majesty's Stationery Office © Crown Copyright

Plate 13 Meathop Moss: excavation of pipe trench

probably reflects a genuinely low level of human activity in the hinterland of Morecambe Bay at this time, as other studies from around its fringes also record a similar story (*eg* Smith 1959; Oldfield and Statham 1963; Wells *et al* 1997) — a situation which is in marked contrast to the West Cumbrian Coastal Plain which saw intense clearance during the Neolithic (*Ch 3*).

The first significant clearance evidence in the north Morecambe Bay area dates to shortly after *c* 2300–1890 cal BC (3690±70 BP, CAR–554; Dresser 1985) with steep falls in arboreal pollen and a rise in pollen of 'cultural indicators' (Wells 1991). However, the evidence for this increased activity is restricted to the Lyth Valley. There are also indications of three small-scale clearances in the same area (Wells 1991), with partial regeneration by birch *(Betula)* occurring following these episodes. The activity is thought to be possibly of *landnam*-type clearances (*sensu* Iversen 1949) being of short duration and rapidly recolonised by trees after abandonment, although there is little evidence of crop cultivation (Wimble 1986; Wells 1991). Previous pollen studies from the mires around the Morecambe Bay area had revealed little dated evidence for Bronze Age human activity, but the data collected by Wimble (1986), as reported in Wells (1991), do indicate anthropogenic activity, bearing out Pennington's (1970, 72) prediction that high-resolution, independently-dated palynological research might reveal evidence of small, temporary clearances in the area at this time.

Together with the results from similarly detailed studies from the northern Fylde to the south (Wells *et al* 1997), the Lyth Valley evidence (Wimble 1986; Wells 1991) is beginning to help build a picture suggestive of low-intensity, but widespread, early Bronze Age human activity around Morecambe Bay. Once more, however, there is a remarkable contrast in the level of human disturbance to the vegetation seen in the north Morecambe Bay/Duddon coastal strip, to that seen in the West Cumbrian Coastal Plain, where activity in the Bronze Age is extensive (*Ch 3*).

Both the Lyth Valley and the Duddon Estuary areas seem to have experienced an episode of clearance *c* 1592–1260 cal BC (3140±70 BP, Car–553; Dresser 1985) and *c* 1158–820 cal BC (2805±65 BP, Car–598; *ibid*) respectively, with open conditions greater than any hitherto recorded. Arboreal pollen is much reduced, particularly elm, ash, and alder; there is also some evidence for cereal cultivation (Wimble 1986; Wells 1991). The increased activity fits into the wider picture of substantial clearance recorded from various parts of Cumbria and to the south of Morecambe Bay in the later Bronze Age. The uplands of western Cumbria are recorded as having undergone extensive deforestation at a period dated to *c* 1620–910 cal BC (3040±140 BP, NPL–124) at Seathwaite Tarn (Pennington 1964), and the south-western fellsides are littered with many structures considered to date from this period (Pennington 1970, 72; Fell 1973, 204; Leech 1983; Quartermaine 1989).

The very end of the Bronze Age saw a return to a more densely forested landscape around the South Cumbrian mosslands as the clearances were recolonised by secondary woodland. This phase is interrupted by indications of minor disturbance to the regenerating woodland cover at *c* 2500 BP (*c* 800–500 cal BC), probably focused on clearance of trees from the richer soils and lasting no more than 100 years. There is little evidence of arable cultivation. Arboreal pollen values recover well after the clearance, led particularly by birch which attains a greater importance than previously (Wells 1991).

Archaeological evidence

The first significant clearance episodes in South Cumbria coincide with a marked increase in the frequency and variety of archaeological material from the north Morecambe Bay area. There are two known concentrations of Beaker burials within this area, at Sizergh (SMR 2508; SMR 2509; SMR 2510) and at Levens Park (SMR 2504), both of which have received antiquarian attention. Two of the Sizergh group (SMR 2509; SMR 2510) were excavated at the beginning of the century when five inhumations were revealed in what appears to have been a cist from SMR 2510; fragments of an All Over Corded Beaker were recovered from SMR 2509 to the immediate east (McKenny Hughes 1904a; 1904b; Fell 1953, 1–5). These burials form part of wider activity, seemingly a barrow cemetery (R Maxwell pers comm). The cairn excavated at Levens Park (Turnbull and Walsh 1996), although this work was carried out under modern conditions, had also been subjected to earlier investigations, which has had a consequent effect upon the interpretation of the site. The excavation revealed a ring cairn with a primary burial associated with Beaker pottery and a pair of flint knives of possible Yorkshire origin (*op cit*, 24). The site clearly had a continued funerary function with possibly two phases of inhumation after the primary burial. The number of inhumations from the SMR 2510 site at Sizergh may suggest that this also occurred there. The Levens Park site mirrors closely the excavated example at nearby Manor Farm, Borwick (North Lancashire), the primary burial of which has been radiocarbon dated to *c* 1740–1410 cal BC (3270±80 BP, Har–5658; Olivier 1987, 180). Elsewhere, a single incidence of Beaker pottery has been found on Walney Island (Clough 1968; SMR 2706) although there may be some doubt about its attribution (Annable 1987).

Aside from the funerary monuments, there are several stray finds of late Neolithic and Bronze Age date from South Cumbria, including a pair of barbed and tanged arrowheads found above the valley mires on Underbarrow (SMR 4108 and SMR 4109) and a single barbed and tanged arrowhead from Gilpin Bridge, within Stakes Moss (Anon 1903; SMR 4067). Several others have been found on Walney Island (Annable 1987, 488). An early Bronze Age flanged axe of late Migdale tradition (*c* 1650–1350 BC; 3400–3100 BP), associated with peat deposits at Roose (Gaythorpe 1897, 446; Fell 1940, 121; Clough 1969, 2; SMR 2603), remains as yet the sole metal find from within a wetland context from the earlier part of the first millennium BC. There is a general lack of characteristically early Bronze Age metal artefacts from the region as a whole, with only two metal finds, that of a single flat axe from Gleaston Castle (Clough 1969, 29; SMR 2329) and an awl from a potential cremation burial at Appleby Hill, Birkrigg (*op cit*, 30; Gelderd *et al* 1914, 468; SMR 2371). One of the more unusual finds which, unfortunately, lacks context, is a Grooved Ware bowl of late Neolithic date from Walney Island (Barnes 1956). This may have been associated with finds of Food Vessels and Beakers from the same area.

Axe-hammers, dated to between *c* 1650 BC and 1250 BC (*c* 3400–3000 BP; Roe 1979, 16), are particularly widespread throughout South Cumbria and do not appear to be restricted to any particular area or environment (Annable 1987, map 28). The widely-accepted petrological origin for the majority of these artefacts, that of Group XV, which was suitable for manufacture by pecking, is centred on the southern Lake District, possibly in the Coniston area; a wide variety of other local stone sources, however, also appear to have been used (Fell and Davis 1988, table 29). As elsewhere, axe-hammers are seldom found in secure contexts, although it has been suggested that they had a more prosaic function than contemporary battle axes, which can occur in burials (Edmonds 1995, 165).

The provenance of almost all of the perforated hammers is vague, as the majority enter the record as stray finds, and it is therefore difficult to draw meaningful conclusions from imprecise distribution data. Typical references, such as 'many hammers and battle axes of different sizes and several kinds of stone...have been found in the Cartmel district' (Stockdale 1872, 251), offer only tantalising glimpses into the possible significance of these implements. The very number of these implements from Furness may suggest a real concentration of activity in the area, although the precise nature of this activity is difficult to determine. Analysis of the finds from North Lancashire suggested that they were principally associated with 'dryland' contexts, in contrast to the contemporary metal artefacts (Middleton *et al* 1995, 205–6). There are, however, several from South Cumbria from marginal wetland contexts, such as that from Egton with Newland, on land adjacent to Newland Moss (SMR 5561); a further example has been found in peaty ground at Colton, at the head of Rusland Pool (SMR 2545). In the far east of the study area, a perforated whetstone was found below '5 feet' of peat within Stakes Moss (SMR 4132) and an axe-hammer is also recorded from Low Park Wood on the limestone of Sizergh (North 1934; SMR 4089). Further north, in the Lyth Valley, a perforated stone implement, possibly a mace head, was discovered near Whitebeck (SMR 4151), whilst perforated stone implements have also been recovered from the immediate moss fringes at Lindale (SMR 2456; SMR 2560). Concentrations of stone hammers having been recorded at Winder Moor, near Flookburgh (SMR 4147), fragments of a stone adze from Grange (North 1934), and 'many hammers as well as celts of brass, copper, and stone found in the Cartmel District particularly at Nab Green' (Stockdale 1872, 250; SMR 4375). Whether such finds represent deliberate deposition or the first marginal wetland settlement is difficult to establish with certainty from the small number of isolated artefacts.

The corpus of finds of mid- to late Bronze Age metalwork from southern Cumbria again derives largely from antiquarian research and suffers from poor provenancing; numerous artefacts are now lost. There is, however, a significant amount of material from wetland contexts, including an axe from 'a peat moss near Cartmel' (Clough 1969) and a dagger recovered from Helsington Moss (SMR 4071), which has close parallels with Irish dirks and daggers dated to *c* 1400–1200 cal BC (*c* 3200–3000 BP; Fell 1940; Fair 1945; Clough 1969). Other metalwork of middle Bronze Age date includes a spearhead from the ridge on Whitbarrow (Anon 1921; SMR 4073) and a remarkable assemblage of finds from Kirkhead Cave (SMR 2415) discovered by local antiquarians, Mr Morris and Dr Barber, excavating in the mid-nineteenth century.

> 'The excavations [were continued] for four feet through the drift, [where] highly interesting relics...turned up, bone implements...the frontal bone of a human skull,... and two stone implements of the rude underground type...pieces of pottery rudely baked...fluted earthenware bead, three bronze rings, one bronze pin, enamelled; bone amulet formed of head of human femur; paalstab [*sic*] in bronze, piece of bronze fibula, bronze tube, one small and one large amber bead, flake of flint embedded in stalagmite, spear-head in bronze, portion of urn carved out of soft stone, large bronze celt and a large quantity of human bones'
> (Baines 1870, 685, quoting Dr Barber).

The objects found within the cave were recorded with little regard to stratigraphy and many have now been assigned dates other than the Bronze Age. However, a recent inspection of some of the pottery recovered from the excavations has identified at least two, plain, wide-mouthed, flat-based buckets of a type allied in form to a number recovered from Walney Island and also from Pilling, which belong to the later Bronze Age and are present from *c* 1500 cal BC to 1000 cal BC (*c* 3200–2800 BP; Gibson 1986, 6; Gilks 1987, 40). Several other areas in the north of England have produced similar cave deposits, such as Elbolton Cave, Yorkshire (Gilks 1973), which appear to be the result of burials over an extended period of time.

The most impressive archaeological structure recorded from the South Cumbrian wetlands themselves must surely be the brushwood corduroy trackway from Foulshaw Moss (SMR 2487) discovered by peat-cutters in the first decade of the twentieth century (Munn Rankin 1910). A stretch of the same trackway was also seen at the nearby Rawson's Moss and was described by Barnes (1904, 207–8) as consisting of:

> 'cross timbers laid side by side on three lines of supporting logs parallel to the direction of the road. The large timbers, some of them 2 feet 6 inches thick, had been split and laid face downwards; the smaller ones were left entire. At short intervals along each side of the road, pointed stakes had been driven deep into the peat...Axe marks were distinctly visible on many of the logs both on the ends and on the split face...the tool used had been about 11½ inches wide...The road is from 15 to 16 feet wide.'

Stratigraphically, the trackway was positioned at the junction between two distinct peat types differing in humification ('the upper and lower *Sphagnum* peats' of Smith (1959)). This was originally interpreted (without the benefit of radiocarbon dating) as representing the *Grenzhorizont*, a recurrence surface considered by Granlund (1932) to reflect a major climatic change to wetter conditions affecting much of North–Western Europe and traditionally dated to *c* 600 cal BC (*c* 2500 BP). The conjunction of this stratigraphical feature with the trackway naturally led to the latter also being ascribed to this date, or, in Smith's words, 'towards the end of the Bronze Age' (Smith 1959, 126).

The crucial humification change central to this assumption was dated independently at Foulshaw Moss by Wimble (1986). The results (Wells 1991) suggest that, in fact, the change is older than previously thought, probably 1592–1260 cal BC (3140±70 BP, Car–553; Dresser 1985) in the middle Bronze Age. The construction of the trackway appears to coincide with a localised change in the wetness of Foulshaw Moss represented by this recurrence surface, presumably formalising existing ways across the moss which became impassable as the mire status changed over time. There is also a coincidence of the trackway construction with an increasing number of clearance episodes in the woodland surrounding the moss (Wells 1991): whether this is a reflection of increased population and settlement, or a reaction to the mosses becoming unproductive, is unclear. It has been argued in North Lancashire that islands within the mosses were an important area for settlement but the surrounding area became unavailable for grazing through increased bog surface wetness (Middleton *et al* 1995, 206); such a process in Foulshaw may have increased the pressure on the surrounding woodland for both grazing and cultivation. This pressure is also reflected on the upland margins with increased settlement of now marginal land (Quartermaine and Leech forthcoming).

The speculation by some authors (Pennington 1970; Oldfield and Statham 1963; Wimble 1986; Wells 1991) that there was a period of relatively slight interference in the landscape in the late Bronze Age and early Iron Age, possibly as the result of climatic deterioration. This is in part reflected in the archaeological record where a steep drop in the recorded material from the late Bronze Age occurs. The late Bronze Age/early Iron Age is so poorly represented in the archaeological record in the eastern part of north Morecambe Bay, for instance, that it appears to be completely empty. The nature of the evidence also demonstrates a marked change from that which preceded it, from the relative abundance of archaeological material which typified the middle Bronze Age, with axes and other weapons represented, as well as ceramic material and extensive evidence for the settlement of formerly marginal land. In contrast, the evidence for this period relies on a few articles of metalwork and a small number of large, yet poorly understood, settlement sites.

The late Bronze Age finds appear to be more restricted in distribution, being confined to areas such as Furness (Higham 1986, fig 3.8). At Skelmore Heads, for example, a hoard of six late Bronze Age socketed axes (*c* 700–600 cal BC; *c* 2500 BP; Clough 1969, 19–21; SMR 2226) was found, and another hoard included a Ewart Park-phase sword (Fell and Coles 1965, 47–52; SMR 2367). There were also further reported hoards, of a spearhead and a number of bronze 'celts' and rings from near Stone Walls at Urswick (SMR 2314), which all date from the late Bronze Age period (Barnes 1963), and of a socketed 'celt' and palstave, together with a ring, from Stone Close, Urswick (SMR 2265). Apart from a relatively rich cist burial at Stainton, which yielded a 'razor' (SMR 2379), a knife from Urswick (SMR 2312), and a spearhead from Piel Castle (Fair 1946b, 175; SMR 2619), these comprise the complete late Bronze Age artefact assemblage for the whole of the South Cumbrian area.

Of these, the most recently excavated site, that of Skelmore Heads (SMR 2248) in Furness (Powell 1963), has revealed at least two phases of occupation, since the prominent stone-banked enclosure was preceded by a timber palisade, but little artefactual material was recovered during the excavations and almost no occupational debris was encountered within the trenches (*op cit*, 14). Parallels to the structural sequence at Skelmore Heads have been found in Scotland and at Huckhoe in Northumberland, which have a suggested late Bronze Age origin (Annable 1987, 255; Higham 1986, 121). It has been argued that the clustering of late Bronze Age metalwork in the vicinity, allied to the Iron Age enclosure, indicates the presence of an oppida-like settlement – a focus of local power exerted over the surrounding area. Although the chronology of occupation is poorly defined, the finds suggest that this may have been established in the late Bronze Age as a palisaded enclosure and continued through the Iron Age (Higham 1986, 121–35).

Excavations at Urswick Stone Walls (SMR 2313), also in the Furness Peninsula, have indicated similar continuity between the late Bronze Age and Iron Age. The excavations on a hut circle revealed a mixed assemblage, including a Bronze Age flint scraper, large quern stones, and a decorated bronze fragment, dated stylistically to 200–100 BC (*c* 2200–2100 BP; Smith 1907, 99; Dobson 1907, 72–94). It is possible that this site constituted an enclosed homestead similar to those found in the east of the county, particularly at Leck and Eller Beck, in the Lune

Valley (Lowndes 1964; Haselgrove 1996, 64; Jecock 1998), and Ewe Close and Waitby to the north (Collingwood 1933b; Higham 1986, 192).

The decrease in human activity indicated palynologically further north and east parallels the notable absence of recognised settlements from the late Bronze Age in these areas. In the upland hinterland of South Cumbria, Pearsall and Pennington (1973) suggested, on the basis of palynological evidence, that population also declined at this time. This conclusion seems to have been borne out by the results from more recent field surveys in South Cumbria, where very few monuments ascribable to the Iron Age have been located (LUAU 1994; 1995). The uplands between 200 m and 300 m OD in South Cumbria seem to have been largely abandoned in the late Bronze Age, leaving the fell slopes littered with a redundant infrastructure of field boundaries and cairns.

Around Morecambe Bay, there is little evidence for significant disturbance to the woodland cover in either the late Bronze Age or much of the Iron Age, the studies of Smith (1959), Oldfield and Statham (1963), and Wells *et al* (1997) paralleling those of Wimble (1986). Similarly, NWWS work in North Lancashire (Middleton *et al* 1995, 70) found little artefactual evidence relating to this period in the area immediately to the south of Morecambe Bay (Over Wyre), despite three seasons of extensive field-walking. Whilst there are real problems in the identification of Bronze Age and Iron Age remains in the North West (Haselgrove 1996), it seems plausible that the retreat from marginal land in South Cumbria and the Morecambe Bay wetlands may have been climatically determined. A shift to colder, wetter conditions around the turn of the second millennium as suggested by, amongst others, Lamb (1977, 373), would help to account for the widespread initiation of podsolisation and peat growth over areas of formerly productive upland. A similar process may have encouraged the abandonment of the Over Wyre mosslands as raised mire systems began to expand and become wetter (Middleton *et al* 1995; Wells *et al* 1997).

Iron Age and Romano-British period

Palaeoecological evidence

The first episode of extensive clearance (*sensu* Turner 1965) and prolonged open conditions recorded from South Cumbria dates from *c* 2400 BP (*c* 700–400 cal BC) in the Duddon Valley and *c* 2200 BP (*c* 350–200 cal BC) in the Lyth Valley (Wells 1991). Initially, there are few indications of arable cultivation although cereal grains are recorded later on; it thus appears that the clearance was increasingly associated with cereal cultivation. Stratigraphical evidence indicates drier conditions on the mire surface at this time and possibly suggests an improvement in climatic conditions (Wells 1991).

An episode of increased clearance is clearly identified in Smith's (1959) diagrams from the Lonsdale mosses, where it is associated with a humified layer in the peat stratigraphy. Smith (*ibid*) originally conjectured that the phenomenon could date to the Romano-British period, as did Pennington (1970, 72), based on its stratigraphical proximity to a radiocarbon date from Smith's putative 'retardation layer' of cal AD 266–680 (1514±100 BP, Q–83; Godwin and Willis 1959). However, Wimble's work (1986), as noted above, indicates that this activity began in lowland South Cumbria rather earlier, in the pre-Roman Iron Age, and continued into the Romano-British period. Recent evidence from Coniston Water, immediately to the north (Pennington 1997, 44), has also yielded additional evidence of this phenomenon with major deforestation beginning just before *c* 110 cal BC – cal AD 120 (1980±50 BP, SRR–1872), suggesting a resurgence of clearance activity across much of South Cumbria. These results are directly comparable to similarly high-resolution pollen and macrofossil data from Fenton Cottage, situated immediately across the Bay from the South Cumbrian mires, which show a parallel trend, namely a major deforestation signal beginning at the end of the first millennium BC (Middleton *et al* 1995, 151, 170; Wells *et al* 1997).

South Cumbria can therefore be seen to follow the trend recorded from most of the recent detailed studies in North West England, indicating a dramatic increase in forest clearance beginning in the late Iron Age. The feature has also been detected in diverse locations throughout the North West, including the Forest of Bowland (MacKay and Tallis 1994), North Cumbria (Dumayne and Barber 1994), and the north side of the Solway Firth (Tipping 1995a; 1995b; 1995c). Comparable evidence has also been found on the upland fringes, such as the episode of upland cereal cultivation from the south–western fells, which is dated to *c* cal AD 210–680 (1560±130 BP, NPL–116) at Burnmoor Tarn and *c* 20 cal BC – cal AD 592 (1750±130 BP, NPL –117) at Devoke Water (Pennington 1970, 72).

At some point at or just after the end of the Iron Age, forest regeneration (and by implication land abandonment) began, possibly a century or so earlier in the Duddon Valley compared to the Lyth. Changes in the local environment, possibly as a result of variations in relative sea level may have encouraged differing settlement patterns in the two areas. Zong (1997), in a study of diatom biostratigraphy from a fossil marsh sequence at Roudsea in the adjacent Leven Estuary, near to the Duddon Valley, has demonstrated that a previously undetected rise in sea-level of >0.5 m OD occurred some time after 800–380 cal BC (2420±80 BP, Beta–88434). It seems possible, therefore, that rising inland water tables resulting from this may have caused waterlogging or flooding in low-lying areas around the Duddon Estuary resulting in abandonment of settlement.

Archaeological evidence

The revised chronology of events in South Cumbria, together with the more detailed data generated by recent studies, means that pollen diagrams from the Lyth and Duddon actually record the pre-Roman establishment of cereal cultivation in the area. Despite this, archaeological evidence for the late Iron Age/early Romano-British period is sparse in South Cumbria. There is a scattering of defended enclosures and enclosed settlement, some of which, perhaps significantly, appears to overlook the low-lying wetlands. These include Skelmore Heads (SMR 2248), where a bank and ditch replaced the earlier palisaded enclosure (noted *above, p 45*), a single defensive enclosed settlement near Banishead Quarry (LUAU 1994; SMR 3219), and hillforts at Helm Crag near Kendal (SMR 2641), Castlehead near Grange (SMR 2446), and Warton Crag in North Lancashire. An apparent contradiction between palaeoecological and cultural evidence is also apparent to the south of Morecambe Bay in the mosslands of Over Wyre where a major deforestation signal dating to the end of the period exists (Wells *et al* 1997), in spite of an almost complete lack of Iron Age archaeology recorded during an extensive systematic fieldwalking programme (Middleton *et al* 1995, 99). However, as Haselgrove (1996, 64) has asserted for Lancashire, the dichotomy between the two strands of evidence may be due to poor archaeological visibility and survival, compounded by problems of dating and a lack of artefacts common to most northern sites.

As with much of the North West, the pattern of rural settlement changes little with the Roman invasion. The greatest change in the Cumbrian archaeological record is the appearance of Roman military works, such as the forts at Watercrook (Kendal: SMR 2078), Lancaster, and Ambleside (Shotter 1997; SMR 1877). There are also records of scattered Romano-British finds across the area, including 'a considerable number of Roman antiquities…found in various parts of the parish [Cartmel] (largely a coin hoard of 524 pieces), but no marks of forts, camps, roads, dikes, or other permanent works' (Baines 1836, 628). Other finds include a tombstone from Lindale (SMR 2458); some finds from Kirkhead cave, including an enamelled pin, an iron axe, a trefoil-shaped fibula, and a coin of Domitian (AD 81–96; King 1974, 196; SMR 2415); and second and third century artefacts from the Dog Holes cave near Haverbrack (SMR 2489) which also contained the skeletal remains of humans, some of whom had apparently been stabbed to death (Benson and Bland 1963; Bland 1994). A site within Levens Park (Turnbull and Walsh 1996, 15), known as 'The Temple of Diana' (SMR 2645), has yielded first-century material with superimposed post-Roman structures. A similar enclosure (the Sizergh Fell Barrow), excavated by McKenny Hughes on Heaves Fell, revealed a skeleton within the corner of one of the walls, interred with a blue melon bead, a bronze fibula and ring, dated to the second–third century AD (McKenny Hughes 1912a; SMR 2490).

Early medieval

Palaeoecological evidence

A period of high arboreal pollen values and low non-arboreal pollen values prevails in all areas, suggesting a recession in activity at and just after the end of the Roman period. Clearance is recorded from Foulshaw following this and, at White Moss, a resumption of activity is suggested by non-arboreal pollen values rising to a maximum by *c* cal AD 690–1000 (1160±60 BP, Car–694; Dresser 1985). Once again the clearance is characterised by the reduction of ash and elm to negligible values but there is little evidence of arable cultivation (Wells 1991). Oldfield (1969), discussing extensive early medieval clearance in the North West, pointed to a contrast between the almost exclusively pastoral activity of the north and more arable agriculture in the south. The evidence from North Morecambe Bay (Wells 1991) suggests that it may be transitional.

In the wider landscape, the peak of early medieval clearance was short-lived, and regener-ation occurred by *c* 800–900 BP (*c* cal AD 1050–1270) although there are indications of very small-scale, but persistent, clearings (Wells 1991). This corro-borates Dickinson's (1975) study from Rusland Moss which records little activity between the late Roman period and putative clearances of the ninth and tenth centuries AD, and also seems to accord well with the sequence of events depicted in Walker's (1966, 94–5) diagram from Ehenside Tarn. Oldfield (1969), summarising the data available from north Morecambe Bay, also constructs a sequence in which fifth- to sixth-century AD regeneration was interrupted by some clearance, then low-level farming activity in the sixth to eighth centuries AD, followed by a distinct regeneration in the tenth or eleventh century. The picture from the south of Morecambe Bay, in Over Wyre, is very similar (Wells *et al* 1997), but better dated. It shows regeneration following the Roman period, and reaching a peak between cal AD 670–990 and cal AD 1047–1280 (1200±70 BP, GU–5156; 820±50 BP, GU–5142). This phenomenon of marked regeneration occurring at some point around the end of the first millennium AD is well known from several studies in North West England, including South Lancashire

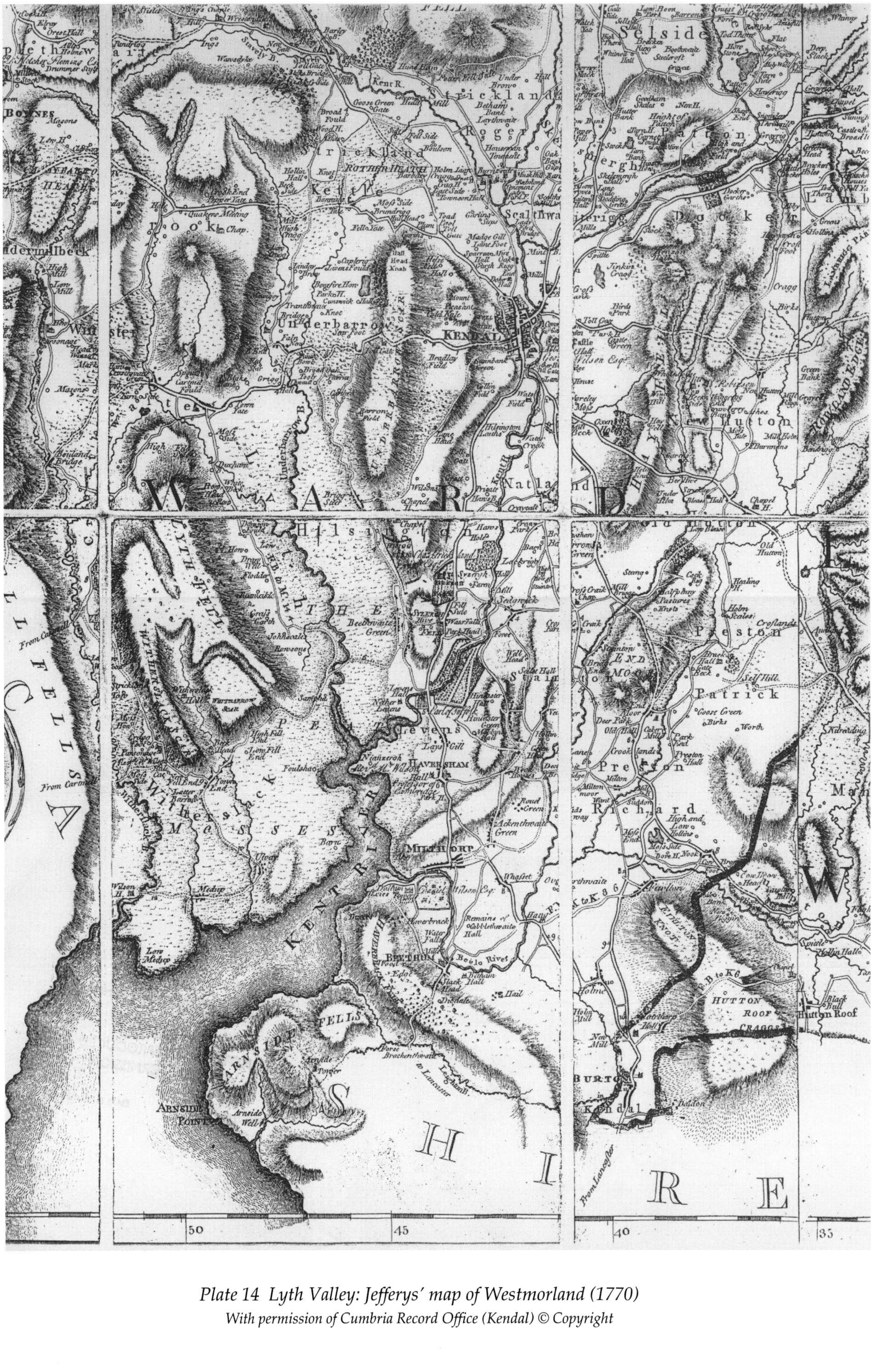

Plate 14 Lyth Valley: Jefferys' map of Westmorland (1770)
With permission of Cumbria Record Office (Kendal) © Copyright

and Cheshire (Birks 1965; Leah *et al* 1997, 154) but particularly from the fringes of Morecambe Bay (Oldfield 1963; Oldfield and Statham 1965).

The regeneration may be interpreted as a reduction of agricultural activities at this time but, if so, the cause remains a puzzle. A climatic explanation appears unlikely, given its timing; at the turn of the first millennium AD Britain and North–West Europe were in the midst of the 'Early Medieval Warm Epoch', the second warmest and driest phase of the Holocene recorded until modern times (Lamb 1977, 374, 435–49). Because of this, Oldfield (1969; Oldfield and Statham 1965, 75) has sought to explain the regeneration as a response to political events, and suggested it represented effects of the so-called 'Harrying of the North' by William I in the years around 1068. However, there are plenty of alternative, or additional, candidates to account for subdued agricultural activities during this period. Following the collapse of the Northumbrian state by AD 876 the area west of the Pennines experienced a period of political instability which ensured it was particularly vulnerable to sea-borne attack and immigration from the Irish Sea (Higham 1986, 317). Any English settlement of Cumbria may have been limited to better agricultural lands, judging by the distribution of characteristic place-names (*ibid*) and, after the removal of any umbrella of protection afforded by Northumbria, coastal Cumbria is unlikely to have been capable of mounting effective opposition to coastal raiding. A series of expeditions sent by English kings between AD 945 and AD 1000 in an attempt to keep Norse and Scottish encroachments in check ultimately failed (Kirby 1962, 86). However, given the geographical extent of the regeneration feature, and the possible length of time it represented (probably at least a century), it may well be unwise to discount other possibilities, such as epidemiological factors.

Archaeological evidence

Archaeological evidence is particularly scarce for the immediate post-Roman period, with only documentary references for a monastic settlement at Heversham, although this has never been located (Bingham 1987), and a tradition that the land of Cartmel, with its native population, was granted to St Cuthbert by King Ecgfrith of Northumbria (Higham 1986, 271). The sole excavated site from this period was inspected in 1911, when trenching in Levens parish revealed four human skeletons, on an east–west alignment (SMR 4068). The excavator interpreted these as Christian, on the basis of their alignment, despite Bronze Age activity in the area. A similar potential early Christian burial has been reported at Roosebeck (SMR 2614) and at Rampside Church a Scandinavian dagger and sword (SMR 2596, SMR 2595) have been found in the graveyard, presumably from a grave; O'Sullivan has suggested the possibly of a long-cist cemetery in the vicinity (O'Sullivan 1985, 23, fig 2.2).

Medieval and post-medieval

Palaeoecological evidence

Whatever the cause of the recession in agricultural activity, non-arboreal pollen values do not rise again in the Duddon area until some time between *c* cal AD 1000 and cal AD 1200 (1000–800 BP), when there is a rapid and consistent build-up of clearance activity (Wells 1991, 39). This is distinct from all those recorded previously by the high values and diversity of arable indicators present, including hemp or hops (*Cannabis* / *Humulus*-type).

By late medieval times the area was extensively cleared with predominantly open conditions. The background to this activity probably lies with the establishment of Furness Abbey in 1127 (Knowles 1940, 74), which promoted clearance of land and agriculture on a large scale. Thereafter, a long period of enclosure and sheep pasturing began in South Cumbria, reaching its zenith in the fourteenth and fifteenth centuries. On the suppression of the Abbey in 1537, the lands were divided up amongst a class of Yeoman farmers (Barnes 1968) which led to further clearance and the establishment of a more mixed agricultural economy. This change may be reflected by the rise in pollen values of cultivated plants (*eg* cereals, *Cannabis* / *Humulus*). The first serious attempts to drain and reclaim the wetlands, a process accelerated by the awards of Parliamentary enclosures in the eighteenth and nineteenth centuries (Pearsall and Pennington 1973, 261), may also have accelerated the clearance activity.

The final changes in the landscape, recorded by the diagrams from Urswick Tarn and Ellerside Moss (Oldfield and Statham 1963), are not recorded widely in the area due to truncation. They show the decline of hemp cultivation in the latter half of the eighteenth century and the beginning of plantations, with pine, in the nineteenth.

Archaeological and historical evidence

The Lyth and Winster Valleys

The mosses of the Lyth Valley have seen both some of the earliest documented agricultural activity on the mosses in Cumbria and also some of the most intensive exploitation of the wetland landscape in the county. Eighteenth-century cartographic evidence (Jefferys' map of 1770; Pl 14) clearly depicts the Lyth Valley as settled along the fringes of the steep wooded hillsides of Whitbarrow and Underbarrow, but with negligible settlement within the clearly marked 'PEAT MOSSES'. However, this belies other evidence for the early use and exploitation of the mosses.

As early as 1241, the mosses were being utilised by Matthew de Redman, who was granted the right of common pasture by William de Lancaster for the land from the 'Sandpol as far as the hard land of Whiteberg' (Farrer 1923–4, II, 118). This pasture would have been seasonal, confined to summer grazing, and would not have had a drastic effect upon the moss itself. However, there is potential evidence for early encroachment and improvement on the mosses dating from a suit of 1315, between the plaintiff Sir Walter de Strickland and Sir Matthew de Redmane [*sic*]. In the suit it was granted that

> 'Sir Walter, his heirs and tenants of Syresherth, Helsington, Brigsteer and Levenys as his right and inheritance common of pasture everywhere in Levenys...and for digging turves, and mowing herbage and rushes in Redker, and to take in heaths and mosses at his or their will'
> (*op cit,* 121).

Whether the tenants took advantage of this right to take in the mosslands is not known, but a later grant dating to 1352 is more illuminating. It states that beasts straying within a moiety of 30 acres of waste at a place called Le Mossebank, in Levens, 'from Medilhoweslak as the fence passes in Le Musse and through Le Musse to the brook of Briggesterewelles and ascending by the same brook to the highway from Levenes to Crossethwayt', would not be impounded, whereas should they cross the highway they would be seized and fines levied (*op cit,* 124). The reference to Le Mossebank certainly implies an earthwork of some sort, whether a drainage dyke or peat-cutting bank is unclear, but when considered with the reference to the fence through the moss, what is clear is that certain areas had been physically divided and enclosed. The differentiation between land on the one side of the highway from the other suggests that land had different status, and that land on the moss was less likely to suffer from cattle trespassing.

A similar arrangement pertained to land called Les Ryddynnges (clearance) and an adjacent area of 28 acres of waste called Wynstremyre which the owners were at liberty to enclose (*ibid*). The ability to enclose the waste at will would appear to have been as a result of a concerted effort on behalf of the Barony of Kendale which issued the grants to improve marginal land without incurring the costs themselves. Similar practices occurred at other wetlands in the country at this time, such as Eyton-upon-the-Weald-Moors in Shropshire, where in the early fifteenth century lands were granted on long leases, rent free for the first eight years whilst the land was cleared (Leah *et al* 1998, 78).

This early practice of defining areas on the mosses and setting about improving them appears to have created a pattern for later activity in the Lyth Valley. Certainly the piecemeal enclosure of waste had continued through to the sixteenth century when an award of 1589 made by, amongst others, Thomas and Nicholas Curwen, with Thomas Preston of Nether Levens and James Bellingham of Overlevens, granted the rights and titles in the wastes, commons, and mosses in Overlevens. In it, it states that both parties 'shall hold and enjoy in common the said wastes etc *now remaining unimproved* and to take common of pasture, turf grass, turbary peats...' (Farrer 1923–4, I, 136). It goes further by declaring that the two grantees were to make presentments to their respective manor courts regarding 'all manner of incroachments, purprestures and improvements' (*ibid*). That there was waste 'remaining unimproved' would hint at a certain amount of previous improvement, although the presentments to the manor courts suggest that such improvements were not to go unchecked.

Early evidence that the mosses may have been over exploited and stripped of peat comes from a court roll dated 1590, where it was stated that '...the jurye aforesayd doo Order that John Clodslye shall translate and make his howse wherein he dwells, into a peatcoote accordinglye as it was buylded...' (CRO(C) D/Lons/L/Kendal Barony/3/3). Although this is not clear evidence for the depletion of the peat at the head of the Lyth Valley, it does indicate that a structure associated with the drying of peat had gone out of use by the seventeenth century, suggesting a certain amount of organised gathering and drying of peat.

Writing approximately a century later, in 1694, Hugh James, agent for James Grahme of Levens, wrote that he had 'gott Graven about 100 day worke of peates and intends to open all the Runners belongs to our moss if it may be drained' (Bagot and Munby 1988, 89). The existence of runners on the moss certainly indicates that the wetlands were managed and, by the collection of 100 days work of peat, that they were exploited intensively.

It is from the seventeenth and eighteenth centuries that the majority of the evidence for the intensive working of the mosses originates. A large number of title deeds exists which outline the rights of turbary that the landowners held on the mosses. An example of such is a lease of a parcel of Hestead Moss belonging to the corn mill at Viver in Heversham to Miles Davis in 1755, the rights of which date to 1636. The lease states that Davis:

> 'also shall and will digg, delve grave and get the turf and peat hereinafter allowed him and them...and also bed and drain the Mosses and Holes from whence got in an Husbandlike

> manner and according to the usual Method of digging bedding and draining Mosses in the like case...'
> (CRO(K) WD/D/He/3/31).

A further stipulation restricted Davis to four days peat work on Black Moss and one day on White Moss which was to be exclusively 'burnt and consumed in the Dwelling House' (*ibid*).

This pattern was repeated well into the eighteenth century when leases and conveyances outlined a rigid and well-organised system of ownership rights on the mosses, which had been separated into individual moss rooms, dales or wands. An indenture of 1711 defined James Hinds' rights:

> '...and also at that Moss or parcell of Turbary ground lyeing in Slake alias Sampoole Mosses containing fourteen yards in breadth or thereabouts lyeing between the Mosses of WilliamWilson's and James Hinds being of the yearly rent of three pence'
> (CRO(K) WD/D/He/3/31/7).

Similar arrangements existed throughout the area, creating a patchwork of landholdings all intensively worked to meet domestic needs.

By the eighteenth century the mosses were providing the locality with peat, which was the preferred fuel for domestic fires, as James informed his master on the Levens Estate in 1693: 'Pray sir let me know if you will have peates gott this yeare, for If I might advise you I would have peates to burne, otherwise you'll destroy your wood' (Bagot and Munby 1988, 79). Equally important was the fact that peat also fired the region's numerous lime kilns. Hugh James fired a lime kiln at Levens in 1694 after having 'gotten a Great many stackes of peates sett up' (*op cit*, 94). Not only would a lime kiln require more peat than a domestic hearth but it would appear that the peats used were larger, as James wrote in August of the same year: 'I have gott all the peats stacked and all the short ones brought home and housed, and hath made two stackes at the Lime Kilne...' (*op cit*, 100).

Peat from the Lyth Valley mosses was also used to supply nearby Kendal with fuel both for domestic use and for the woollen and tanning trade (Bingham 1987, 17). In fact peat was used to such an extent that by 1727 the Kendal Corporation petitioned parliament to be allowed to obtain duty-free coal from Milnthorpe as 'the turf, hitherto their usual fuel, being exhausted, the expense of firing has caused almost the entire loss of trade' (Nicholson and Nixon 1915, 304).

There does appear to be evidence which would support the claim that the mosses were greatly reduced. Some time around 1748, John and his father, Isaac Wilkinson, purchased an iron furnace and forge at Wilson House 'intending to smelt the rich hematite iron ore of Furness with turbary or peat moss, large tracts of which at that time were on every side, nearly, of the furnace' (Stockdale 1872, 210). A canal was excavated 'into the midst of this large tract of turbary', effectively draining the moss (*ibid*). Their attempts to use peat, however, largely met with failure. Various methods were employed for preparing the peat, from using it in its natural state, to compressing the peat into small bricks, and even charring the peat moss itself (*op cit*, 211). The improvement of the moss led to the creation of 1000 acres of arable land, and the first recorded instance of clay mining and brick making in the Cartmel area, and furthermore gave John Wilkinson the notion of improving the whole of Morecambe Bay.

Lyth Valley peat was also used at furnaces other than Wilkinson's ironworks. The use of peat in furnaces was a well-established practice from at least the seventeenth century, when there were repeated prohibitions at the manorial court of Milnthorpe on the taking of peat from the local mosses for use at the Milnthorpe bloomery (Phillips 1977, 21). The most well-known use of peat, however, was at Leighton Furnace in Yealand Redmayne parish, where, in 1715–16, peat was used with charcoal in a ratio of 2:3 (Schubert 1957, 225). Leighton Furnace belonged to the Backbarrow Company, and in the 1730s there were numerous references to that company supplying peat to Leighton from Ulpha Moss, of which it seemed to have owned the peat-cutting rights (LRO DD Mc30/4; Newman 1999, 14). As well as for use as fuel, peat may also have been used for building purposes at the ironworks (Newman 1999, 14).

A further example of the reduction of the mosses occurs in 1799; Margaret Fuller did 'feloniously steal and carry away a cartload of peats valued tuppence' (Bingham 1995, 329). Although tuppence for a cart load of peats would suggest that peat was still plentiful and did not command high prices, the punishment did not appear to fit the crime. She was ordered to be confined in the House of Correction for approximately six months then 'carried in a cart to Blind Beck Bridge [in Kendal], with her body naked upwards from the middle and the words 'A Thief' placed thereon written in large characters and then be discharged' (*ibid*).

In around 1786, John Wilkinson, inspired by his improvements on the mosses at Castle Head, and swept along by canal mania, promoted a scheme to reclaim large areas of Morecambe Bay (Rollinson 1974, 134). The plan was to reclaim 32,500 acres of Lancaster sands, 1600 acres of Ulverston sands, and 4600 acres of Duddon sands, at a cost of £200,000. This was to be achieved by the channelling of the

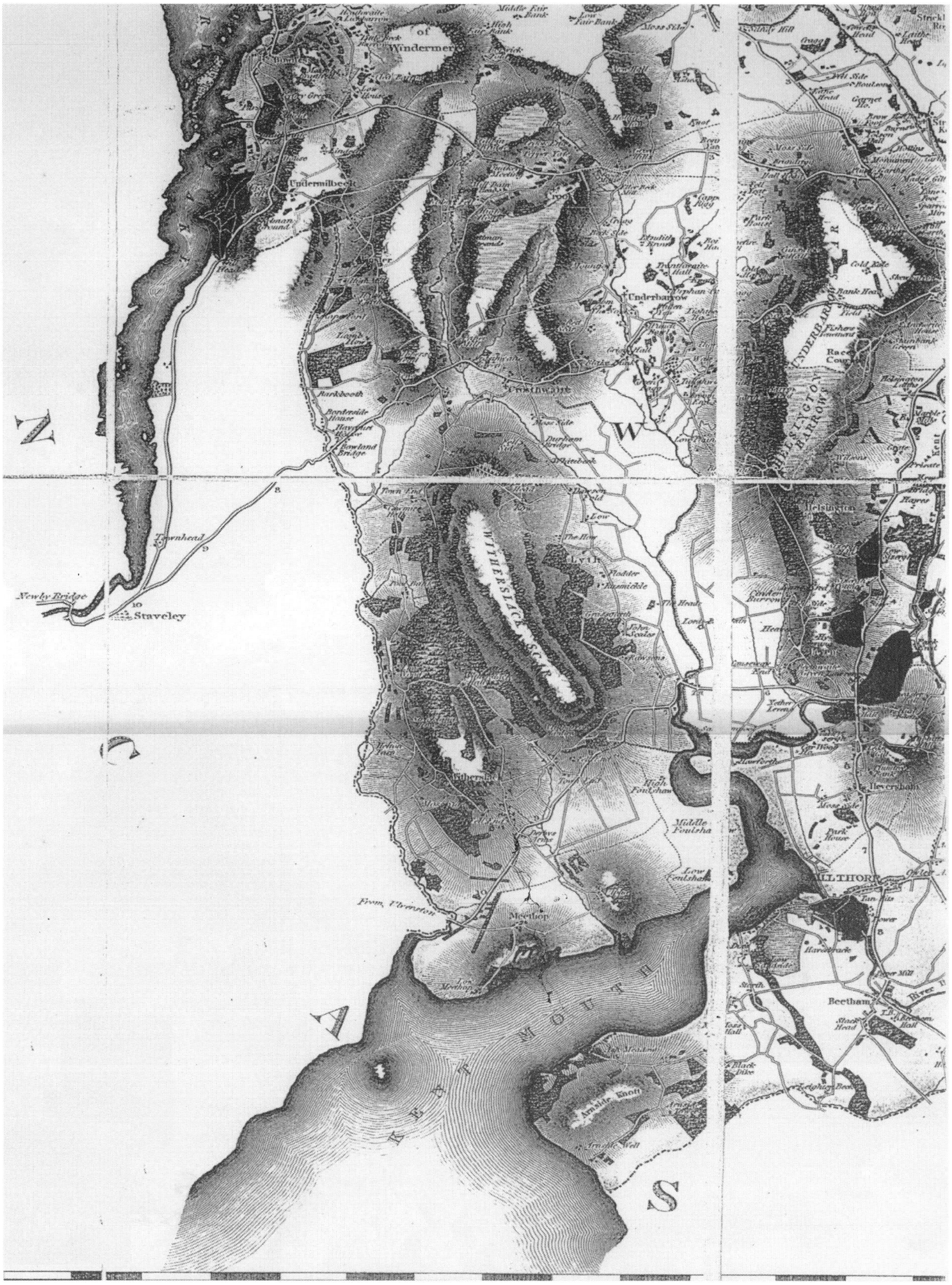

Plate 15 Greenwood's map of Westmorland (1824)

With permission of the Cumbria Record Office (Kendal) © Copyright

Kent, by the excavation of a new channel from Castle Head, across Foulshaw Moss, to join the Kent at Nether Levens, then along the coast to Arnside, to the deep water at Heysham (Hadfield and Biddle 1970, 183). Although the schemes were never enacted, due to what some claimed to be the objections of some 'trifling fisheries' and the refusal of the lords of the manor to relinquish rights to the reclaimed territory, it is a clear indication that, to industrialists and entrepreneurs, the mosses were no longer a barrier to large-scale improvements. When the enclosure movement did reach southern Cumbria in the early nineteenth century, however, the scale of the improvements which were proposed appeared much less ambitious and as a consequence much more achievable.

The extent of the enclosure of the mosses prior to the nineteenth century Enclosure Awards (as shown on Greenwood's map of 1824; Pl 15) has been tentatively estimated, by Steer (1991, 16) to be around 5 % of the valley floor. What is certain, however, is that the mosses received a great deal of attention from the enclosure commissioners. Wood has calculated that, of the 57,746 acres included as part of the awards, 18,956 acres were of moss lands, approximately one-fifth of all the Kendal Ward (Wood 1968, 12). A series of Acts over 40 years, from 1803 to 1843, effectively established the modern landscape of the Lyth Valley through the implementation of a system of drainage which would both drain the mosses and act as a drain for the surrounding 'hard' land. The drainage programme included extensive work raising the river banks and the insertion of two sluices at the east and west of the River Gilpin, which emptied the mosses into the River Kent (*op cit,* 18). Further work included the provision of public roads on the moss to allow people access to their new lands, including a new road across the moss from Levens Bridge to the Turnpike road to Greenodd. This was floated on birch (*Betula*) and heather faggots from the moss and juniper (*Juniperus*) from the adjacent Whitbarrow.

The upshot of the enclosure movement in the Lyth Valley (OS First-Edition map of 1858; Pl 16) was the creation of a large area of land suitable for cultivation and the establishment of new large farms such as Gilpin Bank and Lords Plain Farm, situated in 220 acres of former moss, created by the Levens Hall estate in 1808. The process of change was gradual. After the enclosure and the drainage of the fields, the peat was removed conventionally by a slow process of excavation. Once the peat had been removed, it was found that potatoes could be grown in the first season. Later, after the land had been bedded, it was mainly used for arable, with Kendal market benefiting from the new growing grounds, for in 1825 it was observed '...but a change has taken place, and it is not uncommon to see 200 loads of wheat and a much greater quantity of oats and other meal in the market' (Wood 1968, 37). Later, in 1904, it was remarked that 'instead of a wide expanse of marsh and bog, often supporting little but snipes and wild ducks, there now appears wide fields of golden grain and glistening green crops' (*op cit*, 38).

The desire to improve the area was still strong in the mid-nineteenth century, and although agricultural improvement was not the main aim, it would be a beneficial side-effect. Plans were submitted for the construction of the Grand Caledonian Junction Railway: Stephenson's scheme, submitted in 1837, proposed to link the Carlisle–Maryport line with Lancaster by an embankment across Morecambe Bay from Lancaster to Humphrey Head with the resultant reclamation of approximately the same amount of land as envisaged by Wilkinson (Rollinson 1974, 140). A later scheme was even more ambitious. Hague proposed the construction of a 10-mile-long embankment, from Morecambe to the Furness coast, with the reclamation of approximately 46,300 acres in Morecambe Bay alone. The proposal was to cost £434,131 9s 4d, although the resale of the land at £23 per acre was to recoup three times that amount. Neither scheme was accepted, both being viewed as too expensive (*op cit,* 141). This marked the end of the large-scale reclamation schemes proposed for the Morecambe Bay area, although a railway to Barrow was constructed on more modest terms, between 1850 and 1857 (Marshall and Davies-Shiel 1977, 187).

The rise of arable agriculture, coupled with the availability of better fuel via the railway, had a drastic effect upon what could be described as the peat industry. In 1907, the Vicar of Helsington bewailed, '...whereas half a century ago some 26 peat carts would go to Kendal every day selling from house to house at 2s 6d to 3s a cart, now only one goes occasionally and the price is 7s 6d. Railways have much to answer for' (Bingham 1995, 123). Despite the fall off of peat extraction at the end of the nineteenth century, however, peat was dug on the still unreclaimed mosses during the fuel crisis of the 1940s (Bingham 1987, 17).

Rusland Pool

Stockdale wrote in 1872 that:

> 'When many of the meadows near the little streams in the Cartmel Vale are dug into there is generally met with seven or eight inches of fine loam, and underneath this is a considerable depth of black peatmoss, at the very bottom of which are occasionally found the stems and leaves of the aquatic plants which first began to grow there, preserved quite entire by the

Plate 16 First Edition OS map 6 inch:1 mile (1863): showing Foulshaw Moss and the Kent Estuary

With permission of the Cumbria Record Office (Kendal) © Copyright

> well-known antiseptic properties of the peatmoss or turbary'
> (Stockdale 1872, 578–9).

From this it is clear that the mosses of the Rusland Pool were not subjected to the same intensive exploitation as those of the Lyth Valley had been. Obviously the lack of a major settlement such as Kendal, whose demand for fuel led to the widespread depletion of peat reserves in the adjacent peat fields, had the effect of preserving much of the peat long after the enclosure awards for the Cartmel area.

The medieval occupation and settlement of the mosses in the Rusland Pool area are poorly understood, due largely to the absence of monastic records from the neighbouring Cartmel Priory. The Priory did own land in the surrounding area and presumably used the coastal mosses as sources of fuel and for fishing, but it is not until the dissolution of the monastery in 1536/7 that the extent of the holdings becomes apparent. A petition dated 1641, referring to all the 'lands, tenements, farms, part of the possessions of the dissolved priory of Cartmel, not before granted in any preceding reign subsequent to its dissolution,' outlines the extent of the monastic holdings (Baines 1893, 631). Besides numerous tenements and messuages in adjacent Cark, Flookborough, and elsewhere, the lands included

> 'Saltcoates in the Firth, the mosses of Waythome, Ellerside, Wreke, in the Firth...also turbaries, moors and marshes, lands of Windermore in the said manor, Little Dubbes, Templand Banck, Howlker Banck and Buckburowe Banck, Hampsfell Banck, Cartmell Fell and Cartmell Common '
> (*ibid*).

This was effectively the strip of exceptionally marginal land on the eastern shore of Cartmel Sands which would have been of little agricultural value. There are no other references regarding the use of these lands and it can only be assumed that, like the majority of the lowland valleys in the area, they were used as seasonal pasture, as a source of fuel, and for fishing.

The mosses in the Rusland Valley were, however, under the ownership of the Cistercian house of Furness Abbey. This land was in dispute during the twelfth century between the monks who had been granted the area of the forest of Furness and Waughneia (Walney) in 1127 as its core endowment, and William de Lancaster, Baron of Kendal, who was awarded the lands of Lonsdale and Kentdale in 1150. The land of Furness was defined as 'Elterwater by the valley to Tilberthwaite, and from there by Yewdale to Thurston [Coniston] Water, and by the bank of the same water as far as the Crake, and thence to the Leven' (Rollinson 1967, 77).

Although the Cistercians had a reputation for bold agricultural reclamation, the monks at Furness do not appear to have had much impact upon the mosses of Rusland Pool. Despite undertaking reclamation schemes on Walney Island, including the construction of the Biggar Dyke, they do not appear to have undertaken any schemes within the wetlands. Indeed, there are no explicit references to the way in which the mosses were used during this period, although until 1196 the Barony of Kendal held hunting rights over the area around Rusland. Evidence from Holm Cultram on the Solway (*Ch 4*) has shown that marginal areas were often employed as hunting grounds and, even when settled by farmers with an eye for improvement, it would appear that the importance of the chase was paramount and that areas were left undeveloped.

Evidence of earlier work on the mosses of Rusland is not forthcoming. The sole evidence for wetland management prior to the parliamentary enclosure is from Stockdale, writing over half a century later in the late nineteenth century. He stated that the main watercourses of 'Ellerside, Stribers, Buckbarrow and all the other mosses in those parts were quite *superficial* and universally neglected; whilst the Deer Dyke had actually been allowed to grow up, and was therefore wholly useless' (Stockdale 1872, 328). The precise nature of these watercourses is not clear, nor is the age of the Deer Dyke, which may indicate a continued use of the area as a chase throughout the medieval period and beyond. Certainly, the neighbouring Holker Hall had a park; the Deer Dyke may relate to this early modern establishment (*op cit*, 416).

The later history of the western Rusland mosses is closely related to the construction of the Ulverston Canal and the growth of the railway. The Morecambe Bay reclamation scheme forwarded by Wilkinson in 1786 was to reclaim 1600 acres of land at Ulverston Sands through the construction of a restraining bank, confining the Leven to the shore, with a new channel cut through Plumpton Moss (Rollinson 1974, 135). It has been shown (*see p 51*) that this scheme was not enacted, but the construction of the Ulverston Canal at the end of the eighteenth century did result in some mossland reclamation.

The need for a canal at Ulverston was first advanced in 1791 as a means of providing a safe anchorage from the Levens Sands for the vessels involved in the transportation of slate and iron ore (Marshall and Davies-Sheil 1977, 184). The Ulverston Canal Act was passed in 1793, and stated that the canal was to be cut through Ulverston Moss, which required draining prior to construction. This was achieved by the cutting of a main moss drain, feeding into a back drain running along the northern edge of the canal. As a consequence of the draining operations, the enclosure of the Ulverston Commons and mosses was made easier.

The Enclosure Act of 1799 was carried out in 1813 and included 1109 acres of moss and common (Marshall 1958, 65). The Award of 1813 allows an insight into the uses of the moss prior to the enclosure. It states:

> 'And whereas there are several Mosses or Parcels of Peat Moss marked and mired out, situate within the said Town and Hamlet, the Moss of which hath for time immemorial been conveyed from one person to another, and now held by many different Persons for the sole and exclusive Purpose of digging Peats therein: but the Herbage of Such mosses hath always been enjoyed as Part of the Common of Pasture of and belonging to the said Town and Hamlet of Ulverston'
> (LRO/AE).

Clearly the extraction of peat had been organised prior to the Enclosure Act and the turbary divisions can be seen on the associated enclosure plan. The moss drain post-dates a number of the enclosures in the north of the area, whereas further to the south, towards the canal, the drain defines the edges of the majority of the fields. It is probable therefore that the moss drain incorporated a previous early runner in the south of the area, which was defined by pre-existing turf rooms, hence the erratic course, whilst the northern end of the drain cut through pre-existing fields which were not part of the award, and had been sold by a private contract, in order to include Newland Moss, to the north, in the drainage scheme. This is indicated by a clause in the Award which states:

> '...whereas the Money which arose by the sale of the Herbage and upon and the beneth [*sic*] the several parcels of Peatmoss so marked and meared aforesaid, and the Incroachments which were made within the 20 years before the passing of the said Act, was not sufficient to defray the Costs, Charges and Expenses, incurred in and about the execution of the recited act'
> (*ibid*).

The individual awards were typical of their type, for example:

> 'To Hugh Aulerigg: the Herbage upon and the soil beneth a certain Dale or Parcel of Peatmoss containing in the like Measure one Rood and 24 perches and marked on the said plan 209'
> (*ibid*).

It is interesting to note the area in which rights over the commons extended, since Richard Bell obtained the rights which had formerly belonged to Thomas Ashburner of Urswick some five miles to the south–west. This would suggest that the mosses in the area had been worked quite intensively by people from a large hinterland.

The enclosure of the Cartmel Commons and mosses, further to the east, took place in 1796, comparatively early for the area, and effectively changed the landscape, as the award not only provided enclosed fields but created roads and new bridges. The need for improvement was not solely economic. Stockdale stated:

> 'some persons probably will be inclined to smile when, amongst the many other advantages attendant on the enclosure of the Cartmel Commons is enumerated the entire expulsion thereby from the parish of Cartmel, of that very plaguy complaint 'the ague!'. Often I have heard my late father and other old persons say that previous to the enclosure of the commons in 1796, the ague was one of the most common complaints in this county – as prevalent as it is, or rather once was, in Lincolnshire and other low lying districts, and indeed well might be the case, seeing that in almost every parish there were then extensive morasses, swamps, obstructed streams and much undrained or at any rate, very ineffectually drained, land'
> (Stockdale 1872, 327–8).

Stockdale's father was James Stockdale, an engineer and one of the chief architects for the enclosure of the Cartmel Commons between 1796 and 1809 (Satchell 1984, 88). This may explain the florid embellishments in his description of the Winster mosses and indeed this description appears to be at variance with what the inhabitants of the Cartmel mosslands thought during the time of the awards. A request was received by the enclosure commissioners in 1798 from several owners of moss dales on 'mosses behind Newton Fell' requesting to be purchasers of their moss dales 'or to have the same part of their several allotments'. The commissioners agreed and divided the land into lots which were sold for a total of £1844 (Stockdale 1872, 335).

In 1802, William Wilkinson, a local magistrate, disturbed at the state of the roads that were not being improved as part of the enclosures, compelled the inhabitants of the lower part of Cartmel parish to effect improvements. By 1804, the road between Bigland Scar Meadow and Stribers had been set out but not made, and the road from Stribers across Burnbarrow Moss to Low Wood was considered impassable. Wilkinson considered these to be public highways and not private as they were the only roads 'for the conveyance of peats from the mosses for the adjacent townships of Lower Holker, Upper Holker and Lower Allithwaite...peats being almost the only commodity easily obtainable for fire at this time in this county' (Stockdale 1872, 355). This dispute resulted in the whole parish being fined £2100 and was only resolved in 1825.

The reclamation schemes of the Railway Age, had they been implemented, would have had a profound effect upon the mosses of the Rusland Pool. Stephenson's plan to construct an embankment across Morecambe Bay (*see p 55*) proposed a route from Humphrey Head to Chapel Island off the Furness coast (Rollinson 1974, 140). Hague's massive proposal would also have affected the Rusland Mosses as the scheme was to allow the incoming tidal waters to deposit silt, thus raising the level of the land, enabling cultivation (*ibid*). As has been discussed, these schemes were never enacted.

Little is known of the recent history of the mosses of the Rusland Pool and it must be assumed that, after the Enclosure Award, the area was worked in the same manner as that of the neighbouring mosses in the Lyth Valley.

Duddon Mosses

Records for the early use of the mosses in the Duddon Valley come from the extensive collection of documents concerning the Cistercian house at Furness. The abbey was founded in 1127, on land granted by the future King Stephen of England, after having been relocated from Tulketh in Preston (Knowles 1940, 74). The Cistercian desire for inhospitable and remote locations for their abbeys was particularly well suited to the Furness Peninsula. As their core endowment they were granted the forest of Furness and Walney Island (*see above, p 55*) comprising the whole lordship of Furness, including Ulverston, Walney Island, and Dalton, amounting to 20 carucates of arable land, and by 1292 they had established 11 granges within their lands on the Furness Peninsula (Barnes 1968, 30). The topographical locations of the granges bear little resemblance to those selected by the other large Cumbrian monastic house at Holm Cultram, on the Solway coast; as will be discussed *below* (*Ch 4*), the northern abbey appeared to position its granges deliberately skirting the moss fringes. The positions of the Furness granges display no such preference and they are situated almost randomly throughout the demesne. Later, by 'the course of gradual aggrandisement by the gifts of the various proprietors, the monastery extended its lands and holdings, including notably Angerton and Arnaby mosses on the Duddon (Baines 1893).

The extensive collection of charters within the *Furness Coucher Book* relating to both Angerton and Arnaby Mosses is illuminating in that they outline the important economic role played by the mosses in the Abbey's affairs. Although the southern mosses, along the Winster and Lyth Valleys, had been used from the medieval period onwards, the extent to which they had been settled and exploited can only be extrapolated from inadequate data. Within the Duddon mosses, however, there appears to have been a deliberate and well-orchestrated effort to use the mosses to the best of their ability from the medieval period onwards. Indeed, within the pastoral landscape of the south Cumbrian mosslands, the area around Angerton and Arnaby Mosses produced more archaeological evidence of agricultural activity than the Lyth and Winster Valleys further to the east. This included remnants of medieval/post-medieval ridge and furrow (CU45, CU46, CU47, CU48, and CU52), some of which was very well preserved and, in one case (CU45), was clearly associated with Angerton Hall. Fieldwalking also produced evidence of early industrial processes with the identification of three bloomery mounds and an associated charcoal-burning area on the northern fringe of Heathwaite Moss (CU50).

The position of Angerton Moss, on the tidal reaches of the Duddon Estuary, enabled the monastery to establish fish weirs and salt works and to cut peat as well as pasture their stock. The moss came into the possession of the Furness monks in 1299; previous to that the recorded use of the moss relates to the pasture between *Welpesat* and *Steerspool* belonging to Alan de Kirkby who paid 4s rent yearly to a Simon de Broughton for the right (Farrer and Brownbill 1914, 408). Later, the whole of Angerton Moss was granted to Thomas Skilhare of Dalton by Ralph, son of Alan, for a rent of 5s (*ibid*). Although Angerton Moss was rented by Thomas Skilhare, Ralph maintained his right of turbary over the moss, which was enough for his household consumption and to cover his buildings, and he also allowed Sir John de Kirkby to take 60 cartloads from the moss (*ibid*).

An indenture of 1424 between Robert, abbot of Furness, and Richard of Kirkeby refers to 'a parcell of mosse, more and pasture...callet Angerton Mosse', the extent of which was described as:

> 'begynnyng atte Whelpesatecrag and fro thethyn comand downe by the sand of the water of Doden...and so unto the streme of the watir of Sterespole and so folwyng the streme of the said water...upward un to the downfallyng of Otrepole in to the Sterespole and so folwing Otrepole continuely to the heved of the same Otrepole and fro the heved of Otrepole going streght over the mosse to the Estende of the olde dike extending to Whelpesatecrag and so folwing that dike evyn to the first of the said Whelpesatecrag' (Brownbill 1919, 773).

This land division is still partly visible as the Angerton Parish boundary.

When the moss was granted to Furness in 1299 it would appear that there was already a fishery in existence as agreement between John de Kirkby and Ralph fitz Alan mentions a *piscarium* constructed from large timbers and brushwood (*tinsellum*) at Sterispol (Atkinson 1887a, 321). By 1517, the Doddyng Fysshyng, rented out to John Lampley, knight, was valued at 10s (Brownbill 1919, 603).

There are no explicit references to salt production by the monks of Furness in the Angerton Moss area, although they had a number of salt works elsewhere (*see below*) and there is a reference to a 'Salthus in Furnes' (Atkinson 1887b, xix). There was, however, a grant of 1234, to the Priory of St Bees of land between Slepul where it falls into the Duddon across to the orchard which stands above the Duddon marsh, and from the above orchard going up to the great oak which stands above the Slepul near the great peat moss [*magnum tuberam*] (Wilson 1915, 58). The monks were also granted the right to turbary to fire the salt work [*omnia estoveriasua predicte saline necessaria ubicunque voluerint in magna turbera inter Arnulfpule et Slepul*] (*ibid*).

The extent of the monastic holdings within Angerton Moss is detailed in the *Furness Abbey Rental*, which reveals that in 1537 payment was due from 15 tenants of Kirkby for 40 acres in Angerton Mosse, which was valued at 52s 2d (Brownbill 1919, 603).

The mosses of Broughton were not enclosed until 1847 when the 'Common of Pasture in over and upon the said Commons Waste and Moss lands and to the right of Turbary in the Moss lands as set out by the Moss Liverers of the said manor' were established (LRO/AE). The mosses which were included in this award were Black Moss, Mere Side Common, Middlescough Moss, Wreaks Common, White Moss, and Fox Moss. The system of runners within the mosses, particularly those on White Moss, represent the fossilised turbary rights of the parishioners of Broughton.

There is little evidence for the use of the mosses during the recent past: the construction of the Furness railway was the last significant event which had an effect upon the mosses of the Duddon. The line crosses Herd House Moss and Angerton Moss on the eastern side of the Duddon and Shaw Moss on the west. It was completed in 1850 as the Whitehaven and Furness Junction, effectively joining the West Coast line with the Furness railway (Melville and Hobbs 1951, 38).

Discussion

This chapter has detailed the assessment of the known archaeology and palaeoecology of a large area of Cumbria lying on the northern edge of Morecambe Bay. Some of the wetlands still survive largely intact, although others have been destroyed in the past, such as some of the Lyth Valley mosses. With few exceptions, all of the mosses are in decline owing to past and present exploitation: forestry on Foulshaw Moss, for example, is likely to lead to its desiccation and peat decay. Management works, notably those in the SSSI areas, will lead to the survival of a significant proportion of the mosses, although only those in certain areas with intact (or regenerating) mossland vegetation. The importance of less significant wetlands should not be overlooked, particularly for the more recent periods. An example of the potential of such areas is the trackway from Goldmire, near Dalton-in-Furness, which was initially noted by William Close (West 1805, 9) who recorded 'an ancient road paved with large stones...' discovered in May 1803. Subsequent excavations in 1949 revealed a metalled road built to cross a small wetland in the base of a steep-sided valley. Associated finds appeared to suggest a post-medieval date (Brady 1971, 32–4).

The predominantly pastoral land-use, along with the survival of overlying peats and marine silts, has meant that the archaeological remains are very difficult to detect from surface survey. Even aerial photography has proved difficult to use in such circumstances – such survey in comparable circumstances in North Lancashire, for example, only gave results where there was widespread arable cropping (C Cox pers comm).

The fact that many new sites have not been identified by the survey is significant, and hopeful: it may mean that the wetland sites and relict landscape features lie buried under waterlogged deposits under permanent pasture. The threat to any such sites is thus relatively slight, unless large-scale landscape change is proposed. The only sites which will require future management, and about which archaeologists should express concern, are those within the mosses under the threat of decay. The trackways in Foulshaw Moss, and the palaeoecological work of Wimble (1986), have indicated the potential of such deposits to contain well-preserved archaeological and palaeoecological remains. The threat posed to such peat deposits is all the more insidious as the exact nature of the remains in the mosses is difficult to assess accurately without large-scale disturbance of the deposits: the accurate expression of this potential is one of the key issues facing wetland archaeologists.

The concept of potential archaeological value is also important in the consideration of previous archaeological records. It is clear that many of the patterns visible in the data are products, at least in part, of differences in the visibility of the archaeological record. This is clearly expressed in the Mesolithic period where we are reliant on finds under a barrow (Levens Park), in caves (Bart's Shelter), and in eroding sand dunes (Walney Island). The extent of activity in the landscape indicated by these glimpses of the archaeological record suggests that there are many more sites to be discovered, many of which are likely to be focused around resource-rich areas. Many such sites are likely to be located along the fossil coasts preserved in and under the South Cumbrian wetlands. Similar patterns are present in other periods, where we have glimpses of extensive landscape activity, particularly in the pollen record from later prehistory onwards, which is not accurately reflected in the known number of sites, particularly from wetlands. It is likely, however, that, by the Iron Age, many of the wetlands had developed into full raised mires and, if parallels can be drawn from elsewhere in the North West (*eg* Middleton *et al* 1995, 205–6), they were largely abandoned for settlement and subsistence. There can be little doubt, however, that the coast, with its extensive wetlands, remained important throughout the prehistoric and historic periods both for fishing and for communication. As such, it would have provide an important adjunct to inland activity, particularly in Furness, with its continued importance from the Iron Age into the historic period.

3

WEST CUMBRIAN COASTAL PLAIN

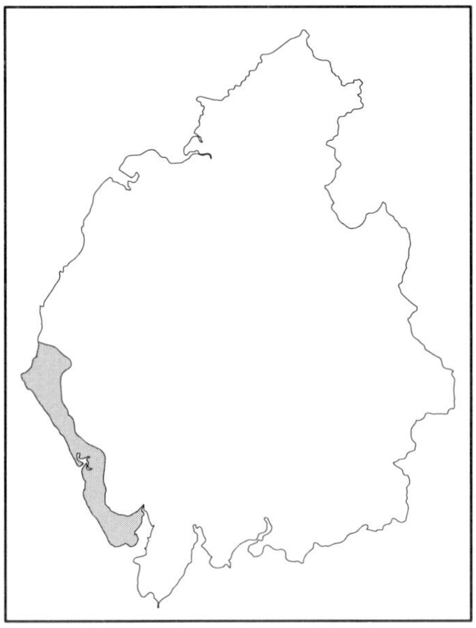

Fig 21 The extent of the area covered by Chapter 3

> 'There can be little doubt that by well-directed inquiry there would be found here yet unearthed a wealth of Neolithic remains... Dispersed over this area are gentle hills, with isolated depressions containing tarns or the remains of similar pools, now boggy places or marshy land suggestive of habitations or lake...dwellings'
> (Watson 1903, 92–3).

Introduction

The West Cumbrian Coastal Plain, as defined for the purposes of this chapter, encompasses a relatively small area of land compared to others discussed in this volume (Fig 21). It is, however, discrete geographically which is recognised by its designation as the '*West Cumbria Coastal Plain*' Natural Area by English Nature (English Nature 1997) and '*West Cumbria Coastal Plain*' Countryside Character Area by the Countryside Agency (Countryside Commission 1998, 25-30; *see* Figs 9 and 10, *p 24*). This, and the importance of this thin strip of land to regional and national prehistory, warrants its discussion as a separate topic.

The plain stretches 40 km from the Duddon Estuary in the south, to St Bees Head in the north, and nowhere is it more than *c* 7 km wide. The plain is bisected at Ravenglass by the Rivers Irt, Mite, and Esk which form a small estuary fringed by sand dunes. Further north the Rivers Calder and Ehen reach the sea near Sellafield. The seaward part of the plain in this vicinity is marked by a series of former raised beaches — the results of mid-Holocene sea-level changes. For the most part the coast is low-lying, with shingle beaches and low boulder-clay cliffs, but at St Bees Head New Red Sandstone outcrops to form sea-cliffs (Pl 17).

Geology and soils

Most of the coastal plain between St Bees and Ravenglass is characterised by deep, well-drained coarse loamy brown earths, intermixed with gleyed brown earths and brown sands (Wick 1 Association): these derive from fluvio-glacial deposits and drift of variable stoniness (Jarvis *et al* 1984, 302). The lower reaches of the River Ehen and areas north–east of Beckermet and east of Seascale are distinguished by extensive areas of freely-drained, medium (occasionally coarse) sandy soils (Newport 1 Association). On the dunes of the Esk Estuary, deep, well-drained sandy soils of the Sandwich Association predominate; south of the estuary extensive areas of seasonally waterlogged loams of the Clifton Association have developed on till deposits.

Archaeological and palaeoecological survey

Archaeology

Extensive investigations by a series of workers over the past half-century have shown that the West Cumbrian Coastal Plain was a focus of late Mesolithic and early Neolithic activity. The initial work of Nickson and Macdonald (1955), who first discovered

Plate 17 Boulder clay cliffs showing peat section at St Bees

early prehistoric lithic assemblages at Drigg, through the extensive fieldwalking of the Cherrys during the 1960s and 1970s (Cherry 1963; 1965; 1967; 1969; Cherry and Cherry 1973), has led to the integrated studies of landscape change around the Esk Estuary of the 1970s and 1980s (Bonsall *et al* 1986; Tipping 1994).

The area has continued to excite archaeological enquiry and generate a rich source of data. At the same time palynological research, sometimes associated with archaeological endeavours, sometimes undertaken independently, has shown the West Cumbrian Coastal Plain to have been one of the earliest areas of the British Isles to have been extensively and permanently cleared for agricultural purposes. Palaeoenvironmental and archaeological information is also available with which to interpret events in the later Mesolithic, Neolithic, and Bronze Ages. Earlier and later periods (with the exception of the Late Devensian), however, are still poorly served by data from both strands of research.

The North West Wetlands Survey field strategy followed that used elsewhere (*see above p 16–8*) and had the primary aim of assessing the peat resource by rapid field survey (Fig 22). Since the predominant land-use was permanent pasture it was envisaged that a programme of fieldwalking would not have been successful, and, in the light of the work of the Cherrys over many years, investigating each arable or ley grassland field that arose, it was considered that fieldwalking arable would be a duplication. The only exception to this was a field near Ehenside Tarn, which was walked to cast some light on the previous work at the site.

Palaeoecology

Owing to time constraints, and the amount of palaeoecological work already undertaken on the West Cumbrian Coastal Plain, no new palaeoecological work was carried out in this area as part of the NWWS field programme. Brief visits were, however, made to a number of small basin mires near to Ehenside Tarn to assess the possibility of organic preservation. The survey has made extensive use of the work of Walker (1966), Pearson (1962), Pennington (1965; 1970; 1975), Coope (1994), and Tipping (1994), for information on the Late Devensian, later Mesolithic, Neolithic and Bronze Ages, in the compilation of this chapter.

Wetland distribution

The majority of the West Cumbrian Coastal Plain mosses are located in close proximity to the coastline, often occurring to the landward side of the coastal sand dunes, for example, Williamson Moss (Fig 23), which is close to the archaeological site of Eskmeals (Tipping 1994). However, many wetlands have been destroyed by cutting and only relict fen peats remain from these former valley and floodplain mires. Church Moss and Silver Tarn are examples of a suite of basin mires which developed in kettle holes after the last glaciation; these are also a feature of south-east Cumbria, Cheshire (Leah *et al* 1997) and Shropshire and Staffordshire (Leah *et al* 1998). There are only a few lowland raised mires in West Cumbria and these occupy the narrow strip of land between the coastal plain proper and the Cumbrian Mountains. This mire type is represented by Black Moss (Pl 18), the most westerly example of a lowland raised mire in England (English Nature 1985). Acid valley mires are also found in this narrow strip and Hallsenna Moor (Pl 19), now a poor fen, is an example of these (Ratcliffe 1997, 58). At a slightly higher altitude is the well-preserved, large basin mire of Distington Moss, which represents the most extensive peat land in the area and supports rare plant species.

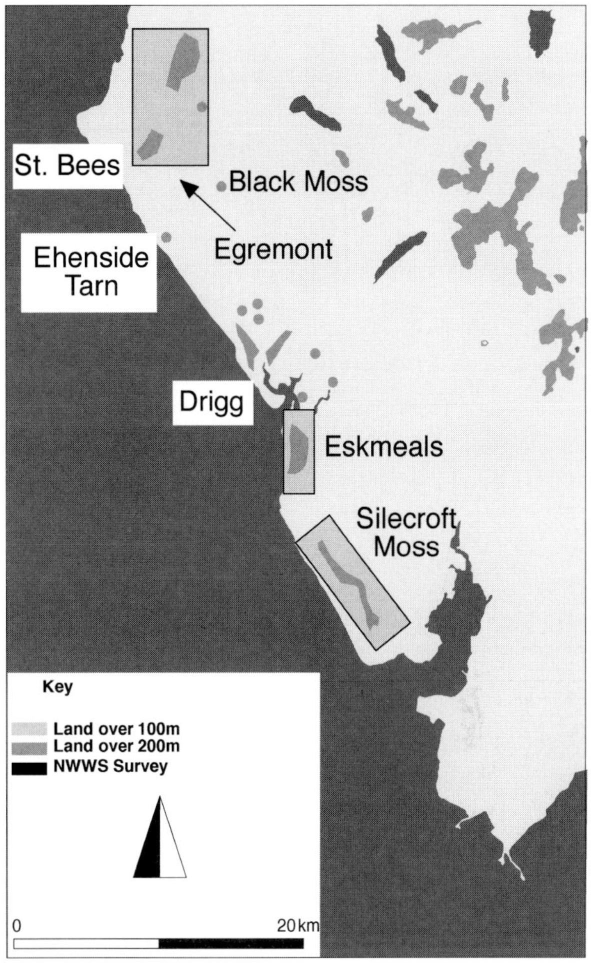

Fig 22 West Cumbrian Coastal Plain: areas of archaeological survey

Plate 18 Black Moss showing its limits

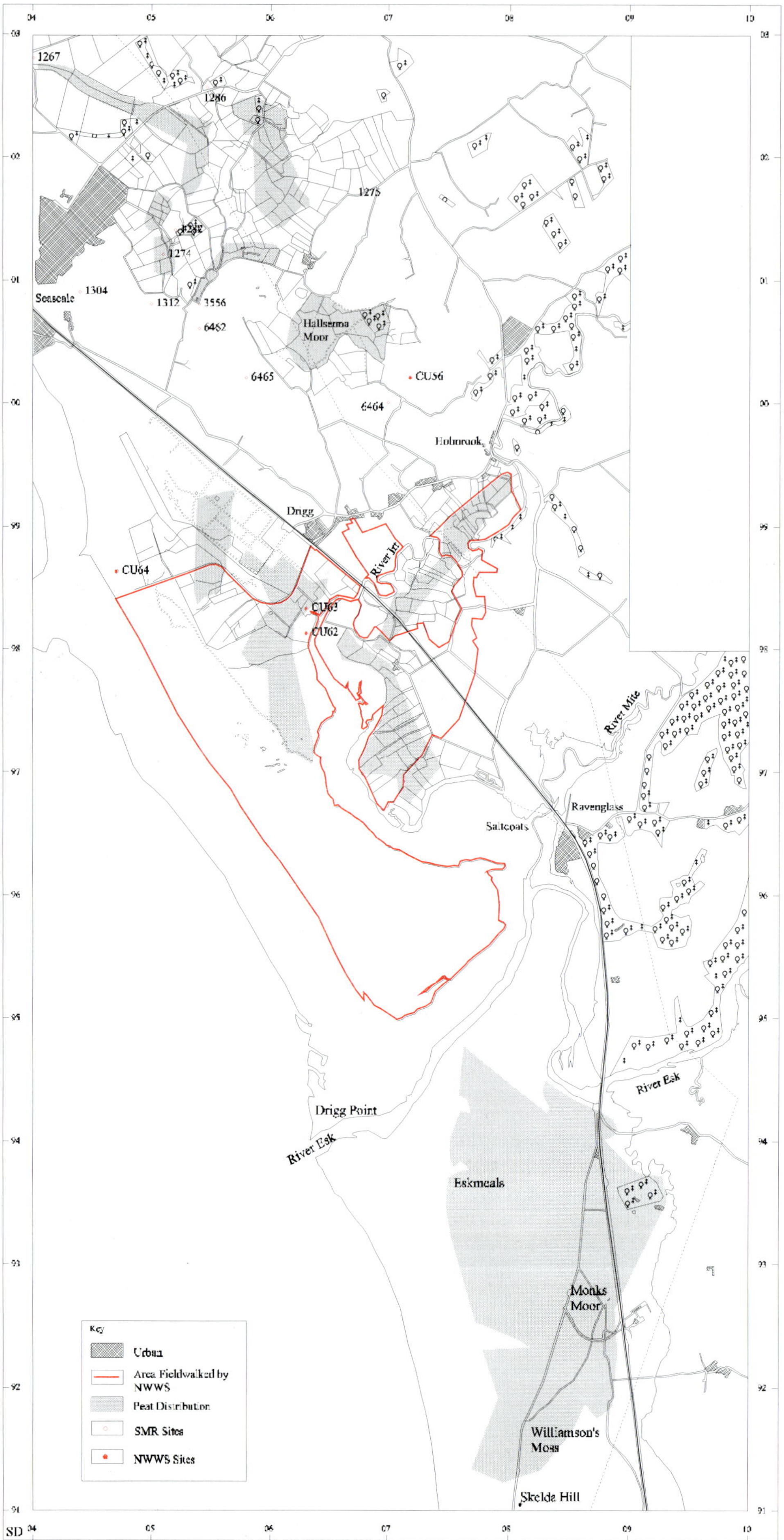

Fig 23 Eskmeals: predicted peat survival and actual area surveyed (Scale: 1:50,000)

Based upon the 1975 Ordnance Survey 1:10,000 maps with permission of the Controller of Her Majesty's Stationery Office © Crown Copyright

Plate 19 Hallsenna Moor: a small intact valley mire

The fertile coastal plain has inevitably been the focus of many of the larger settlements within the county of Cumbria, and some of these have a long history of industrial development, such as Workington and Whitehaven. Several of the county's major communication routes link these settlements, with consequential detriment to the wetland areas. Similarly, mining activity and arable farming practices have taken their toll on former mosses. Moresby Moss, for instance, has been destroyed by open-cast coal mining, and Brownbank Moss is truncated by the B5344, the construction of which undoubtedly had some negative impact on the site. Nevertheless, some of the wetland areas have to date survived relatively intact with only peripheral damage.

Landscape Evolution

Late Palaeolithic

Palaeoecological evidence

The west of Cumbria is fortunate in possessing a palaeoecological signature of events during the last stages of the Devensian glacial period, which is recorded in a classic section in cliffs south–west of St Bees (NX 962 116; Pl 20). An organically-rich sediment layer is intercalated between Quaternary drift deposits which, although noted for some time (Smith 1912; Eastwood *et al* 1931; Trotter and Hollingworth 1932), were first subjected to palynological study in the 1950s (Walker 1956; 1966). These studies revealed vegetation changes attributable to the well-established tripartite sequence characteristic of the climatic oscillation at the end of the Devensian glaciation (Late Devensian (I–III)). A rise in tree and shrub pollen marks the earliest stages of the Windermere Interstadial between biostratigraphy dominated by herbaceous pollen characteristic of colder, open conditions; recent studies at the site suggest that the latter phase of the Windermere Interstadial (LDe II) can be dated to 11,806–11,196 cal BC (11,500±120 BP, Birm–647) and 11,419–10,888 cal BC (11,180±120 BP, Birm–649).

Palaeoecological work at St Bees was extended in the early 1960s to include study of the sub-fossil coleoptera remains (Pearson 1962). Further study of these assemblages by Coope and Joachim (1980) led to an enhanced understanding of environmental conditions in West Cumbria during this period and analyses of the ecological tolerances of the beetles identified in the stratigraphic sequence has added considerable detail to the thermal conditions during the Late Devensian in West Cumbria. In a recent reappraisal of the evidence, Coope (1994, 88) suggests that mean July temperatures at the site were probably *c* 15–16 °C during most of the LDe II (Windermere

Plate 20 St Bees: cliff section showing Late-Glacial peats

Interstadial) thermal optimum, falling to *c* 10 °C during the following LDe III (Loch Lomond Stadial) phase. The reliability of this model of the Late Devensian oscillation, with its abrupt, rapidly changing pattern of temperatures, has recently been enhanced by its similarity to climatic curves for the same period derived from high resolution Greenland ice core data (Johnsen *et al* 1992) and other sub-fossil coleoptera studies (Walker *et al* 1993).

Mesolithic and early Neolithic

Palaeoecological evidence

Currently, there is no palaeoenvironmental evidence for the early Holocene on the West Cumbrian Coastal Plain. In contrast, the nature of the environment during the later Mesolithic / early Neolithic transition is better understood, largely as a result of recent work in the Esk Estuary.

All Mesolithic activity on the West Cumbrian Coastal Plain was played out against the backdrop of a coastline influx. For example, between *c* 5970–5480 cal BC (6810±130 BP, Hv–6209) and 5613–5240 cal BC (6495±95 BP, Hv–5210) a relative sea-level rise of some +2 m OD had produced a perimarine fringe *c* 1–1.5 km further inland than at present at Eskmeals during which a series of shingle ridges was deposited. These began to build coastal barriers which helped to isolate inland basins from marine influence, encouraging the development of lagoon and estuarine environments (Huddart *et al* 1977, 142; Bonsall *et al* 1986, 178).

Recent palynological studies at Eskmeals have focused on two sites: Williamson's Moss (Pl 21), and the infilled channel immediately to the east which contained a possible Mesolithic structure (Tipping 1994). Williamson's Moss had originally yielded evidence of Mesolithic vegetation disturbance in earlier studies (Pennington 1975) but the latest research adds much greater detail and a more precise chronology to the sequence of events.

Plate 21 Landscape of the Irt Estuary

Tipping's work (*op cit*) at Eskmeals, undertaken in conjunction with extensive archaeological survey and excavation, has shown that, after the isolation of the lake basin from direct marine influence at *c* 5000 cal BC, a fen carr began to develop around it. Not long after *c* 4780–4470 cal BC (5760±60 BP, SRR–2658), the first indications of disturbance to the carr, possibly anthropogenic, are discernible in the form of a slight opening up of woodland. Following this, like several other palynological sites in Cumbria (Oldfield 1963; Walker 1966; Pennington 1975), Williamson's Moss produced evidence for a double Elm Decline. Tipping (*op cit*) has demonstrated that an exceptionally early primary decline occurred in the area, dated to *c* 4458–4047 cal BC (5440±70 BP, SRR–3065). Although the decline at this point is small, it is statistically significant, and other species, such as hazel and oak, also seem to have been affected. The occurrence of *Plantago lanceolata*, and the fact that mineral inwash affected the lake sediments during this time, indicating localised soil erosion, strongly suggest anthropogenic influence. Palynological work from the channel to the east of Williamson's Moss containing a possible Mesolithic platform (*op cit*, 126) also records the primary Elm Decline and, given that the radiocarbon date for this event from Williamson's Moss appears to match closely that of the timber of the platform itself, there appears to be strong circumstantial evidence that the activity at the site is connected with clearance in the wider landscape.

These findings appear to amplify signals of early 'primary Elm Decline' disturbance recorded from Barfield and Ehenside Tarns, situated 4.5 km and 12 km respectively along the coastal strip to the south and north-west. At Barfield Tarn, Pennington (1970; 1975) recorded traces of small forest clearings dating to *c* 4457–3825 cal BC (5340±120 BP, K–1057), while at Ehenside Tarn, Walker (1966) registered the presence of similar reductions in arboreal pollen

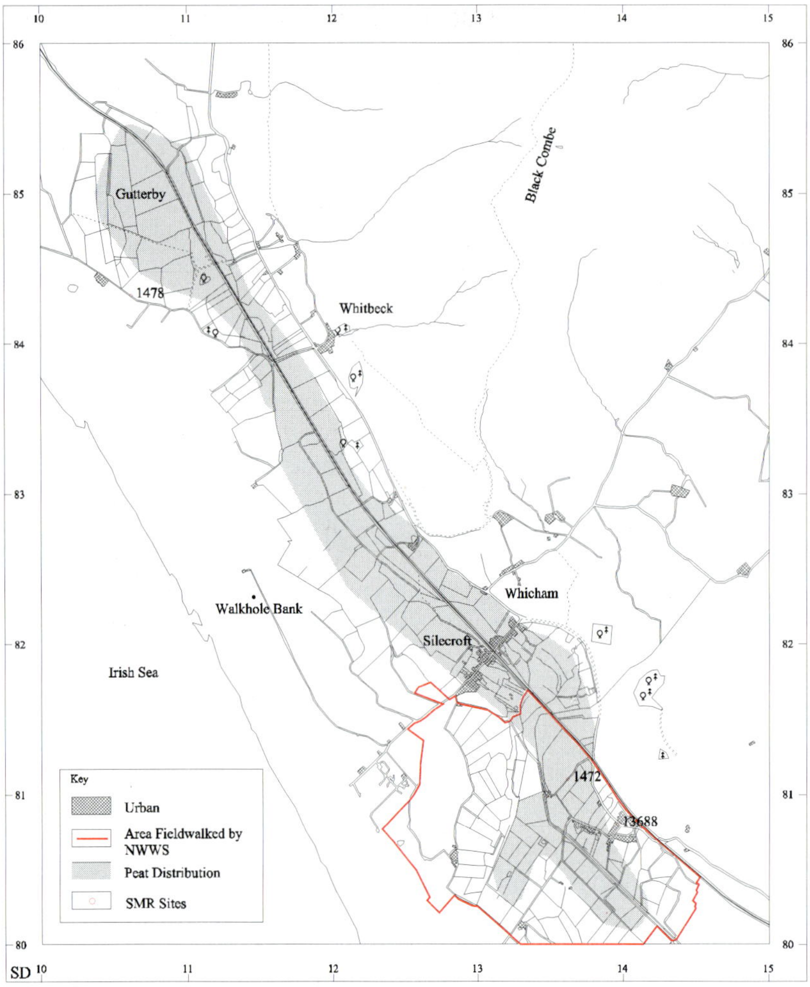

Fig 24 Silecroft: predicted peat survival and actual area surveyed (Scale: 1:50,000)

Based upon the 1977 Ordnance Survey 1:10,000 maps with permission of the Controller of Her Majesty's Stationery Office © Crown Copyright

prior to the main Elm Decline. The wide distribution of this phenomenon seems to support the suggestion that the West Cumbrian Coastal Plain was a focus of Mesolithic settlement and it has even been speculated that these initial falls in arboreal pollen represent the effects of 'small-scale assarting' (Tipping 1994, 126).

Nevertheless the clearings were undoubtedly of limited impact for by *c* 3990–3650 cal BC (5010±70 BP, SRR–3067) partial woodland regeneration was reclaiming the open areas around Williamson's Moss. The recession in activity was short-lived and the 'main' Elm Decline followed, dated to *c* 3893–3381 cal BC (4850±80 BP, SRR–3068), from which the tree species failed to recover fully. At Barfield Tarn the main (secondary) Elm Decline was accompanied by reductions in other tree pollen, especially oak, and an increase in cereal pollen. These changes in the composition of the pollen sum occur at the point when the sediment lithology changes from an organic mud to a pink clay. This strongly suggests that considerable clearance of woodland for pasture and cereal cultivation occurred at the same time as the Elm Decline, with the consequence that soil erosion around the tarn was greatly increased (Pennington 1975).

Similar activity is also represented at Ehenside Tarn where the chronology of the sequence relies on proxy dating of palynological features by comparison with radiocarbon-dated analogues elsewhere (Walker 1966). Based on this, the indications are that major clearance may have begun here slightly later than at Barfield, the earliest cereal pollen occurring at the same time as the beginning of extensive and permanent clearance around Ehenside, possibly between *c* 2900 cal BC and 2600 cal BC (*c* 4300–4100 BP; Pennington 1970, 69). Taken together, the evidence from these three sites indicates that the West Cumbrian Coastal Plain began to be permanently cleared of forest during the early Neolithic, making it one of the earliest recorded regions in the country to suffer this fate (Bell and Walker 1992, 165).

Archaeological evidence

Evidence for Mesolithic settlement on the West Cumbrian Coastal Plain is almost exclusively confined to raised beaches of former coastlines. The exceptions to this pattern, such as Fleswick (SMR 17827) and Rottington (SMR 1179), are located on the cliffs north of St Bees (Cherry and Cherry 1983, 8). On the coasts the distribution of sites is widespread, from lithic assemblages in dunes at Pow Beck, St Bees (SMR 1182), to more subtle signs of activity at Walkhole Bank (SMR 18908) near Silecroft (Fig 24; Cherry and Cherry 1987a). The tight focus of the sites to the coasts, and to estuaries in particular, may suggest that they were positioned to take advantage of the greater biodiversity of estuarine environments with the larger number of potential food resources compared to other coastal habitats (Bonsall *et al* 1994, 93). In addition, the major sites of Williamson's Moss (SMR 13940) and Monk Moors (SMR 1391) would also have been ideally positioned to exploit pebble flint sources from coastal shingle bars.

Nearby basin fens may have provided other resources to supplement those of the coast. Thus, the sites at Eskmeals (Fig 23) might have been positioned to exploit both the adjacent marine environment and that of the inland wetland of Eskmeals Pool, while those at St Bees are similarly situated between the coast and the ancient alder carrs of the Pow Beck (Fig 25). Furthermore, it is conceivable that dune-slacks (which have left a record of their former presence in the peat deposits intercalated within the coastal dune deposits) and fens to the rear of the dunes offered a similar suite of conditions at Drigg. It is also possible to view the poorly-defined Walkhole Bank coastal site (SMR 18908) in this light, positioned as it is in close proximity to both the coast and the mires of the Silecroft Beck.

The lithic assemblages of the Mesolithic sites share a number of physical and typological similarities, notably the predominance of beach pebble flint which has resulted in small artefacts of variable quality. The most common implements are geometric microliths and scrapers, with core axes and bevelled tools lacking (Bonsall *et al* 1994, 96). Cherry and Cherry (1984, 14) have suggested that the similarities in the assemblages from St Bees (SMR 17803; SMR 17804) and Nethertown

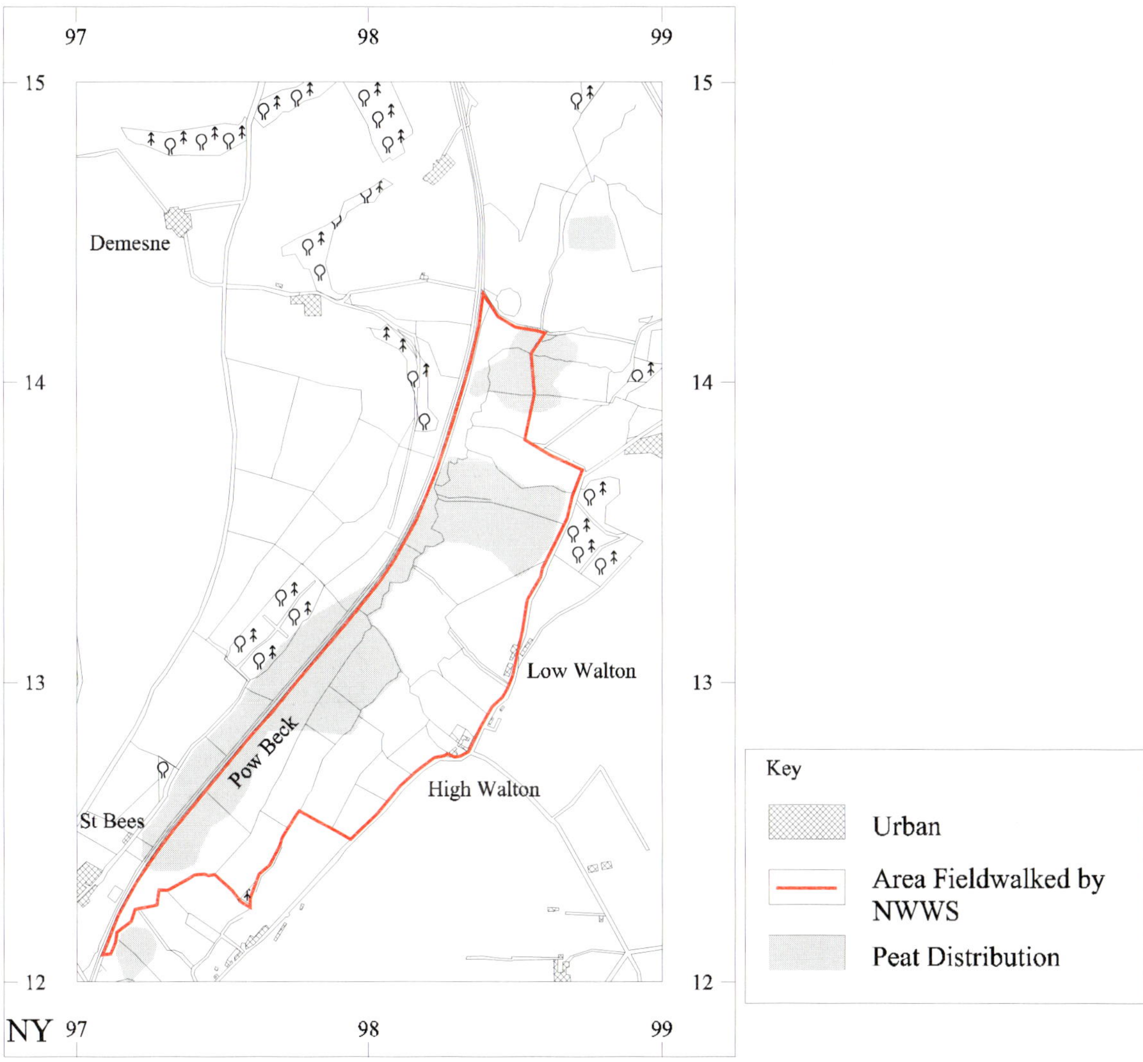

Fig 25 St Bees: predicted peat survival and actual area surveyed (Scale: 1:25,000)

Based upon the 1988 Ordnance Survey 1:10,000 maps with permission of the Controller of Her Majesty's Stationery Office © Crown Copyright

(SMR 6475; SMR 6481) are so marked that in each case it may be the work of a single knapper, and that the scatters may represent a single family / hunting unit.

Within this widespread Mesolithic activity there is a concentration of settlement in the Esk Estuary and it is this area which has received the most archaeological attention. Detailed fieldwork has identified *foci* of activity on Skelda Hill at SD 082 912 (SMR 1386; SMR 13946; SMR 13949), Williamson's Moss (SD 085 917; SMR 13940), Monk Moors (SD 090 924; SMR 1391; SMR 13950–13953), Eskmeals Pool III (SMR 13956), Eskmeals IV and V (SMR 13957), situated near Langley Park, and a further site at Eskmeals near Newbiggin (SMR 1387). All of these were initially identified as surface lithic scatters, some of which were subsequently excavated. Judging from the range of radiocarbon dates from Williamson's Moss and Monk Moors, obtained from hearth charcoal — *c* 5970–5360 cal BC (6752±156 BP, BM–1216) to *c* 1252–910 cal BC (2859±49 BP, BM–1386), which are statistically consistent (T′ = 4.9; v = 4; T′ (5 %) = 9.5; Ward and

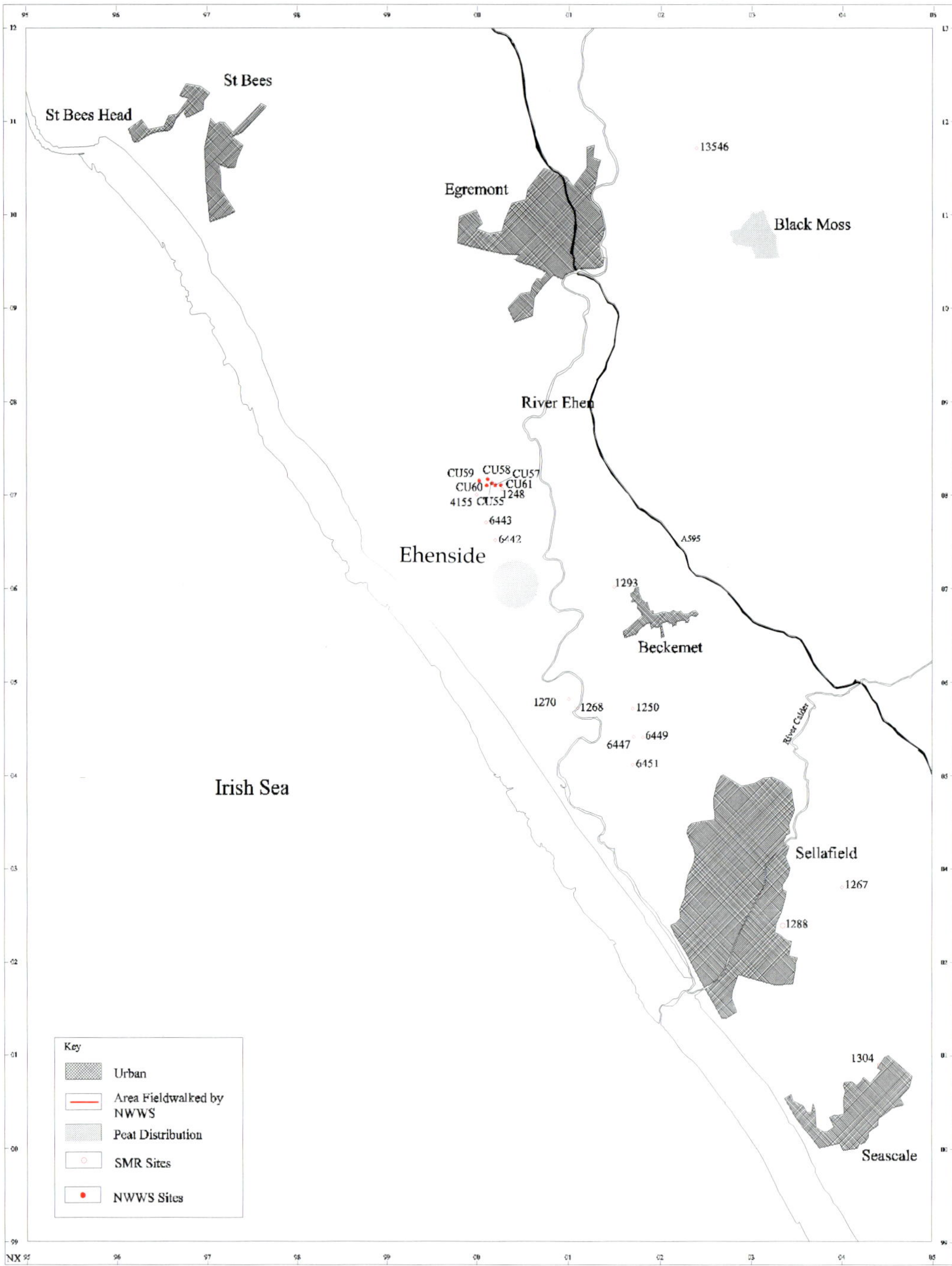

Fig 26 Ehenside Tarn: map showing the location of the tarn and associated finds (Scale: 1:75,000)

Based upon the 1982 Ordnance Survey 1:10,000 maps with permission of the Controller of Her Majesty's Stationery Office © Crown Copyright

Wilson 1978) — some of the sites have multiple phases of activity stretching into the Bronze Age (Burleigh and Matthews 1982; Bonsall *et al* 1994).

The most detailed site specific studies have been concentrated on Williamson's Moss (SMR 13940). This basin on the raised beach had been subject to phases of marine sedimentation but, by *c* 5473–5074 cal BC (6320±85 BP, Hv–5227), it became isolated from the sea by the development of shingle ridges, allowing an initially mildly-brackish water lake to develop which subsequently became a freshwater one (Huddart *et al* 1977, 142). It was around the fringes of this lake, and subsequent fen carr, that Mesolithic activity appears to have been concentrated: small excavations on the margins of Williamson's Moss led to the recovery of over 34,000 pieces of flint, chert, and tuff (including over 600 tools) from a single 125 m^2 area (Bonsall *et al* 1994, 97).

More ambiguous are the organic 'structures' radiocarbon-dated to the fifth millennium cal BC (4756–4246 cal BC, 5555±40 BP, UB–2545; 4667–4370 cal BC, 5650±50 BP, UB–2546; 4530–4166 cal BC, 5520±85 BP, UB–2712; 4490–4169 cal BC, 5500±70 BP, GU–1664; 4510–4047 cal BC, 5480±90 BP, UB–2713; (based on an average of five dates from samples of birch bark (Bonsall *et al* 1994, 99), which are statistically consistent (T' = 4.9; v = 4; T' (5 %) = 9.5; Ward and Wilson 1978)), which were excavated within an infilled channel immediately to the east of Williamson's Moss (Bonsall *et al* 1994, 97–99, areas E1 and E2). These consisted of oak branches criss-crossed beneath well-humified peat containing brushwood (*op cit* 1994, 98), and it has been conjectured that they represent the floors of buildings, although no diagnostic artefacts were discovered in association with them nor were conclusive traces of woodworking displayed (*ibid*). A similar suite of evidence from Church Moss, Davenham in Cheshire, was demonstrated by more extensive excavation to have been produced entirely by natural forces (Howard-Davis and Buxton forthcoming). Nevertheless, these possible structural remains are thought to be contemporary with the undeniable evidence for Mesolithic occupation in the immediate vicinity.

On the basis of the available evidence it is difficult to assess whether such a high density of artefacts reflects an actual preference by Mesolithic groups for the Esk Estuary, or whether it may be simply a reflection of the fact that excavation has tended to be concentrated on the most promising flint scatters. The limited excavation of a superficially-similar surface scatter at St Bees golf course (St Bees Site VIII; SMR 17803), however, yielded significantly fewer artefacts (Bonsall 1983, 13), and certainly seems to go some way towards bearing out Bonsall's theory that Williamson's Moss and Monk Moors were the *foci* of the Mesolithic groups, with the hinterland being exploited to a lesser extent.

It has further been suggested that the scale of the activity in the vicinity of Eskmeals may indicate that the area was permanently occupied during the late Mesolithic (Bonsall 1981, 469), perhaps as a result of the year-round abundance of food sources in the area (*eg* fish, gulls, gamebirds). It seems likely, however, that any change to a more sedentary way of life would have had a marked effect upon the surrounding landscape and that it should be possible to detect this signal in the palaeoenvironmental record. Such a signature may be represented by the extensive clearance activity recorded around the Elm Decline (*see above, p 67–8*).

Ehenside Tarn (Fig 26)

Much of the early Neolithic archaeology of lowland Cumbria has been defined in terms of Ehenside Tarn (SMR 1248; Pl 22), perhaps the most famous, and yet enigmatic, prehistoric site on the West Cumbrian Coastal Plain. Large numbers of artefacts and other items were discovered during the draining of a small tarn in 1869, which were noted by Rev S Pinhorne (Fair 1932, 57) and recorded by Rev Kenworthy (quoted in Coles and Coles 1989, 37–9). These included charcoal, nuts and leaves, wild animal bones (including deer and cattle), flint and stone artefacts, substantial amounts of pottery, axe-marked wood, and two wooden paddles.

In 1871 the Society of Antiquaries sent R D Darbishire to research the site, which led to the (for the time) systematic excavation of a small part of the tarn edge (Fig 27). Crucially for our understanding of the site, the results were published very promptly and clearly (Darbishire 1873), thus allowing the reinterpretation and re-examination of the site (Fair 1932; Piggott 1954, 295–9; Walker 1966, 82–116; Coles and Coles 1989, 37–9; Bradley and Edmonds 1993, 136). During the reclamation works and subsequent excavations, the large number of finds included (Fig 28):

- a wooden bowl fragment;
- three wooden 'clubs', one decorated;
- two possible wooden 'fish spears';
- a plain bowl and Peterborough Ware pottery;
- a stone saucer-quern;
- stone *polissoirs;*
- two stone axe-blade rough-outs;
- a polished stone axe blade, in a wooden haft;
- a further wooden haft;
- a stone 'pot boiler';
- the horn and metacarpel of *Bos longifrons.*

Plate 22 Ehenside (Gibb) Tarn today

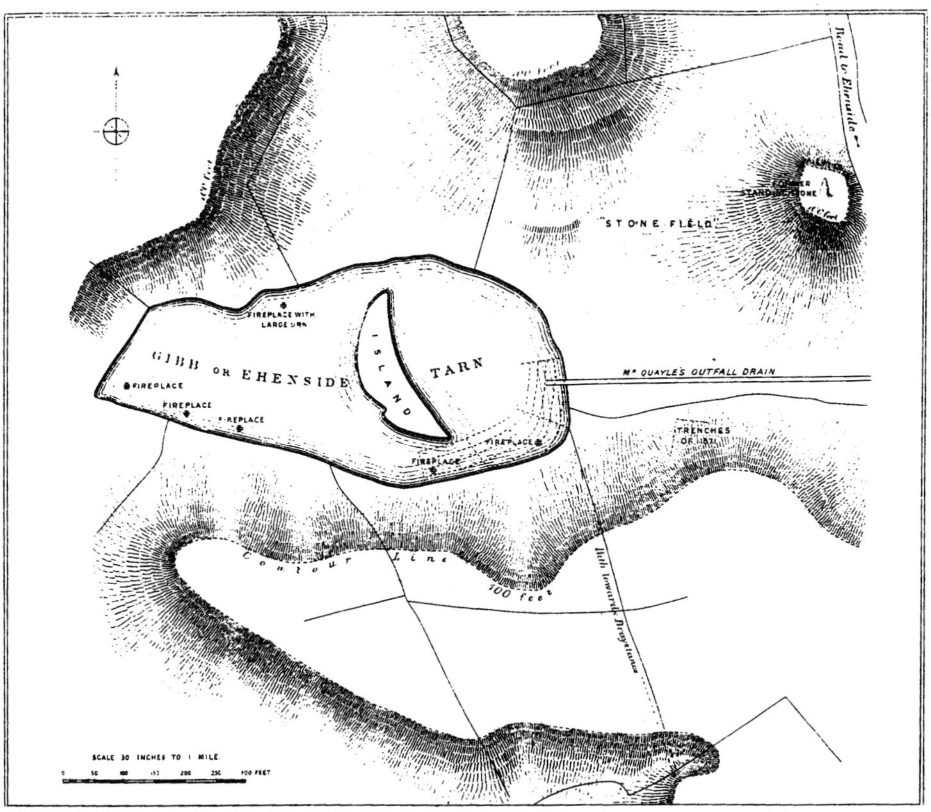

Fig 27 Plan of Ehenside Tarn as excavated (from Darbishire 1873)

Of these, the stratigraphic position often can be identified, all from in and around the 'Forest-bed', and some of the finds were directly associated with a series of 'hearths' around the tarn. Re-excavation of the Forest-bed in 1957 (Walker 1966, 91) suggested that it may have been an artificial platform. In addition, a rotary quern and Crambeck Ware of Romano-British date were found in 1871, as well as a dug-out canoe which may actually have been wood from a mill which burnt down in 1830, the remains of which were dumped in the tarn (Fair 1932). The subsequent excavations of 1957, adjacent to the site of the earlier work, produced two worked flints, an unidentified wooden object, a possibly worked piece of wood, and a fragment of medieval pottery (Walker 1966, 91–2). A further barbed and tanged arrowhead is also known from the site (Fell 1967, 230).

The polished axes appeared to suggest that the material from the tarn was Neolithic in date, although subsequent finds analysis (Fair 1932) and radiocarbon dating appear to suggest several phases of activity. The radiocarbon dates were all undertaken early in the development of the technique and thus do not have the precision of more recent measurements (Table 4, Figs 29 and 30). Calibration of the results, including C–462, the sample which may have an unreliable context (Walker 1966, 102), shows a wide spread from the late Mesolithic to the middle Bronze Age.

Fieldwalking of the area in the vicinity of Ehenside Tarn during the present survey produced additional material (CU55, CU57, CU58, CU59, CU60 and CU61). Whilst many of these were isolated pieces of waste flint, two burnt blade fragments (CU58) were recovered, and also two scrapers (CU59; Fig 60, *p 154*). The diagnostic pieces amongst the assemblage clearly suggest a Neolithic date, with CU58 perhaps the earliest recognised.

The wide date range of the material assemblage, along with the manner in which the site was discovered and excavated, makes interpretation problematic and it is now impossible to associate finds directly with the stratigraphy and the dates. There may also have been a certain selectivity in the material retained and reported upon; flint implements, for example, are mentioned by Kenworthy (Coles and Coles 1989, 37) but not by Darbishire (1873).

Owing to the uncertainty of dating there is now doubt over the provenance of the worked wooden implements, for which the site is best known and it may be that the finds were not a discrete assemblage but derived from several periods of activity around

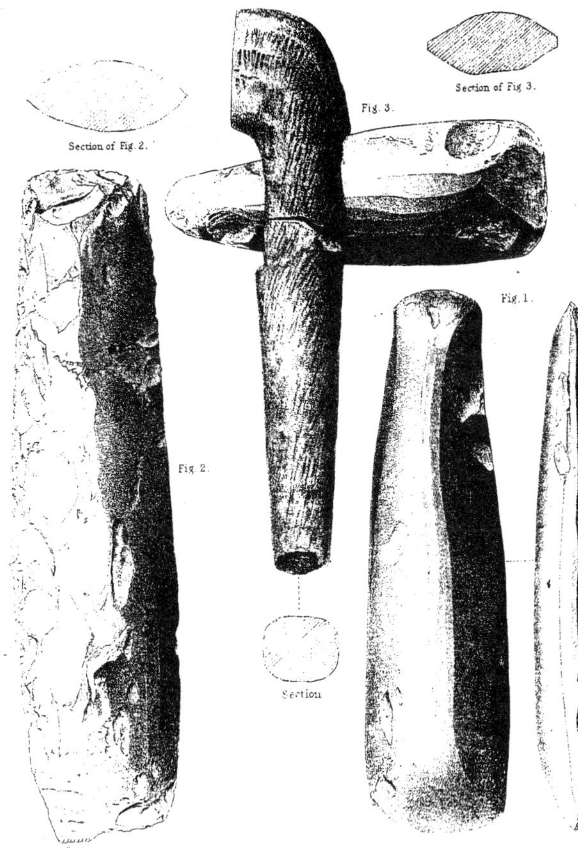

Fig 28 Finds from Ehenside Tarn (from Darbishire 1873)

the tarn. The assemblage, however, is unique and comparative examples for most of the types have yet to be discovered. Sadly, all but two of the organic finds, a bowl fragment and a decorated spatula (both in the British Museum), are now lost (C Earwood pers comm), perhaps because of the interesting method of preserving them adopted after discovery (Darbishire 1873).

The presence of rough-out, part-polished, and finished axes in association with *polissoirs* denotes the site's relevance to the production of axes in Langdale (Bradley and Edmonds 1993, 144). It should be noted, however, that polished axes with *polissoirs* are known from other wetland contexts, such as Roose Moss in Furness (SMR 2605; *see Ch 2*) and Pilling Moss, Lancashire (Middleton *et al* 1995, 57); in both cases it is assumed that the finds were deliberately deposited. The presence of plain bowl pottery would place the earliest demonstrable phase of activity within the early Neolithic (Herne 1988), although closer dating is not possible. The site does show continuity between the late Mesolithic and

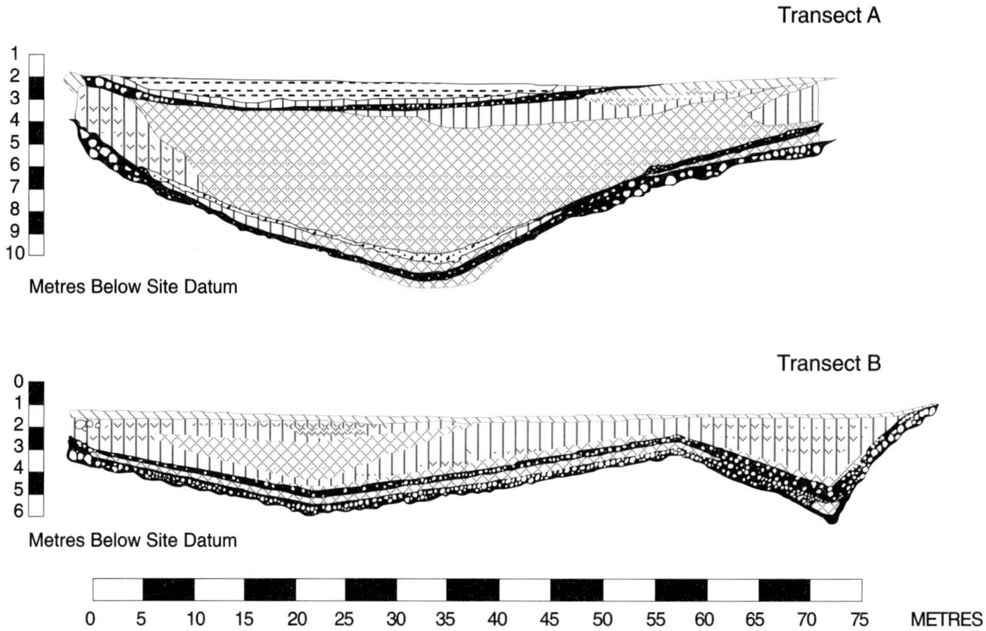

Fig 29 Stratigraphy of Ehenside Tarn (from Walker 1966)

early Neolithic in terms of the palaeoecological record (Walker 1966) and similarities exist between the wooden structures from Ehenside and those purported at the late Mesolithic site at Eskmeals (Bradley and Edmonds 1993, 136). This supports the wider evidence for continuity between the last hunter-gatherers and first farmers, such as the use of volcanic tuff in Mesolithic assemblages (Cherry and Cherry 1984) from St Bees (SMR 1182), Rottington (SMR 1179), Nethertown (SMR 6475), and Ravenglass (SMR 3556; SMR 17914).

As elsewhere in the county, the most important indicator of Neolithic activity is taken to be the polished stone axe which has been used as the defining artefact of the period. The apparent occurrence of axes in wetland contexts, such as at Ehenside Tarn, led to early theories that the west coast of Cumbria was settled by lake dwellers (Watson 1903). Subsequent chance finds and concerted archaeological survey has revealed a broader pattern of axe distribution, with an approximate total of *c* 35 rough-out and finished axes recovered from a wide variety of topographic locations throughout the West Cumbrian Coastal Plain. The distribution is wider than that for Mesolithic material, although axes cannot be taken as a proxy indicator of subsistence activity.

Flintwork from throughout the area continued to be characterised by industries based on beach pebble

Lab Ref	Radiocarbon (BP)	Cal date range (2σ)	Reference
C–462	4964±300	4500–3040	Arnold and Libby 1951
BM–68	3530±150	2300–1520	Barker and Mackey 1961
Q–303	4051±115	2910–2290	Godwin and Willis 1960
Q–303	4125±115	3016–2407	Godwin and Willis 1960
Q–303*	4088±106	2916–2362	Godwin and Willis 1960

**Weighted mean (T′ = 0.1; v = 1; T′ (5%) = 3.8; Ward and Wilson 1978)*

Table 4 Radiocarbon dates from Ehenside Tarn

flint, although surface material of earlier Neolithic date is scarce. Leaf-shaped arrowheads, however, have been recovered from the sandhill sites at Drigg (SMR 1396; SMR 3561), fringing Williamson's Moss (SMR 13930; SMR 13931), and apparently from St Bees (Annable 1987, table 8). These locations appear to coincide with areas of late Mesolithic activity, although the reasons for this coincidence are unclear.

Late Neolithic and Bronze Age

Palaeoecological evidence

The available palynological evidence indicates that occupation of the area was continuous during this Neolithic/Bronze Age transition, with arable farming playing a major role in the agricultural economy of the coastal strip. Around Ehenside, Walker (1966) reconstructed a series of putative clearance and regeneration episodes spanning the whole of prehistory. The relatively low resolution of the original data, however, and the lack of a robust dating control, ensure that it is impossible to assign a firm chronology to these episodes, or indeed to characterise their actual nature. Despite this, in general terms it is probably safe to assume that human interference with the vegetation continued unabated throughout prehistory, and increased dramatically in the early to mid-Bronze Age, perhaps sometime *c* 2000–1500 cal BC (*c* 3700–3200 cal BP). The picture from around Barfield Tarn presents much the same picture with arable farming continuing throughout most of the prehistoric period (Pennington 1970, fig 17). Further down the coast at Drigg there is also evidence for marked forest clearance activity during the third millennium cal BC which appears to be directly related to artefactual evidence (Pennington 1965; *see below, p 77*).

Archaeological evidence

Late Neolithic activity on the West Cumbrian Coastal Plain is more visible within the landscape than that from earlier periods, notably in the form of stone circles and large ritual monuments. The artefactual assemblage is, however, difficult to distinguish from later assemblages.

The distribution of the circles is particularly striking with a concentration of (now lost) monuments around Kirksanton (SMR 1472; SMR 1478), Lacra (SMR 1460; SMR 1474; SMR 1476; SMR 1480; SMR 4845; SMR 5351), and Gretigate, near Gosforth (SMR 1292), together with a reconstructed example at Greycroft, near Sellafield (SMR 1288). A stone circle was also formerly known to exist to the west of Egremont (SMR 1198). The chronology of these monuments is unclear and contrasting dates have been obtained from the west Cumbrian circles; the Greycroft circle, when excavated, revealed a fragment of polished axe within the packing of one of the stones, which has led to the monuments being closely linked with axe production and exchange (Fletcher 1957). Further excavations, however, revealed the presence of a burial cairn with a cremation, associated with other Bronze Age artefacts (Waterhouse 1985, 64). Typologically, the circle at Greycroft (SMR 1288) belongs to the middle period of circle building, *ie* the late Neolithic to early Bronze Age (*c* 2670–1975 cal BC; *c* 4150–3650 BP; Burl 1997, 19), although several phases of activity are present, suggesting that the monument was in use for several centuries. Similar longevity was also noted at Lacra, where Circle A surrounded an early Bronze Age cremation burial, and at Circle D a Collared Urn also indicates Bronze Age use of a Neolithic monument. Lacra D also contained the remains of stone avenues (Waterhouse 1985, 46–52, fig 3.9).

The lithic assemblage of the late Neolithic period is closely associated with that of the early Bronze Age, primarily focusing on the production of flakes from beach flint, and consequently the two are difficult to separate. Transverse arrowheads occur within earlier flint scatters, such as those from Drigg (SMR 1398), Eskmeals (SMR 1399), and a pair from St Bees, perhaps signifying one aspect of late Neolithic landscape use. One of the St Bees examples occurs in close association with both an extensive Mesolithic site (St Bees VIII; SMR 3562) and with rare finds of chalk flint, perhaps from Antrim or chalk deposits under the Irish Sea (Cherry and Cherry 1996).

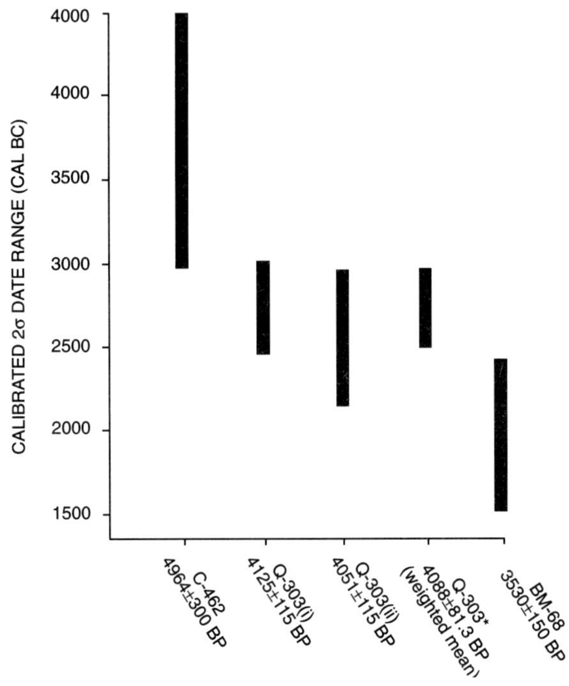

Fig 30 Ehenside Tarn: graph showing the range of calibrated radiocarbon dates

Unlike the neighbouring Morecambe Bay fringe (*Ch 2*), the evidence for Bronze Age settlement of the western coast relies almost entirely upon the lithic assemblages and on the survival of numerous monuments on the adjacent upland slopes. The almost complete absence of Bronze Age metalwork, and the lack of excavation, precludes the establishment of a detailed chronological sequence.

The evidence of clearance and burial cairns on the upland margin suggests an expansion of permanent or semi-permanent settlement during the Bronze Age. The internal and / or external pressures leading to this move up-slope are difficult to discern but the evidence from Langdale points to the extensive use of the uplands for seasonal grazing throughout the Neolithic (Bradley and Edmonds 1993): the more permanent settlement may reflect the formalisation of this previously transitory activity and the establishment of land 'ownership'. The Beaker sherds from Mecklin Park, Santon Bridge (Clough 1968, 2; Annable 1987, 601), may suggest that the upper limit of settlement, denoted by burials, was already moving upward by the late third millennium cal BC. Similar processes may have been underway in the lowlands, despite the fact that the settlements, and field systems, seen in stone on now-marginal uplands, do not survive, although the large amount of lithic material recovered by fieldwalking, allied to the sketchy palaeoecological data, would suggest that much of the lowland was settled and the coast exploited. Fragments of a lowland funerary tradition have been recovered at Herding Nab, Seascale, comprising a Bronze Age 'clay burial urn with small bones' (Fair 1943, 50; SMR 1266) with other burial traces known from Ravenglass (SMR 5347) and, away from the coast, at Holmrook Hall (Bradley and Edmonds 1993, 51; SMR 1277).

Significant sites have been identified south of Beckermet (SMR 6447), with flake-based assemblages containing both beach and chalk flint (Cherry and Cherry 1984, 7). Similar material has also been found at Walkhole Bank (SMR 18908), and at both Drigg (SMR 1398) and Eskmeals (SMR 13935), where the dunes appeared to have been a focus of activity (Cherry 1963; 1965). Indeed, the first archaeological artefacts recorded from Eskmeals were from Bronze Age flint-knapping sites which were discovered on raised gravel beaches (Fair 1936; 1937; Spence 1937; SMR 1399). These assemblages are all characterised by the poor quality of the raw materials which, with the exception of the Beckermet site, are derived from beach flint. As in earlier periods, this has resulted in

Plate 23 Drigg: section through burnt mound eroded by the sea

a constrained assemblage of material, with scalar cores and discarded struck beach pebbles present at Walkhole Bank (SMR 18908).

In spite of the poor raw material, more finely-worked artefacts also occur, notably barbed and tanged arrowheads which are spread throughout the area, particularly on the coastal sites: examples have been found at Ehenside Tarn (Fell 1967, 230; SMR 1248) and two from Williamson's Moss (Cherry 1969; SMR 1248), which were augmented by hearths dated to 2179–1926 cal BC (3665±40 BP, UB–2568) and 2480–1890 cal BC (3755±105 BP, BM–1396; Bonsall *et al* 1994, 99). Monk Moors also yielded a possible hearth (SMR 1391), radiocarbon dated to between 2883–2403 cal BC (4050±55 BP, BM–1385) and 2456–1705 cal BC (3650±120 BP, BM–1395), and a pit with charcoal dated to *c* 1258–900 cal BC (2860±50 BP, BM–1386).

Barbed and tanged arrowheads (SMR 16924; SMR 1400; SMR 1396) have also been recovered from the Drigg dunes where, in 1960, an intercalated organic band was recorded from a seaward-facing section which had been exposed by erosion (Pl 23; Cherry 1982; SMR 4300). Within the organic deposit was an extensive concentration of burnt granite stones which was interpreted as a 'hearth', although they may be part of a burnt mound (M Leah pers comm). A large quantity of timber, interpreted as the 'ground framework of a rectangular, possibly square, structure built on the peaty surface' (Powell and Corcoran in Cherry *op cit*, 4), was also identified. It comprised an horizontal timber, approximately 4 m in length, which appeared to have been laterally split, although there was no other evidence for working. It lay with the inner section facing upwards and was associated with a flint core, perhaps of Bronze Age date. An associated vertical peg measured 0.05 m in diameter, although the full length was never established.

Palynological investigation of the Drigg organic section (Pennington 1965) suggested that it represented a two-stage accumulation of raw humus plus sand, thought to have formed on an emerging raised beach immediately after the Elm Decline (*ie* <*c* 3500 cal BC; *c* 4700 BP). Subsequent radiocarbon dating of charcoal associated with the burnt stones produced dates of 2456–2039 cal BC (3780±55 BP, UB–905) and 2900–2507 cal BC (4135±55 BP, UB–906; Cherry 1982, 5). When the organic band began forming the immediate area appears to have been damp and dominated by alder woodland. The level of the stratigraphy coincident with the flint core's position, however, also marked a change to drier conditions, characterised by birch, oak, and hazel. Above the flint level there is also clear evidence of forest clearance with steep rises in the percentage representation of grasses, bracken, and open ground herb indicators, such as *Plantago*, *Urtica*, *Rumex*, and Caryophyllaceae.

Isolated metal and stone finds are rare on the coastal plain; Bronze Age metalwork includes a 'copper battle axe' from Rainors or Bolton Wood (SMR 1285) and a stop-flanged axe from Town End, Santon Bridge (Fair 1943, 52). Stone axe-hammers are similarly scarce, particularly in comparison with South Cumbria; five are known, including examples from Ehenside (SMR 1248), Winscales (SMR 1054), and Silecroft (Cross 1950). Whether these are settlement-related rather than intentionally deposited (Edmonds 1995) is unknown.

Iron Age and Romano-British period

Palaeoecological evidence

Palaeoecological evidence relevant to more recent archaeological events is scarce on the West Cumbrian Coastal Plain. The only palynological data derive from the undated upper sections of Walker's (1966, 92–105) diagrams from Ehenside Tarn (which are complicated by an hiatus in one profile) and can give only an approximate indication of land-use during the Iron Age and Romano-British periods. Nevertheless, it is possible to discern that a relatively stable regime of continuous agriculture was characteristic much of this time.

Archaeological evidence

Evidence for Iron Age activity on the West Cumbrian Coastal Plain is also scanty and contrasts with the Furness evidence (*see Ch 2*). Only Eskmeals has produced artefacts of possibly early Iron Age date from a lowland context, consisting of a pair of blue beads found in association with an earlier flint scatter (Cherry 1963, 50; SMR 1399). Such limited evidence is insufficient to prove continuity of habitation of the sand-hill sites from the Bronze Age (*op cit*).

In the earlier part of the nineteenth century, a well-preserved male body was recovered from Seascale Moss (Turner 1989, 21; SMR 5330). The preservation of the body was such that only skin, nails, and hair remained, which:

> 'gave the hands an appearance of a pair of fine leather good: the nails still continuing to the fingers. The left ear and feet are quite perfect, part of the scalp had hair on it, and the chin still showed a vestige of beard...A walking-stick of hazel was lying by his side'
> (*ibid* quoting *The Cumberland Paquet* for 3 June 1834).

The fact that the body does not survive and that it was discovered a long time ago means that it is very difficult to date. By analogy with other dated examples from both Britain and the rest of Europe, an Iron Age or Romano-British date is likely, although some, such as that from Scaleby Moss (*Ch 4*), may be of Bronze Age date (van der Sanden 1996, 71–89). The presence of apparently ritually killed bodies from this period is well known from North West Europe's wetlands (Stead *et al* 1986) and Turner (1989, 23) considers it likely that the Seascale body was part of this tradition.

To the south, at Whicham, a double-ditched enclosure (SMR 13688), interpreted as a bivallate hillfort, has been found by aerial photography (Higham 1986, 130). This site, which has not been tested archaeologically, may be similar to other enclosures around the Central Fells, such as those at Skelmore Heads and Warton Crag and the enclosed settlements at Urswick (*see Ch 2*). Other than their generalised importance in the expression of power over the landscape, however, these sites cannot, as yet, be directly related to lowland settlement and wetland exploitation.

Evidence for the Roman use of the West Cumbrian Coastal Plain comprises military remains and scattered finds. The dearth of rural settlement evidence continues from the Iron Age, in marked contrast to the extensive evidence from the Solway Plain (*Ch 4*). None of the present evidence suggests that wetlands were used or exploited. Roman military activity was focused on Ravenglass (SMR 1378) where a fortlet, established during the early Hadrianic period, was re-modelled into a larger fort during the mid–Hadrianic period (Shotter 1993, 36). The fort is generally believed to have formed the terminus of an extended coastal defensive system to protect the Solway (Potter 1979), although doubt has been expressed about this view (Bellhouse 1989).

Scattered finds are concentrated at Drigg and Eskmeals where there appear to have been small encampments within the sand-hills; a site at Drigg revealed a carved pebble head (SMR 6498), thought to date to the second century (Cherry and Cherry 1985, 5), whilst several sites at Eskmeals have evidence for a local iron manufacturing industry with associated fourth-century pottery (Cherry 1966, 55; SMR 1399). There have also been several coin finds, including a hoard of *denarii* from Braystones (Shotter 1990, 188; SMR 1261), a first-century *denarius* of Nerva from Halsenna (SMR 1275), and a fourth-century *solidus* of Theodosius from Muncaster (SMR 4006).

Medieval and post-medieval

Palaeoecological evidence

Indications of more intense clearance episodes are noticeable in the uppermost stratigraphy of Ehenside Tarn which is likely to relate to the post-Roman era (Walker 1966, 80). The earlier of these may be early medieval in date although a later clearance episode is probably post-medieval: the dating remains highly speculative. A peak of *Humulus/Cannabis*-type pollen occurs immediately before this clearance episode, suggesting the cultivation of hemp (*Cannabis sativa*) in the vicinity, which was previously thought to date to the seventeenth or eighteenth centuries AD (*op cit*, 111) in other parts of Cumbria, but recent work on Glasson Moss (*Ch 4*) and at Littlewater (LUAU 2000) has indicated a much earlier date range.

Archaeological evidence

The early medieval period, as elsewhere in the North West, is poorly recorded in the archaeological record, with no excavated settlement sites which can be dated to the period and few chance finds. The West Cumbrian Coastal Plain, however, has produced a remarkable collection of pre-Conquest stone crosses from a large number of sites (Bailey and Cramp 1988, figs 2 and 3), which attest to the presence of early churches throughout the region. Crosses which can be dated to the eighth to ninth centuries, with Northumbrian attributes, have been identified at Beckermet St Bridget (SMR 1259), Irton (SMR 1271), and Waberthwaite (SMR 3984), and of these the cross at Irton is regarded as one of the finest examples of ninth century sculpture in the country (*op cit*, 116–7). The number of sites producing sculpture with Scandinavian motifs of tenth and eleventh century date is much greater, with pieces not only from Irton, Waberthwaite (SMR 3984), and Beckermet St Bridget (SMR 1259), but also Beckermet St John (SMR 1262), Muncaster (SMR 4723), Haile (SMR 4758), and St Bees (SMR 1193). The greatest assemblage has been recovered from Gosforth, where the finest example of Scandinavian sculpture, the Gosforth Cross (SMR 1272), still stands in its original position in the churchyard. This masterpiece clearly demonstrates the attempt to link Scandinavian myths to the Christian creed, and this theme is continued on other sculpture from the site, including a number of hogback tombstones (Bailey 1980). It is thus clear that a vibrant population continued to exploit the coastal plain, although their use of the wetlands, and indeed of so much of the area, continues to elude archaeological research.

As with most of Cumbria, the first recorded use of the West Cumbrian coastal mosses was by the large monastic houses. In the twelfth century Furness Abbey was granted land at *Meles* (Millom) and Kirksanton:

> *'a Sandscard super mare usque in siket in Selvernmire et sic per siket de Selvernemire* [sic] *usque in siket inter duas Layriggas et sic per*

eundem siket usque in Helevepul et sic per filum aque Helevepul descendendo usque in mare...'

[at Sandscard by the sea up to the syke in Selvernmire and so over the syke in Selvernmire up to the syke between the two Layriggs and so along the syke to Hellpool through the winding water of Hellpool descend(ing) into the sea]
(Brownbill 1916, 576).

A similar land grant of 1279 also mentions the mosses around Kirksanton:

'apud le Sandscard et sic directe transeundo versus boream usque Tequitmos et de Tequitmos descendendo usque fossatum subtus Coklayk et sic sequendo dictum fossatum usque Helyepol et sic sequendo Helyepul descendendo usque in Doden...'

[at the Sandscard cross directly opposite the wood as far as Tequitmos and from Tequitmos descend(ing) as far as the ditch below Coklayk and thus along the said ditch as far as Hellpool and then along Hellpool descending as far as the Duddon]
(*op cit*, 537).

The Abbey *ie* Furness was also granted land at Monk Foss, where they had fish ponds. Excavation of the ponds in the nineteenth century revealed that the area was chosen for its low-lying, wet aspect and, after the removal of peaty ground, the diversion of a stream created three ponds (Collingwood 1900, 118–9).

Other monastic houses with holdings of moss lands in the area included the Priory of St Bees, which was granted land (*c* 1240–50) at Holegill Bridge at Hall Foss, south of Bootle, containing Barfield Tarn (Wilson 1915, 361–2), and a similar grant (*c* 1230–40) was made to the Hospital of Jerusalem, which gained three tofts of land between Whicham and Silecroft, close to Hellpool (Wilson 1915, 298).

These particular grants concerned rights of pasture and there are no explicit references to rights to improve or take in the mosses. The mention of the sykes and ditches, however, would imply that a certain amount of control and division of the mosses was taking place, but not perhaps a full-scale drainage operation.

The sole reference to turbary rights concerns the grant of land by Matthew de Redmayn to Furness Abbey at Drigg and Carleton in the mid-thirteenth century. The grant outlines the privileges *'...in bosco et plano, in pratis et pasturis, in turbariis, salinis, piscariis'* [in the woods and fields, meadows and pastures, and to turf-cutting, salt production, and fishing] (Brownbill 1916, 538). The only explicit reference to the cutting and use of peat was associated with the grant to Furness Abbey in 1220 of a saltern at Lairwath which comprised:

'unam salinam juxta Lairwath...cum quatuor acris terre arabilis eidem saline proximo adjacente...et hac presenti carta mea confirmavi turbariam sufficientem ad sustentationem ejusdem saline ...'

[a saltern near Lairwath... with four acres of arable land adjacent to the saltern and with the present charter...I confirm sufficient turbary to maintain the saltern]
(*op cit*, 531).

Similar grants have been noted on the Solway Plain (*see Ch 4, p 137*).

Whilst a number of references exist to monastic holdings of moss lands, there appears to have been very little improvement of mosses during the medieval period. An inspection of the holdings of the large monastic houses at the Dissolution indicates that the monasteries no longer had any possessions on the coastal plain: only the Priory of St Bees had a parcel of land adjacent to the Priory called Petegroves, presumably along the Pow Beck, although there is no specific reference as to what purpose this was put to (Wilson 1915, 449, 598).

All references from the post-medieval period relating to the mosses of the West Cumbrian Coastal Plain relate to the rights of turbary laid down in the manorial courts (Dilley 1972, 198). The common right of turbary was of significant value to the medieval and post-medieval Cumbrian farmer as it provided the main fuel supply in a largely treeless landscape. There were, however, restrictions on the amount of peat that could be removed; on Drigg Common, for example no peat could be dug before 26 April in 1768 (Dilley 1972). Similar restrictions were in place at Bootle in 1688, when no peat could be won until after 10 May.

Peat was also won from other, less obvious, sources. Jefferson, writing in 1842 regarding the coastal encroachments of the past, noted:

'at low-water extensive plots of vegetable soil or peat-moss are visible, from one of which, two or three years ago, an inhabitant of the parish, named Mandle, dug several cart-loads, which, as an article of fuel, was found to be far superior to the peat commonly in use in the neighbourhood'
(Jefferson 1842, 109).

Plate 24 Greenwood's map of Cumberland (1823) with permission of Cumbria Record Office (Carlisle) © Copyright

It was not until the Enclosure Movement of the late eighteenth and early nineteenth centuries that the small tarns and mosses came under the eye of the improving agriculturists. Even then, however, most of the smaller mires were ignored due to their unsuitability for agriculture. Of the Enclosure Awards for the area, only three dealt with large areas of moss: those of Egremont, Eskmeals, and Drigg (Pl 24). Of these awards, only one, that of Drigg in 1828, improved an area of moss into agriculturally viable land (CRO(W)YSPC/16/7). The award created a number of fields within an encircling road system, at the centre of which were several 'ancient inclosures', which were effectively surrounded by low-lying marginal, estuarine silts and peat. The enclosure drained Bell Moss Tarn and brought smaller moss deposits into cultivation.

The other awards dealt with the enclosure of the common lands of Egremont and Eskmeals which included Black Moss, east of Egremont, and Williamson's Moss, although in both cases the mires remained unenclosed. After the Egremont Award, Black Moss remained an important fuel supply for the town and the 16 acres were divided between 54 individuals (CRO(C) QRE/1/30; Pl 25). The only non-fuel related allotment involved the incorporation into the northern edge of the moss of a mill race. Similarly, Williamson's Moss was left largely unaltered by the Eskmeals Award, the moss not receiving an apportion number and remaining in private hands (CRO(C) QRE/1/120).

In contrast, small-scale drainage works were undertaken to reclaim some of the mosses for agriculture. One of the most completely documented of these is Ehenside Tarn where drainage revealed numerous prehistoric and Romano-British finds (*see above, p 71–6*). The work was undertaken by the local farmer in 1869 who improved the moss by cutting an outfall drain to the east to take water from the enclosed basin. By 1871 the tarn was drained and the peat had turned into a 'solid, heavy, peaty mass' (quoted in Coles and Coles 1989, 39). Subsequently, this peat surface was covered with an even depth of boulder clay to provide a firm surface (Walker 1966, 86). The reclamation was not entirely successful, however, as two pools subsequently developed in the former tarn basin, now known as Gibb Tarn and Silver Tarn.

This example illustrates the small-scale enterprises that must have been undertaken in many parts of the county to bring small tarns and mosses under cultivation. This may have been particularly worthwhile on the coastal plain, given the value of the land around the mosses for agriculture and the low population density which may have lessened pressure on every moss for use as fuel. Domestic-scale peat-cutting did not die out until the 1950s (Rollinson 1981, 121) when labour for the task was difficult to find. Some cutting is still undertaken in South Cumbria on a very small scale (*see Ch 2*; Pl 26), although, in the long term this will cease and the skills of a once-common rural craft will be lost.

Discussion

This chapter has outlined the known evidence of settlement of all periods from the West Cumbrian Coastal Plain. This small strip is topographically very distinct and, through a combination of circumstances, has provided abundant evidence for prehistoric activity of national importance. Critically, it has enabled the late Mesolithic and early Neolithic periods to be elucidated, eras for which there is little evidence nationally, particularly in the key transitional phase.

Our understanding of these periods developed from a fortuitous set of circumstances, the first amongst which was the discovery of Ehenside Tarn and, critically, its exploration and publication by the Society of Antiquaries of London (Darbishire 1873). The quality of this work has enabled the site to be re-assessed on several occasions (for instance, Coles and Coles 1989) and has allowed it to have a pivotal role in the appreciation of both the potential of wetland archaeology and the understanding of the lowland component in the Neolithic axe production and dispersal process (Bradley and Edmonds 1993).

Subsequent to this, the twentieth century saw systematic programmes of both palaeoecological and archaeological work, Pennington's study (1970) of, amongst others, Barfield Tarn illustrating pre-Elm Decline ('Mesolithic') clearance activity. Walker's study (1966) of the Cumbrian lowlands amplified this work and attempted to relate mire sequences together to aid the understanding of landscape development through time. Interwoven through this was the long-term study by the Cherrys of the archaeology of the Plain, which identified many previously-unsuspected lithic scatters confirming and amplifying the intensity of human use of the landscape. Their survey work was followed by intensive interdisciplinary studies (Bonsall *et al* 1994), notably around the Esk Estuary, which identified with finer resolution human interaction with the environment.

Because of these important pieces of work, our understanding of the Plain in the earlier prehistoric period is apparently more complete than others parts of lowland Cumbria. This chapter has demonstrated,

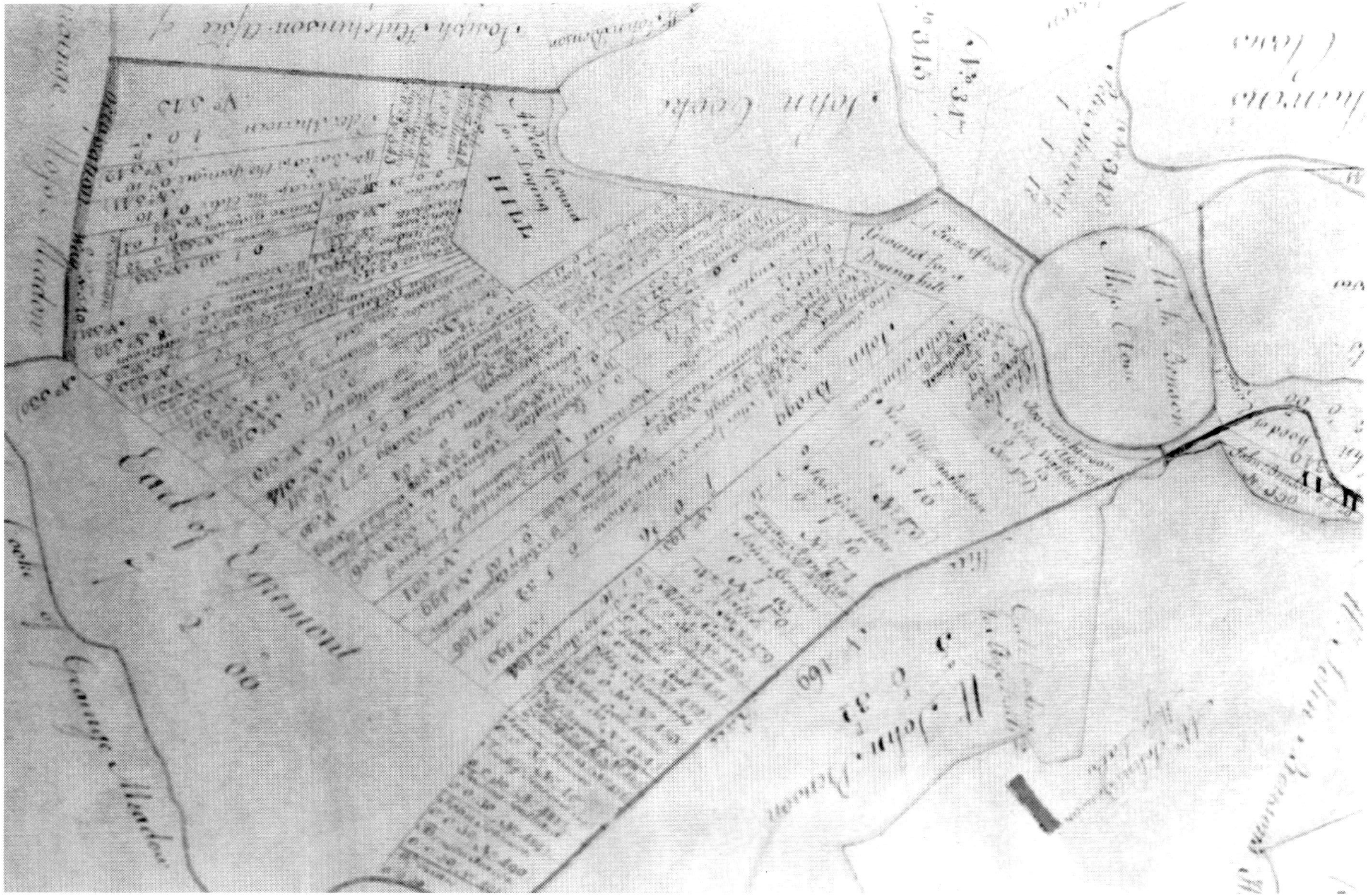

Plate 25 Black Moss: extract from the Egremont Enclosure Map (1783) with permission of Cumbria Record Office (Carlisle) © Copyright

Plate 26 Peatcutting at Whitbarrow Moss during the 1960s (Lancaster University Library)

however, that there are still large gaps in our knowledge, notably concerning periods more recent than the Bronze Age, and the use of wetlands in the past.

Small wetlands are scattered throughout the Plain and have been a commonplace to local communities throughout the Holocene. There can be little doubt that they were exploited; the evidence from Ehenside Tarn, whatever its failings, does indicate the attraction of small tarns over several millennia. The bog body from Seascale Moss (Turner 1989) suggests a later use for such areas of water for ritual deposition. Aside from this coarse-grained, empirical evidence, we currently lack the fine resolution data which can identify how different wetlands were used and how that use changed. Evidence from elsewhere in the NWWS survey area indicates that, for example, fens were more readily exploited than raised mires (*eg* Leah *et al* 1997) and 'dry phase' mires were exploited until they became raised mires (Middleton *et al* 1995, 195–6).

One of the reasons we do not have this fine detail is the problem of access to the archaeological remains. The Cherrys' work has shown that archaeology can be identified as long as the small areas of arable are investigated as they arise each year. The mosses, however, are seldom cultivated and so the remains within them are never disturbed. Even where mosses were reclaimed, such as at Ehenside Tarn, the peat was covered with Boulder Clay to provide a solid surface which has effectively masked the underlying archaeology from surface survey. The work of NWWS in the vicinity of Ehenside Tarn, however, has yet again reinforced the need for repeated fieldwalking, whenever fields are ploughed.

One of the themes running through this chapter is the intensity of the early settlement of the Plain which is reflected in both archaeological remains and palaeoecological sequences. There can be little doubt that the mosses (and the coastal deposits which were critical to past settlement) retain substantial amounts of well-preserved archaeology from all periods; characterising that archaeology is another matter. The known record presents only glimpses of the true potential.

4

THE NORTH CUMBRIAN PLAIN

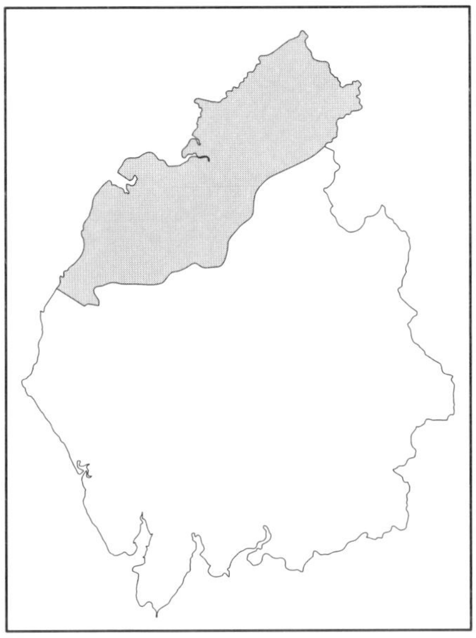

Fig 31 The extent of the area covered by Chapter 4

> The level upon which the trunks lay, was a little below that of high tides; and from 8 to 10 feet below the surface of the ground they were embedded in; which, excepting the superficial soil, is a soft blue clay, having the appearance of marine alluvion [*sic*]; ... Although the precise period at which this forest fell is not ascertainable, there is a positive proof that it must have been long prior to the building of the wall of Severus [Hadrian's Wall], because the foundations of that wall passed obliquely over it, and lay 3 or 4 feet above the level of the trees; all of which were of oak, and several of them above 4 feet in girt [*sic*]
> (Chapman 1832, 117).

Introduction

The North Cumbrian Plain is defined here as the low-lying land (below *c* 200 m OD) north and west of the Lake District massif, stretching from St Bees in the south to the Northumberland border in the east, and including parts of the Eden Valley south-east of Carlisle (Fig 31). The area encompasses all of *Countryside Character Area 6: Solway Basin* (Countryside Commission 1998, 19-24) and its coastal fringe forms the Solway Coast Area of Outstanding Natural Beauty (Countryside Commission 1995; *see* Fig 9, *p 24*). For both of these landscape categorisations, descriptions have been compiled to which the reader is referred for more details of landscape form and 'character'. It also falls within the *Solway Basin Natural Area* defined by English Nature (1997; *see* Fig 19, *p 24*).

Most of this region is mantled by a deep accumulation of glacial till forming a gently undulating landscape of low ridges, intersected by a predominantly south-west to north-east trending drainage system. Extensive areas of salt-marsh fringe the south bank of the Solway north of Silloth around to the Scottish border. Land-use is dominated by pasture, although significant areas of arable are present around the Abbeytown Ridge, east of Silloth.

Geology and soils

Over all of the survey area, the solid geology is masked by Devensian tills upon which are soils chiefly of the Clifton and Brickfield Associations. The former, which are seasonally-waterlogged, form the characteristic soil type north of Carlisle to the Scottish border, and also much of the plain to the south and west of the city, stretching in a broad swathe from the Eden Valley to Maryport (Lawes Agricultural Trust 1983). Further south and west, the Brickfield Association is characterised by seasonally-waterlogged fine loams and clay-rich surface water gleys which have developed over drift deposits derived from sandstone and shales (Jarvis *et al* 1984, 123).

Other major soil types include the well-drained loams of the Wick Association developed in river systems, notably around the River Ellen and west of Carlisle. The Rockcliffe Association, derived from marine alluvium, is present along the Solway coast, while small pockets of well-drained sandy loams of the Newport Association are found on the coast, and

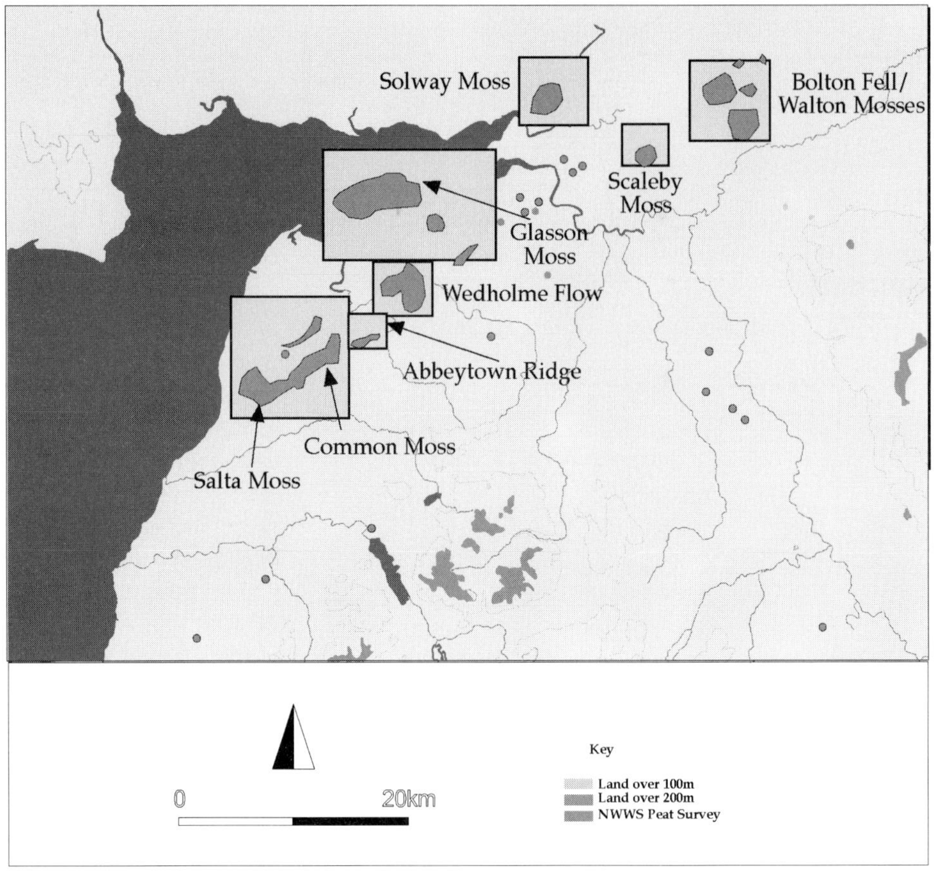

Fig 32 North Cumbrian Plain: areas of archaeological survey

around Pelutho. South–west of Carlisle, a significant area of Salwick Association loams occurs (*ibid*).

Archaeological and palaeoecological survey

Archaeology

The field survey strategy adopted on the North Cumbrian Plain (Fig 32) differed from that adopted elsewhere in the county. The reasons were twofold: the presence of (for the county) significant amounts of arable on the Abbeytown Ridge, which permitted systematic artefact recovery, and the number of the mires where peat is extracted commercially, again allowing systematic survey of cut faces .

Systematic fieldwalking was undertaken on the sandy soils which are regularly cultivated. Relatively large areas of arable, particularly on the Abbeytown Ridge, were surveyed to characterise the ploughsoil archaeology of contiguous areas of landscape (*see rear of volume*, Figs 33–6). This narrow ridge, stretching north–east from Salta Moss across to Wedholme Flow, reaches heights in excess of 40 m OD and, as such, forms a significant topographic feature defining the southern boundary of the Solway Plain. Smaller areas were also fieldwalked south–east of Silloth and east of Allonby to determine the pattern of artefacts elsewhere. Compared with other areas of the North West, however, the total area of arable was restricted and gave a picture of a relatively small part of the Plain as a whole.

With some exceptions, most of the cultivation was undertaken on a rotation which meant that not all fields were available in any one season. To overcome this, a strategy of short, annual visits was adopted between 1991 and 1994 to walk the available fields. This was usually, but not always, undertaken in tandem with assessments of intact mosses and reclaimed mossland under pasture in a similar manner to those in the remainder of the county.

The methodology of the survey of the cut-over mosses has been presented in Chapter 1 (*p 16*). The work was undertaken on four occasions during the project, between 1991 and 1995, although a rapid assessment was additionally carried out on a trial basis in 1989 before the formal start of the survey phase of the project. A baseline survey was made in 1991 of Wedholme Flow, Bolton Fell Moss, and Solway Moss when each was surveyed in detail, including the examination of each ditch section. Sample profiles were also recorded such that the rate of loss over time could be measured. These sections were not regularly revisited during the project, however, largely because the rate of change was found to be insignificant, for reasons relating to the manner in which the moss was exploited. The data on the sections have been stored in the project archive and are available for future consultation.

In subsequent years (1993–5) each of the mosses was re-visited annually, but only those areas actively cut that year were inspected. For each moss a rapid assessment was made of new workings to identify remains exposed in freshly-cut sections and disturbed peat. Conditions were found to vary greatly between and within the mosses, reflecting differing modes of working. A significant part of the project also involved maintaining a liaison with the peat companies who worked the sites to ensure that finds were reported promptly and their staff understood the archaeological value of the mires. On two occasions the value of this liaison was shown by finds being discovered by peat workers and reported to project staff, leading to excavations of the material. The two excavations, of a cow 'head and hooves' and a sheep at Solway Moss, were undertaken in 1991 and 1992. Reports on both are presented later in this chapter (*p 123–36*).

Another strand to the methodology was the assessment of sites on the wetland edge where development may reveal peat stratigraphies. A single visit was made to Salta in 1990 where a house was being built on the edge of the moss; no finds were revealed, although the line of the wetland edge was confirmed. This strand of the work was superseded by PPG16 (DOE 1990) which placed such work on a commercial footing.

Archaeological research on the North Cumbrian Plain has, historically, focused upon the Roman settlement of the area, and particularly on the complex series of frontier works which culminated in Hadrian's Wall, the western end of which was characterised by Jones (1982) as the Solway Frontier. The Wall itself tended to grasp the imagination and attention of the antiquaries of the nineteenth century, and much archaeology was interpreted in terms of its relationship with the Wall. A good example is the intertidal forest between Glasson and Kirklands, the 'Sunken Forest of the Solway' (Daniels 1978, 341), which was discovered during excavation for the Carlisle Canal between 1818 and 1832. 'There is a positive proof that [the forest fell] long prior to the building of the Wall, because the foundations of the Wall passed obliquely over it and lay 3–4 feet above the level of the trees' (Chapman 1832, 117). The wood, which was buried in anaerobic clay in the intertidal zone, was so sound that it was 'used with other timber in forming the jetties at the outlet of the

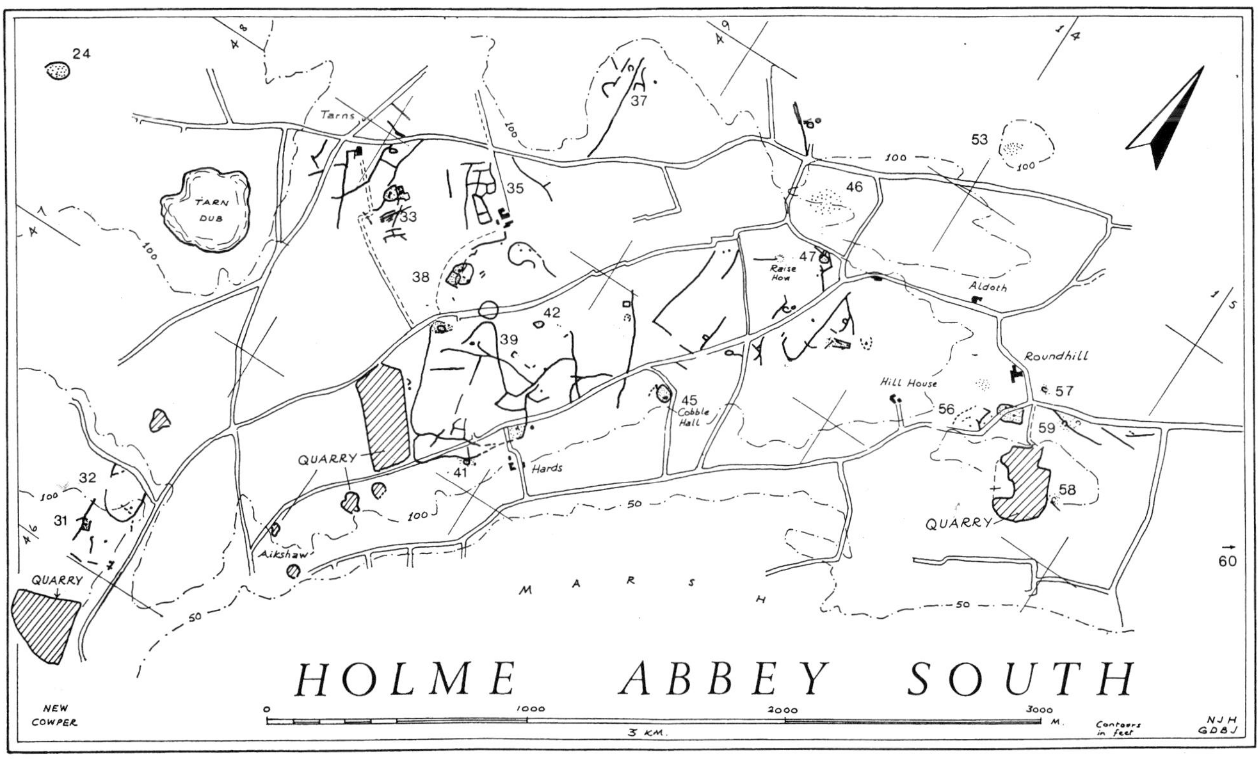

Fig 37 Sites on the Abbeytown Ridge identified by aerial photography (after Higham and Jones 1975)

Plate 27 Biglands Bog: aerial view (© R H Bewley)

Plate 28 Aerial view of modern peat cutting on Wedholme Flow (© R H Bewley)

Plate 29 Solway Moss: palaeoecological coring

canal into the Solway, and the President's chair of the Society of Antiquaries [of Newcastle upon Tyne] was made of it' (Bruce 1867, 302–3).

This emphasis has created not only a Romano-centric bias in the archaeological record, but a bias which is further skewed towards Roman military monuments as opposed to civilian and native settlement. Aerial photographic reconnaissance from the mid-1970s to the present has gone a long way towards correcting this (Fig 37), with the identification of over 200 previously unrecorded sites on the North Cumbrian Plain (Higham and Jones 1975). This has in turn encouraged studies of landscape use through integrated programmes of excavation and survey allied to aerial reconnaissance (Bewley 1986; 1994). The importance of such work lies in the establishment of the North Cumbrian Plain as a major area of prehistoric and Romano-British activity in Northern England (Higham 1986, fig 5.1), and one of the few in the county where such remains are susceptible to aerial photography (Pls 27 and 28).

Palaeoecology

The NWWS palaeoecological survey focused on mires where detailed archaeological work had been undertaken, in particular Solway Moss (Pl 29), which had previously seen little work. Transect coring was undertaken to place both the 1991 cow and 1992 sheep excavations in their stratigraphic context (*see*

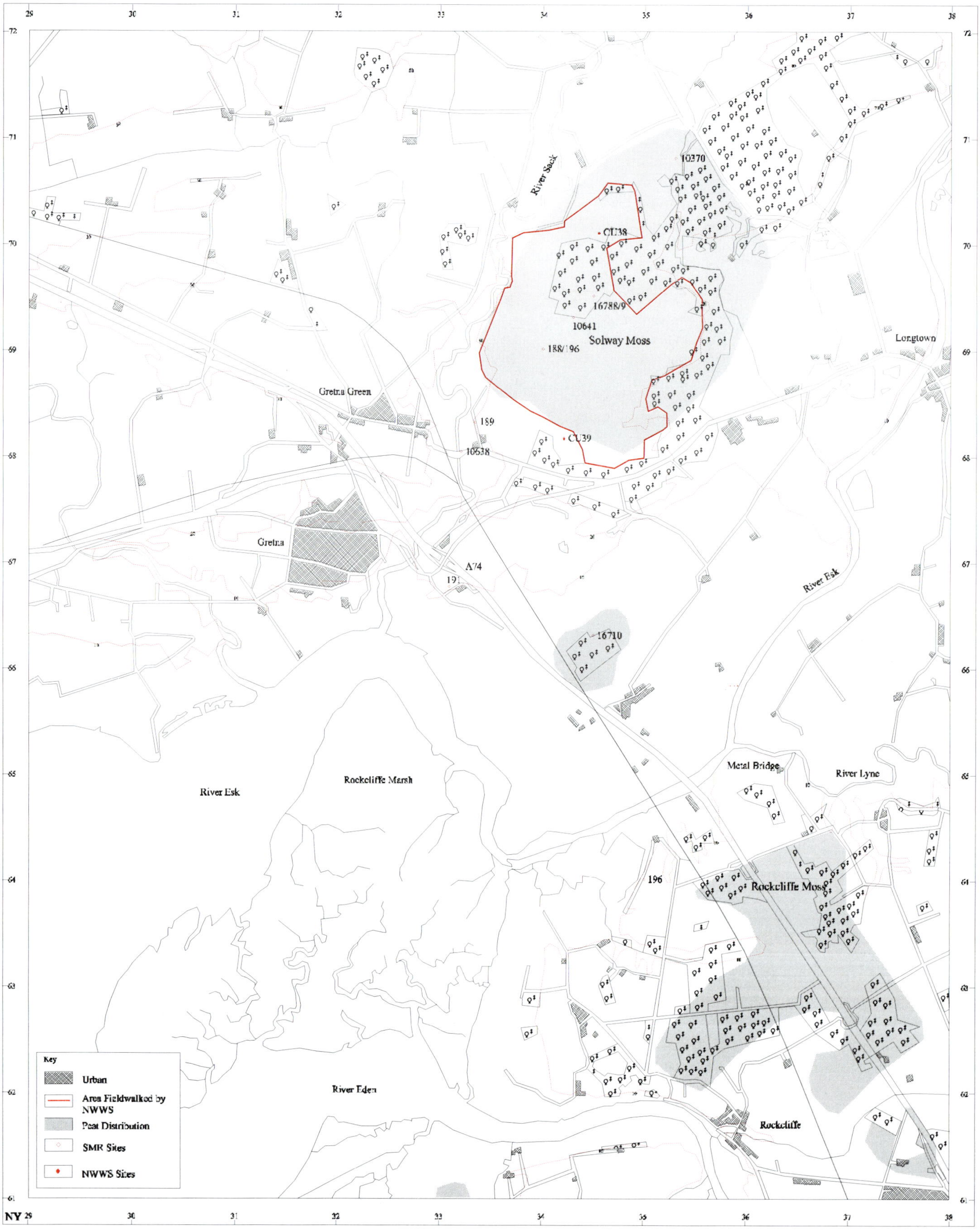

Fig 38 Solway Moss: peat extent and actual area surveyed (Scale: 1:50,000)

Based upon the 1988 Ordnance Survey 1:50,000 maps with permission of the Controller of Her Majesty's Stationery Office © Crown Copyright

Plate 30 Solway Moss: topography

below, p 123 onwards) and was allied to detailed pollen and macrofossil studies of dated stratigraphies. Additional stratigraphical studies on Bolton Fell Moss, Walton Moss, Glasson Moss and Solway Moss, to provide evidence for mire development, was completed as part of Paul Hughes' research for his PhD thesis at Southampton University (1997), and he has also provided a brief summary of his work for this chapter.

Much of our knowledge of the vegetational history of North Cumbria derives from the work of Walker (1966) who mapped the vegetational history of the area in *The Late Quaternary History of the Cumberland Lowland*; this remains the most important source of data for the environmental history of the area. Because the study was undertaken over 30 years ago, however, there are limitations in the extent to which the data can be used to reconstruct human impact on the landscape, since they are of relatively low resolution by modern standards and mostly lack absolute dates on which to base a firm chronology for the biostratigraphic events recorded in many of the pollen diagrams. Because of this, anthropogenically-related events evident in the diagrams can only realistically be viewed in broad, qualitative terms, and so Walker's original detailed sub-regional 'Cumbrian' zonation is disregarded in the treatment of the data below. Within the area encompassed by this chapter, Walker's dataset is relevant only to the first part of the Holocene, truncation of the stratigraphy by hand-cutting at his four main study sites (Bowness Common, Abbot Moss, Scaleby Moss, and Oulton Moss) having limited their post-Bronze Age palynological record.

Fortunately, most recently published palynological studies from the North Cumbrian Plain have concentrated on post-Bronze Age sediments (Barber 1981; Barber *et al* 1993; Dumayne and Barber 1994), allowing these later periods to be viewed in general terms. Recent studies at Walton, Glasson, Solway, and Bolton Fell Mosses by Hughes (1997 and this volume) have added some detail to the earlier periods in the north–eastern part of the county. In addition to these studies from the Cumbrian side of the Solway, evidence from the high-resolution palaeo-environmental studies of Tipping (1995a; 1995b;

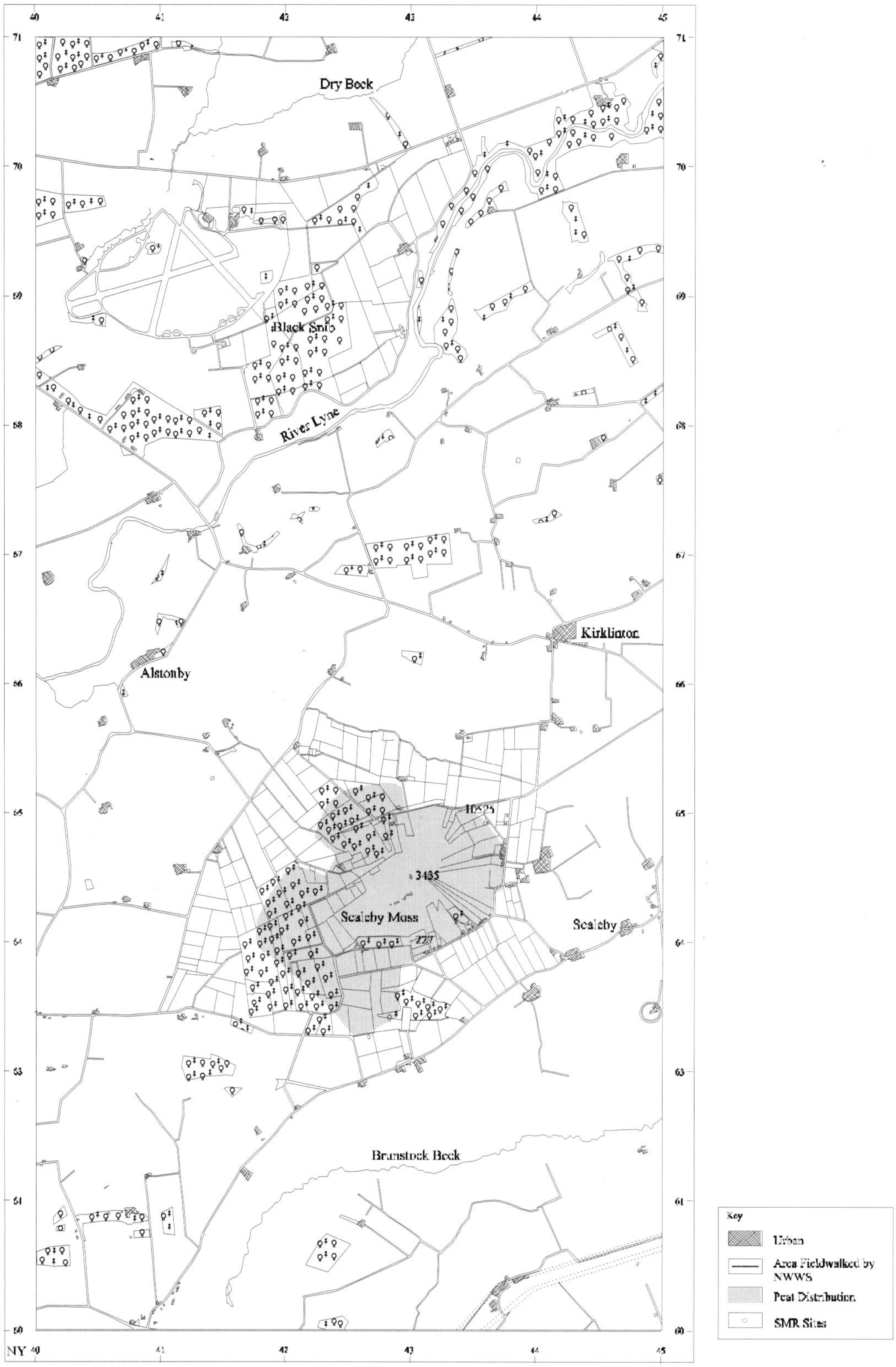

Fig 39 Scaleby Moss: peat extent and actual area surveyed (Scale: 1:50,000)

Based upon the 1988 Ordnance Survey 1:10,000 maps with permission of the Controller of Her Majesty's Stationery Office © Crown Copyright

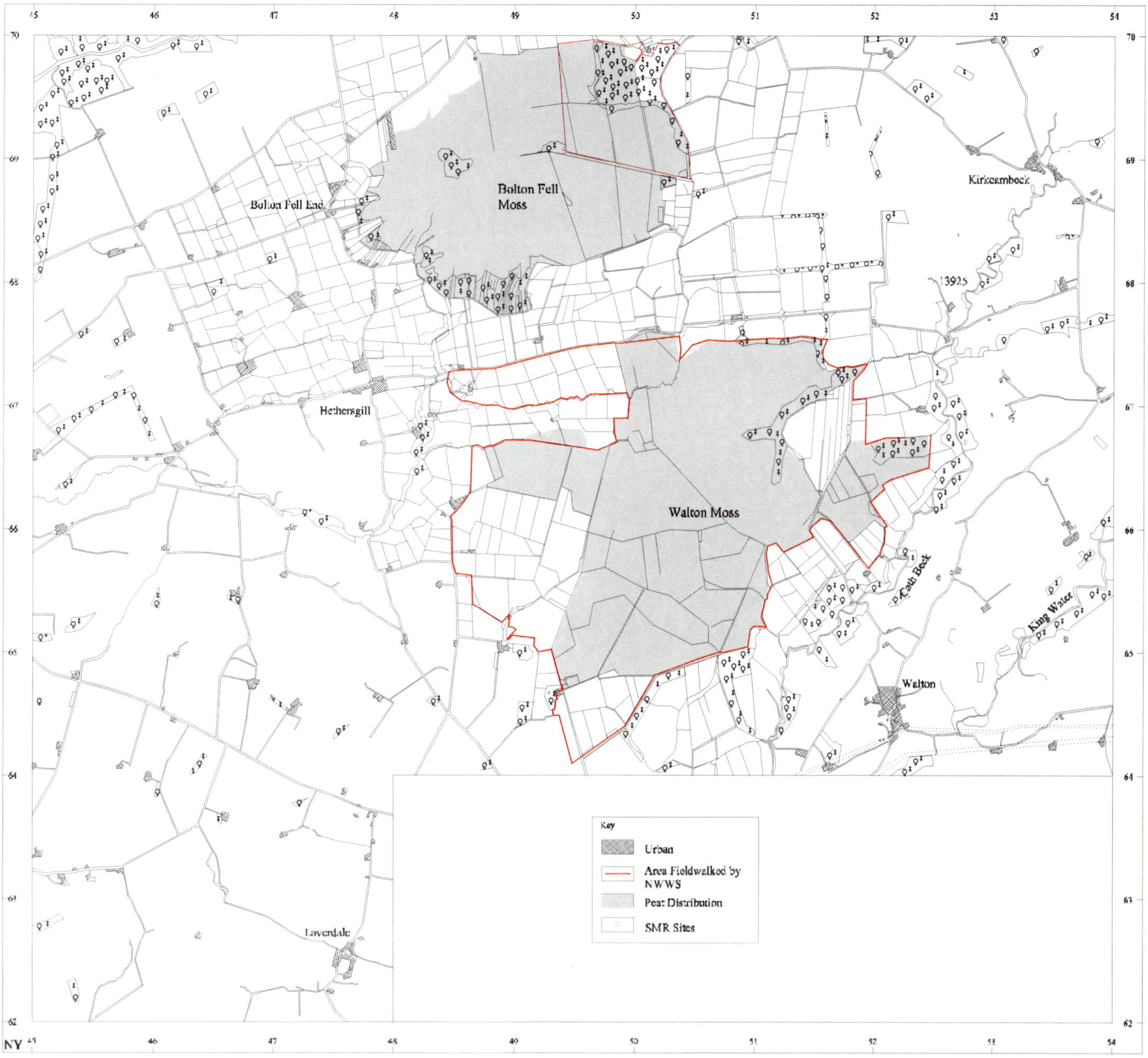

Fig 40 Bolton Fell/WaltonMoss: peat extent and actual area surveyed (Scale: 1:50,000)

Based upon the 1988 Ordnance Survey 1:10,000 maps with permission of the Controller of Her Majesty's Stationery Office © Crown Copyright

1995c) from Burnfoothill Moss on Scottish coast are also instructive.

Wetland distribution

Two major mire types are prominent on the North Cumbrian Plain: extensive raised / intermediate mire systems clustered around the Solway coast; and shallow fen peats of former valley and floodplain mires. The major raised and intermediate mires include Solway Moss (Pl 30), Wedholme Flow, Bowness Common, Glasson Moss (Pl 31), Scaleby Moss, Oulton Moss, Abbot Moss, Drumburgh Moss, Bolton Fell, and Walton Moss (Figs 33–6, 38–40). The reclaimed valley and floodplain mires are characterised by Altcar Association soils and occur in a thin band running south–west along the lower stretch of the River Waver south of Abbeytown (Fig 33). In

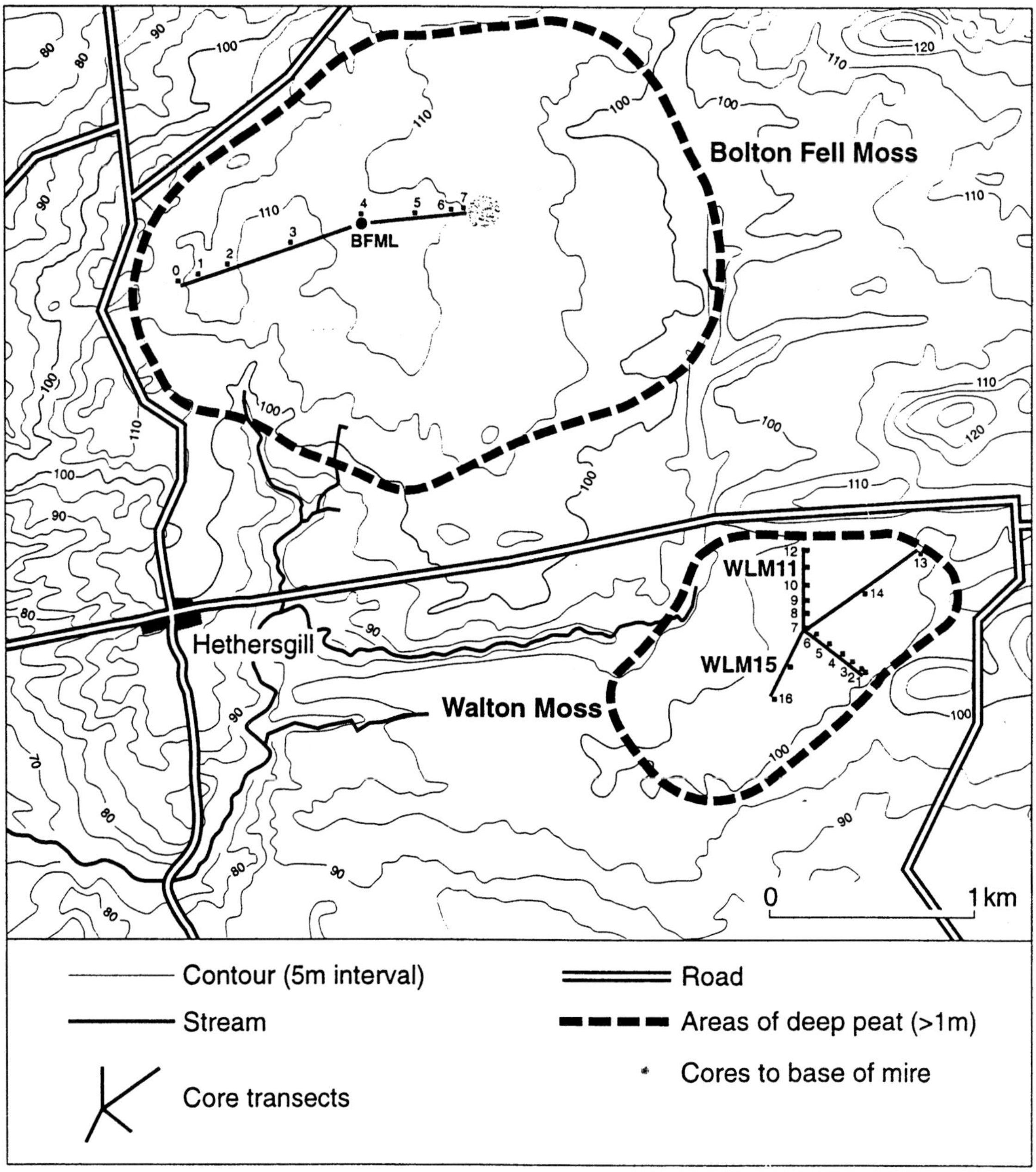

Fig 41 Bolton Fell/Walton Mosses: location of coring transects (from Hughes 1997)

addition, small basin mires also occur, such as that at Biglands Bog, and small raised mires (*eg* Dubbs Moss) between Cockermouth and Eaglesfield.

Small-scale peat-cutting has badly disturbed the upper stratigraphy of many mires, in particular Scaleby, Oulton, and Dubbs Mosses, but others, in more remote locations, have seen only peripheral damage, notably Walton Moss, Glasson Moss, and Bowness Common. Commercial peat-cutting has affected large portions of Wedholme Flow, Bolton Fell Moss, and Solway Moss, although some undisturbed stratigraphy does survive on each mire.

The four major mire systems

Following on from survey work by Walker (1966), four of the raised/intermediate mires (Bolton Fell Moss, Walton Moss, Glasson Moss, and Solway Moss) were investigated in detail to provide information on their origins and the sequences of local vegetation development. These were selected for study since they are four of the largest mires where a complete Holocene record still exists in the region and the work formed part of a wider project to examine the Fen Bog Transition (Hughes 1997). The field stratigraphies produced at the four sites provided the context for a comparative study of the vegetation changes leading to the transition from fen to raised bog in North Cumbria (*ibid*). Results of macrofossil analyses at Walton Moss are reported in Hughes *et al* (in press).

At Walton Moss (Fig 41) and Solway Moss (*p 103–5*), the field stratigraphy was investigated using two crossed transects of boreholes, whereas at Bolton Fell Moss (Fig 41) and Glasson Moss (*p 99*) single transects were used. The mire stratigraphy was described using the Tröels-Smith (1955) system of sediment notation.

Walton Moss (Figs 42 and 43)
Walton Moss (NY 504 667) is located 15 km east of Carlisle and 1.5 km north of Walton village, which lies astride Hadrian's Wall. The moss covers 150 ha and remains almost completely intact, having suffered only minor hand-cutting at the edge of the rand and on thin peats surrounding the moss (Barber *et al* 1998). The surface vegetation of the mire is classified as NVC M18 (*sensu* Rodwell 1991), a *Sphagnum*-rich lowland raised mire type. The plant communities are dominated by *Sphagnum cuspidatum, S magellanicum, S papillosum* and *S capillifolium* var *rubellum*, accompanied by *Eriophorum vaginatum, Eriophorum angustifolium, Rhynchospora alba, Vaccinium oxycoccos* and *Narthecium oosifragum*. Walton Moss is particularly striking because the surface topography is very subdued when compared to similar raised mires in Northern England and Scotland which tend to display a pronounced hummock/hollow surface form. Consequently, higher hummock species such as *Calluna vulgaris* are rather poorly represented on the mire.

Peat formation began at *c* 8300 cal BC (*c* 9200 BP) at Walton Moss in a shallow basin located 300 m from the northern margin of the modern mire. The lowermost organic horizon is a stiff detritus peat containing species typical of mesotrophic fen conditions. This community was quickly replaced by fen carr, principally composed of *Betula*, followed by a more open poor fen containing *Sphagnum palustre* and *Sphagnum* sect *Acutifolia*. After 1 m of peat accumulation the mire succession became oligotrophic at *c* 7850 cal BC (*c* 8800 BP), marking the point at which the growing surface became fully separated from the influence of groundwater seepage. The rather brief phase of fen development at Walton Moss was succeeded by a community composed of species indicative of deep or fluctuating water-tables on raised mires, which persisted for more than 2000 years before being replaced by pool edge and then deeper pool species. The change to wet conditions occurred at *c* 5850 cal BC (*c* 7000 BP). This wet shift has been recognised at other sites in Cumbria (Hughes 1997) and falls within the time frame suggested by Korhola (1994; 1995) as a major period of mire initiation and expansion in Scandinavia.

At the time of the marked change to wet conditions in the centre of the basin there were also major changes on the fringe of the mire. Shortly before *c* 5880 cal BC (*c* 6600 BP) the raised mire overtopped the edge of the confining basin, and increased water flow across mineral ground to the south of the main basin led to fen initiation and expansion in this area. After approximately 1500 years the fringing fens developed into ombrotrophic mire forming one continuous peat dome across the whole peat complex (Hughes *et al* 2000).

Bolton Fell Moss (Fig 44)
The southern edge of Bolton Fell Moss (NY 495 695) is located 1 km north-west of Walton Moss. A 30 ha section of the original mire surface remains intact in an area known as 'The Reserve', but large parts of the remaining peat mass have been cut down to early-Holocene strata and in some places the underlying till has been exposed. Barber (1981; Barber *et al* 1994) described the vegetation of the site in the 1960s and observed a marked deterioration in the vegetation over the following decade as a result of industrial peat mining altering the drainage. Although the mire communities were originally similar to those found at Walton Moss, the present cover has become very hummocky and species such as *Calluna vulgaris, Eriophorum vaginatum*, and *Molinia caerulea* are more extensive than they were three decades ago, but a range of *Sphagnum* still thrive between the hummocks.

The stratigraphic survey carried out by Walker (1966) suggested that Bolton Fell Moss was 4–5 m deep. Subsequent commercial peat depth surveys demonstrated that the moss achieved a maximum depth of 11 m in an area to the west of the geographic centre of the modern mire. After several decades of commercial hand-cutting followed by industrial-scale milling of peat the remaining uncut reserve area, which is located on the deepest peat, is now 10 m deep, and thus the whole peat mass appears to have shrunk by as much as 1 m since the 1960s as a result of drainage. The shrinkage can be seen in the stratigraphic diagram (Fig 44) which traverses the reserve: Cores 0 to 4 no longer record the form of a distinct peat dome.

Peat began to accumulate at Bolton Fell Moss in an arc-shaped basin shortly before 8432–8099 cal BC (9305±45 BP, SRR–4945). The pioneer community, dominated by *Salix*, was replaced by *Betula* carr

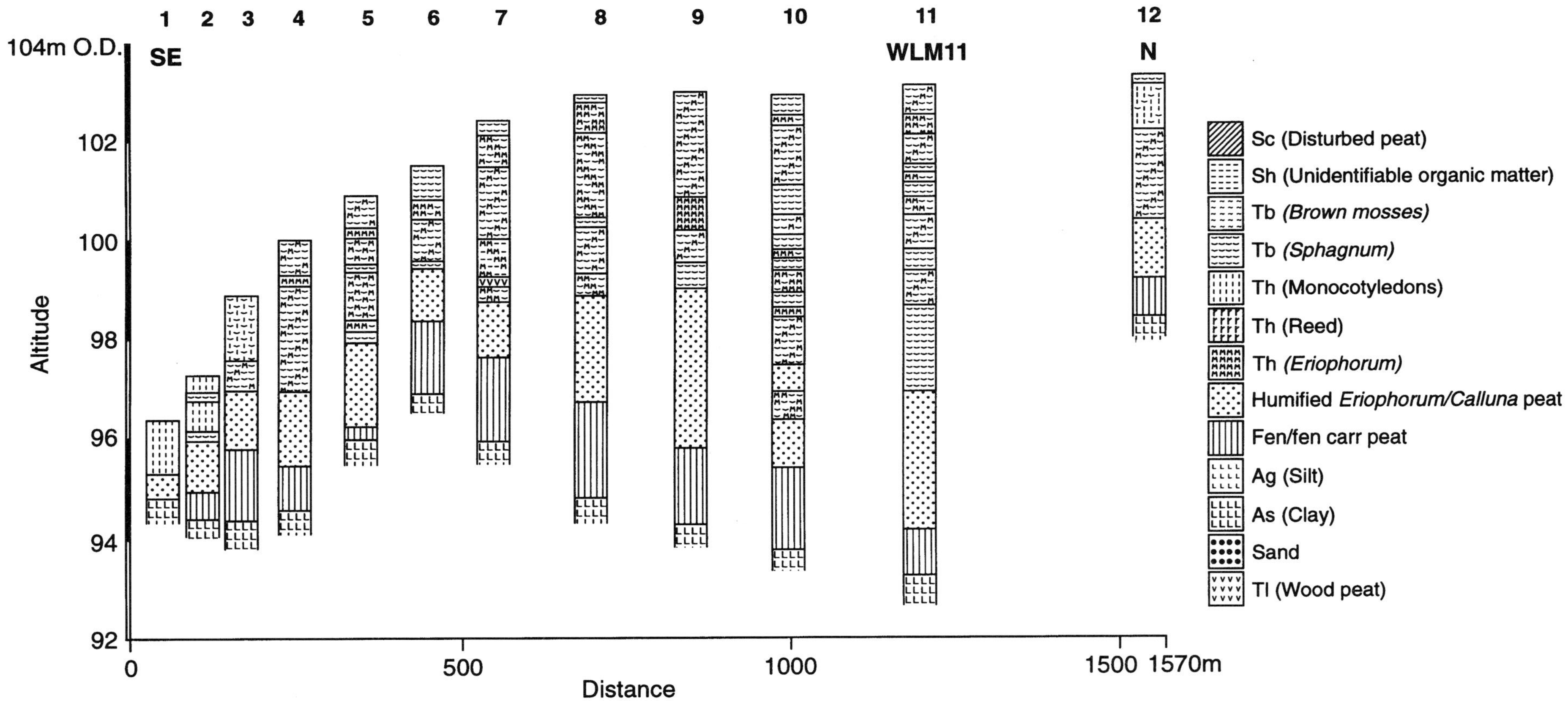

Fig 42 Walton Moss: Transect A (from Hughes 1997)

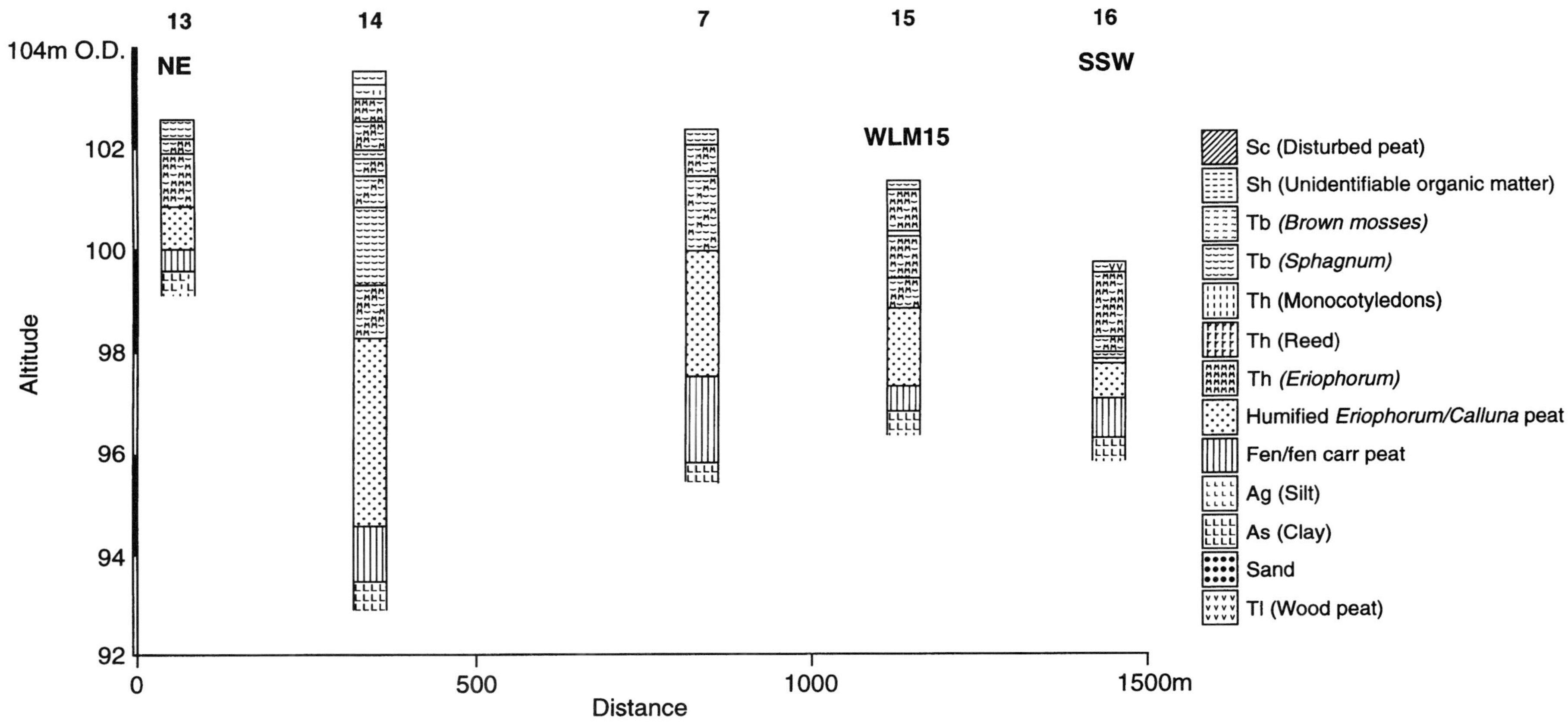

Fig 43 Walton Moss: Transect B (from Hughes 1997)

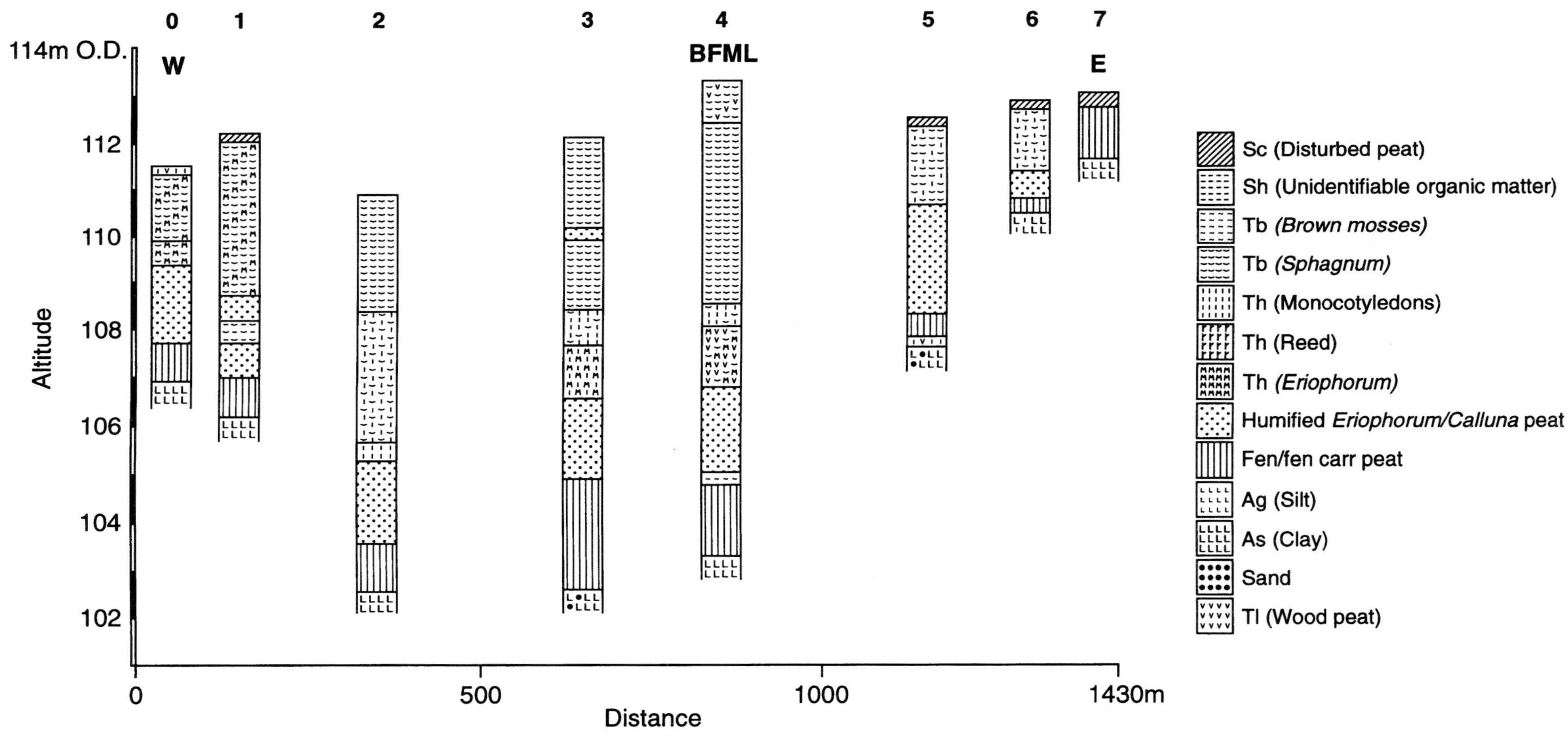

Fig 44 Bolton Fell Moss: field stratigraphy (from Hughes 1997)

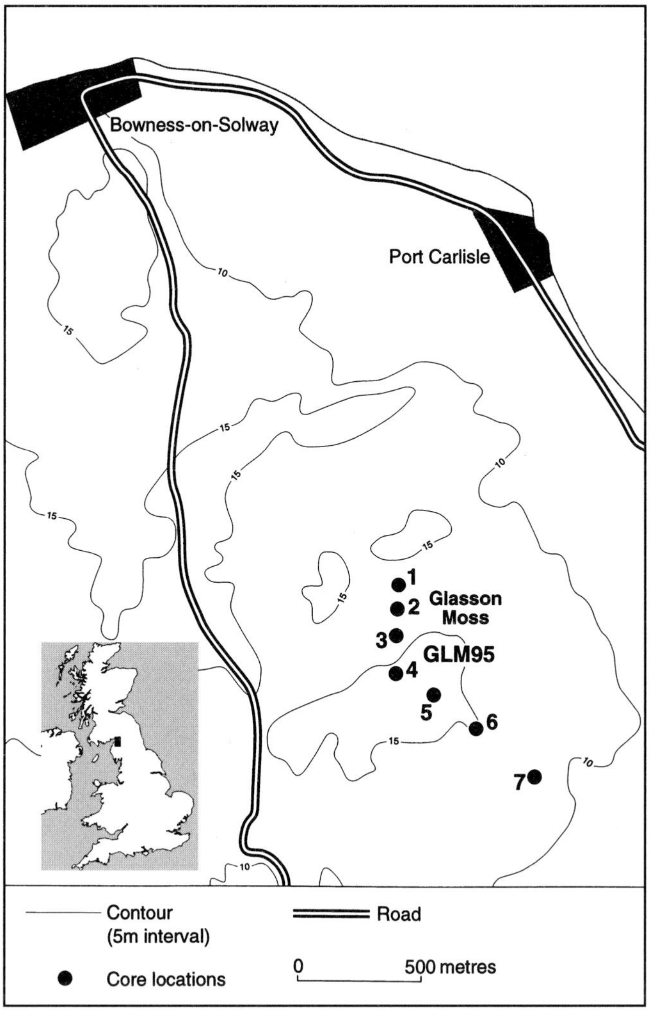

Fig 45 Glasson Moss: location of coring transect (from Hughes 1997)

woodland after no more than a couple of centuries. At *c* 7650 cal BC (*c* 8700 BP) the woodland was succeeded by a succession of fen mosses indicative of mesotrophic conditions, at the core site marked BFML on Figure 41. At 7561–7427 cal BC (8470±50 BP, SR–4942), macrofossil results show that *Eriophorum vaginatum/Calluna vulgaris* mire rather similar to that described at Walton Moss displaced the poor fen vegetation. The *Eriophorum vaginatum*-dominated phase lasted for approximately 3000 years, ending with the colonisation by *Rhynchospora alba*, a pool edge and mud bottom species. *Sphagnum imbricatum*, the main peat former on raised bogs in the mid-Holocene, was seen at core site BFML at a date of 3496–3104 cal BC (4575±45 BP, SRR–4937), marking the establishment of lawn conditions: this stratigraphic change can be traced for 400 m in the vicinity of the BFML coring site. Thereafter, the stratigraphy of Bolton Fell Moss demonstrates a series of striking alternations between dominance by *Sphagnum imbricatum* and dominance by *Sphagnum* section *acutifolia* and monocotyledons indicative of hummocks. These fluctuations have been attributed to changes in effective precipitation by Barber *et al* (1994). The replicability of the proxy-climate record at Bolton Fell Moss has been demonstrated by Barber (1981) from open sections and by Barber *et al* (1998) from multiple short cores.

Glasson Moss (Fig 45)

Located 0.5 km from Glasson village and 0.5 km from the south bank of the Solway Firth, Glasson Moss (NY 238 603; Pl 31) has a maximum peat depth of close to 10 m. The surface of the mire ranges from between 9 m and 14 m OD, which places the basal peat deposits within a few metres of the modern mean high water mark in the Solway Firth.

An extensive peat depth survey carried out by English Nature (F Mawby pers comm) suggests that Glasson Moss began to develop in a number of basins between a set of low hummocks which may be drumlins. Macrofossil core GLM95 (Figs 45 and 46) suggests that the basal deposits formed in shallow open water which filled with reedbeds. In common with the previous three sites, a dry *Eriophorum/Calluna* mire formed the first oligotrophic community at Glasson Moss, but in marked contrast to the other sites there was a clear return to poor fen at *c* 7750 cal BC (*c* 8800 BP), after approximately two centuries of raised mire development. The renewed poor fen phase is characterised by a major increase in the percentage of *Sphagnum palustre* and a marked reduction in the frequency of oligotrophic species, but after approximately 0.30 m of *Sphagnum palustre* peat accumulation, representing several centuries of mire

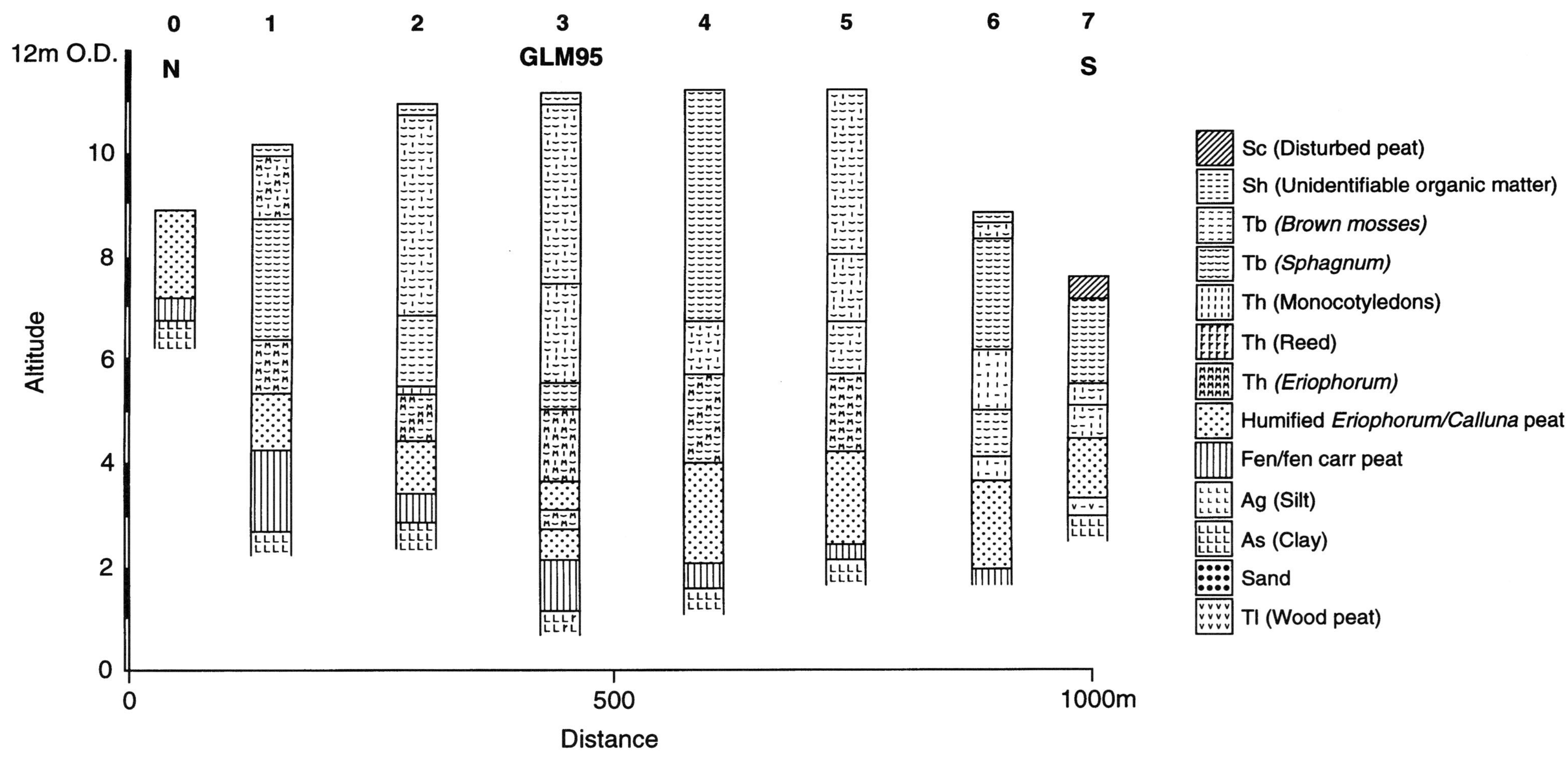

Fig 46 Stratigraphy of Glasson Moss (from Hughes 1997)

Plate 31 Glasson Moss

development, *Eriophorum vaginatum*-dominated raised mire returned to the GLM95 core site.

The alternation between fen and oligotrophic species during the early-Holocene coincides with a time period when sea-levels were rising rapidly in the Solway Firth (Haggart 1989). After *c* 5550 cal BC (*c* 6600 BP), the rate of sea-level rise began to reduce and after *c* 3050 cal BC (*c* 4500 BP) the trend reversed towards falling sea-levels. The *Sphagnum palustre* phase within the raised peat was probably a response to the backing up of groundwater as the base-level increased.

The stratigraphy of the *Sphagnum*-rich peats overlying the early cotton grass mire suggests that raised mire pool formation began in the sixth millennium BC and that there has been a series of alternations between hummock and pool phases on the mire surface. The upper 3 m of peat was investigated by Stoneman (1993) who identified a number of distinct wet shifts that have been interpreted as climatic deteriorations.

Solway Moss (Fig 47)
Solway Moss (NY 345 688) is located 2.5 km from the north coast of the Solway Firth and 1.5 km from Gretna Green on an interfluve between the Rivers Esk and Sark. The uncut surface of the mire lies at approximately 20 m OD, underlain by pale green and red sandy till of Devensian age. Much of the moss has been stripped of vegetation and milled for commercial peat extraction, removing the top 4–5 m of peat. Small areas of uncut mire remain but these mainly support very hummocky vegetation and *Pinus sylvestris* trees, and in places the till has been exposed. Deep drainage ditches also cross the mire, frequently penetrating to the underlying till.

The stratigraphic sections for Solway Moss (Figs 48–50) show that the basal topography undulates and it is probable that the mire began as a series of smaller fens formed in shallow kettleholes. At the macrofossil and pollen core site, SOLA2 (Fig 47), located close to the centre of the present mire, 0.30 m of deposits containing a rich variety of fen herbs was recovered, which are characteristic of Late-Glacial material (Hughes 1997). A further set of macrofossil cores, SOLA1–SOLA5 (Fig 47), also uncovered Late-Glacial material containing the remains of *Betula nana* (dwarf birch), whilst examination of the deeper drainage ditches showed that the ground between the basins was initially colonised by *Betula* (tree birch). Paludification of the *Betula* woodland commenced during the first

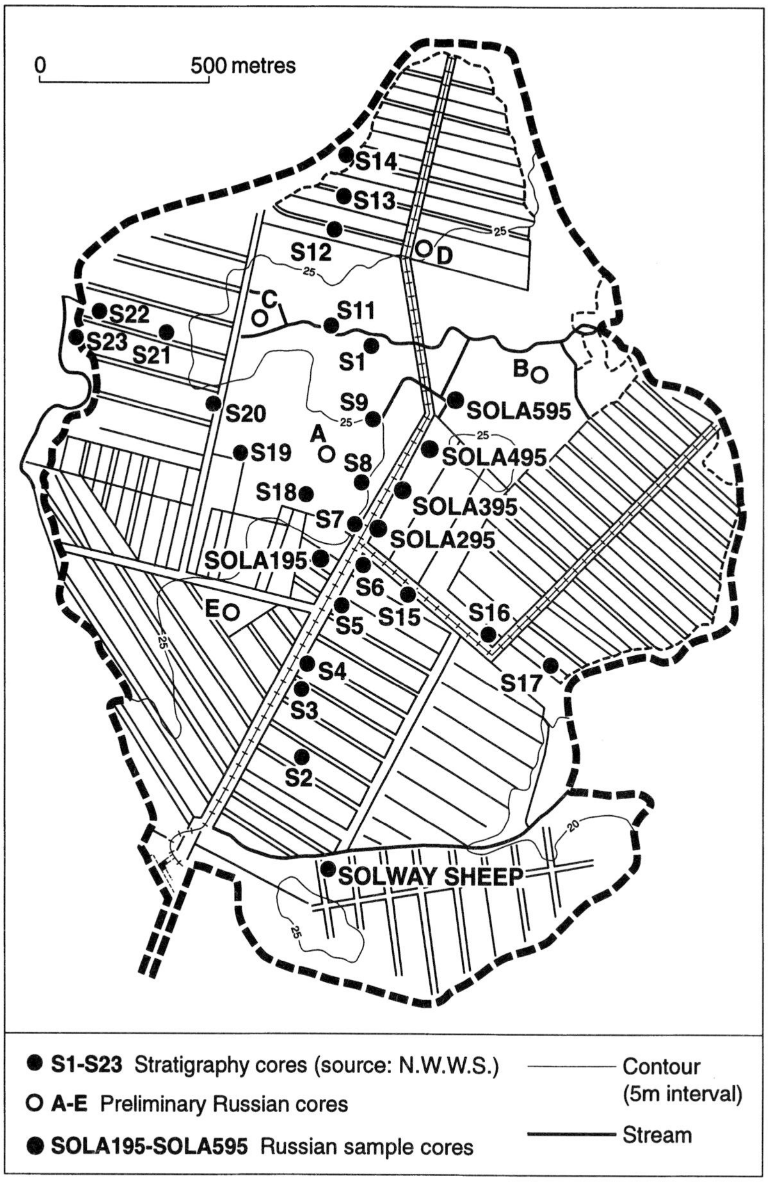

Fig 47 Solway Moss: location of coring sites (from Hughes 1997)

millennium of the Holocene (tenth to eleventh millennium BC) as the fens expanded beyond the confines of their original basins, and they subsequently developed a cover of carr woodland containing a mixture of *Salix* and *Betula.* At core site SOLA2, the fen and carr phases persisted for approximately 1500 years before being replaced by an open oligotrophic community containing *Sphagnum* sect *Acutifolia* and small quantities of *Aulacomnium palustre.* Following the *Sphagnum* phase, the mire changed to a vegetation type indicative of deep or fluctuating watertables on raised mires, dominated by *Eriophorum vaginatum* and *Calluna vulgaris.* This community persisted for a little over 2000 years before *Rhynchospora alba* and *Sphagnum imbricatum* arrived, marking the estab-lishment of a high water-table. Much of the mid- to late-Holocene *Sphagnum*-rich peat has been removed from the mire, accounting for the very uneven surface of the cross-sections presented in Figures 48 to 50. Further work is required to recover a detailed macrofossil record for the late-Holocene at this site.

Summary

The stratigraphies of all four mires indicate that the pioneer raised bog vegetation was relatively dry and hummocky in character. The appearance of soil fungi, such as *Cenococcum geophilum,* in the macrofossil assemblages from this phase suggests that the mire surfaces were aerated on many occasions. This form has been recorded in many mires of different ages and may reflect a distinct development phase when the newly-formed raised mire was incapable of maintaining a permanently high water-table

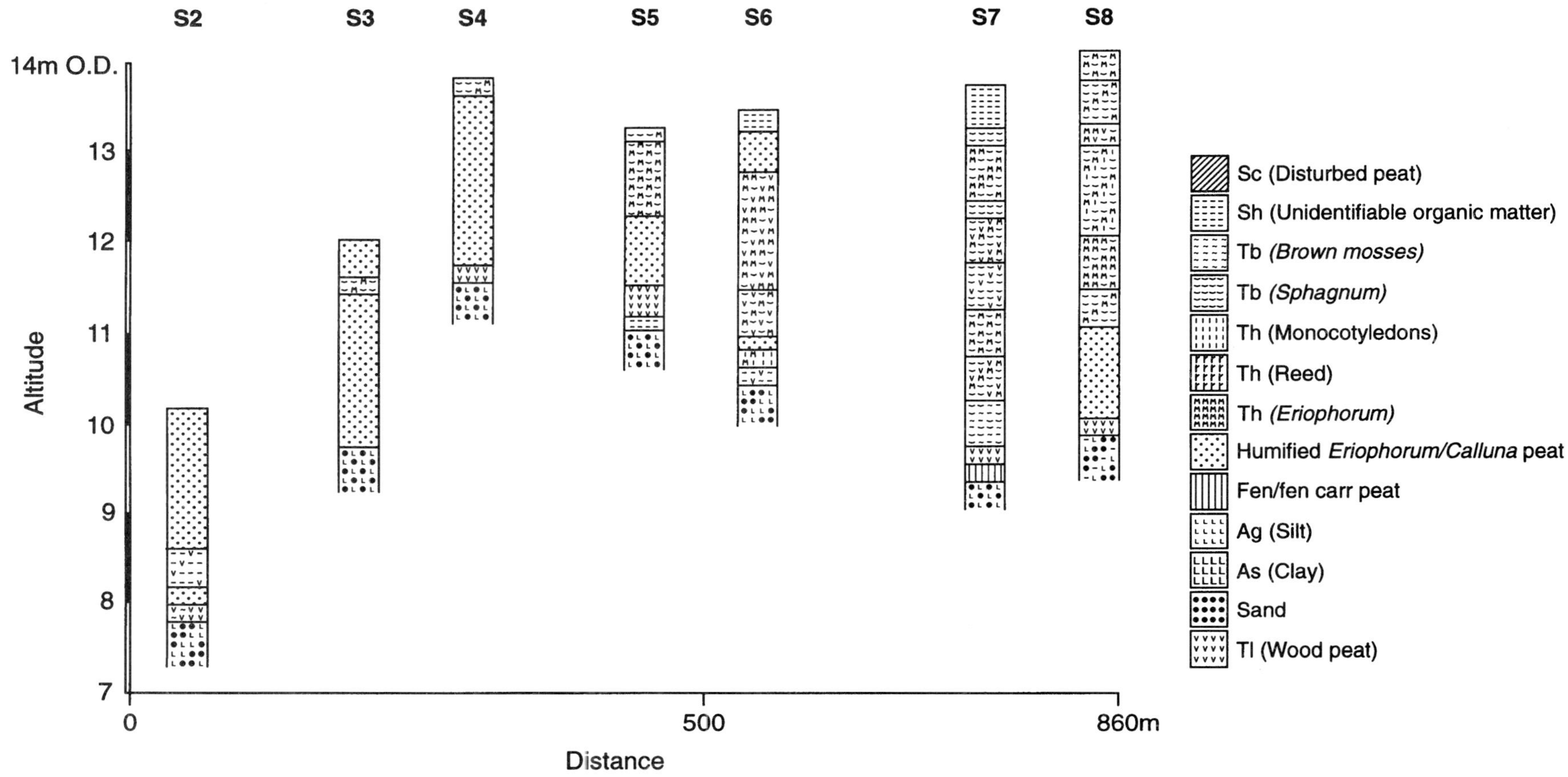

Fig 48 Solway Moss: Transect 1a (from Hughes 1997)

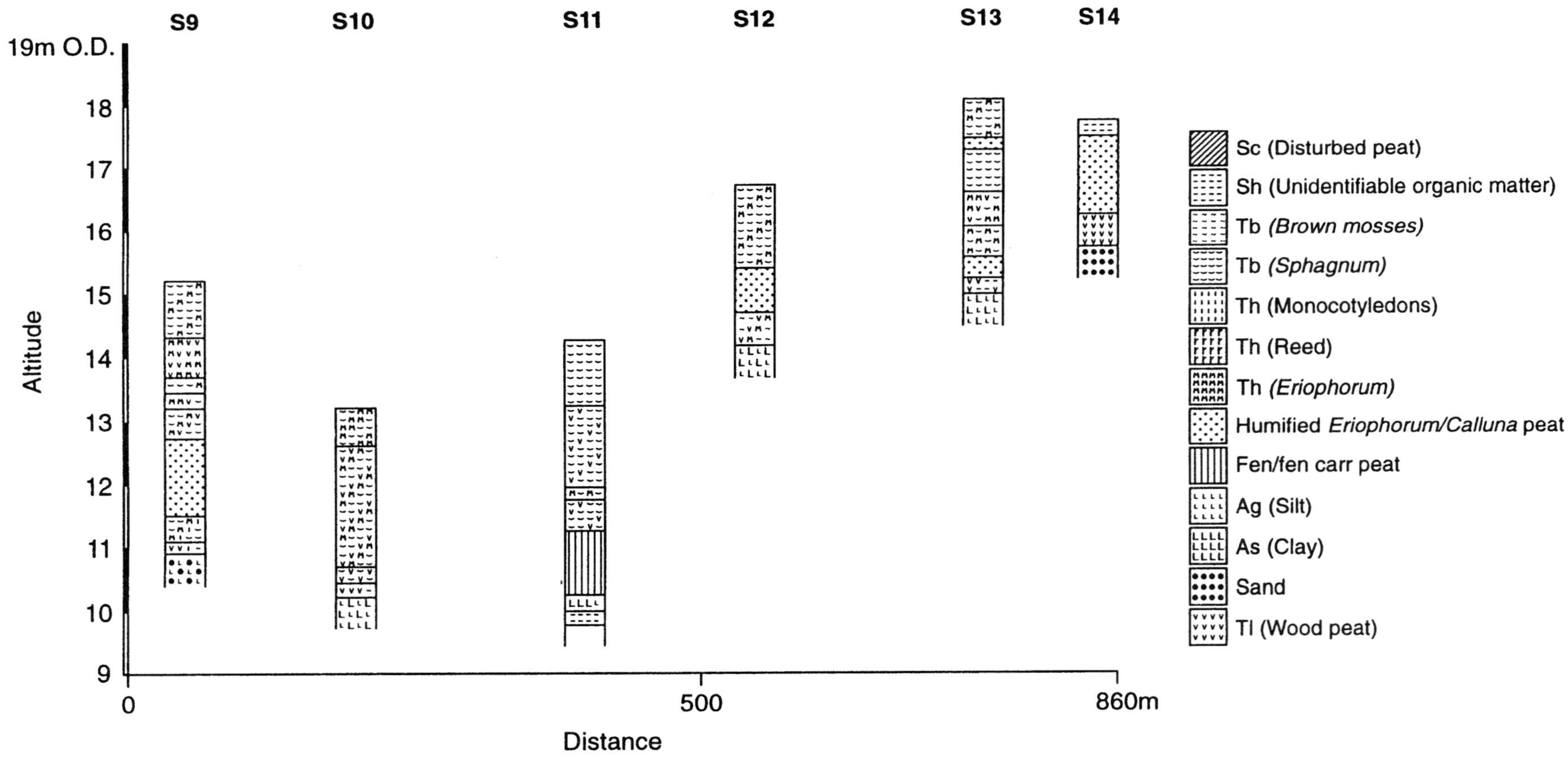

Fig 49 Solway Moss: Transect 1b (from Hughes 1997)

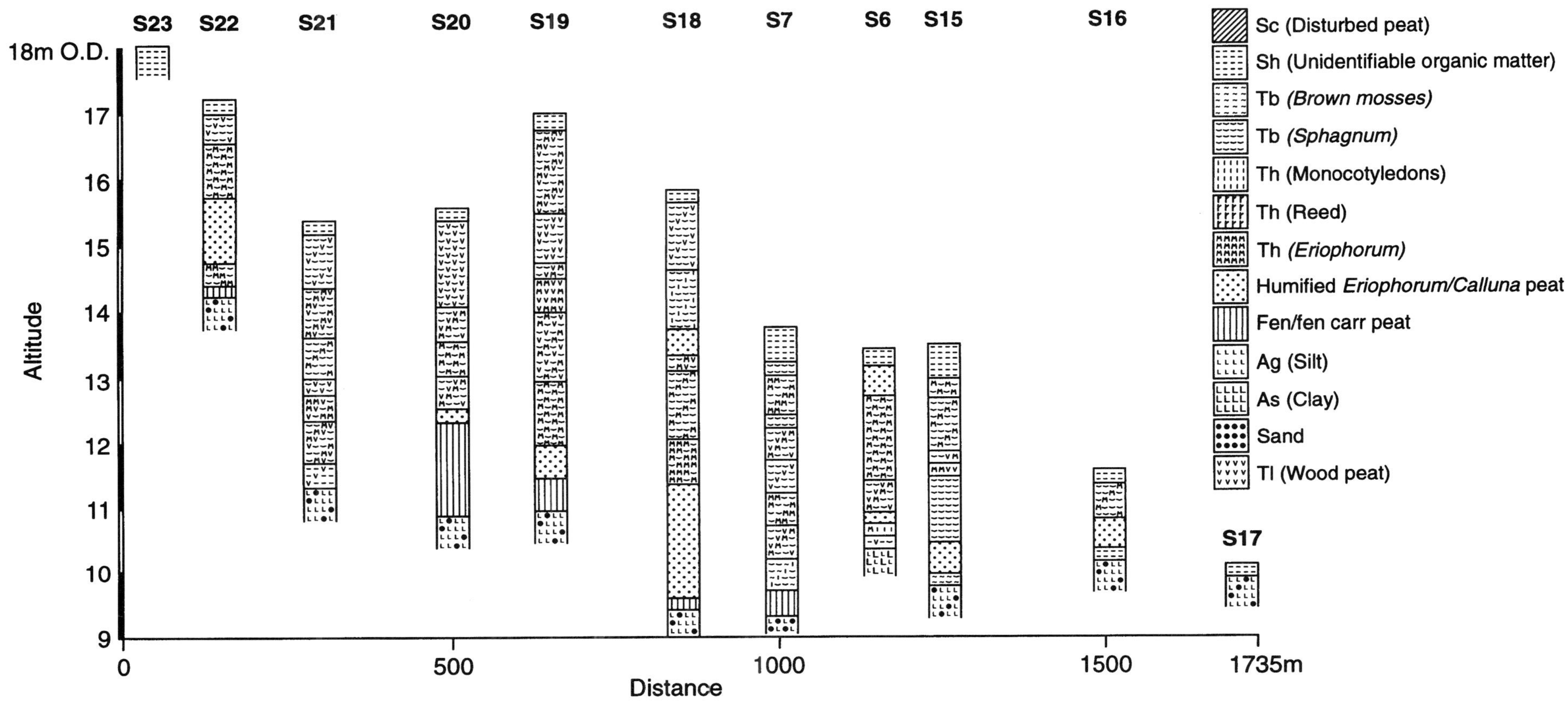

Fig 50 Solway Moss: Transect 2 (from Hughes 1997)

throughout the year. Alternatively, the *Eriophorum*/*Calluna* mires of the early-Holocene may reflect a period when climatic conditions were not favourable for the development of oceanic *Sphagnum*-rich vegetation. These themes are explored further by Hughes *et al* (in press).

Landscape Evolution

Late Palaeolithic

Palaeoecological evidence

The first evidence of climatic amelioration in the North Cumbrian Plain is represented by pollen identified in the organic muds laid down in the deep basins of Moorthwaite, Scaleby, and Oulton Mosses, during the Windermere Interstadial (Walker 1966, 149), the Walton, Bolton Fell, Glasson, and Solway data also show this (Hughes 1997). The end of this milder interlude is dated to 11,092–10,155 cal BC (10,160±207 BP, Q–147) at Scaleby Moss (Godwin and Willis 1959). Soil solifluction around the basin indicates the cold conditions of the ensuing Loch Lomond Stadial, which probably ended sometime between 10,542 cal BC and 9028 cal BC (10,160±193 BP, Q–152; Godwin and Willis 1959). However, summer conditions were not severe enough to deter the growth of aquatic vegetation completely. In the wider landscape, the vegetation was predominantly open and characterised by herbaceous flora throughout the Late Devensian, with trees and shrubs playing a very minor role. Pine (*Pinus*) and (tree) birch (*Betula*) both increased their representation substantially during the Windermere Interstadial, along with juniper (*Juniperus*), before declining again during the colder conditions of the following Loch Lomond Stadial (Walker 1966, 154, 158). No archaeological evidence has been found to suggest this landscape was exploited by human groups.

Mesolithic and early Neolithic

Palaeoecological evidence

Sea-levels

Mesolithic activity in the Solway area was played out against a background of changing relative sea-levels, which may have had profound effects on the nature of settlement and subsistence. To illustrate this, in southern Scotland early Mesolithic sites are located on raised beaches, indicating an earlier, higher sea reach associated with a period before the main marine transgression. The late Mesolithic material tends to be found on the present coastline and is thus later than the main marine transgression (Cressey and Toolis 1996, 465). The marine transgression has been dated (at Redkirk Point and Newbie Cottages, Annan) to between 7490–6604 cal BC (8135±150 BP, Q–637) and 4460–3970 cal BC (5360±116 BP, Birm–200), with the maximum inundation in the eastern Solway Firth occurring at *c* 6000 cal BC (7150 BP, Jardine 1980) before a slight regression and then a further transgression 3 m above the present level, at *c* 2500 BC (*c* 4000 cal BP; Lloyd *et al* 1999). On the southern shores of the Solway, where more limited work has been carried out (Walker 1966, Huddart *et al* 1977, Lloyd *et al* 1999), a mean sea-level between *c* 4.5 m and 6 m higher than present affected Bowness Common between 5820–5630 cal BC (6850±60 BP, Hv–6208) and 5298–4340 cal BC (5875±220 BP, Hv–6207). Further south, sediments at Wedholme Flow register the effects from the same transgression when the developing moss was inundated by way of the Rivers Waver and Wampool between 5970–5550 cal BC (6870±95 BP, Hv–5228) and 4893–3640 cal BC (5385±280 BP, Hv–4713). At Crosscanonby, the transgression occurred between 5970–5480 cal BC (6810±130 BP, Hv–6209) and 5613–5240 cal BC (6495±95 BP, Hv–5210) at altitudes between +3.71 m and +4.44 m OD. The relatively early end to full marine influence recorded there, compared to Wedholme Flow, is probably due to the formation of a sand and shingle barrier on the raised beach which protected the basin, in which the sedimentary record was contained, from marine influence. Recent work on the southern shores at Boustead Hill and Drumburgh Moss suggests that at the latter the relative sea level remained high and fairly static after *c* 4000 cal BC (*c* 6000 cal BP) with minor fluctuations before the final regression at 730–338 cal BC (2430±60 BP, B–103261; Lloyd *et al* 1999).

A further, less-pronounced, marine transgression than that recorded in the Mesolithic may have affected part of the North Cumbrian coast in the early Neolithic, the final stages of which are recorded from Pelutho, south of Silloth and dated to 3934–3370 cal BC (4845±100 BP, Hv–4418; Huddart *et al* 1977, 139–142).

Vegetation

At the opening of the Flandrian stage the North Cumbrian Plain was colonised by a 'park tundra' vegetation (*sensu* Iversen 1954), firstly of juniper (*Juniperus*) and willow (*Salix*), followed by birch (*Betula*) and then hazel (*Corylus*) to produce an open landscape with small copses isolated amongst grassland and herbaceous communities. Simultaneously, hydroseral development in waterlogged basins led to sedge swamps becoming sufficiently acid for *Sphagnum* mosses to begin colonisation.

By 8425–7576 cal BC (9009±194 BP, Q–161) the record from Scaleby Moss (Walker 1966, 42) shows that hazel, which had been present in small amounts for the previous thousand years or so, had expanded to become a major woodland-forming species in the area. In turn, elm (*Ulmus*) and oak (*Quercus*) became predominant, the elm (*Ulmus*) establishing itself slightly ahead of oak (*Quercus*). Finally, in common with most of lowland England, alder (*Alnus*) expanded its cover rapidly between *c* 6500 cal BC and *c* 6000 cal BC (7700–7150 BP), the beginning of the alder rise being dated at Scaleby to between 6457 cal BC and 5888 cal BC (7361±146 BP, Q–167; Godwin and Willis 1959).

By *c* 6000 cal BC (*c* 7150 BP) the vegetation of the North Cumbrian Plain had developed into 'an attenuated form' of the forest covering much of lowland England (Walker 1966, 179). Pine was much less common than further south and east, and lime was absent, due to its dependence on a mean July temperature of *c* 16°C necessary for the production of fertile seed (Pigott and Huntley 1981). Climatically North Cumbria was either beyond the range of lime, even during the mid-Holocene thermal maximum, or the necessary temperatures were not maintained long enough for lime to migrate that far north. Overall, Walker (*op cit*) considers that the vegetation of North Cumbria at this time had much in common with that of Northern Ireland, being relatively impoverished and subject to a more oceanic climatic regime.

The clearest palynological indications of possible human interference with the natural vegetation in the Solway area during the early Holocene come from the Scottish side of the estuary. In an episode dated to *c* 5850–5750 cal BC (*c* 7000–6900 BP; based on interpolated radiocarbon dates), Tipping (1995b) records a small peak in microscopic charcoal from Burnfoothill Moss which coincides with the appearance of pollen from clearance indicators such as *Pteridium* and *Plantago*, and the fire indicator *Melampyrum* (Simmons 1996, 119). Tipping (*op cit*) considers this association to suggest anthropogenic forest disturbance, and, in terms of magnitude and species assemblage, it resembles the putative Mesolithic clearance activity recorded from the south Cumbrian border at Thwaite House Moss (Middleton *et al* 1995, 187). The presence of *Melampyrum* has also been demonstrated to be common in episodes of Mesolithic forest clearance recorded from the North York Moors (Simmons 1996, 89–102).

Given the evidence for probable Mesolithic disturbance from Burnfoothill Moss, situated only 15 km from the Cumbrian coast, it seems very likely that, with the application of high-resolution studies, similar palynological signals would be detected in North Cumbrian mire sediments of suitable age. At present, however, the dataset available is probably inadequate from which to discern the limited disturbance events which are characteristic of this period. Work on the North York Moors has shown that most Mesolithic disturbance events have been detected at 'sites...at which a fine sampling interval of only 1 cm or 2 cm [*sic*] has been employed, suggesting that a wider sampling interval may sometimes not reveal small, short-lived events of this kind' (Simmons and Innes 1987, 392; Wells *et al* 1999).

Nevertheless, there are hints of possible anthropogenically-induced disturbance from late Flandrian II contexts in some of Walker's (1966) diagrams. Simmons and Innes (1987, fig 1), for instance, considered that the presence of microscopic charcoal in sediments of Flandrian II age at Oulton Moss, together with grains of *Plantago*, *Artemisia*, and Chenopodiaceae and a rise in the pollen of *Corylus*, *Fraxinus*, and *Hedera*, strongly suggest the effects of human interference around *c* 5000 cal BC (*c* 6100 BP). Similarly, in the Bowness Common diagram, peaks in indicators such as *Calluna*, *Rumex*, *Plantago*, *Urtica*, and *Artemisia*, together with rises in *Corylus* and *Betula* and a fall in *Quercus* values during late Flandrian II, may be a signal of anthropogenic activity in the surrounding landscape (J Innes pers comm).

In addition to the palynological record, it is possible that an indirect signature of human activity may be represented by the widespread presence of macroscopic charcoal in North Cumbrian mire stratigraphy of Mesolithic age (Hughes 1997, figs 5.4–8, 6.8–11). The NWWS survey at a site on the southern margins of Solway Moss in the vicinity of the sheep remains (*see p 130–6*), for example, demonstrated that carbonised plant material can be seen to occur regularly through the lower peat stratigraphy (Fig 51). The record suggests that fire disturbance was continuous from at least 6080–5790 cal BC (7080±50 BP, GU–5277), when peat accumulation began in fen carr, reaching a peak at 5240–4940 cal BC (6180±60 BP, GU–5315), which coincides with a sudden drop in the frequency of wood fragments and the first appearance of *Eriophorum*, indicating a switch from fen to acid bog, or poor-fen (*Appendix 6*). This corresponds to a sharp decline in alder and birch pollen, suggesting that carr woodland may have been destroyed by fire during the Mesolithic: whether the fires were deliberately started or as a result of natural processes is unclear (*Appendix 6*). This may have resulted in the establishment of acid mire communities for, after a brief recovery of wood fragments following this episode, these disappear completely as a cotton-grass and heather-dominated bog develops. The frequency of burnt plant remains also declines after this point but, nevertheless, they remain an almost constant feature of the profile until the upper stratigraphy, dated to 3350–2890 cal BC (4400±80 BP, GU–5313; Huckerby and Wells 1993).

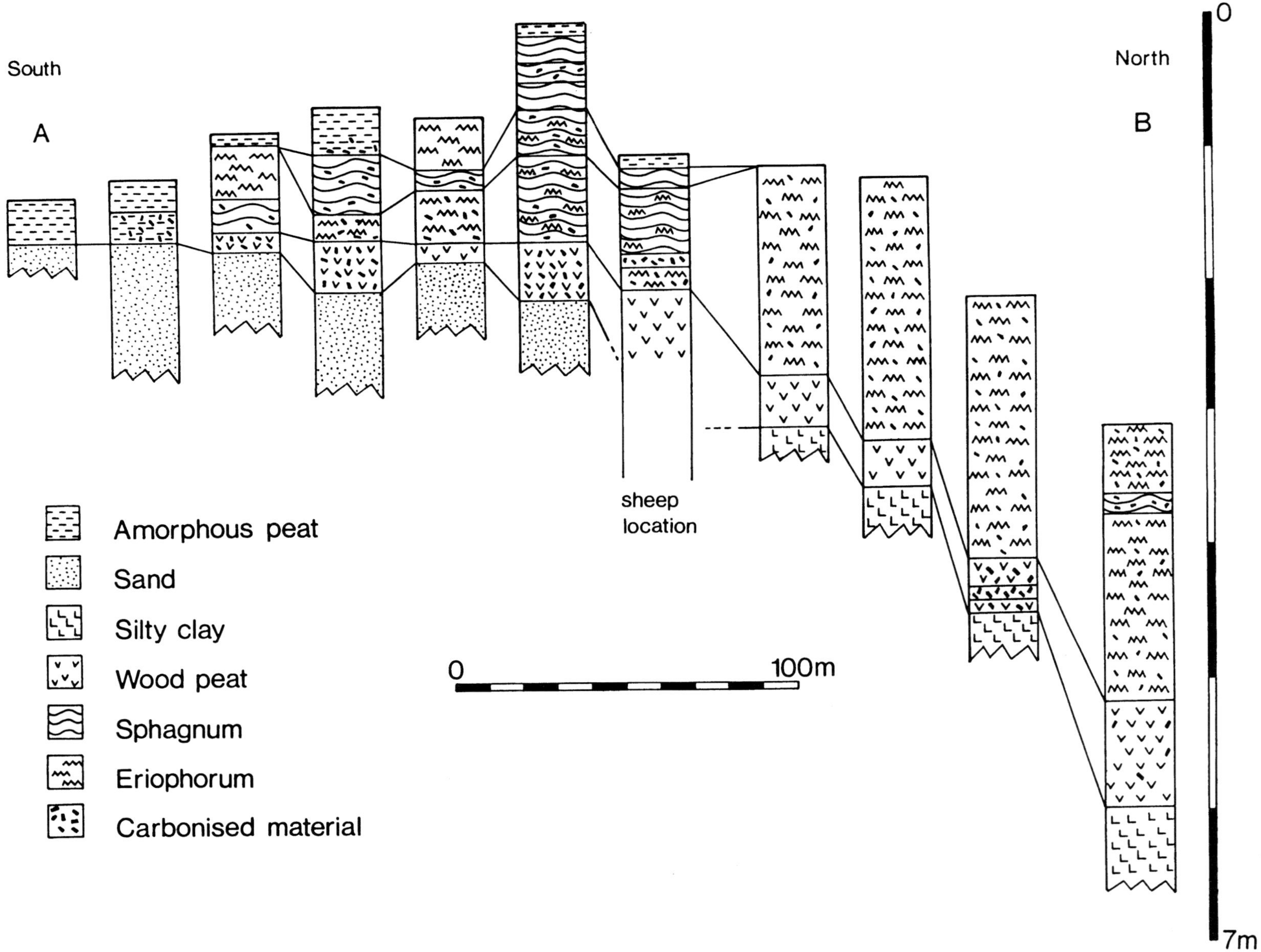

Fig 51 Solway Moss: stratigraphic transect in the vicinity of the sheep remains

Plate 32 Charcoal lens in the perimeter dyke onSolway Moss

A vivid demonstration of the signature left by these ancient fires on Solway Moss was revealed during survey and excavation there. Systematic survey of cleaned dyke sides on the southern and western edges of the moss in 1991 revealed a continuous spread of charcoal at the interface between the mineral soil and the peat (Pl 32), the comminuted charcoal occurring in small lenses spread throughout the deposit as far as it could be traced. The limit of the deposits was set purely by the available cleaned dyke side; it is likely that the charcoal occurred around the whole moss, although this would need to be established.

During the excavation of the sheep from the moss in 1992 (*see below, p 130*) the carbonised 'ghost' of the lignified parts of a complete plant of heather were revealed. The find suggested that the burning of upstanding ericaceous vegetation on drier hummocks had occurred, similar to the effect produced during modern controlled 'muirburn' where the leggier heather is 'clipped' by the passing fire-front as it sweeps over the ground (Ratcliffe and Oswald 1988).

It is not possible to assign a definite cause, or causes, to these apparently widespread fires which so evidently afflicted the mires during the Mesolithic. Natural, lightning-set fires, with a frequency and efficacy enhanced by a more 'continental' climate during the early Holocene thermal optimum, is a very real possibility (Huntley 1993, 211). Human agency, however, remains a possible cause of at least some of the major burning events recorded from this period (Simmons 1996, 129) and it might be expected that this is more probable in the wet environment of a mire. Plausible incentives for hunter-gatherer groups to ignite fen vegetation certainly exist, and have been proposed by, amongst others, Mellars (1976), and Simmons (1996, 180). These include encouraging the development of grass- and sedge-dominated communities in order to provide collecting and traversing areas for game, along with clear sight lines along which to aim weapons; similar strategies are well documented among ethnographically analogous communities, such as Native Americans (Lewis 1977; Lewis and Ferguson 1988). It is not inconceivable, therefore, that some of the charcoal from Solway Moss could represent a signature of early attempts to manage the wetland landscape.

The earliest definite clearance activity recorded directly from the north Cumbrian lowlands appears to be associated, in classical fashion, with the mid-Holocene Elm Decline, which is dated to *c* 4520–3990 cal BC (5421±130 BP, Q–120) at Scaleby Moss (Walker 1966, 43), and to just before 4036–3780 cal BC (5110±60 BP, GU–5275) at Solway Moss (Fig 88; Huckerby and Wells 1993). No evidence for burning is associated with the episode from Scaleby, nor does there seem to have been any cereal cultivation there. However, reductions in arboreal pollen, particularly oak and elm, occurred, and disturbed ground indicators such as *Plantago* (plantains), *Rumex* (sorrels) and *Artemisia* (mugwort) were present. Grasses also increased at this time, suggesting more open ground, while percentages of pollen from hazel (*Corylus*), ash (*Fraxinus*), and occasionally birch (*Betula*), can be seen to rise, perhaps in response to the opening up of the woodland canopy (Walker *op cit*, 196). Clearings appeared to have been sustained for some time, with bracken growing in importance as shrub clearance was undertaken. Woodland regeneration reclaimed the open ground in due course although there are hints that small-scale disturbances may have affected the secondary woodland in the vicinity of Bowness Common (*op cit*, 199).

The work of the NWWS at the site of the sheep remains (*see Appendix 6*) suggests that the picture from Solway Moss is slightly different and it is possible that cereal cultivation was being undertaken during the early Neolithic (immediately prior to *c* 3340–2707 cal BC (4370±90 BP, GU–5314)). This is suggested by the present survey's record of a cereal-type pollen grain in peats at the same stratigraphic level as a significant increase in mineral matter occurs (*Appendix 6*; Huckerby and Wells 1993), implying input from wind-blown sources, possibly as a result of exposure of nearby soils by woodland clearance (Vuorela 1983). Another cereal-type grain has been identified by NWWS from Solway Moss peats in the same vicinity and dated to *c* 4036–3780 cal BC (5110±60 BP, GU–5275). These two observations may represent the earliest cereal cultivation in North Cumbria, pre-dating the earliest previous records by *c* 1500 years (Walker 1966).

These initial inroads into the wildwood were undertaken during an apparently relatively dry climate, according to recent work on proxy climate signals provided by peat stratigraphy in North Cumbria. This has suggested that the early Neolithic (from *c* 4314–4006 cal BC (5315±35 BP, SRR–4631), to *c* 3080–2780 cal BC (4320±50 BP, SRR–4811) was generally characterised by relatively dry conditions, with perhaps a switch to wetter conditions around 3600 cal BC (*c* 4750 BP), but with a very dry phase around 3200 cal BC (*c* 4550 BP; Barber *et al* 1994). The significant change to a very wet and possibly also cold climatic phase at *c* 3080–2780 cal BC (4320±50 BP, SRR–4811) is likely to have had serious implications for prehistoric agriculture.

Archaeological evidence

Although it may be possible to suggest tentatively from the palaeoecological evidence that there may have been some human management of the landscape (*see above*), artefactual evidence for Mesolithic activity on the North Cumbrian Plain is limited to a single late Mesolithic triangular microlith from Carlisle (Caruana and Cherry 1994) and a possible early Mesolithic convex-edged blade of trapezoid section from Crofton, north–east of Wigton (Fell 1985). The present survey, however, produced four separate findspots of Mesolithic/early Neolithic material (CU1, CU6, CU7, and CU20), although some was in assemblages that also included later finds. Whilst most was waste from tool making, a backed blade was identified (CU1). Some bone harpoon points (SMR 4538) have also been found in wasting peat deposits at Crosby-on-Eden (Hodgson 1895; Annable 1987, 43). The finds demonstrate the definite presence of groups with which to associate the tentative palaeoecological evidence, and suggest that further Mesolithic finds are likely in the future, particularly in view of the potential attraction of the coast and wetlands for hunter-gatherers and early farmers, which has been demonstrated on the West Cumbrian Coastal Plain (Bonsall 1981).

The relative lack of material from the North Cumbrian Plain is perhaps best explained by the poor visibility of the archaeological remains, rather than an absence of activity. This is shown by the extent of the evidence from the Scottish side of the Solway, for example, at Redkirk Point near Gretna, where Mesolithic activity has been revealed through coastal erosion of raised beach deposits: the small early Mesolithic assemblage there included a flint blade, and a hearth which provided two radiocarbon dates of 7045–6614 cal BC (8000±65 BP, UB–2445) and 7193–6470 cal BC (7935±110 BP, UB–2470; Masters 1981). The extensive remains from Luce Bay in Galloway (Morrison 1981) were also found on raised beaches and areas clear of cycles of deposition and erosion, and it may be the presence of such features which determines the visibility of sites. Bewley (1994, 54) has, however, noted that, in Cumbria and southern Scotland, the Mesolithic material tends to occur only on south– and south–west-facing coasts, which may have been favoured for settlement.

The potential of the area for prehistoric material is shown by Bewley's work at Plasketlands (1993; SMR 607). Until this excavation, Neolithic domestic structures were unknown in Cumbria and most of the cropmarks identified in the area were assumed to be Iron Age or Romano-British in date. The site at Plasketlands, near Mawbray, was partially excavated (Pl 33) in order to establish its date and function (*ibid*). It was shown to comprise a large sub-rectangular ditched enclosure, with what has been described as an 'annexe' of large post-holes on its north–eastern side. This was interpreted as defensive in nature, due mainly to the presence of the large ditch, and the massive timbers which were employed in the construction of the 'annexe', each estimated at approximately 0.60 m in diameter. Despite the lack of Neolithic artefacts, three radiocarbon dates, of 3970–3525 cal BC (4940±90 BP, GU–2573), 3775–3380 cal BC (4810±60 BP, GU–2571), and 4032–3720 cal BC (5090±60 BP, GU–2572) from charcoal within a posthole established early Neolithic origins for the site. These dates appear to coincide approximately with the first definite clearance activity recorded on the North Cumbrian Plain.

This site is clearly of the greatest importance in our understanding of the first farmers in the area, yet the lack of artefactual material and the morphology of

Plate 33 Plasketlands: excavation under the direction of Bob Bewley

the site had not suggested early origins. It is likely that there are other such sites awaiting detection, at present assumed to be of later date.

As with the remainder of the county, evidence for Neolithic settlement largely rests on the distribution of polished stone axes. It has been estimated that there have been approximately 100 axe finds from the Solway Plain area (Bewley 1994, 56) but few of these have been found in reliable contexts. The exceptions are two reported to the project during fieldwork (CU17; Fig 52, and CU33; Fig 58) where the location of the finds has been established (Middleton 1991b; 1993c); and an axe flake found during fieldwalking (CU28 (Fig 59, 14); Middleton 1993c). These were all recovered from a relatively confined area on the Abbeytown Ridge which produced material of all periods during the NWWS survey.

Of the earlier chance finds, around a quarter of the total were found in locations in close association with mires in the Solway Plain. Notable concentrations occur at Salta (SMR 4882), and Mawbray (SMR 4524; SMR 4880–4883; SMR 6474), all of which overlook areas of wetland. A further concentration of axes from Beckfoot, north of Mawbray, was backed by deposits of fen peat (Bewley 1994, 56). Other axes occur as isolated finds in broadly similar locations (Fig 34): at Anthorn (SMR 377), and Allonby (SMR 651) on the coast, and Chapel Moss (SMR 637), Foulsyke (SMR 638), Dub Beck (SMR 652), 'The Hill' (SMR 375), and Mireside (SMR 376) skirting the valley mires inland. Rough-out axes have been found at Mawbray (SMR 4524), Kirkbride (SMR 396), and Allonby (SMR 651).

The wetland associations of some of these axes is notable, although it should be said that in many areas of the Solway Plain almost any findspot is going to be near to a wetland. Of those axes with a secure context, that from Pelutho (CU17 (Fig 52)) is from a precise wetland context — peat to the north of the Abbeytown Ridge. Since the axe was found during ploughing the context of the find is not fully secure, although there can be little doubt that it came from the peat. Another is the polished stone axe from Solway Moss which was found during peat cutting ('6 ft deep') complete with haft (Evans 1897, 151–2); these, and other chance finds mentioned *above*, may suggest a continuation of the pattern of axes from wetlands noted in other parts of the county (*this volume, Chs 2 and 3*). This is further discussed in *Chapter 5*.

One significant group of material of generalised Neolithic date was found during fieldwalking on the southern slopes of the Abbeytown Ridge, above

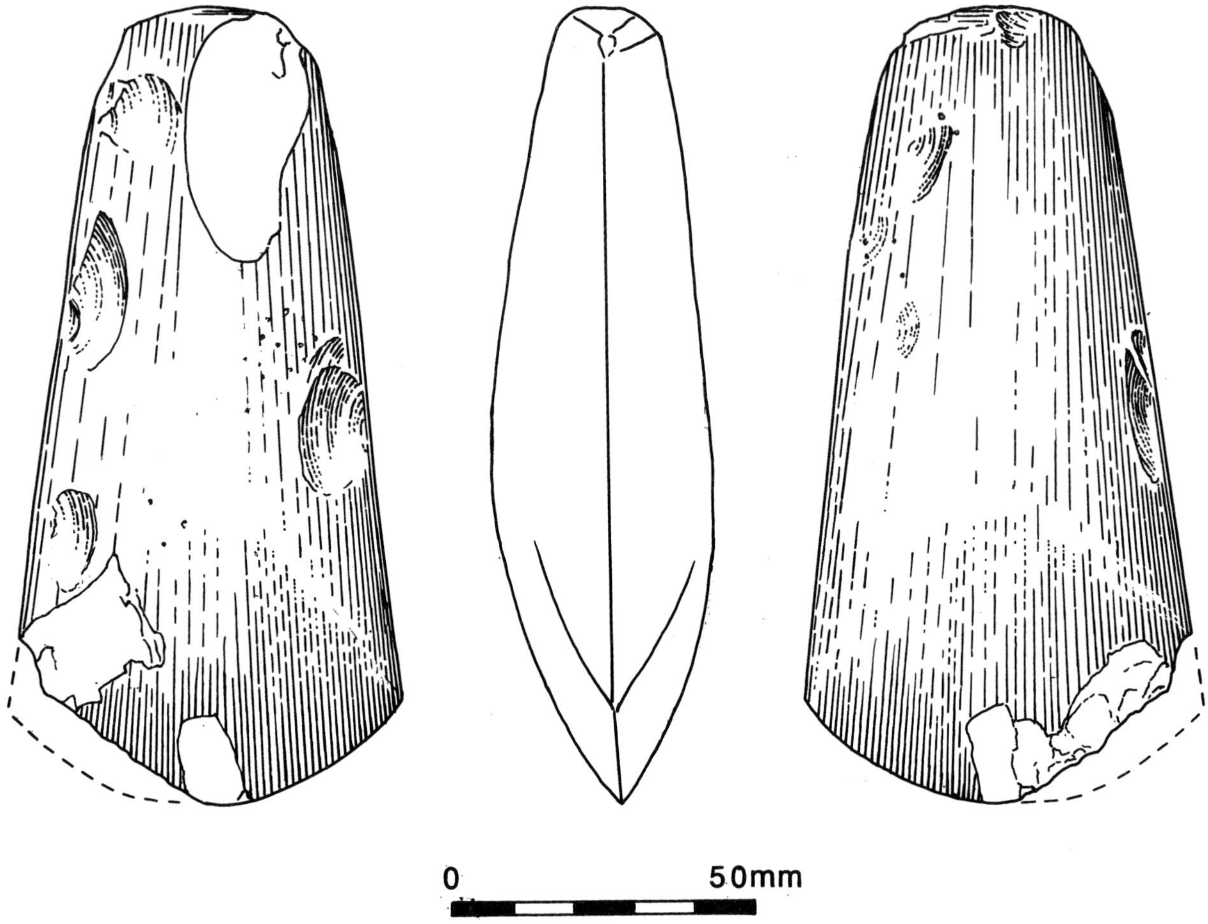

Fig 52 Polished stone axe from Pelutho (CU17)

Common Moss (CU4). Amongst waste material there were both preparation and utilised flakes, as well as an end scraper and a point. In addition, a single leaf arrowhead (CU18) was found in association with waste pieces further to the east along the ridge near Southerfield Hall.

In order to examine the possibilities for Neolithic exploitation of the North Cumbrian Plain, Bewley (1994, 72–3) has undertaken a Territorial Exploitation Analysis of some of the Neolithic sites, including Plasketlands (SMR 607), Hill Farm (SMR 9723), Mawbray (SMR 605), and Beckfoot (SMR 3183–3186, 3026). The results indicate that the area around the sites contained a diverse range of land-types including potential arable land, pasture, and sources of wild resources, such as the coast and wetlands. The analysis provides a range of potential resources, although these are the ones that are available in the modern landscape; the landscape will have been very different in the Neolithic, particularly in light of relative sea-level change, wetland development, and land-use over the past 6000 years.

Late Neolithic and Bronze Age

Palaeoecological evidence

Palynological evidence for early Bronze Age cereal cultivation on the North Cumbrian Plain is currently restricted to the records from Bowness Common, where cereal-type pollen is recorded for the first time at *c* 2000–1500 cal BC (*c* 3700–3200 BP), although the dating is very tentative. At the same time a major increase in clearance activity is recorded from Abbot Moss, although cereal cultivation there appears to have been of minor importance (Walker 1966, 199). Woodland around Oulton Moss also received a series of sustained attacks between perhaps *c* 2000 cal BC and 1500 cal BC (*c* 3700–3200 BP) with cereal cultivation gaining in importance as time went on. At Scaleby Moss, too, a major increase in clearance activity seems to have begun in the early Bronze Age, growing in intensity toward the middle of the second millennium BC.

Archaeological evidence

Many of the polished axes mentioned *above* are of likely late Neolithic date (Bradley and Edmonds

1993), although the only closely dated evidence for late Neolithic activity is from Brownelson, where fieldwalking over a large cropmark revealed a selection of flint and chert artefacts (Caruana 1989).

For the early Bronze Age, despite a significant increase in clearance activity and initiation of cereal cultivation in the pollen record, archaeological evidence is scarce. Scattered finds, such as a broken fragment of a jet bead from Brownelson (Caruana 1989), appear to characterise the record. This was supported by the findings of the present survey, which discovered some ten separate sites of the period scattered along the Abbeytown Ridge (CU3; CU5; CU11; CU15; CU19; CU29; CU30 (Fig 59, 15); CU32; CU35, and CU42, *(Appendix 1)*). Of particular note was a flake knife (CU8 (Fig 59, 12)) from south of Tarns Dub, and another (CU30) approximately 1 km to the north–east, in addition to a thumbnail scraper amongst the largely earlier assemblage from Lowsay Farm (CU1 (Fig 59, 1)). Three axe-hammers (SMR 4522; SMR 4884; SMR 4885) have been recovered from near Mawbray and another three are known from Bowness-on-Solway (SMR 172; SMR 174; SMR 176). Single examples have been found south–west of Cocker-mouth (Jackson 1906; SMR 176) and near Silloth (Anon 1917). The general absence of finds may reflect the lack of systematic survey over most of the Plain, although the NWWS survey on the Abbeytown Ridge suggests that settlement sites are either not represented by large scatters of flint, or are located lower in the landscape.

Aerial reconnaissance has revealed numerous sites of archaeological potential although most are not dated and have not been tested by excavation. Bewley (1994) has analysed the morphology of the sites and suggested that 21 of the circular cropmarks can be interpreted as barrows. One such at Aldoth (SMR 16932) has been excavated and yielded an upturned collared urn containing a cremation (*op cit*, 61–2). The site at Ewanrigg (SMR 3692), located south–west of the Plain near Maryport, also indicates the potential of the area. It was discovered by fieldwalking and proved to be a Bronze Age cremation cemetery surrounding a cist burial, which has been dated from 2470 cal BC to 1520 cal BC (2470–1760 cal BC, 3700±120 BP, HAR–7077; 2454–1880 cal BC, 3700±90 BP, HAR–5961; 2290–1760 cal BC, 3640±90 BP, HAR–5959; 2170–1704 cal BC, 3570±80 BP, HAR–7072; 2012–1630 cal BC, 3470±70 BP, HAR–5960; and 1890–1520 cal BC, 3400±70 BP, HAR–7071; Bewley 1992; Bewley *et al* 1992). The site is important for two reasons: firstly, its full potential was not evident purely from the fieldwalked remains and the surface evidence was only present because the site was on a sandy promontory and was being badly eroded. Secondly, it indicates the longevity of use of some areas of landscape; the radiocarbon dates suggest that burials were being interred over a period of about 940 years. Analysis of the radiocarbon results, however, shows that the span of the burials is skewed towards a shorter distribution, 150–700 years at 68% and 0–1100 years at 95% confidence limit.

The evidence for Bronze Age use of the wetlands is scanty, although some metalwork is known, the most famous of which is the Salta Moss rapier (Pl 34) which was found during ploughing at Edderside

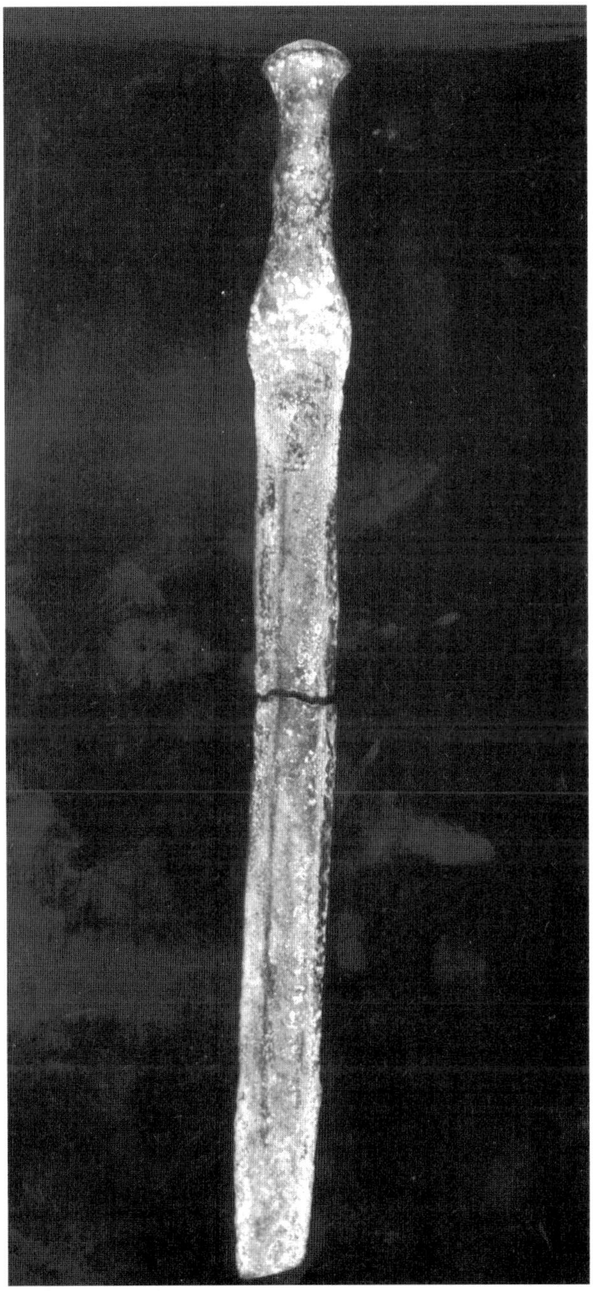

Plate 34 Salta Moss rapier (Tullie House, Carlisle Musuem and Art Gallery)

(Coles 1961; SMR 4521). It was cast as a single piece with a distinctive pommelled hilt and was probably manufactured locally. The find was poorly provenanced and so its exact context cannot be directly confirmed, although there can be little doubt it came from the moss. Other metal finds not directly related to mosslands include a late Bronze Age bronze spearhead from Caldbeck, on the banks of the River Caldew at Lynewath (Anon 1915; SMR 901). The use of small wetlands for the deliberate deposition of metalwork is suggested elsewhere in Cumbria by the find of a late Bronze Age cauldron in a small mire in Ravenstonedale near Kirkby Stephen to the south of the area (Turnbull 1995). A stone axe-hammer from Solway Moss (SMR 198), which was found during peat-cutting, may suggest that stone implements, in addition to those of metal, were also deposited in wetland contexts (*cf* Edmonds 1995).

Past records of wetland finds and sites are often difficult to interpret and present significant dating problems. One such is the 'palisade' found during peat cutting on Bowness Common in 1903 (Hodgson 1904). It comprised a single row of roundwood stakes of various spieces, including oak (*Quercus*), birch (*Betula*), hazel (*Corylus*), and possibly alder (*Alnus*). Each was *c* 50 mm in diameter and *c* 1 m long with one third of the length driven into the sand beneath the peat, overlain by *c* 0.6 m of peat at the level created by peat cutting. Most of the stakes had been sharpened to a point with 'a heavy tool having a clean cutting edge' (*op cit*, 211), although some retained trimmed roots; stones in the peat suggested a securing mechanism. In the area seen by Hodgson only a single line of stakes was visible, although several such structures, most with double lines of posts, were known from other parts of the moss.

The lightness of the stakes would suggest that they could not have supported a superstructure nor were they part of a defensive structure. The most plausible explanation is that they represented a fence or palisade dividing the moss edge into discrete parcels; the occurrence of such structures in other areas perhaps suggests a formal organisation of the landscape. If the structure is Bronze Age (and this must be open to question), the closest parallels may lie with the land boundaries of the fen-edge at Fengate or the reaves of Dartmoor (Pryor 1982; Fleming 1988). With such fragmentary evidence, interpretation must always be speculative, although the suggestion of landscape organisation cannot be ignored, if only because it is the first such recovered from the Cumbrian wetlands. It is worth noting also that this site remains much as it did in 1903 when it was discovered, and thus re-excavation of the 'palisade' may be very worthwhile if only to date this potentially important structure.

Another aspect of wetland use is represented by the bog body found during peat-cutting in Scaleby Moss in 1845 (Turner 1988; Turner and Scaife 1995, 112–13; SMR 3435). The dating of this to the Bronze Age is based on the description of the 'dark peats' in which the body was found *c* 3 m below the surface, although Bronze Age bodies are known from Britain and other parts of Northern Europe (van der Sanden 1996, 76–7) and such a date cannot be discounted for the Scaleby find. What this means in terms of landscape use is difficult to determine, although it may reflect the marginalisation of at least some of the mires away from the main areas of settlement and subsistence.

Iron Age and Romano-British Period

Palaeoecological evidence

In common with many other parts of North West England, the first part of the first millennium cal BC in North Cumbria is distinguished by a quiescence in the anthropogenic 'noise' discernible in pollen records. Tree cover remained extensive but there is evidence for low-level agricultural activity occurring in the lowlands. This includes indications of cereal cultivation in the vicinity of Bolton Fell Moss and the continuous curves of the pollen from species indicating cleared ground such as *Plantago* and *Rumex* (Barber 1981, 112). This points to a picture of a steady but low level of agricultural activity in the North Cumbrian lowlands during this time.

By way of contrast, in common with the rest of Cumbria and North West England, a major expansion of forest clearance and agricultural activity seems to have occurred in North Cumbria during the final centuries of the first millennium cal BC; this appears to coincide with the possible dating of many of the cropmark enclosures on the Plain (Bewley 1994, 63). The pollen record from Bolton Fell Moss shows dramatic falls in arboreal pollen beginning before cal AD 4–322 (1860±60 BP, Hv–3085; Barber *et al* 1993, 43, fig 17); at Walton Moss intensive clearance started before 111 cal BC–cal AD 83 (2000±45 BP, SRR–4531), before declining at 8 cal BC–cal AD 139 (1925±40 BP, SRR–4530; Dumayne and Barber 1994). Across the Solway at Burnfoothill Moss, Tipping (1995b) records a similar process beginning before 157 cal BC–cal AD 76 (2015±45 BP, SRR–3752), while the pollen diagram from Midgeholme Moss, situated immediately to the north of Hadrian's Wall at Birdoswald, also shows a late Iron Age clearance initiated before 355 cal BC–cal AD 125 (2040±80 BP, OxA–2325; Innes 1988). An intriguing consequence of this widespread deforestation is the possibility that during the Roman period there may have been few trees older than 250 years in age available for construction in the area around Carlisle. Indeed, a detailed analysis of dated timbers from the fort at Carlisle (*Luguvalium*) has shown the majority to have started growing in the first or second centuries BC (McCarthy 1995). The relatively open landscape conditions initiated in the late Iron Age appear to have been maintained, and perhaps expanded, during the Roman period in North Cumbria, as seems to be the case in other parts of the county, such as east Cumbria (*Appendix 7*).

In the vicinity of Hadrian's Wall (Pl 35), it was originally argued that the initiation of the major forest clearance episode was actually related to the Wall's construction, the impetus being provided by the timber requirements of the military infrastructure (Davies and Turner 1979; Barber 1981, 113). This assertion has subsequently been repeated many times (Dumayne 1993a; 1993b; 1994; Dumayne and Barber 1994) but, nevertheless, has always rested on data with relatively poor dating control (Dumayne *et al* 1995). The arguments have been challenged as a consequence (McCarthy 1995; Huntley 1999) and a revised view of the evidence has been formulated which suggests that sites close to Hadrian's Wall fit into the pattern of clearance initiated in the late Iron Age for agrarian purposes. Huntley (1999) has recently demonstrated this, summarising her conclusions in the chart reproduced in Figure 53. This confirms that there is no direct correlation between the initiation of the Roman period in the area and woodland clearance; on the contrary, it is clear that the story varies. At Bolton Fell Moss, clearance synchronises relatively closely with the building of Hadrian's Wall, while at Glasson Moss it occurs somewhat later, whereas Walton Moss seems to demonstrate Iron Age clearance. Site-specific work has tended to support the idea of a patchwork of different landscapes before the Wall was built. A palynological study of ditch fills from Vindolanda, for example, indicates that deforestation occurred before *c* AD 85 (*c* 1900 BP; Manning *et al* 1997), whereas recent work at Appletree and Birdoswald has concentrated on buried ground surfaces below the Turf Wall, which are particularly useful in giving a snapshot of the environment immediately before the construction of the Wall in the AD 120s. Birdoswald showed a damp, alder-dominated woodland, while only 1.4 km to the west, at Appletree, a grazed moorland environment existed on land which had been cleared long before the Romans

Plate 35 Hadrian's Wall, east of Birdoswald

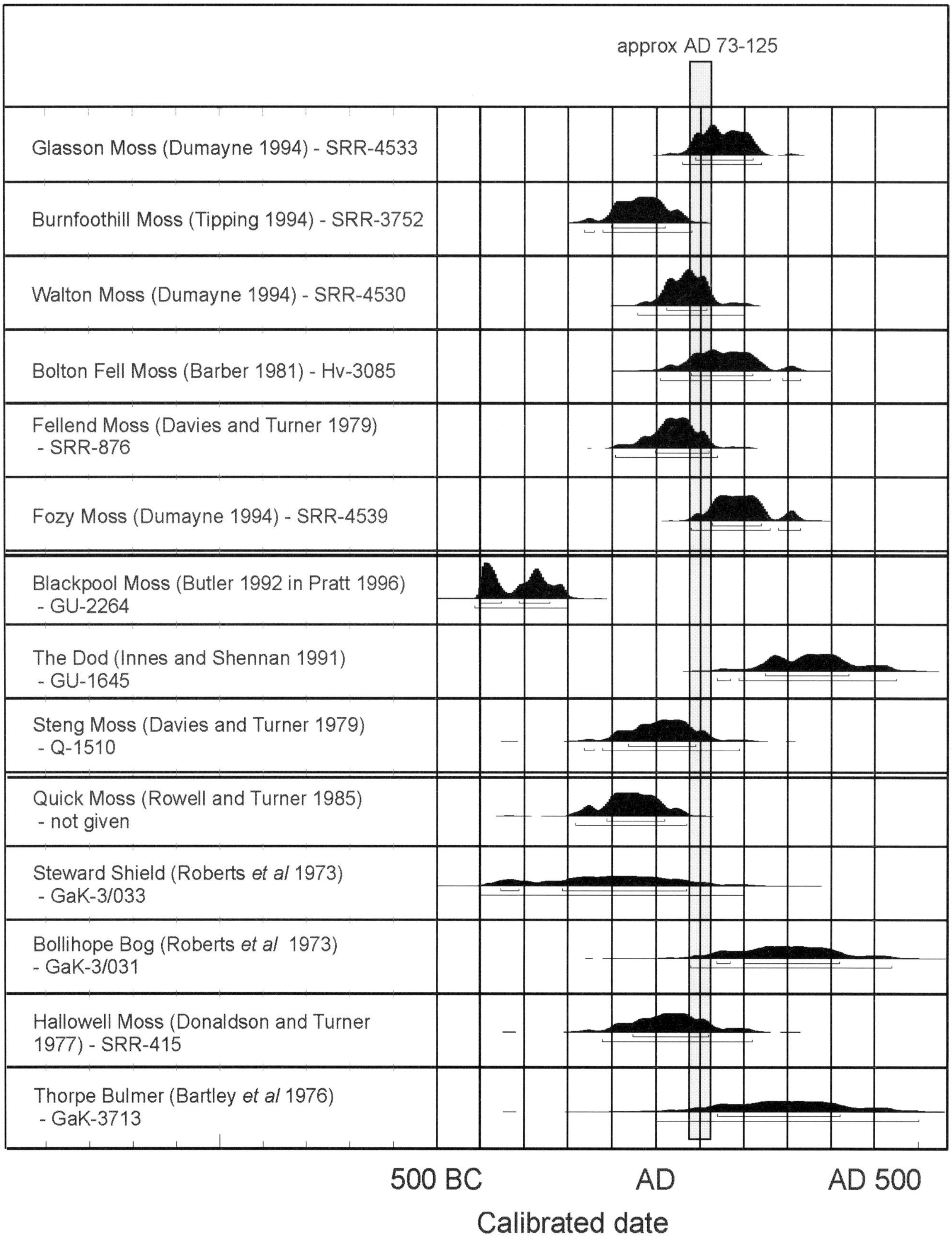

Fig 53 Probability plots of calibrated dates to 2σ of major clearance of woodland in the Hadrian's Wall area (after Huntley 1999)

arrived (Wiltshire 1992; and in Wilmott 1997, 25–37). Further to the west, in the low land around Carlisle, it is clear that pre-Wall cultivation took place, attested by the presence of ard marks beneath the counterscarp of the Wall ditch at Tarraby (Esmonde-Cleary 1997, 415); this evidence parallels the situation to the east, between Wallsend and Carrawburgh in eastern Northumberland (Huntley 1988). Such studies are particularly useful in giving detail of local conditions against the sort of regional picture available from the palynological studies of wetlands, and future approaches designed to explore both sources have great potential. Across the Solway in southern Scotland, the late Iron Age clearances persisted into the Romano-British period, and appear to have intensified until after *c* cal AD 150 (*c* 1850 BP), by which time little woodland remained (Tipping 1995b). The frequency of fires also seems to have increased during this phase.

Archaeological evidence

The lack of evidence for forest clearance in the early–mid-Iron Age appears to be supported by an extreme scarcity of archaeological material of early- to mid-Iron Age provenance from the Solway Plain (Bewley 1994, 63). Aside from the hillforts on the southern edge of the Plain at Carrock Fell (SMR 2973) and Swarthy Hill (SMR 609), dated to *c* 500 BC (*c* 2450 BP; Bewley 1992), the only site definitely ascribable to this period remains that discovered beneath the Roman milefortlet at Rise How (SMR 4239), where excavation revealed an Iron Age burial with pottery (Bellhouse 1984). Nevertheless, as has been pointed out by McCarthy (1995), archaeological fieldwork in the area has not been intensive, in addition to which is the added difficulty of identifying Iron Age sites because of their essentially aceramic character in the region (Bewley 1986). On the basis of current knowledge, however, the clear picture which seems to emerge from both strands of evidence is of a sparsely-populated landscape in the north Cumbrian lowlands for the first part of the Iron Age.

Bewley (1994, 63) notes, however, the large number of undated prehistoric enclosures on the Solway Plain, detectable where conditions are suitable for cropmark development. Many of these may be of Iron Age date, being the physical manifestation of large-scale landscape exploitation evident in the palynological record (*see above, p 114–5*). Hints of such dating have been given outside the study area by the excavations of a bivallate enclosure at Dobcross

Plate 36 Prehistoric and Romano-British enclosures at Wolsty Hall (NY 105 512) on a sandy ridge rising out the surrounding fen and bog (© R H Bewley)

Plate 37 Romano-British farmstead overlying an earlier prehistoric settlement (NY 251 573). This site lies on the very gentle slopes above Fingland Moss (© R H Bewley)

Hall, Dalston (SMR 698), where an Iron Age site appears to have been re-used in the Roman period (Higham 1981). On the Solway Plain, small-scale excavations suggested the Roman re-use of a prehistoric circular enclosure at Boustead Hill (SMR 3386), and a possibly similar situation at Fingland (Bewley 1986; SMR 3303; Pl 36), at the Ewanrigg settlement (Bewley 1992; SMR 791), and at Edderside (Bewley 1998; SMR 605). If the circular enclosures are prehistoric in origin, in contrast to the Roman rectangular and square examples (Bewley 1994; *cf* Bewley 1998, 114), their ubiquity would indeed suggest a widely-settled landscape in the Iron Age (Pl 37). The invisibility of such prehistoric enclosures to surface survey, however, has been amply demonstrated by several programmes of work (Bewley 1986; Caruana 1989), including the present survey.

Despite the paucity of securely-dated sites, the evidence suggests that individual enclosed farmsteads were the normal settlement pattern which appears to have existed across (at least parts) of the Solway Plain by the end of the first millennium cal BC. The economy of such settlements is likely to have been mixed, with possibly a strong element of seasonal stock movement; such a pattern of settlement and subsistence has been noted in other parts of northern England where archaeological visibility and survival are better (Cunliffe 1991, 275–9; Haselgrove 1996).

Another use of wetlands is demonstrated by infrequent occasional finds. The bog body from Scaleby Moss (discussed *above, p 114*) may be Bronze Age, although the deposition of such bodies is an established feature in North Europe throughout prehistory. The importance of some mires is suggested by metalwork which may have been deposited intentionally; the clearest example is the Bewcastle cauldron (SMR 94) of suggested second-century date, found during peat-cutting (Davenport 1996). This joins the scatter of such material from mires across the North West (Middleton *et al* 1995; Middleton *et al* forthcoming).

The Roman period in North Cumbria is, as already remarked, virtually synonymous with Hadrian's Wall and its predecessors, and it is worth considering the use which the Roman frontier works made of the mosslands, and how far the relationship between them may be used towards a strategic interpretation of the frontier.

As is well known, the first frontier line on the Tyne-Solway isthmus was the Stanegate road, upon which a series of forts was located, including Corbridge, Vindolanda, Carvoran, and Carlisle. To the west of Carlisle, an extension of the Stanegate has been recognised, the western terminus of the Stanegate being at the point where the River Wampool flows into the northern arm of the bay known as Moricambe. Here, at Kirkbride, a Trajanic auxilliary fort (Bellhouse and Richardson 1982) is consistent in date with the rest of the Stanegate forts and is considered to be the terminal fort of the system. The line of the Stanegate has been traced from Kirkbride eastwards as far as Fingland Rigg by aerial photography (SMR 4677), and has been excavated on Fingland Rigg itself (Jones 1989). Although not physically traced eastwards of here, it is probable that the route continued to one of the three forts at Burgh-by-Sands (Fort I: SMR 4395, Fort 2: SMR 415, and Fort 3: SMR 6486). Though these forts are poorly understood, especially in terms of dating, the line of the Stanegate suggests that the fort known as Burgh-by-Sands I (SMR 4395) might be the Stanegate installation. Fort I overlies a ditched palisade running east–west across the site, and also a circular tower of the early second century; at least one similar tower has been identified on Fingland Rigg (Jones 1982, 284–5; 1989; SMR 6485). This suggests that the Stanegate frontier is of at least two phases in this sector, while the later enlargement of Fort I hints at a third.

The Stanegate line is carefully planned to take a route over a low ridge some 3 km south of the Solway coast, deliberately keeping to the south of the major mosslands of Bowness Common, Drumburgh Moss, and Glasson Moss. There can be little doubt that the mosslands of the Bowness and Cardurnock peninsula were the major determining factor in the selection of this line for the Stanegate, since the Burgh–Kirkbride route certainly represents the best line for a road. As it is relatively high, it avoids the necessity for complex engineering works of the kind often used to carry Roman roads over boggy ground (Chevalier 1976, 90), and is beyond any threat from coastal innundation. Watch towers like those at Burgh and Fingland Rigg could command views covering the low-lying coastal mosses, the Solway itself, and across to the northern coast. In all of these respects the Stanegate line is superior to that taken by the later line of Hadrian's Wall.

The builders of Hadrian's Wall took the coastal route from Burgh-by-Sands, along the coast to Bowness, and then by way of a western extension consisting of a palisade, turrets, and milefortlets (Jones 1976) around the Cardurnock peninsula to Moricambe. The actual line of the Wall is questionable at some points; the debate over whether the Wall ran across Burgh Marsh, or was diverted to the south of it, has run since Horsley (1732, 156) took the latter view. Bruce (1867, 92) was inclined to agree with Horsley, while admitting the want of evidence. Hodgson (1840, 302) considered that it crossed the marsh, and it took the excavation work of Haverfield (1900) to demonstrate that the postulated southern routes had no foundation. A map of the course of the Carlisle Canal in this area, drawn up in 1818 (Chapman 1832), shows 'stony vestiges' west of Burgh-by-Sands leading into the marsh and this may be have been part of the Wall. Generally, however, Chapman acknowledges that there were no traces of the Wall found in the work for the canal. It is clearly possible that the Wall was destroyed either by innundation from the sea, or by a bog-burst such as that which took place on Solway Moss in 1771 (*see below p 141*). The scale of the engineering required to construct the Wall along this coastal route was revealed at Milecastle 79 (Richmond and Gillam 1952, 27–28; SMR 165), where the milecastle and the Turf Wall running east towards Port Carlisle and west towards Bowness were constructed upon a substantial sea-bank originally some 2 m in height from the natural ground surface to the base of the Wall. The bank was built of alternate layers of peat or turf and gravel and the excavators thought it likely that a similar bank might have crossed Burgh Marsh. This extraordinary construction, which is much overlooked in general literature on Hadrian's Wall, would not have been necessary had the Stanegate route been followed.

It seems to the present writer (T Wilmott) that the effort expended in taking the coastal route must have been prompted by a very powerful motive. In many ways the coastal route was less obviously advantageous than that of the western Stanegate. The available views were not as good as on the ridge, from which, as Wooliscroft (1994, 59) has implied, perfectly adequate watch could be maintained on the Solway. In addition, the amount of effort required to preserve the Wall from inundation was substantial, and the land to the rear of the Wall consisted of a virtually continuous tract of mossland. It is argued here that the Wall from Burgh-by-Sands onwards should be seen functionally as part of the same concept as the extension around Cardurnock in giving an alternative route for the frontier to Moricambe, and that a reason needs to be sought for the enclosure of all land up to the coast in this area. In fact the choice of the littoral route seems to inform the current debate on whether the Wall was a non-defensive barrier intended to control movement (*pace* Breeze and Dobson 1976, 143) or whether it was a military barrier with a defensive capability from its inception (*pace* Daniels 1979, 360), a debate summarised recently by Bidwell (1999, 31–5). If the former, then the better Stanegate route would surely have been followed.

The only explanation for the enclosure of the coastal mosses is military; it was to prevent the use of the Cardurnock/Bowness peninsula as a bridgehead for potentially hostile forces from across the Solway, with the Wall itself continued to the headland at Bowness, providing a formidable barrier along the south bank of the Firth, which could easily be forded. From Bowness westwards, where fording is possible but more risky (Jones 1982, 286), a palisade with milefortlets and turrets was considered to be adequate. It is well known that the coastal sites west and south of Bowness were not re-occupied after the withdrawal from the Antonine Wall. Bidwell (1999, 34) has recently emphasised the view that these installations were established to meet a specific military threat which later receded due to either diplomacy or warfare, rendering coastal defence redundant. It is suggested that this threat also resulted in the alignment of Hadrian's Wall from Burgh to Bowness, which was retained as part of the continuing system in the post-Antonine period (Wilmott in prep).

As already demonstrated with reference to the sea bank along the Solway shore, Roman engineering was capable of adapting to difficult terrain in order to secure an objective. This is visible on the line of the Wall in North Cumbria at two other points. At Birdoswald (SMR 343), the primary line of the Wall was placed along a ridge between Midgeholme Moss to the north and a small basin mire to the south. The natural advantages of the site in terms of its proximity to the Irthing river crossing at Willowford, and the commanding views which it has in all directions, made it ideal for a fort. The topographic disadvantage of the bog to the south was disregarded, as this was drained, truncated, and sealed in advance of the construction of the fort (Wilmott 1997, 42). At White Moss and Blea Tarn, further to the west, the *Vallum* to the south of the Wall was continued through the moss in a built-up form, having four mounds and a vestigial ditch, rather than the usual substantial ditch between two mounds (Haverfield 1895; SM26085).

Medieval and post-medieval

Palaeoecological evidence

After *c* cal AD 300 (*c* 1750 BP), palynological trends from north of the Solway demonstrate that partial woodland regeneration occurred (Tipping 1995b). There has been speculation that this could be associated with a gradual diminution of Roman control in the area after *c* AD 350, leading to a reduction in economic activity (Higham 1986). This, however, contrasts with the view posited by Turner (1979), who, on the basis of radiocarbon-dated pollen diagrams from four sites on Hadrian's Wall, argued that a high degree of political and economic stability persisted for some time after the Roman withdrawal, a view shared by Barber (1981, 114) on the basis of palynological data from Bolton Fell Moss. One interpretation of these apparently conflicting palynological signals might be that they reflect radically contrasting local responses to the major political events of the fourth and fifth centuries, with economic activity retrenching to areas of security provided by military infrastructure. However, as Tipping (1995b) has cautioned, the imprecision inherent in radiocarbon dating and the limits in resolution of most pollen datasets ensures that most attempts to correlate palynological data with discrete historical events must remain, at best, speculative.

In broad terms, it is clear that *c* cal AD 600 (*c* 1500 BP), renewed anthropogenic activity was underway north of the Solway in the vicinity of Burnfoothill Moss, at a time of well-documented (for the period) political importance of the area, such as the establishment of Kentigern's See at Hoddom (*ibid*). South of the Solway, agricultural activity also appears to have continued, albeit at a low level, in the vicinity of Bolton Fell Moss, but at *c* cal AD 714–980 (1170±50 BP, Hv–3084), there are indications of a general recession in agricultural activity with a rise in arboreal pollen and a decline in grassland (Barber 1981, 115), and it is conceivable that this coincided with a sustained period of wet conditions, which resulted in widespread pool formation on the moss. Nevertheless, it seems likely that some cereal cultivation continued to be undertaken.

In the Border Region, the effects of the 'medieval Warm Epoch' (*sensu* Lamb 1977) are possibly reflected in the palaeoclimatic signal from peat stratigraphies which have been studied in detail. North of the Solway Tipping (1995a) records that mire surface conditions at Burnfoothill Moss appear to have been relatively dry from the end of the Roman period until perhaps *c* cal AD 1200 (*c* 850 BP). At Solway Moss and Bolton Fell Moss, after the phase of wetter conditions described *above*, there is evidence for a drier phase which affected the mire surfaces, and this has been dated at Bolton Fell Moss to between *c* cal AD 1100 and cal AD 1300 (*c* 900–600 BP; Barber 1981, 191, fig 76; Stoneman 1993). A short-lived (perhaps *c* 150 years) deforestation episode seems to have occurred in the vicinity of Bolton Fell Moss during the earlier part of this period, with indications that 'secondary' woodland was being cleared. However, this episode was immediately followed by a rapid decline in agricultural activity and forest regeneration (Barber *op cit*, 115).

The later medieval period was witness to a complete reversal of the climatic trend, with a change to colder, wetter conditions (Lamb 1977). In North Cumbria the effects of the late medieval climatic deterioration may be recorded in the stratigraphy of Bolton Fell

Moss between cal AD 1258–1395 (680±50 BP, Hv–3078) and cal AD 1298–1429 (560±35 BP, Hv–3081), when the mire surface became dominated by pool development, indicating a 'phase-shift' to much wetter surface conditions (Barber 1981, 191). At the same time, the palynological record appears to indicate the last extensive regeneration of woodland recorded from the environs of the moss, with Gramineae values declining to pre-Roman levels. This, together with an apparent abandonment of cereal and hemp cultivation, suggests that climatic conditions did not favour human settlement in the area at this time, although it has been speculated that political and epidemiological events (such as Scots border raids and plague) may also have played a part in discouraging agricultural activity (*op cit*, 116–17). Across the border at Burnfoothill Moss, Tipping (1995a) also registers a shift to wetter mire surface conditions, based on a multi-proxy array of mire wetness indicators. Although absolute dating for this level of the stratigraphy is unavailable, it has been estimated that the shift may have occurred slightly later than at Bolton Fell Moss, perhaps at *c* cal AD 1550 (*c* 400 BP). However, it is difficult to be precise about the timing of the climatic conditions thought to be responsible for these features of mire stratigraphy. This might be due to factors such as possible lags in response time peculiar to the individual hydromorphological systems governing particular mires, and the often non-synchronous nature of the responses of different proxy indicators employed to detect such shifts.

Overall, it is probably best to view the available evidence in broad terms, as reflecting the effects of a climatic deterioration which began to make itself felt seriously in North Cumbria during the fifteenth and sixteenth centuries AD — the first part of the 'Little Ice Age' (*sensu* Grove 1988). Despite the problems which a wetter, colder climate must have inevitably introduced, and the apparent contradiction with the early medieval evidence, Barber (1981, 118–9) believes that the palynological signals from Bolton Fell Moss indicate that after the initial decline in agricultural activity at the beginning of this phase, an increase in woodland clearance around the mire soon occurred, together with a return to cultivation of cereals and also hemp. It is, perhaps, significant that such an upturn in cultivation also coincides with the development of more settled conditions along the Anglo-Scottish border from the seventeeth century onwards (Jones 1969).

It is possible that one spectacular consequence of the latter stages of the climatic deterioration in North Cumbria was a bog burst recorded from Solway Moss in the early winter of 1771 (McEwan and Withers 1989). The Solway 'Irruption', as it was described at the time by Thomas Pennant (1774), is just one of a series of bog bursts in North Western raised mires which are clustered around the centuries generally thought to relate to the Little Ice Age. Its immediate precursors included Chat Moss, Greater Manchester, in 1526 (Leland, quoted in Crofton 1902; Hall *et al* 1995), White Moss, also Greater Manchester, in 1633 (*ibid*), and Pilling Moss, North Lancashire, in 1744/5 (Sobee 1953, 153). Other famous bog bursts in the British Isles also cluster around this period, notably those of Charlesville near Limerick, Eire, in 1691, and Flanders Moss near Stirling in 1629 (McIntire 1941).

The mires would probably have been reaching their greatest extent at this period, just prior to the commencement of large-scale drainage and reclamation operations. Increases in precipitation, which characterised this era, are more likely than at any other time in the historic period to have tipped the hydrological balance of certain systems into instability, causing these spectacular catastrophic failures of the mire structure. This certainly seems to have been the case at Solway Moss, as it is known that a major regional storm affected southern Scotland and northern England on the 16–17 November 1771 (Pennant 1774). It seems that this extreme weather event may have caused the moss to pass a threshold of hydrological stability, causing a mass movement of liquified peat out of the north–eastern sector of the mire — an area, perhaps significantly, where early peat-cutting had been concentrated. The ensuing peat flow resulted in the temporary loss of *c* 500 ha of farmland and the death of some cattle (McEwen and Withers *op cit*).

Nevertheless, within a few years the land had been reclaimed from the ravages of the flow by a process known as 'hushing'. This involved the construction of conduits from artificial reservoirs which led into the River Esk. The powerful artificial streams thus created were used as a kind of liquid conveyor belt into which the dried crust of bog-burst debris was dumped using an army of labourers (McIntire 1941). From these small beginnings, when the moss had the capability to disrupt the infrastructure of its surroundings, not only would the former fields be recovered, but the moss itself would eventually begin to be reclaimed for agriculture, until eventually the advent of peat harvesting in the twentieth century came to threaten the mire's actual existence.

Hints of the later agricultural history around Bolton Fell Moss are also present in the palynological record. Further clearing of scrub occurred, probably in a period relating to the late eighteenth and early nineteenth centuries, and the first plantations of windbreaks in the newly enclosed landscape of the period are discernible in the rise of pollen from pine

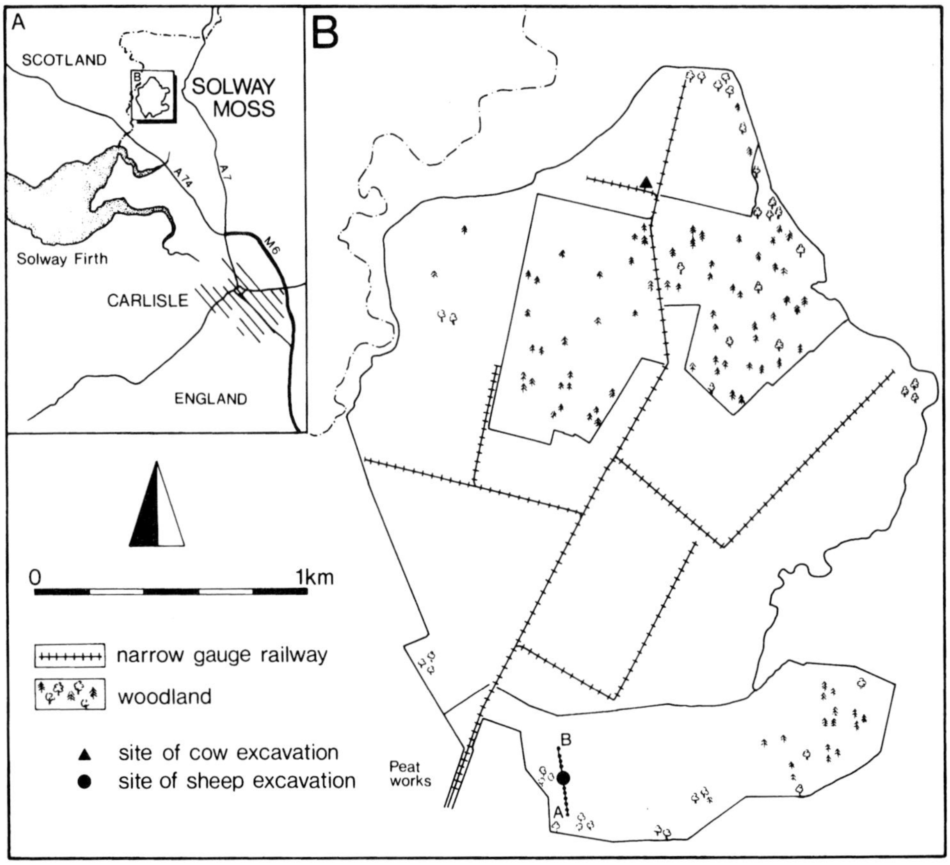

Fig 54 Solway Cow: location of cow and sheep finds

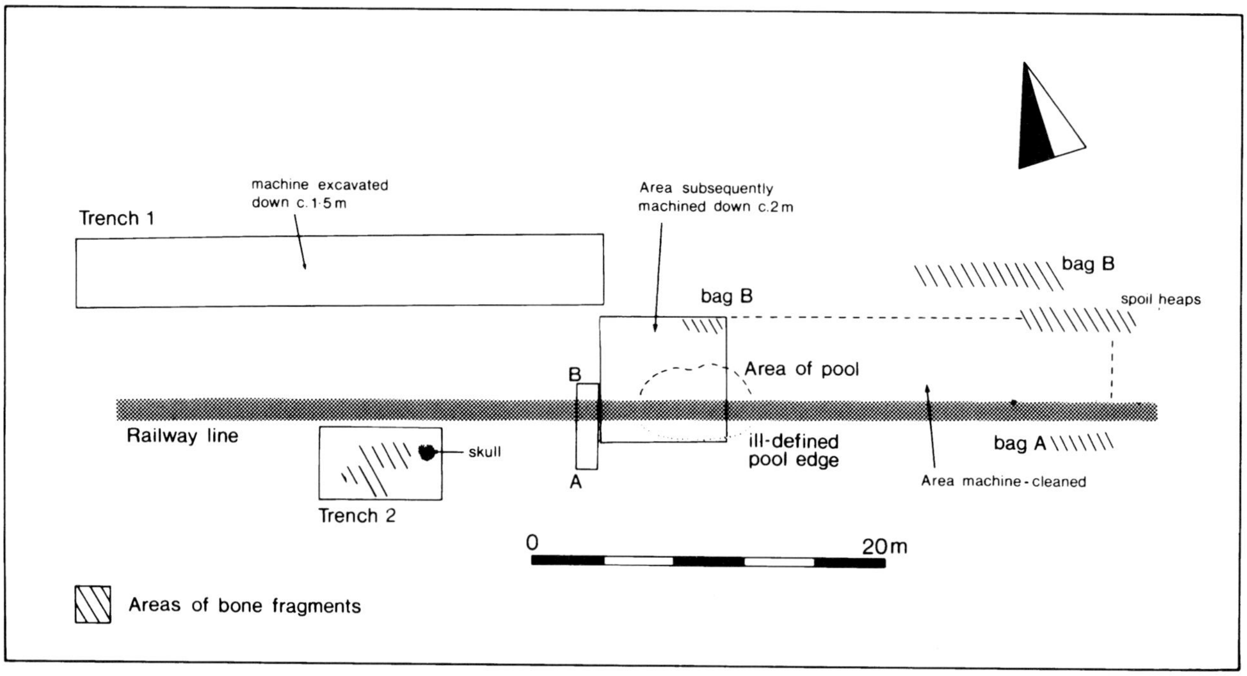

Fig 55 Solway Moss 1991: excavation plan of area around the cow finds

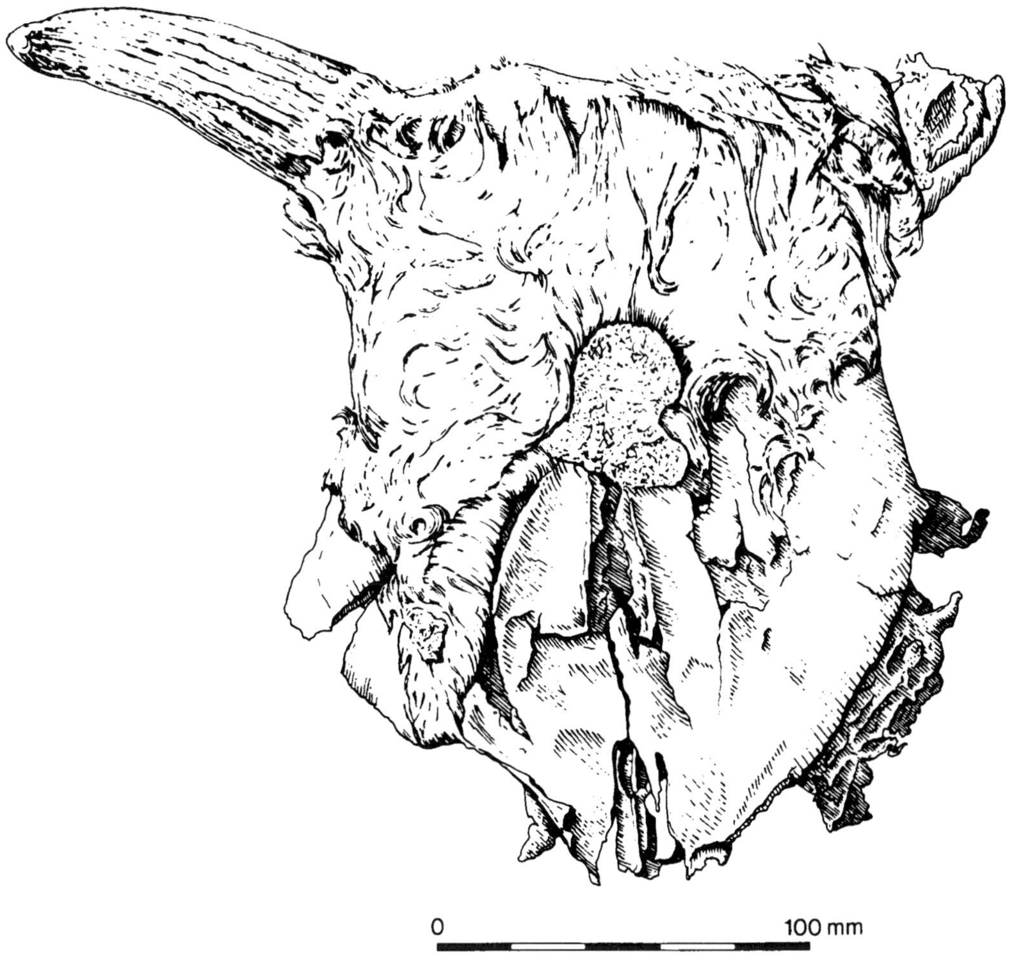

Fig 56 Solway Cow 1991: skull

(*Pinus*), elm (*Ulmus*), beech (*Fagus*), ash (*Fraxinus*), and spruce (*Larex*) (Barber 1981, 119).

Archaeological evidence

The archaeological evidence for the early medieval and later use of the mosslands has two elements: excavation data from two investigations in Solway Moss, and information from documentary sources which has been augmented by a programme of work undertaken by the University of Bournemouth at Glasson Moss. This is reflected in this section of the chapter.

Solway Cow

The discovery and excavation of the remains of the Solway Cow (SMR 16788) took place in July 1991, during routine commercial peat extraction (Figs 54–5). A baulk of peat was being levelled when the machine driver noticed a cattle horn protruding out of the levelled surface, which proved to be part of a substantial portion of a cattle skull that still had hornsheaths covering its horncores, as well as considerable amounts of skin and hair on the frontal regions between the horns. The partial skull was lifted and kept to one side, the findspot marked, and a halt called to the levelling operations by the staff of the peat works who then contacted NWWS staff.

Although the partial skull was comparable in size to that of a calf of modern 'improved' breeds, the discoverers noted the presence of several growth rings on the hornsheaths indicating that it derived from a mature adult animal. It was, in fact, comparable in size to that of 'Celtic shorthorn' cattle, the small (waist high) domestic cattle that were ubiquitous throughout Britain and continental Europe from the Iron Age to the medieval period, and which were particularly characteristic of the Iron Age. Figure 56 illustrates the original find of the skull, without its hornsheaths refitted onto the horncores. The position of this find, at a depth of 1 m from a surface that had not recently been cut, suggested that the skull might have been of prehistoric date. A large, peat-filled dent in the centre of the frontal part of the skull suggests that someone pole-axed the animal immediately before it died. It was dubbed the Solway Cow in deference to the remains of 'Lindow Man' (Stead *et al* 1986).

Plate 38 Solway Cow 1991: the finds in situ

Given the high potential of peat deposits to preserve organic tissues and the possibility that this partial skull indicated the presence of a complete cattle carcass buried in the peat, a small-scale excavation programme was initiated, funded by English Heritage separately from the main survey programme. The project aimed to retrieve the complete carcass *en bloc* to enable detailed laboratory excavation and analyses to take place. The several aims for this programme were:

1 to understand the location of the finds in the peat stratigraphy;
2 to understand the environmental context within which deposition occurred;
3 to identify the extent of the remains and recover them for detailed laboratory analysis;
4 to practice a quick-alert response for potential finds of human bog bodies;
5 to facilitate detailed scientific analyses of soft tissues of 'Celtic shorthorn' cattle, which are well known from their bones, but almost unknown for their soft tissues, hair colour and type, diet, and potential for DNA studies;
6 to test the relative dates of the find (the carcass) and its enclosing sediments (the peat) using detailed radiocarbon dating;
7 to test the comparability of radiocarbon measurements taken from different body tissues of a single individual (in this case, bone, hoof, and skin).

Plate 39 Site of excavation of the Solway Cow

Fig 57 Solway Cow: schematic outlines of the two individuals thought to be present, with the recovered material highlighted

The cattle remains are archived in the Centre for Archaeology, Portsmouth.

Diligent searching of all areas of peat recently disturbed by the levelling machine was undertaken in conjunction with hand- and machine-excavation of areas situated in the immediate vicinity of any findspots of cattle remains (whether these were *in situ* in undisturbed peat, or in disturbed deposits (Pls 38 and 39)). Details are given in Middleton (1992b). In the undisturbed peat, where the horn had been noticed protruding from the levelled surface, further finds of portions of the same skull were made, together with one metatarsal (foot) bone. Finds tended to be recovered from concentrations, and the presence of more horncores indicated that at least two individuals were represented. Their similar states of preservation suggested that they might have derived from a single deposit, and analyses of how the levelling machine operated confirmed that their spread was consistent with this hypothesis. Further work by trenching across the peats led to the conclusion that the original findspot of the skull of the Solway Cow was not, in fact, *in situ*, but was, itself, material that had been disturbed in 1976 during earlier peat extraction activities. Although some further cattle remains were recovered, it was clear that a complete carcass did not survive *in situ* in the peat.

The concentrations of cattle finds centred on an area that was well known to the current peat-cutters because of its exceptional wetness. A trench cut around the wet area to a depth of 0.5 m showed it to consist of relatively unhumified *Sphagnum cuspidatum* peat bounded by a ring of oxidised *Eriophorum* peat, representing an ancient pool on the mire surface, and, although other areas of *Sphagnum* were present around and beneath this pool, none were either as large or as well defined. Given their distribution (after disturbance), the cattle remains are considered to have derived from this wet area, although this cannot be proved.

The skeletal material derived entirely from the heads of two mature individuals, plus various foot bones that very probably derive from either or both of these two animals. Figure 57 shows the outlines of two individuals, with a schematic representation of which parts of the body are represented skeletally. Since there are fewer foot bones than would be expected from two four-footed individuals, it is not possible to state categorically that two individuals are represented by feet, but there are definitely two skulls, both horned.

One of the skulls, probably that of the original find, is very probably from the same individual as a

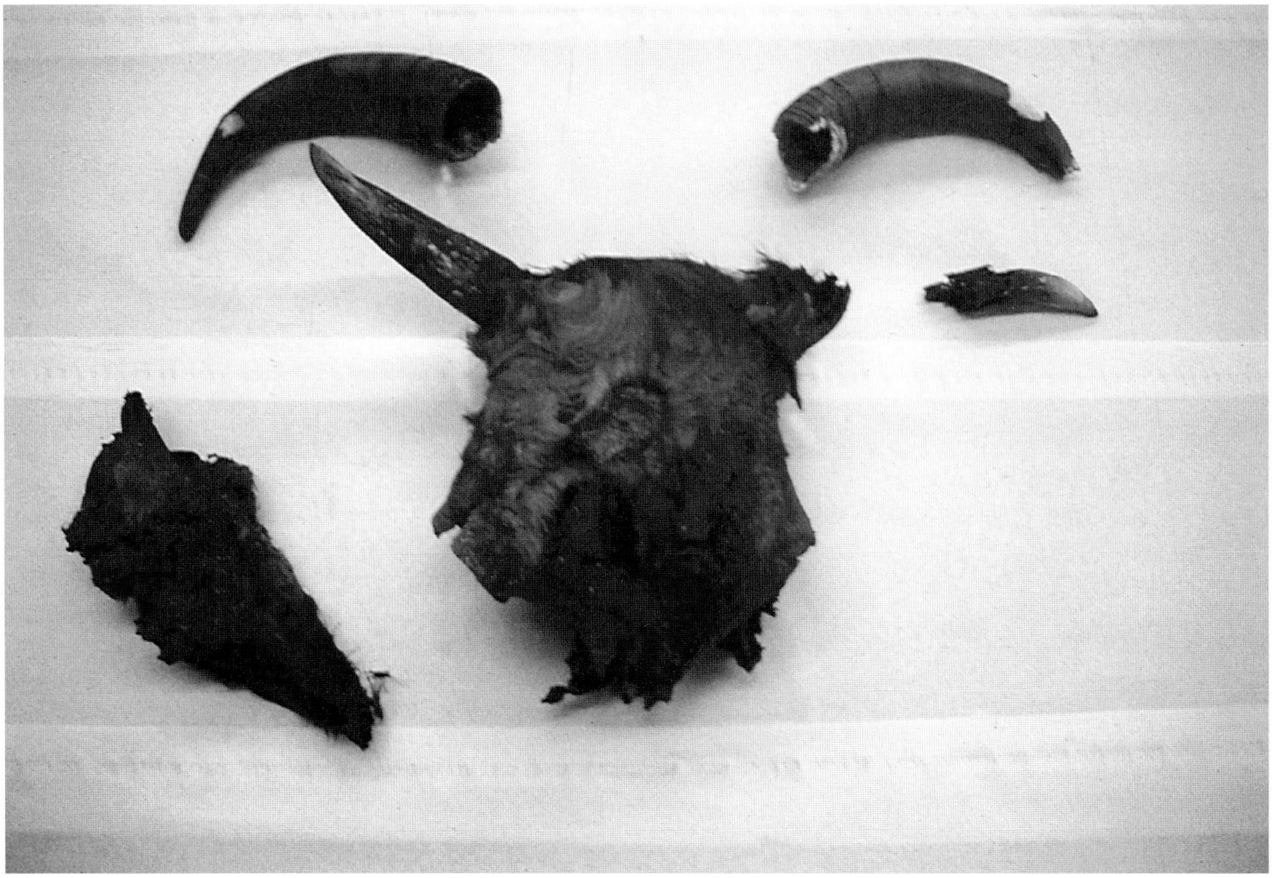

Plate 40 Solway Cow: skull and fragments

mandible and a hyoid bone (which is situated in the throat), suggesting that it was deposited in the bog as a complete head. This is the skull that appears to have been pole-axed (Pl 40). Although neither skull can be completely refitted from the fragments recovered, it is clear that both had short horns (of similar sizes and shapes), and that they were complete from the back of the skull right down to the muzzle. Due to the lack of length measurements (required for comparisons with breadth measurements), it is not possible to suggest whether the skulls derive from males or females. The shapes of the horncores are unlike those of bulls, but could derive from either castrate males (sometimes referred to as 'oxen') or females: the texture of the horncores, the stages of tooth eruption and wear, the presence of minor periosteal alterations of the bone alongside the tooth sockets (indicative of low grade gum disease), and the numbers of rings on the hornsheaths of the Solway Cow skull, all indicate that both of these cattle were mature adults, probably in the region of eight to ten years when they died.

One of the upper jawbones is congenitally missing the first adult premolar (P2), although this tooth is represented by a socket from which the tooth has fallen out, post-mortem, in another upper jaw and in the mandible (lower jaw). A relatively high incidence of this trait is thought to reflect a restricted gene pool, and the trait occurs in a significant minority of Romano-British cattle in this region (*see*, for example, Stallibrass' (1993) analysis of cattle from Carlisle). The incidence is much lower in medieval cattle, but records for prehistoric and early medieval cattle are too few to be useful in comparisons.

In addition to the head bones and the hornsheaths belonging to the Solway Cow, there are several fragments of hooves (the horn-like covering of the lowermost toe bones), and various pieces of preserved soft tissue thought to be skin or some form of partially decomposed body tissue. Several of the skull and jaw fragments are partially covered with skin and hair. Most of the hair is a gingery colour, but around the muzzle area of the maxillae (upper jaw bones) and the mandible (lower jaw bone), the hair is much darker brown in colour. The hair on the skull of the Solway Cow, between the horncores, is quite wavy (like a fringe). The survival of keratinous material (the 'finger-nail'-like hornsheaths and hoof fragments, and the hair) is restricted to particular burial conditions, such as peat bogs and arid environments, and the recovery of this type of material is rare in Britain, for any species.

All of the non-head bones derive from cattle feet: metapodials from the main part of the foot; toe bones (including the bones that are surrounded by the hooves during life); and some very small bones (the sesamoids) that fit behind the toes. The epiphyses (ends) of the foot bones are all fused, indicating (an) adult or sub-adult age(s) at death, and the asymmetrical splaying demonstrated by the distal metacarpal has been linked in a study of modern individuals with weight, which can be mimicked by heavy work during life (Bartosiewicz *et al* 1993). This may add support for the interpretation that these cattle were mature when they died (or were killed) and were not in their prime for providing tender meat.

When the bones were catalogued, no cutmarks or toothmarks were observed on any of the exposed surfaces of the bones, although some areas were obscured by coverings of peat or soft tissue. The bones do not give the appearance of having been butchered for meat in any way, nor do they appear to have been within reach of any scavenging dogs that might have found the presence of soft tissue and bone rather appetising. Modern studies of scavenging canids (*eg* Stallibrass 1986) have demonstrated that they usually attack the muzzle and anal region of a complete carcass first, yet none of the muzzle bones (*ie* premaxillae, maxillae and symphysis region of the mandible) show any signs of attack.

The *Sphagnum* peat infilling the putative pole-axing hole in the centre of the skull of the Solway Cow was left *in situ*. The edges of this hole appear to be ancient rather than modern, and its position and size are certainly consistent with pole-axe damage. Given the absence of any substantial timbers or hard objects such as stones in the vicinity of the cattle remains (which might have caused similar damage by accident or design close to the time of death, or during post-mortem disturbance), it seems quite probable that this animal was stunned or even killed by pole-axing. Radiography might be able to confirm this interpretation, but no further archaeozoological analysis of the bones has been undertaken. Pole-axing is often used to stun an animal before it is killed by having its throat cut.

Apart from some unworked pieces of roundwood, which may be a 'natural' component of the peat, there were no cultural items associated with the cattle remains (*ie* there were no items of worked wood or leather, no ceramic, glass, or stone artefacts, and no stone manuports). Although the states of preservation varied between bones and fragments (and, indeed, even within a single fragment in some cases), this distribution of head and foot bones cannot be explained by the relative robusticity of different skeletal elements. The bones of the elbow joint, the sockets of the shoulder-blade and hip, the lower portion of the hind leg and the small bones of the wrist and ankle are all as dense as those recovered (Lyman 1984), and singularly more robust than skull bones, and yet none were found. Although the wrist and ankle bones are small, they are not as small as the sesamoids, some (although not all) of which were found, and the other robust bones are quite large and unlikely to be overlooked. Differential preservation and / or recovery, therefore, cannot be cited to explain the skeletal distribution represented by the material recovered. Neither can disturbance by peat-cutting machines be regarded as a very credible explanation: to decapitate a complete carcass so neatly that nothing between the head and feet of one individual is retrievable (despite the diligent searching) is unlikely. To decapitate two such carcasses seems infinitely improbable.

In summary, the vertebrate remains recovered from these investigations comprise the partially skeletalised heads of two cattle, including horn sheaths, skin and hair, similar remains from the feet of one, possibly two cattle, and some unattached soft tissue, at least some of which is cattle hair, and, probably, more skin. If these finds are a good indication of what was present in the peat prior to disturbance in the late twentieth century by modern peat-cutting equipment, they appear to indicate that the head and feet of one, and at least the head (and quite probably the feet as well) of a second individual were deposited in or beside a pool in the bog. Given the presence of skin and hair as well as the skeletal material, the heads and feet may well have still been attached to each other by the hides of the animals, both of which were mature, probably working, cattle.

The presence of damage to the skull of at least one of the individuals, suggestive of pole-axing, and the lack of obvious explanations for the body part representation that do not require some sort of selective deposition, suggest a degree of human involvement in the deaths and / or deposition of the cattle remains. These do not appear to have been cattle who wandered into the bog and died accidentally. The lack of any 'material culture' remains and the location of the deposit raise the possibility that the cattle had been involved in some non-utilitarian activities.

The survival of keratinous material provided a rare opportunity to study the hair structure and colour of 'Celtic shorthorn' cattle which were, at the time of their discovery, expected to be of Iron Age or Romano-British date.

Given the darker colour around the muzzle area of several fragments, and the overall ginger colour of the remains, it was suggested that the animal may have been white with dark 'points' (*ie* white with dark or red ears and muzzle, a combination sometimes referred to in Celtic legends and mythology as having particular significance (Ross 1974, 389)). Macroscopically, it is not possible to tell whether a ginger colouration is due to a natural (life-like) colour, or to post-burial staining by tannins in the peat.

Ryder (1994a) studied samples of hair taken from the skull of the Solway Cow for their fibre type and their colouration, and the following section is based upon his report. Three tufts of hair were taken as samples: one from the top of the skull (which was ginger in colour), and two from a maxilla (one coloured ginger, the other dark brown-black). The diameters of 100 hairs in each tuft were measured, together with the percentage of hairs with a central medulla (a cavity that is found in coarser hairs), and the percentage of hairs with natural pigment. The density of pigmented granules, where found, was also recorded.

Ryder (*op cit*) has identified two types of cattle coat on the basis of the distributions of hair diameters: 'primitive', which has a greater proportion of finer to coarser hairs, and which persisted until after the medieval period, and 'modern' which has a greater proportion of coarser hairs, and which has been identified from the Bronze Age onwards. The Solway Cow samples indicate a 'modern' type of hair. The actual hair diameters, averaging about 40–50 µm, are typical of mature rather than younger cattle, agreeing with the interpretation based on the skeletal material. The hair from the top of the head was wavy, and may indicate that the whole of the coat was shaggy, although it could just indicate a 'fringe'. The hair on the muzzle was straight, but head hair is often different to that on the rest of the body. The presence of tapering tips of hairs indicated new growth but, since the head hair tends to be replaced continually (unlike the body hair), this cannot be taken to indicate a season of death.

The samples contained some debris including soft tissue material, some of which could be identified as muscle tissue although none was positively identified as skin. Relatively few of the hairs had a medulla, and most contained pigment of a brown, tending towards tan, colour. The dark brown/black sample from the muzzle had a greater proportion of densely pigmented hairs; there were no white hairs. This pattern of pigmentation rules out the possibility that this animal was white with dark points; rather, it was a reddish-brown colour with a dark muzzle. Ryder (1994a) noted that four of the hairs in the sample from the top of the head had a more reddish pigmentation, and suggested that the animal might have had a decorated red 'topknot' when it died.

Consequently, duplicate samples of hair from the top of the head and the dark brown part of the muzzle were submitted for tests for organic dye, tannins, and iron compounds, using absorption spectrometry, thin-layer chromatography, and chemical tests. The following section is based on the report by Walton Rogers (1994) of Textile Research in Archaeology.

The tests for organic colorants (with or without mordants) failed to detect the presence of any red colorant in either sample, but did suggest the presence of a yellow or brown substance in both samples. This appeared to be in a greater concentration in the dark-brown muzzle sample. Thin-layer chromatography systems designed to characterise yellow colorants and tannins (Walton Rogers and Taylor 1991) detected pyrogallol, a common tannin, in both samples, but no other colorant was detected. Since iron compounds such as iron-rich clays (including ochre) have been used for millennia to obtain an inorganic red colorant, the samples were also tested for the presence of iron, in conjunction with control samples of modern wool. These chemical tests are highly sensitive, but failed to detect any iron or iron compound in the samples.

It should be stressed that the original four hairs detected by Ryder as having unusual pigmentation have not been tested. Walton Rogers (*op cit*) suggested that the animal's coat may have been rather patchy in the way that it took up staining from tannins from the peat, and that the red colour noted by Ryder may have been the more natural colour of the coat, in an area less affected by overstaining after burial.

Given the disturbed nature of the finds, it was not possible to test the contemporaneity of the 'carcass' with its enclosing peat deposits, but radiocarbon determinations were sought on three samples from the cattle remains themselves, both to test the consistency of radiocarbon dates taken on different tissues of a single individual, and to ascertain whether or not the expected date of Iron Age/Roman Period was correct. A sample of hoof material (OxA–3640), a toe bone (OxA–3641), and some soft tissue, thought to be skin, that immediately overlay the base of the skull of the Solway Cow (OxA–3642), were submitted to the University of Oxford Radiocarbon Accelerator Unit. There is marked similarity amongst the measurements (Table 5) and the three results are statistically consistent ($T' = 0.1$; $v = 2$; $T'(5\%) = 6.0$; Ward and Wilson 1978); the calibrated range of the weighted mean is cal AD 684–947 (1203±43 BP, OxA–3640 + OxA–3641 + OxA–3642).

These dates indicate that the remains of at least one of the individuals (represented by the skull of the Solway Cow), if not both, were deposited in the early medieval period, and may suggest that the pool in

Lab Ref	Material	Catalogue No	$\sigma^{13}C$ (%)	Radiocarbon age BP	Calibrated date range (95% confidence)
OxA–3640	hoof	S1	–22.9%	1185±70	cal AD 670–1056
OxA–3641	toe bone	S10	–22.4%	1210±80	cal AD 660–1056
OxA–3642	soft tissue ?skin adhering to bone	S16	–21.2%	1220±75	cal AD 660–980
*(OxA–3640–3642)				1204±43	cal AD 684–947

**Weighted mean (T′ = 0.1; v = 1 T′ (5%) = 3.8; Ward and Wilson 1978*

Table 5 Radiocarbon dates for samples from three different types of body tissues taken from the Solway cattle remains

the bog existed at that time. During this period, the region was under some Scandinavian influence, but the degree to which Celtic/Britonic, Roman, Northumbrian and Scandinavian cultures and settlers interrelated is poorly understood. Thus, any interpretation of these cattle remains must be based on what is known of them with relative certainty or, at least, must make use of circumstantial evidence that has some compatibility with the facts.

Accidental deaths have been ruled out because of the selective presence of restricted parts of the carcasses and the apparent pole-axing of at least one of the individuals. The material is unlikely to represent rubbish disposal, given its location 300 m into a bog (*ie* well beyond the usual 'toss-zone' of sites bordering wet ground) and the apparent association of hide with the head and foot bones. Although heads and feet can comprise primary butchery waste, hides are usually too valuable to be thrown away. Even if these hides were judged to be of poor quality or unwanted, there is still the question of why they might have been transported into a bog and left in or beside a pool of open water. A non-utilitarian explanation appears to be the most likely.

Discussions of how the cattle remains might relate to religious beliefs and practices have already been presented by Middleton (1992b) and Stallibrass (1996). The following section draws on both of these and adds further suggestions and references.

Ironically, any attempt to ascertain whether these cattle remains indicate 'Celtic survival' or 'Scandinavian influence' with regard to the people presumed to be responsible for their deposition is hampered by the fact that cattle and watery places were held to be of great significance by both cultural communities (Davidson 1988). Roman influence is more likely to have adjusted to local beliefs rather than to have overprinted them.

Piggott (1962) conducted a detailed review of finds of 'head and hoof' burials in Eurasia, in which a human body appeared to have been wrapped in a hide (often of cattle, but also of other species such as horse or bear) although, in most instances, the nature of the burial conditions ruled out the survival of the hide itself. On Solway Moss, there is no indication whatever of any human skeletal remains amongst the finds (although Piggott (*op cit*) did find occasional instances in which burials were made adjacent to, rather than wrapped up in, the hides). Nor were the burials typically associated with wet places.

Alternatively, offerings of a variety of items (of organic or inorganic materials) into wet places were common in both Celtic and Scandinavian cultures, where a breach into the underworld through natural features such as caves and open water (*ie* pools, lakes, and rivers) or through man-made shafts including pits and wells, could be regarded as a means of contacting chthonic forces. The reasons for making contact were multitudinous, including gifts, sacrifices, attempts to elicit good harvests or knowledge. Boyer (1991, 289–90) emphasises the importance placed upon wet places in Odinism, and upon sacrifice and open-air ritual in Germano-Norse paganism. Coles and Coles (1989) draw attention to the large number of bog finds of all types from Denmark and Germany dating to the fifth and sixth centuries, in a period preceding the impact of Scandinavian culture in Cumbria (Higham 1986).

Davidson (1988) mentions early Irish, Welsh, Scottish, and Icelandic records of people wrapping themselves in cattle hides and lying in remote places in order to receive foreknowledge. What happened to these hides afterwards is not mentioned, but she also makes reference to the practice of placing the hide (with or without the head and feet) of a sacrificed animal on a pole outside the house in which the sacrifice took place (the rest of the animal being eaten in a feast).

Placing hides on poles is also mentioned by Piggott (1962) and the Larousse *Encyclopaedia of Mythology* (1959) in various contexts connected with ritual beliefs across Eurasia up until the last century. On a slightly more practical, though probably also superstitious note, Henderson (1911) refers to a post-medieval Scottish practice of disposing of hooves, and sometimes the head, of a diseased animal in an out-of-the-way place to prevent further infection of livestock. The reasoning behind this seems to have had more to do with making an offering than with any genuine understanding of hygienic disposal of noxious organic waste.

Summary

The apparent association of the remains of two cattle, which died during the early medieval period, with an area of open water some distance from the edge of a large lowland raised bog does appear to reflect some ritual significance. However, it is not possible to ascertain precisely what this significance was.

The material, consisting of heads, hide and feet, may well represent the remains of sacrificed animals, although the cattle concerned were by no means prime meat animals, and were probably mature beasts that had fulfilled various utilitarian functions during their lives (such as providing milk, traction, offspring, and dung). The animals were probably a reddish brown (tan) colour in life, and at least one of them probably had darker points, at least around the muzzle. The bringing of their remains to the pool in the bog may indicate a desire to contact chthonic forces, and they could have been used in divination rituals.

The lack of any indications of chewing by dogs may suggest that the hides were thrown directly into standing water, or suspended above it on poles until they dropped in. Alternatively, they may have been beyond the reach of scavengers simply by being brought so far out into the bog.

Whether the remains of the two individuals were deposited simultaneously or sequentially cannot be ascertained. Either seems possible, given that 'special places' often retain their significance through folk memory for considerable periods of time.

Solway Sheep

The discovery and excavation of the remains of the 'Solway Sheep' (SMR 16789) took place in June 1992, under similar circumstances to those of the Solway Cow the previous year, excepting that the sheep remains were found towards the southern end of the moss (Fig 54 p 122). This report draws upon the interim reports by Middleton (1992b) and Wells (1992) on the excavation and site stratigraphy, respectively,

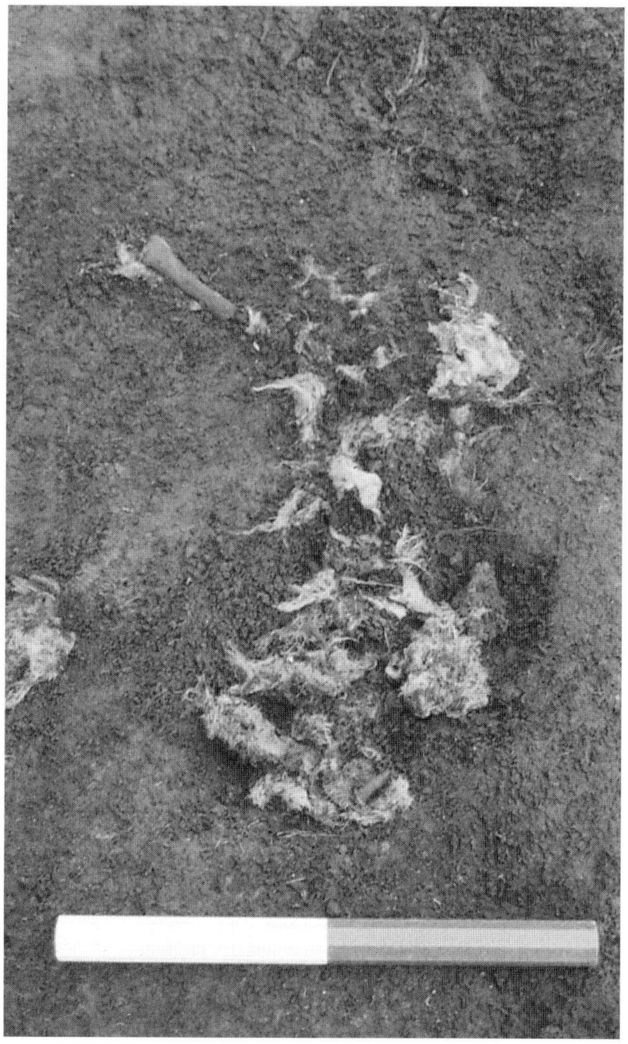

Plate 41 Solway Sheep in situ

and on a detailed pollen report by Huckerby and Wells (1993). The original description of the remains of the sheep (Stallibrass 1993) is reproduced here. Since then, an unpublished detailed report on samples of fleece has been produced by Ryder (1994b) which is summarised below. The aims of the studies were similar to those for the Solway Cow. Previous survey on the moss edge near the findspot had suggested that the peats in which the remains were contained were of prehistoric date; there was thus an initial expectation that the faunal remains would be contemporary.

The initial finds, exposed by a peat extraction machine, included considerable quantities of fleece and a few bones, including ribs and leg bones (Pl 41). These finds were spread around some material that remained *in situ*. During a two-day excavation programme, the surface of the surrounding area and the adjacent spoil heaps of the peat extraction machine were searched, and a block of peat centred on the *in situ* remains was lifted (Pl 42). This block was sent to the Department of Archaeology at the University of

Durham, where it was excavated by Jenny Jones (conservator).

No cultural material was found during the excavation, nor were any 'unnatural' features observed. The peat appears to have remained undisturbed until the extraction machine cut away the overlying peat and revealed the sheep remains.

The location of the findspot was approximately 300 m from the edge of the moss, in peats dominated by *Sphagnum* sect *Cuspidata*, which indicates a wet mire surface with numerous pools. A stratigraphic transect was taken from the southern border of the moss, traversing the position of the sheep and continuing right across to the northern edge of the peat, and a core was obtained immediately adjacent to the *in situ* sheep remains. Detailed palaeoecological analyses were carried out on this core, and the results are presented in *Appendix* 6.

The general stratigraphy of the moss was revealed by the transect, which indicates that beneath the layer of *S* sect *Cuspidata* peat are layers of *Eriophorum*-dominated peats, sometimes overlying *Polytrichum*-rich peats which, in turn, overlie woody peats above the natural sands and silty clays. Carbonised plant material occurred regularly through the profiles, possibly reflecting human interference with, or control of, the natural vegetation, and a patch of what appears to be *in situ* carbonised remains of a *Calluna* plant was revealed during the field excavation of the sheep remains. The macrofossil and pollen evidence from the palaeoecological core provide more detail and dating for the peat adjacent to the sheep remains. Radiocarbon dates were provided by conventional facilities at the SURRC laboratory (*see below, Appendix 8*).

The woody peat began to accumulate shortly after *c* 6500–6000 cal BC (*c* 7700–7200 BP), when fen carr became established. Fire disturbance appears to have been continuous from the beginning of this phase until it reached a peak at *c* 5000 cal BC (*c* 6100 BP) when much of the woodland component of the peat disappeared. It was replaced by *Eriophorum* and *Calluna*-dominated mire vegetation. Traces of fire persist, but at less-frequent intervals throughout the core, only disappearing in the topmost levels. After *c* 4500–4000 cal BC (*c* 5700–5200 BP), *Sphagnum* sect *Acutifolia* became a prominent component of the *Eriophorum*-dominated mire, which continued until a brief interlude dated to *c* 3300–2700 cal BC (*c* 4500–4150 BP) when a pool dominated by *S* sect *Cuspidata* formed. Later, there was an abrupt change from the *Eriophorum*-dominated mire to one characterised by an abundance of *S imbricatum* in the topmost 50 mm of the core.

Plate 42 Solway Sheep during lifting

On the basis of these palaeoenvironmental studies, it appears that the area was fen carr until it was, at least partially, cleared during the Mesolithic period. It then became a lowland bog, gradually becoming wetter although still subject to clearance by fire from time to time. These studies have also provided evidence of early cereal cultivation (cereal-type pollen is recorded) in the early Neolithic, immediately prior to 3340–2707 cal BC (4370±90 BP, GU–5314). The sheep remains were associated with an horizon of *S* sect *Cuspidata*, indicative of a wet mire with pools, and dated to the Neolithic / early Bronze Age.

The remains of the sheep comprise fleece and a few immature bones, mostly coming from the disturbed peat. The block removed for excavation under laboratory conditions retained considerable quantities of fleece, but relatively few bones: there is no need to postulate the presence of more than one individual. The preservation condition of the bones varies considerably but, unlike the skeletal remains of the Solway Cow, various parts of the body are represented, including ribs, upper limb bones from both the front and the hind limb (humerus and femur), and a lower limb bone from the front limb (radius). The unfused epiphyses (ends) of the femurs indicates that the animal was still young, and the unfused tuberosity at the distal end of the humerus may indicate that the animal had not quite reached one year of age.

The incompleteness of the skeleton can be explained by factors relating to preservation and recovery; the peat extraction machine may have removed or destroyed some of the bones beyond detection and recovery. Small bones, in particular, would have been difficult to recover, and the variable preservation states of the collected material indicate that some bones may have decomposed *in situ*. Immature bones (which are incompletely mineralised, and which have a more open structure than mature bones) are particularly susceptible to diagenesis.

The overall size and shape of the bones indicate that the animal was not a primitive, prehistoric type. Although some increases in sizes and weights took place in the Roman period, the Solway Sheep remains most closely resemble relatively recent livestock of the post-medieval or modern periods. Given the juvenile nature of the bones, which had not completed their growth, together with their state of preservation, it is not possible to estimate the absolute size of the animal.

The fleece is of an off-white to gingery colour, but this may be due to staining by tannins from the peat. Early sheep tended to be coloured, but white sheep are known from the Roman period onwards, and became common in the post-medieval period.

During the excavation in the laboratory, two short (*c* 300 mm) cores of peat were taken immediately adjacent to (*ie* within 50 mm of) the *in situ* fleece, one from the east side and one from the west. They were taken very approximately from the same level and were roughly 0.5 m apart spatially: these were used for radiocarbon dating of the enclosing deposits. In addition, one of the ribs and some fleece were submitted for dating of the sheep itself. All samples were submitted to the Oxford University Radiocarbon Accelerator Unit. The results are shown in Table 6. There is a marked similarity amongst the measurements on different body tissues (fleece and bone) and the two results are statistically consistent ($T' = 0.1$; $v = 1$; T' (5%) = 3.8; Ward and Wilson 1978).

The two measurements on the peat samples are, however, statistically different ($T' = 37.2$; $v = 1$; T' (5%) = 3.8; Ward and Wilson 1978). Given that they were removed from the block of peat after it had, itself, been removed from the ground, with no absolute heights above Ordnance Datum available for either sample, and that they may not have been placed at similar heights since they had to avoid the remains of the sheep, this apparent discrepancy may

Lab Ref	**Material**	**σ ^{13}C (%)**	**Radiocarbon age BP**	**Calibrated date range (95% confidence)**
OxA–4215	fleece	–24.7%	265±95	cal AD 1440–1955
OxA–4216	rib bone	–22.7%	230±75	cal AD 1470–1955
(OxA–4215–4216)*	sheep tissues		243±58	cal AD 1640–1955*
OxA–4213	peat west-side	–24.9%	3840±85	2569–2039 cal BC
OxA–4214	peat east-side	–26.3%	4595±90	3622–3040 cal BC

* *Weighted mean for sheep tissues (OxA–4215 and 4216) ($T' = 0.1$; $V = 1$ T' (5%) = 3.8; Ward and Wilson 1978)*

Table 6 Radiocarbon dates for samples from two different types of body tissues taken from the Solway sheep remains, and two samples of adjacent peat

be due to sampling error (*ie* different peat strata may have been sampled). Alternatively, the peat itself may have had a variable surface, particularly if it consisted of hummocks and hollows. This aspect of peat bog dating is not yet fully understood.

The peat dates taken immediately adjacent to the *in situ* sheep remains are in accord with the overall dates provided by the palaeoecological core for the formation of the *S* sect *Cuspidata* peat during the Neolithic / Bronze Age. The surprise provided by the AMS determinations is that the sheep remains, so closely associated with the Neolithic/Bronze Age peat, are firmly dated to the post-medieval period. The calibrated range of the weighted mean is cal AD 1470–1955 (243±58 BP, OxA–4215 + OxA–4216).

There are no obvious reasons to discount any of the radiocarbon dates. It would appear that either the sheep remains had sunk into earlier deposits, or that they were deposited in an open area within the surface of the peat, possibly filled with water, that only became infilled with peat at a later date. Since the profile had been truncated by the peat extraction machine that caused the remains to be discovered in the first place, there was no opportunity to date the peat overlying the sheep remains.

Middleton (1993d) suggested that the sheep may have fallen into old peat workings, but there were no obvious signs of weathered edges of any former cuts through the peat, and any hollow in the mire surface may have been a natural one between hummocks of *Sphagnum* or other vegetation. Traditional hand peat-cutting methods leave a trench into which turves from the vegetated surface of the mire are placed. In an ill-drained mire, this trench turns into a 'soup' of old peat which appears to lose its structure and stratigraphy, particularly in winter. The sheep may have strayed into one such cutting, becoming trapped at the bottom on the trench within the structureless peat, which subsequently dried out and became exposed during peat-cutting operations. The presence of such cuttings, being an obvious danger to livestock, would suggest that the moss itself was not grazed, although it is likely the cut-over and 'dry' edges were, and it is from this zone that the sheep strayed.

Unlike the remains of the Solway Cow, there were no indications of any human involvement with the deposition of the sheep carcass: there were no artefacts, no cultural features, no cutmarks, no mysteriously absent skeletal parts. It would appear that a young sheep fell into a hollow (possibly flooded) in a peat bog and died at some time in the last 500 years. There is no clear indication of season of death but, given the likelihood that it was born in springtime, as are most modern domestic sheep, and the possibility that it was just under one year old when it died, it is quite likely that the animal died in the winter time. This is a period when its body fat reserves would have been low, food supplies would have been meagre, and weather conditions may have been poor. It is a time when young, immature animals are particularly at risk.

Because of their organic, biodegradable composition, ancient fleece remains are rarely found. Fleece of any pre-modern date, therefore, is of interest in studies of changes through time associated with evolution, domestication and selective breeding.

During the post-medieval period, documentary, pictorial and osteoarchaeological data indicate that considerable changes were made to the types of sheep extant in Britain. Some differentiation may already have been made during the medieval period, between sheep populations bred in different regions or counties, but the fifteenth to nineteenth centuries saw major developments in response to needs and desires for new types. Some selective breeding programmes probably had specific end-products in mind, such as the provision of larger quantities of meat (for the developing urban centres), or greater quantities and differing types of fleece (for textile manufacture and trade). Other changes to livestock types may have been more accidental, or due to curiosity and experimentation during a period when scientific enquiry became fashionable. However, the data are sparse, and possibly are biased towards records that had most chance of survival and discovery. Details of when and where specific changes occurred are extremely rare, and there is little indication as to which were ephemeral and which were critical for later developments.

It would seem likely that the Solway Sheep was part of normal stock left to range on the open area adjacent to or on the moss. It seems far less likely to have belonged to highly valued experimental breeding stock, which might be expected to have been kept closer to home for better control over breeding and feeding. The Solway Sheep remains, therefore, provided a substantial amount of well-dated fleece from an apparently 'ordinary' animal from North West England, dating to the post-medieval period. An additional bonus is the fact that (unlike textiles) the fleece has not been treated in any way, which means that complete staples (locks) are available for study. Ryder (1994b) undertook a detailed analysis of the fleece type, and the following section is based on his work.

Two staple samples were taken from clumps of fleece that appeared to retain natural groupings of fibres (*ie* they were not tangled or teased apart), and a subsample of 100 fibres was taken from each. Staple lengths and fibre diameters were measured, and the percentages of fibres with a central medulla (indicative of coarser fibres) were recorded. Fibres with natural pigment were recorded as having a low, medium, or high density of pigment.

Firstly, the staples were confirmed as deriving from sheep not goat, since they lack the very fine underwool fibres seen in goats, but do have the lattice shape typical of sheep in the medullae of the hairy fibres.

The two subsamples of Solway fleece are very similar in terms of percentages of medullated fibres, but are very different in percentages of pigmented fibres. The density of the pigmentation also varies, and may indicate that the sheep was piebald rather than self-coloured.

The staple lengths are long: 180 mm and 255 mm, much longer than those that were typical of the primitive sheep that survived until the end of the medieval period (with staples in the region of 60 mm). Staple length varies over the body of the sheep (longest on the breeches and shortest on the shoulders), and modern fleece analysis takes samples from the mid-side in order to provide a standard 'average' position. Since the fleece had become detached from the carcass, it is not possible to suggest which part of the body these samples may have derived from, but they fall within the ranges seen in modern longwoolled breeds (which have a Medium-type fleece), and also in particularly long examples of Hairy-fleeced breeds, such as the Scottish blackface (of the type currently used to provide wool for carpet or tweed manufacture).

Several factors suggest that the Solway samples derive from a Medium- rather than a Hairy-type fleece. The percentage of medullated fibres is very low compared to those seen in Hairy fleeces; the ranges and means of the fibre diameters of the two subsamples are very similar, but the means themselves are unusually high for a Hairy type. Apart from a minor second peak, the distribution of the diameter measurements is virtually symmetrical. This is also unusual for a Hairy-type fleece, which tends to have a skewed distribution of fibre diameters. These traits are all far more similar to those seen in Medium-type fleeces. Given the length of the staples of the Solway sheep, this Medium-type fleece would be identified as a longwool. The shallow wave of the Solway staples accords with the first fleece from a long-woolled sheep (*ie* the fleece grown in the first year of life, prior to shearing).

Ryder (1994b) compared the results of his tests on the Solway fleece samples with data that he had previously obtained from yarns spun from late eighteenth century Longwools. These yarns were used to make textile samples for the Kendal Pattern Book, which is dated to about AD 1770 (Ryder 1990). The Kendal data show that late eighteenth century Longwool fleeces differed quite significantly from those of late twentieth century (*ie* modern) Longwool fleeces, but Ryder found considerable similarities between them and the Solway fleece. He concluded, therefore, that the Solway sheep was probably an early form of Longwool.

Historically, the Longwools are thought to have originated in the English Midlands, particularly the eastern counties of Leicestershire and Lincolnshire, whence they spread into the East Riding of Yorkshire, eastern County Durham, and Northumberland by the close of the eighteenth century: famous types include the Teeswater and Wensleydale Longwools. Solway Moss, on the western border of England and Scotland, is a considerable distance away from this 'core' area of their development. The earliest historical reference to the crossing of Cheviot sheep from the Border region with Leicester Longwools (to create the Border Leicester breed) is AD 1746 (Ryder 1983, 514).

Longwools, generally, were at their most popular during the eighteenth century and became relatively widespread. Although it is possible that the Solway Sheep could be an early (*ie* seventeenth century) example of a Longwool, found further north and west than the core area suggested by contemporary historical sources, it is more likely that it is an eighteenth century example, located within the zone that was 'colonised' when the type was becoming increasingly popular.

Summary

The most reasonable interpretation of the Solway sheep remains is that they are the remnants of the body of a young post-medieval animal that strayed onto the moss at some time in the last 500 years. It probably died of natural causes during its first winter, when weather and feeding conditions can be poor. The remains were associated with a particularly wet and hummocky part of the moss, and the young sheep could have fallen into an old peat cutting, pool, or hollow filled with peaty mud, water or even snow. The animal's fleece indicates that it was of a Longwool type that is known from yarn samples of the eighteenth century but which is no longer extant due to subsequent selective breeding programmes. The sheep may have been piebald in colour (unlike modern examples of Longwools which are usually all 'white', again due to selection by breeders).

Summary of the Solway Moss animal remains

The two finds, the Solway Cow and the Solway Sheep, were discovered within a year of each other, in the same moss, but the dates and interpretations of their remains are very different. Whilst there appears to be no satisfactory 'natural' explanation for the nature and location of the cattle remains, which are considered here to represent some form of ritual or superstitious deposit, the sheep remains appear to have become incorporated within the moss without any intervention from people. Both cases do, however, point to the place of the moss within the landscape. The cattle remains may suggest deposition in a prominent pool within the moss and away from everyday activities. The sheep may have died by straying into an old peat-cutting on the moss which would not have been part of the 'everyday' agricultural landscape. This, and the documentary evidence (*see below*) may suggest that the mosses were only exploited for peat on their margins and were not a core part of the settlement system.

The dates of the faunal remains are very different, both from each other (*ie* early medieval and post-medieval), and from the enclosing (prehistoric) peat deposits. These results have serious implications for the interpretation of materials found within peat bogs. It is clear that the nature of the peat formation (hummocky peat, well-stratified woody peat, etc), and the history of past land-use and moss exploitation, must be taken into consideration before any apparent association of peat and enclosed material is accepted. It would be best practice to test any such apparent associations (as here) by the application of separate dating tests to both the enclosing peat and the enclosed items.

Both sets of faunal remains provided radiocarbon measurements that are statistically consistent between the different types of organic tissue sampled (in the case of the cattle remains: skin, hair and bone; in the case of the sheep remains: hair and bone). These results are encouraging and permit greater confidence in the comparison of dates from vertebrate remains that consist of different types of tissue. Both sets of remains have also provided useful information regarding the development of livestock types in the historic period, particularly with regard to hair and fleece types that no longer exist, but which were the fore-runners of some of our modern breeds.

Glasson Moss

The following section is based on the work of Margaret Cox and others (Cox *et al* in press) to whom the authors are grateful for permitting the material to be included in this volume.

Following a fire in 1976 several circular and linear features, probably drains, became evident on the surface of Glasson Moss, the margins of which were characterised by the growth of *Calluna* in contrast to the *Sphagnum*-dominated mire surface (Pl 43). The cause of these features was not clear, and human action was suspected. A programme of field survey and aerial photography (Pl 44) was undertaken by the University of Bournemouth in 1996/7, which aimed to elucidate the origin and purpose of the features. The results suggested that natural pools on the moss were being used for retting hemp in the early medieval period. Radiocarbon dates suggest that this activity started earlier than cal AD 670–960 (1220±60 BP, WK–5479) and finished by cal AD 1060–1290 (790±60 BP, WK–5478), although *Cannabis* appears to have been cultivated in the area until the mid-seventeenth century (Dumayne 1992). It is suggested that the hemp stems were placed in the pools to ret for about six weeks with the water supply being controlled by three linear drains and, it is presumed, sluices. Mire pools may have been used since the process would have polluted water sources near settlement and in fields.

The work is important for a number of reasons. Firstly, it demonstrates the importance of an integrated approach to the investigation of wetland features. It also shows the value of archaeological work for the elucidation of processes which are seldom evident in the documentary record and which may indicate a significant use for wetlands, particularly large raised-mire systems, for which there is now little evidence, whether archaeological or documentary.

Documentary studies

The early medieval period in Northern Cumbria has, like all other parts of the county, very little concrete evidence to provide a framework of events into which such archaeological finds as those described *above* can be set. The area seems to have been subject to a number of influences in the centuries between the end of Roman Britain and the Norman takeover of Carlisle in 1092, place-name evidence indicating not only a continued British-speaking population, but also Northumbrian and Scandinavian influences, as well as a seemingly renewed British influence, perhaps as a result of the increasing power of the Kingdom of Strathclyde from the ninth century (Kirkby 1962; Jackson 1953). It is clear that Carlisle continued to be a place of some significance following the formal withdrawal of the Roman administration, with documentary references to activity within the area of the Roman town in the later seventh century (Colgrave and Mynors 1940), including the foundation of a nunnery, to add to a slowly growing archaeological record of activity there at least into the fifth century (McCarthy 1990). The adoption of Christianity is perhaps the most clearly demonstrated event of the period, there being a large number of pieces of stone sculpture from the area, those of Northumbrian origin clustering around Carlisle, whilst those containing Scandinavian motifs have a

Plate 43 Glasson Moss: growth of Calluna *on* Sphagnum*-dominated mire surface highlighting linear feature (© M Cox)*

Plate 44 Glasson Moss: aerial photograph of irregularities of surface vegetation interpreted as evidence of hemp retting (© C Cox)

wider distribution, with a relatively dense spread immediately to the south of the main wetland expanses of the North Cumbrian Plain, creating an arc from Bromfield in the north–east to Workington in the south–west (Bailey and Cramp 1988, figs 2, 3).

The Abbeytown Ridge

The Abbeytown Ridge is endowed with possibly the most comprehensive assemblage of documentary material relating to the late medieval and post-medieval exploitation and enclosure of the lowland mosses for the whole of Cumbria. This is due largely to the ownership of much of the mosses in the north of the area by the Lonsdale Estate, and by the ownership and control of the mosses around Abbeytown by the monks of Holm Cultram and subsequently by the Crown.

The foundation of the Cistercian abbey of Holm Cultram, in 1150 (Grainger and Collingwood 1929, 92), is probably responsible for the initiation of the permanent settlement and exploitation of the fringes of the wetlands of the Solway Plain. Prior to the foundation of the monastic house, the area appears to have been neglected by the post-Conquest surge in the colonisation of the marginal lands taking place throughout the rest of Cumbria (Winchester 1987, 38). Indeed, a study of the place-names surrounding the abbey reveals that very few originated before its establishment and that those which were contemporary generally indicate a lack of agricultural activity (Donkin 1978, 104). The only possible indication of a long-established community is the name Holm Cultram itself, which has been linked with a 'mansio' belonging to the See of Lindisfarne named 'Culterham' in AD 854 (Symeonis Dunelmensis 1868). The name appears to incorporate a British element, *culter*, meaning 'narrow piece of land, isthmus' and -*ham* an Old English settlement reference (Armstrong *et al* 1950, 288). The only reference for the pre-foundation occupation of the area comes from the 'so-called' Foundation Charter of 1150 which states that the area was formerly a royal hunting ground (Grainger and Collingwood 1929, 91).

The inhospitable location of Holm Cultram appealed to the ethic of austerity required by the Cistercians in their statute of 1134 (Canivez 1933) and the siting of the monastery in an area of wetland would appeal even more so. A significant number of Cistercian houses were sited close to marsh or fenland and many were also given tracts of unreclaimed marsh and fen as gifts. The royal charter confirming the abbey's lands, dating to 1189, outlines the extent of the wetlands belonging to the abbey. The land began:

> 'by the beck under Kyrkebride between the outer dyke of the monks and the vill of Kyrkebride as it falls into the Wathepol [Wampool]; up by the same beck outside the said dyke to Cockelayc following the junction of solid ground and moss; thence straight up to the middle of the moss between Waytheholm [Wedholme] and the island St Lawrence [Lawrenceholme]; thence across the moss and wood to Antrepot; down by the Waver to its meeting with the Cromboc [Crummock Beck]; up by the Cromboc to the place where the stream from Wytheskeld falls into the Cromboc; up by that beck to Wytheskeld; straight west to the sike [Holme Dub] that goes round Midelrig [Mealrigg] on the north and west, and falls into Polneuton [Black Dub]; down by Polneuton to the sea; thence along the shore and up by the Wathepol to the place where the said beck under Kyrkebride falls into the Wathepol'
> (Grainger and Collingwood 1929, 73).

This core of land was supplemented by gifts from numerous other parties, some of which were also on the mosses, particularly the grant by Richard I, who gave up all claim to the wastes outlined in the previous grant, and that of Richard fitz Osbert, who granted:

> 'all Eyntrepot towards the west along the Waver, as far as the place called Sceplending [Ship-landing]; and from Eyntrepot northwards between the wood and the moss as far as the dyke leads to the monk's moss'
> (*op cit* 1929, 28).

These references to the construction of the dykes generally suggest that they performed a boundary function. They were generally banked earthworks, thrown up to demarcate the monastic possessions, and not ditches, and thus cannot be considered as an act of deliberate land drainage; one such earthwork is still visible to the immediate east of the abbey. The sole reference to any groundworks which may have had a remedial effect upon the wetlands was the recorded excavation of a watercourse at Mawbray. Some time earlier than 1195, Adam fitz Ketil de Neuton granted that 'the monks' watercourse from Polneuton [Black Dub] towards Mayburg may be made over his land' (Grainger and Collingwood 1929, 68).

By 1175 the abbey had erected a total of five grange farms within the bounds of their Holm Cultram lands. The placing of these granges displays a certain focus towards the mosses: of them, those at de Ternis (Pelutho), The Old Grange (in the Southerfield/ Cowfold area), and Mayburg (Mawbray) were all adjacent to mossy areas. Indeed, of the two later granges in 1185, Arlosk (Newton Arlosh) was situated adjacent to Wedholme Flow (Baxter 1914, 274–5).

The role of the mosses within the agricultural concerns of the monks is unclear. There are no explicit references to the monastery using the mosses during the medieval period apart from the obvious assumption that they used peat for fuel; sheep were grazed at Mawbray although no reference to the moss is made, but it is likely that the more accessible mires were employed seasonally as pasture for their stock. Further lands were granted in which peat supplies may have played an important role in the economic concerns of the abbey. For instance, in 1240, the Register records that Thomas de Multon exchanged a salt works that Demann de Burstead held of the abbey for a salt works at Drumburgh (Grainger and Collingwood 1929, 3). Other grants of land for salt works give more detail of the working practices. A grant by Robert Brus to Melrose of a salt work at Tordiff, in Scotland, comprised peat hearths for the work and common pasture (*op cit*, 34). Between Kirkbride and Sevill Cote, Holm Cultram held 21 individual salt works, each with its share of peat moss (McIntire 1942, 5).

It was not only the monks of Holm Cultram who were producing salt along the Solway. A grant of peat bogs [peatheres], at Burgh, was made by Ralph de la Ferte to Lanercost Priory, for their *salina* (Todd 1997, record 158). Wetheral Priory also owned two salt works at Burgh from the mid twelfth century, one of which was later exchanged for another salt works at Drumburgh, along with the rights to turbary in Drumburgh (McIntire 1942, 3). The Priory of Carlisle also had four *salinae* between Burgh and Drumburgh (*op cit*, 4).

Despite the mosses appearing to have been a focus for the siting of the granges of Holm Cultram, their importance can be over-stressed. There appears to have been no concerted effort to improve the wetlands in the area, possibly because of the substantial grants of land better suited to grazing, including lands away from the monastery, particularly at Lazonby, and their 'lawn' illegally assarted in Inglewood Forest (Grainger and Collingwood 1929, 124). The only evidence for improvements taking place on marginal land in the area is associated with the, at least part, reclamation of the marshes probably for the superior turf, as Dugdale records that the abbot of Holm Cultram permitted certain benefactors of the abbey '*seminare et imbladare mariscum quantum pertinet ad dictum dominiam infra fossatum de Burgo*' (Dugdale quoted in Donkin 1978, 119–20). Another reference to Burgh indicates that the reclamation was not solely for pasture, as there existed '*duas acras terrae arabilis cum pertinentiis in marisco de Burgo*' (*ibid*).

It can be seen therefore that the monks effectively looked out from the moss, to their extensive lands within the Royal Forests and elsewhere and whilst undoubtedly the moss fringes were used for pasture, the vast open bog, *magnam mossam*, would have claimed more stock than it fed. In fact, there is evidence that the mosses were to be kept intact and unimproved.

Although technically not within the bounds of the Royal Forest of Allerdale, the Crown was certainly keen that the monks of Holm Cultram were subject to its laws. It is from the records of the Royal Forest that a vivid picture of the extent of the impact the monks had on the mosses can be gained. A petition dated 1292 from the monks of Holm Cultram requested the right to disafforest the island. The jury concluded that:

> 'If the island were disafforested it would be a loss to the King, and a nuisance to the forest of Inglewode, by causing destruction and damage to the deer in many ways. For there are two marshes thick with alders which join at the said island, namely, Brimselmire and Swalebymire, and these marshes extend from the island right up to the great covert of the forest, so that hinds and other of the King's deer can come and go under the covert and the main cover of the forest as far as the island, and back again; and there is another marsh there called Ellerby. So that the King's deer commonly frequent and go about in these marshes, especially about mowing-time, and all the deer which frequent those marshes go upon the island to the grass and wood contained within Holme Cultram [*sic*], namely Leaholme, Bronewra [Brunshaw Moss], Aykesom [Aikshaw], Kyngesetemire [Kingside]. And if the said island were deafforested, these deer would be hunted and taken...so that the King's venison which went about in those marshes would be destroyed, especially the great stags. And if those marshes were destroyed, the whole forest of Allerdale would be destroyed in consequence...they say that the island is distant two leagues from the main covert...and they say that two towns, namely Dundraye (Dundraw) and Blencogou lie directly between the island and the great covert'
> (Parker 1905, 58).

These restrictions imposed by the Crown may have had an effect upon the long-term settlement of the mosses. Certainly, at the time of the jurors' report the mosses were uncultivated and unreclaimed. Indeed the Bowness peninsula, the land between the Eden and the Wampool approximating to the Barony of Burgh, was included as part of the Allerdale forest by Henry II, which may also have had an effect upon any plans to settle and instigate improvement schemes (*op cit*, 45). It would be wrong therefore to state that the monks set about draining and improving the mosses wholesale, as Denton suggested when he wrote that they 'turned all into arable, meadow and pasture' (Denton 1887, 60).

There is little other direct evidence for the use of the mosses during the medieval period, but at the Dissolution a survey of the lands gives an impression of how the monastery had used its mosslands. Swaby Mire was then a freehold lease valued at 5s for 15 acres (Grainger and Collingwood 1929, 163), and it is possible to assume that the moss was always regarded as inferior grazing, similar sized areas of meadow elsewhere being valued at 8s for four acres and 13s 4d for seven acres (*ibid*). There was obviously a differentiation between the quality of the various mosses also. In a survey of the area after the manor reverted to the Crown in 1573, the Common Moss 'where the monastery did grave peats' was valued at 23s 4d and let to John Borrowdale, whereas Wedholme Moss, let to John Chambers, was valued at 6s 8d (*op cit*, 169). By the time the *Register of Holm Cultram* was edited, in the early years of the twentieth century, the value of the land was priced at 4s (*ibid*); thus the value and use of the moss was most obviously much higher in the early post-medieval period.

The study of the post-medieval exploitation of the mosses on the Solway Plain is assisted by a number of surveys carried out in the sixteenth and seventeenth centuries. These outline a particularly backward and isolated area which was not open to change. In the survey of 1638 of the Lonsdale lands of the Barony of Burgh, the surveyor bemoans the essentially medieval landscape, stating:

> 'there land lyeth very much intermixed in very small parcels and if it could bee laied [*sic*] to gather it would bee much better for The tennants but I find the Freehoulders noe whitt willinge to exrange [*sic*] and I think that it will hardly bee Brought to pass'
> (CRO(C) D/Lons/L5/2/41/53).

That the local populace was reticent regarding improvement and change is noticeable. There appears to have been no encroachment onto the mosses in the Solway area; in fact, the same surveyor noted that 'the comons and Mosses such are very spacious and large and where the tenants and Freehoulders gett very good Turfe and Peate' (CRO(C) D/Lons/L5/2/41/53), whilst the surveyor of Holm Cultram declared that in 1573 'there are no common to spare to be improved' (Grainger and Collingwood 1929, 169).

It was not until the mid-eighteenth century that large-scale enclosure and improvement schemes were initially proposed. The Lonsdale Estate, which owned extensive land in the Holm Cultram area, enlisted the services of the agricultural reformer Robert Maxwell who after inspection of the lands submitted his report in 1750. In it he detailed numerous improvements which could be made throughout the better areas of agricultural land; of the mosses he wrote:

> 'With respect to Wead Holm, it being almost totally a very deep Spungy Moss, or Peat Earth; The same, without any material Difference, as Locher Moss by Dumfries, and I having in my Esay [*sic*] on the Nature, qualitys, and Different Methods of Improving Moss, and sowing grass seeds, and in my Memorial concerning Locher Moss aforesaid...Exhausted my knowledge of this soil... I chuse rather you referr a you to them, than to say the same things over again'
> (CRO(C) D/Lons/W4/16).

Whether the essay on land improvement was consulted by Lowther is uncertain; what is certain, however, was that the task of reclaiming and improving Wedholme was never enacted, the work obviously being too large or not cost effective.

While the Lowther Estate appeared to have been reluctant to invest in improving the Wedholme Flow area, they were quite prepared for others to initiate improvement schemes on the Lonsdale Estate lands at their own costs. In 1755 a proposal to enclose the commons of Holm Cultram was put forward by Edward Stephenson to the Court Leet but could go ahead only with Lowther's consent. On hearing that the tenants were favourable he set up a public notice declaring his intention. The area to be affected was primarily the Stank and Acredales (the old town field); part of Colt Park and 'the Mosses without any rent' (CRO(C)/D/Lons/W26 HC). A petition to Parliament was made by a number of tenants, copy holders, and freeholders in the same year protesting that:

> 'the said Comons Comonfields Moors Mosses and Marshes, Except some small parts thereof, are very poor Cold Wet and Barren Land, mostly inclinable to Moss, the greatest part whereof is not capable of being Drained Fenced and Improved, but at an Expence, Exceeding what the Inheritance for such Lands would Sell for when so Inclosed Drained and Improved, Seeing there is not any Coal, Limestone, or other Matter, within the said Lordp. or within Several Miles thereof wherewithall to make any Improvements'
> (CRO(C) D/Lons/W26 HC).

This was given as the primary reason for the objection to the enclosure of the commons, but an inspection of the Petition reveals that Stephenson had 'lately by threats and other pretences' coerced local landowners to enter into an agreement to enact the enclosure of the commons and had allotted and assigned to himself the Stank or Abbots Fishpond, the only profitable improvable land. Whilst the tenants:

> 'would all of them have been as ready to have promoted the said Division, as the said Edward Stephenson or any others who have Executed the said Articles, had such Division been offered and been put upon a fair just and Equitable Footing and could such Comons Moors Mosses Marshes and Wastegrounds have been Inclosed Drained, and Improved, at any Moderate or reasonable expence'
> (CRO(C) D/Lons/W.26 HC),

it is uncertain whether the mosses could have been drained and improved. The Petition was upheld in the Houses of Parliament and the enclosure was never enacted.

Despite the lack of evidence for the improvement of the mosses, they were worked intensively and closely regulated through the manor court and the Sixteen Men of Wedholme, a quasi-local parliament established by Elizabeth I, after the abbey holdings had reverted to the Crown. The regulations imposed by the manor court and the Sixteen Men were intended to restrict the wholesale stripping of the mosses for domestic fuel. This was achieved by regulating when, how, and for what purpose the peats were put to when excavated, although possibly the most important restriction that the manor court could impose related to the areas from whence the peats could be dug. Doubtless during the medieval period there were regulations regarding the permitted extraction areas, but it is not until the sixteenth century onwards that these restrictions become clear. An early reference from the Manor Court of Holm Cultram illustrates the nature of these restrictions when it decreed that:

> 'for graving of flacks no man shall grave any before he shall discover his need to ye Court...and thereupon ye Steward to appoint four sufficient men to be sworn to view his need and to appoint a place where he shall grave, and how many shall be taken'
> (Grainger and Collingwood 1929, 206).

It is later in the same manor that it is possible to see these restrictions becoming more tangible on the ground. In a dispute over a moss at Holm Cultram, in 1622, it was ordered that:

> 'every of them shall grave from pearche to pearche or by line...in the said mosse...accordinge unto pearches or mere stones appoynted or agreed upon by the said partyes...shall have hold occupye and enjoye ther severall partes or porcons of the said mosse'
> (CRO(C) HC/Chambers quoted in Dilley 1972, 150).

The right of turbary was particularly important in the Holm Cultram area. Due to the constant threat of coastal erosion at Skinburness, from an early period it was established that the timber stocks of Wedholme Wood were to be used exclusively for the maintenance and upkeep of the sea dyke, which was in existence in the late sixteenth century, although it appears to have been rebuilt more recently (Fletcher and Miller 1997). This was rigorously overseen by the 'Sixteen Men of Holm Cultram'. The preservation of timber stocks (that washed ashore was even prohibited from use as domestic fuel) must have put a great onus on the right to extract and consume peat. Therefore the importance of the definition of rights as outlined above was long lasting.

The Enclosure Awards brought about by acts of Parliament in the area from the early nineteenth century onwards were notable in that despite enclosing and improving over 3000 acres they avoided large areas of moss. Wedholme Flow, Hangingshaw Moss, and Bowness Common were not improved, although the rights to turbary within them were firmly established. Indeed, the only activity in the Wedholme Flow area was the establishment of the turf dales at Moss Side and Newton Arlosh (CRO(C) QRE/1/148).

Whilst the need for domestic fuel obviously shaped the enclosure of the mosses in the Wedholme Flow area, the prime aim of the Enclosure Movement was to bring more land into cultivation and to reorganise the long-established land ownership system, which characterised the late medieval open-field landscape. It would be wrong to say that the Enclosure Awards of the early nineteenth century were the first concerted effort to improve the land on the Solway Plain – the earliest proposal to improve the Airedales and Stank area of Holm Cultram has already been mentioned – but they formed the first large-scale programme. Previous to this all enclosure had been piecemeal; undertaken in the north and east, particularly around Bowness, encroachment into the mosses had taken place at Rogersceugh, Cocklicks, and Akesbams (Armitage nd, 6). The evidence for encroachment at Rogersceugh comes from the Lowther Estate survey book when, in 1742, the building there was described as in 'bad repair'; a plan of mid-eighteenth century date depicts a ruin surrounded by a ditch located centrally within Bowness Common (Messenger 1975, 334). By 1765 the ditch had become an inner field boundary, with a larger elliptical boundary enclosing a former moss called Lingy Part (Dilley 1991, 367). Other encroachments occurred at Cardurnock and undoubtedly elsewhere.

Armitage points out that the majority of the Enclosure Awards relating to the Solway Plain dealt specifically with waste land, as by the 1800s land better suited to agriculture had already been enclosed (Armitage nd, 14). Therefore, in 1810, the award for Bowness Common dealt with land to the west of the parish, approximately 1500 acres being subject to improvement, the majority moss land. Strips of fields varying between 60 and 155 acres were established, which were later sub-divided (*ibid*). The remainder of the common was divided into two large fields of 169 and 272 acres (Pl 45). Similarly, this pattern of large enclosures on the mosses was repeated at Glasson where the moss enclosures were conspicuous amongst the earlier enclosures of two to eight acres, with moss fields measuring from two acres near Glasson, increasing to 30–40 acres near the centre of the moss. The same occurred between Anthorn and Rogersceugh, where area was divided into plots of 60–177 acres, likewise at Cardurnock where plots of 100–190 acres were created (*op cit*, 19). The same situation was occurring at Holm Cultram where, despite the need for peat for fuel, which resulted in the enclosure of numerous small strips for peat cutting, the field layouts at Common Moss, Cowfold Moss, Bellwater Moss, Bromfield Moss, and Kelsick Mire were established as a result of the 1800 Enclosure Award (CRO(C) QRE/1/148).

The enclosure of the mosses resulted in a pattern of channels being cut through the peat fields in order to improve the land. With the subsequent drainage of the mosses the land, certainly on the Bowness peninsula, was well suited to grain crops. Those new fields which were imperfectly drained were used as meadow pasture. This appears to have continued into the twentieth century, although grain is no longer grown on the Solway.

The establishment of commercial peat-cutting on Wedholme Flow and Solway Moss is the last major event to have shaped the mosses on the Solway Plain. Cutting was established on Wedholme Flow in the early 1900s by the British Midland Peat Company for use as animal litter, with operations ceasing in the early 1930s (Coles 1995, 61). Peat-cutting was resumed again in 1948 by the Cumberland Moss Litter Company, which mechanised peat extraction during the 1960s. Extraction has continued on Wedholme Flow by Fisons Horticulture Ltd and most recently by Levingtons.

East of Eden

The peat mosses to the east of the River Eden have a long history of neglect, much of the area being referred to during the unstable medieval and post-medieval periods as The Debatable Lands, due to its unenviable position on the turbulent and shifting border between Scotland and England. The vulnerability of the area to Scottish attacks and guerrilla banditry rendered much of the area unsettled. Even land which was suitable for agriculture was often neglected and it is therefore of no surprise that the large tracts of impenetrable bogs remained so intact into the eighteenth and nineteenth centuries.

There appear to be no specific references to the mosses receiving attention from farmers or potential improvers in this area prior to the early modern period. Bolton Fell Moss, then in Scotland, is first referred to in 1384 (Armstrong *et al* 1950, 91). Similarly Solway Moss, historically called Sollom Moss, fluctuated between Scotland and England and was the scene of a skirmish in 1542 between a hastily gathered Scot's force and 'two valiant captains, Thomas Dacre and Jack Musgrave' and a hundred horsemen (Nanson 1886, 259). The political instability no doubt hampered peaceful and enterprising settlement within the area. It is also possible that the area was kept deliberately as a no-man's land by both kingdoms and even by the fugitives, moss troopers, and reivers who sought its protection, although whether this was the case and the moss was deliberately unsettled is not certain. That it was used as a *cordon sanitaire* is clear.

The Survey of the Barony of Gilsland, undertaken in 1603 on the assumption of the land by Lord William Howard, permits the first clear picture of the extent of the mosses in this area. The area of Bolton Fell Moss and Leap's Moss which fell into the Barony of Gilsland (the neighbouring Walton Moss was part of the Lanercost Estate and thus not surveyed) was described as:

> 'a great piece of comon pasture & moare called Bolton Fell and Leapesmosse: extendinge from Houghton Crosse & Levenholme on the weste to Kirkamocke on the easte: & lieth betwene the severalles of Kirkacamocke, Creakrope, the Dormansted & diverse other tenemnts on the north: and Salmane grounde, Mothersike, & the middle part of Leapesmosse where it parteth from Hethersgill groundes to the south' (Graham 1934, 14).

The whole area totalled 2010 acres. The description of the southern boundary of Bolton Fell Moss is interesting as it also demarcated the boundary between Gilsland and the Dacre family's Lanercost

Plate 45 Solway Plain: from Greenwood's map of Cumberland (1823). By kind permission of Cumbria Record Office (Carlisle) © Copyright

lands. A description of the boundary is also found in the Survey. It is recorded as:

> 'begininge at Houghton Crosse it proceedeth eastward through the middeste of Leapsmosse unto Leapesburne: & then in a mossye bottome called Mothersike unto Clughki... & still eastward by Salmeane groundes unto Camocke...'
> (*op cit*, 4).

These bounds correspond exactly to an existing drain which cuts through Leap's Moss, immediately to the north of Leap's Rigg. There is a tradition of excavating dykes to define land boundaries in the area and this dyke may be seen to be for that purpose as opposed to serving a deliberate drainage function. However, the excavation of such a dyke, falling into a stream, would enable surplus rainfall to be carried away, thus preventing any dangerous build-up, and would also facilitate the small-scale extraction of peat for domestic use.

The need for drainage in the area was amply illustrated in 1771, when, as described *above* (*p 121*), Solway Moss experienced an 'eruption'. The event was attributed to the retaining wall of peat being breached by the moss on a particularly stormy night. The description of it by William Gilpin makes an interesting narrative and merits a full quotation here:

> 'The enormous mass of fluid substance moved slowly on, spreading itself more and more as it got possession of the plain. Some of the inhabitants, through the terror of the night, could plainly discover it advancing like a moving hill...One house after another, it spread round – filled – and crushed into ruin; just giving time for the terrified inhabitants to escape. Scarce anything was saved, except their lives: nothing of their furniture: few of their cattle. Some people were even surprised in their beds and had the additional distress of flying naked from the ruin...This dreadful inundation, though the first shock of it was most tremendous, continued still spreading for many weeks, till it covered the whole plain — an area of five hundred acres; and like molten metal poured into a mould, filled all the hollows of it, lying in some parts thirty or forty feet deep, reducing the whole to one level surface. The overplus found its way into the Esk; where its quantity was such as to annoy the fish, no salmon, during that season, venturing into the river. We were assured also that many lumps of earth, which had floated out to sea, were taken up some months after at the Isle of Man'
> (quoted in McIntire 1941, 11).

The threat of disaster was obviously recognised due to the existence of the bank to retain the wall of peat. It was stated of the mosses by William Gilpin that:

> 'The substance of it is a gross fluid, composed of mud and putrid fibres of heath, diluted by internal springs, which arise in every part. The surface is a dry crust, covered with moss and rushes; offering a fair appearance over an unsound bottom – shaking under the least pressure'
> (*op cit*, 2).

This overburden of peat was very effectively removed by the 'hushing' process. The work was supervised by John Wilson, whom contemporaries described as a:

> 'truly great man... [who] was debased by one of the most degrading of vices – he was a confirmed drunkard: and ebriety in him was not softened by any of the usual pleas that are offered in execuse [*sic*] of it. He did not drink to drown care, nor to promote hilarity; but merely to gratify an animal appetite'
> (Hutchinson 1794, 543).

Despite being cleared of the mossy overburden and after having been made safe, Solway Moss did not elicit any interest from agricultural reformers or improving landlords. Even the great agricultural reformer, Dr Robert Graham, who lived nearby in Arthuret at the time of the inundation, did not feel moved to try to improve the land by drainage, and used the moss instead for fowling (McIntire 1941, 10).

By the time of the Enclosure movement, the mosses were still open and unreclaimed. The Bolton Fell Moss. Enclosure Award of 1800 (CRO(C) QRE/1/88) shows the moss land as being divided up into large fields and was not sub-divided into turbary plots, as Black Moss at Egremont had been ((CRO(C) QRE/1/30; *Ch 3*)). The lack of turbary plots on the Enclosure Award, however, does mean that the area was not used for peat extraction. The characteristic long thin plots of turf rooms are depicted on the OS First Edition map of 1865. These peat rooms have been extended into the moss in the period between the 1865 and the 1977 coverages, particularly in the Cocklet Hall Farm area.

Evidence for the scale of the domestic peat workings in the area has been recorded in a local newspaper article detailing the discovery of a bog body in peat at Scaleby Moss. In the *Carlisle Journal* of 7 June 1845, it was reported that a George Hogg, whilst peat-cutting '... came to the remains of a human skeleton buried about eight or nine feet beneath the surface, and closely embedded in the lowest stratum of black peat' (Turner 1988, 1). Peat extraction at a depth of nine feet implies that Scaleby Moss was intensively worked and this was no doubt due the close proximity of Carlisle.

From the late nineteenth century peat-cutting was carried out commercially on Bolton Fell Moss, particularly on the eastern and northern parts of the moss. Hand cutting was carried out by Boothby Peat Co between 1956 and 1968, whereupon block cutting machines were introduced. Peat extraction is still carried out commercially today. Similarly, Richardson's Moss Litter Company began excavation on Solway Moss in 1898.

Discussion

There are clear links, both geographical and archaeological, between the West Cumbrian and North Cumbrian Coastal Plains; the distinctions made are artificial. In terms of their archaeological records, the two areas share common issues, the most significant of which is the importance of visibility and its effect on the archaeological record. The number of known sites is skewed to those areas of landscape where cropmarks show well, or there is arable land to fieldwalk, or the proximity to Hadrian's Wall, which has long been a key factor in the region's archaeology. These factors have led to researchers and survey programmes, including NWWS, focussing on certain areas, notably the Abbeytown Ridge, thus reinforcing the pattern. It has also meant that much of the Plain's archaeological record has been interpreted in terms of Hadrian's Wall and the general assumption that it was its construction that led to the first large-scale settlement and landscape exploitation.

As with other parts of the Cumbrian landscape, there are certain finds that indicate the potential of the area for settlement throughout prehistory and into the post-Roman period. Recent work examining cropmarks (Bewley 1993) has revealed an early origin for a number of enclosures, suggesting that there may have been widespread settlement in the Iron Age, if not earlier. The pollen record has indicated the long history of humanly-induced vegetational change across the landscape, which markedly increases in the Iron Age with extensive, and permanent, forest clearance.

The North Cumbrian Plain differs from other parts of Cumbria and the North West by virtue of the excellent survival of many of its mires. Well-preserved peat deposits may still mask finds and well-preserved landscapes, which are difficult to find with current technology, and this makes assessment problematic. This point will be discussed more fully in the next chapter.

The evidence for large-scale wetland use through all periods is, thus, less clear, although what emerges are glimpses of a much more extensive hidden archaeological record. Because the mires have not been disturbed extensively in the past, scattered finds have been fewer than in other parts of the North West but they are, however, significant. The stone axe from Solway Moss, the sword from Salta Moss, and the bog body from Scaleby Moss, all suggest that mosses were being used in different periods in prehistory.

Later use of the mosses is suggested by the Solway Cow and the retting pits on Glasson Moss. The significance of the close correlation in date between the evidence of retting from Glasson Moss and the Solway Cow, however, is difficult to distinguish. It is likely to be the coincidence of a long-lived process and a single event which happen to have produced similar dates but, in the absence of any early medieval archaeological record from around the mosses, it is difficult to place these events in context. Future work may wish to consider the use of the mosses in this period and whether there was a renewed interest in them as an exploitable area of landscape, either economically and/or symbolically.

The documentary evidence expands this knowledge, particularly in terms of peat-cutting, which was clearly of the first importance both for salt production and domestic use. The documents do indicate that the use of the mosses by local people was not as straightforward as has been assumed in the past, with mosses closest to settlements being utilised first through common rights. The example from Wedholme Flow clearly shows that control was exercised over the use of the mosses, if only to allow the King's deer a safe haven. The documents also indicate that the mosses were clearly the last areas to receive attention for agricultural improvement; even the monks of Holm Cultram, Cistercians who elsewhere had 'turned all into arable, pasture and meadow' (Denton 1887, 59), used the mosses in this region only for peat, primarily to fuel salt production.

This chapter, particularly the work at Solway Moss, has highlighted starkly the essential irony of wetland archaeology; finds are made where they are most at risk, and where there are people to record them. The North Cumbrian Plain contains the only substantial areas of large-scale peat extraction/harvesting in the North West. It is a process which shows up the challenges facing the management of the wetland archaeological resource; should cutting cease, to leave the archaeology undisturbed, or should it continue with monitoring to ensure that the archaeological and palaeoecological resource can be characterised to inform the management of other mires?

Questions concerning the management and use of the mires identified in this chapter will be considered further in *Chapter 5*.

5

SYNTHESIS AND DISCUSSION

Introduction

Previous and ongoing wetland surveys have shown the importance of not only revealing and identifying the rich cultural resource that wetlands represent, but also of action to conserve some of the resource for future generations. The projects have been initiated in the light of landscape changes which, although variable, have posed immediate threats to the archaeological resource. These have included peat-cutting and agricultural drainage in the Somerset Levels, and agricultural drainage and peat wastage in the East Anglian Fens and the Humber Wetlands; these all involve the irrevocable loss of peat deposits. Approaches to active management have differed between projects to reflect these local circumstances. In the Somerset Levels some areas were identified as being of National or Regional importance with a prohibition on disturbance (Brunning 1999; Coles 1989); in the Fens, spot Scheduling was complemented by a series of assessments to identify management options, although the very scale of the issue in this area was, and remains, problematic (Hall and Coles 1994). In the Humber Wetlands, management prescriptions have been variable and localised, reflecting the systematic manner of the survey and publication (Van de Noort and Etté 1995).

For Cumbria, the previous chapters have described the extent to which the resource has been revealed and identified. These aspects will now be assessed at county level, followed by a discussion of future research and management needs. This latter is particularly significant in the light of the international importance of the mosses for their wetland flora and fauna.

Fieldwork constraints

Traditionally, archaeological management solutions have depended upon the identification and assessment of the resource and the conservation of a significant sample, sometimes associated with the presentation of a small number of sites to a wider public. In the Cumbrian wetlands there are two problems with this approach: firstly, the resource is often difficult to identify as it is usually buried and, if not, will be under threat from desiccation and in need of active management, usually by excavation. The second problem is that of land-use and the visibility of the archaeological record; Cumbria is predominantly a pastoral county, although there has been significant ploughing in the past when mixed farming systems were more common. Earthworks have only localised survival, notably on the upland fringe in areas of semi-natural vegetation, and flattened sites are very difficult to detect either by ground or aerial survey. The exception to this is the western end of the North Cumbrian Plain, where several fortunate circumstances, including the occurrence of free-draining soils, have led to the detection of a concentration of sites from the air. These conditions, associated with the proximity of Hadrian's Wall and related coastal defences, have led to a focussing of archaeological work in the area (*eg* Higham and Jones 1975; Bewley 1994).

Despite the apparent site visibility 'problems' in the pastoral areas, Cumbria is fortunate in having a rich archaeological tradition, through which have developed appropriate site detection methods. Allied to this has been a long history of palaeoecological research which, from a relatively early date, was closely related to the identification of human interaction with the landscape. Thus a rather variable archaeological record has developed with a good knowledge base in some areas, notably the Abbeytown Ridge, parts of the West Coast, and Furness, with other areas being a blank. A part of the reason for the apparent absence of archaeological remains in some areas is the wetlands themselves, which are particularly well-preserved over much of the county when compared to other areas of the North West. The good survival of peat deposits, from vestigial remains to intact mires, has meant that ancient land surfaces, sites, and finds remain undetectable by traditional methods (Pl 46).

Owing to these fieldwork constraints, and the obvious need for management action to protect archaeological sites and palaeoecological sequences, an approach is

Plate 46 Silecroft: reclaimed mire now supporting silage and pasture

required that considers both the known and the unknown archaeological record. In this chapter some emphasis has been placed on the concept of archaeological potential as a guide to management (Middleton 1999). The term 'archaeological' is taken in this chapter to include the wider cultural heritage potential, not just archaeological sites and artefacts but also the palaeoecological record.

Importance of mires

The importance of the wetland archaeological record in Cumbria is undoubted. The number of chance wetland finds made in the past — the Scaleby Moss bog body, Ehenside Tarn, and the Foulshaw Moss trackway to take three examples — and the important palaeoecological sequences indicate that there is a wealth of data yet to be discovered within surviving peats and silts. Whilst the preservation of archaeological material *in situ* must be the first option in any scheme of management, it is perhaps ironic that we may never realise the true potential of mire archaeology if mosses are protected from damage and disturbance in the future, as the archaeology will remain undetected. Major advances in geophysical technology should be awaited, which may allow organic finds and deeply buried sites to be mapped accurately from the surface without disturbing them.

The palaeoecological importance of the mires is, perhaps, more established than that of the site and artefact record; past work in the county helped to establish the methodologies and procedures for the discipline nationally and so the potential of the mires has been widely recognised. It is notable that Plantlife's *Commission of Inquiry into Peatlands* emphasised the high value of mires as peat archives (Plantlife 1992, 2, 20). This does not mean, however, that the management of the mires should take palaeoecological data for granted. The coring work of NWWS in the county and elsewhere in the North West has established not only the significance of prehistoric sequences but, perhaps more importantly, that of the Roman, medieval, and post-medieval deposits. These are important for the study of the environment of Hadrian's Wall (*eg* Dumayne and Barber 1994) and may also aid the investigation of climate change (*eg* Barber 1981). The upper stratigraphy of mires has been vulnerable to past

Plate 47 Woodland development on the central area of Solway Moss

peat-cutting (particularly for fuel, animal bedding, and litter), desiccation, and disturbance by tree roots and burrowing animals (Pl 47). It is thus of critical importance that where whole sequences exist, they are protected *in situ* with particular emphasis on intact surfaces. In a wider context, the very number and variety of mires within differing landscape contexts means that the palaeoecological resource present in the county is, when taken as a whole, of considerable significance for our understanding of both micro- and macro-scale environmental change.

Although the palaeoecological value of the mires is well established, the wetlands are perhaps best known for their flora and fauna and for the potential of restoration of damaged areas. The advent of the Peat Campaign in the early 1990s raised awareness of the vulnerability of peat bogs to change and the international importance of lowland mire habitat. The Cumbrian mires were important in this campaign since there were large areas of mire still intact, many were protected, and three were still cut commercially. Cumbria contains a significant proportion of England's mire resource which is reflected in the great proportion of largely intact mires designated as Sites of Special Scientific Interest (SSSIs), National Nature Reserves (NNRs), and Ramsar sites. In Cumbria, mires support a number of important habitats including Lowland Raised Bog and Fens, identified under the Biodiversity initiative (UK Biodiversity Steering Group 1995).

Management

The importance of the wetland flora and fauna of Cumbria's lowland mosses is recognised both by their conservation designations, such as SSSIs, NNRs, and by the EC Directive on *Habitats and Species* (92/43/EEC), particularly for Glasson Moss, Wedholme Flow (Pl 48), Bowness Common, and Drumburgh Moss, all on the Solway Coast. The importance of the mires is reflected in the recent effort by English Nature and the Cumbria Wildlife Trust to introduce appropriate management of the mire surfaces, including re-flooding to restore wet vegetation, scrub clearance, and habitat creation for particular species as part of the Lowland Peatland Project (Coles 1995, 61–5). These measures have aimed to raise water tables by retaining rainfall, thus encouraging the re-

Plate 48 View of Wedholme Flow from Sleightholme

growth of mire communities and the suppression of scrub, particularly birch.

There has, however, been no such action directed specifically at the protection of wetland archaeology and palaeoecology in the county. The reasons for this are complex, shaped by the largely invisible and thus unquantified nature of the resource and the fragmentation of data sources relating to past human activity. It is to be hoped that, in bringing together the wetland data, this volume will promote recognition of the need for the management of wetland archaeology, which can be activated both by working with other conservation organisations and through separate, archaeologically-led, initiatives.

Landscape history

In Chapters 2–4 in this volume, a wide-ranging background has been sought to place wetland data in its historical context and to identify the role of the lowland wetlands of the county, more specifically peatlands, in the human past. This section draws together the archaeological and palaeoecological data, highlighting the major trends in the development of the wetlands in the county and assessing the extent and intensity of human exploitation.

Palaeolithic

The main evidence for the Palaeolithic exploitation of the county comes from the limestone of South Cumbria where evidence suggests that numerous caves were used during the late Upper Palaeolithic. There is no evidence for a human presence during the Late-Glacial in other parts of the county, and this may be a reflection of hostile climatic conditions until the climatic amelioration from *c* 9000 cal BC (*c* 9500 BP). The presence of cultural and faunal material in caves is largely an accident of survival, a fragment of a wider occupation and use of the landscape. The interpretation of this evidence, however, is hampered by the sketchy nature of much original excavation data and problems over the interpretation of cave stratigraphy at Kirkhead (Gale and Hunt 1985; 1990; Salisbury 1986; Tipping 1986). More recent excavations, whilst apparently revealing some level of Late-Glacial activity, await full publication. At Kirkhead, a single radiocarbon date and pollen data may suggest activity during the Loch Lomond Stadial, and the position of the site could indicate its use as an early hunting stand overlooking the large plain which now lies under Morecambe Bay. The site at Poulton-le-Fylde in North Lancashire (Hallam *et al* 1973) suggests that wetlands and areas of open water would have been of importance across this plain as places where animals may have congregated. Sadly, the number and extent of these wetlands is unknown although, given the low elevation of the landscape, and the number of wetlands north and south of Morecambe Bay, it is likely that they were a significant component of the landscape. The archaeological evidence is too fragmentary to suggest modes of wetland use although, given the importance of large game in the Late-Glacial, it is likely that wetlands were used extensively.

It is unlikely that the location of Palaeolithic cultural material can be predicted within wetland contexts, and thus its management is problematic. It is similarly difficult to characterise such a scattered and poorly preserved dataset; in the light of the Poulton material, however, the possibilities of chance finds should not be discounted. Management action could be concentrated where cultural material may occur in the lower levels of mires, that started to form in the Late-Glacial, adjacent to known areas of activity. In this light, the mires of South Cumbria may be priorities, although it is possible that some, at least, may not have peats dating to earlier than the Neolithic, due to periodic marine inundation; however, the threats to such deep stratigraphy, it could be argued, may be relatively slight, though boreholes in the vicinity and chemical change could cause severe damage. Stratigraphic work throughout the county has revealed that Late-Glacial sequences are almost ubiquitous across the landscape within small basin mires and kettleholes. Whilst their prevalence does not deny their importance, any archaeological finds of the period are by definition of national importance and therefore a regime of monitoring the condition of the resource must remain a priority.

Mesolithic/early Neolithic

The first evidence we have for the use of wetlands in association with settlement comes from the Mesolithic period, notably from work on the West Coast at Eskmeals, which has identified substantial activity adjacent to a dynamic wetland environment (Bonsall *et al* 1991). Fieldwalking in the area has identified several scatters of lithic material which probably represent contemporary activity along a fossil shoreline. The visibility of the surface material, through rotational arable cultivation and sustained fieldwork by the Cherrys (*Ch 3*) and Bonsall and his team, has led to the recognition of the importance of these sites, and the work at Williamson's Moss has indicated the role of natural resources in attracting repeated settlement to favoured locations. The Esk Estuary would have possessed a ready supply of fish and other marine, estuarine, and riverine resources, as well as providing access to other parts of the landscape, notably the Esk Valley and neighbouring uplands.

The associated palaeoecological evidence indicates that settlement was associated with woodland clearance around a 'double' Elm Decline, which was locally initiated at *c* 4458-4047 cal BC (*c* 5440±70 BP, SRR–3065). The association of woodland clearance, ruderal weed pollen, and mineral inwash indicates a pattern of local landscape management, and it has been suggested that this was associated with long-term, perhaps permanent, settlement, along the coast (Bonsall 1981). There is, however, little precise correlation between the palaeoecological and archaeological evidence to associate individual clearance episodes with site occupation.

A relatively widespread use of the landscape is suggested by the proxy record of charcoal found in many mire sequences. Although the nature of the activity represented is difficult to determine, and natural causes are likely for some of the charcoal, the ubiquity of burning within wetland sequences in the North West, and elsewhere in the country, would suggest an anthropogenic origin for a proportion at least. The evidence from wetlands may have been the result of burning fen vegetation to improve the chance of attracting and catching herbivores; in Cumbria, for instance, the evidence from around the southern edge of Solway Moss may represent several small, but undated, episodes of burning on the wetland edge. The importance of the charcoal is that it may indicate widespread and perhaps large-scale landscape management in the late Mesolithic, and wetlands may have been exploited as an important part of this.

Several problems exist, however, in using charcoal as an indicator to conclude that the pattern of human activity established for the Esk Estuary can be extended to the rest of the Cumbrian lowlands. The extensive resources would not have been available uniformly down the coast, suggesting that activity may have been concentrated in the estuaries and within sheltered bays and inlets; the fragmentary evidence from Walney Island may indicate one other such location. Bewley (1994, 54) has suggested that survival conditions for sites and artefacts on north- and north–west-facing coasts may have been different from those on coasts facing south–west. For example, the material from Levens Park and the caves of South Cumbria, along with the sketchy material from the North Cumbrian Plain, suggests some coastal activity, although its nature is difficult to determine and, in the light of the substantial rise in sea-levels during the late Mesolithic, it is clear that many sites will now lie beneath coastal deposits. A clue to the nature of settlement may be derived from the material that occurs on the raised beaches on the Scottish shore of the Solway, where relatively dense scatters of Mesolithic material have been found, whilst finds from later in the period are concentrated 9–15 m above present sea-level on a raised beach formed at the time of the maximum high sea-level (Morrison 1981, 154). However, although these may give an insight into the nature of possible buried Cumbrian material, care should be taken in supposing that there was coastal Mesolithic activity alongside each Cumbrian maritime wetland.

There are similarities in the palaeoecological record between the West Coast and the Arnside / Silverdale area in the south of the county, particularly activity around the time of the Elm Decline (Middleton *et al* 1995, ch 6). There are, however, problems in extending the model elsewhere in the absence of evidence. It remains to be seen whether this results from factors such as a lack of appropriate studies and, indeed, of deposits to study, or whether it is more complex, arising perhaps from sea-level changes which have drowned many of the favourable areas where repeated occupations would have produced the required signatures in the palaeoecological record. Those sites that have survived are in higher parts of the landscape favourable for settlement (such as Little Hawes Water, Taylor *et al* 1994), and are dated to the end of the Mesolithic after the last significant rise in sea-level in the fifth millennium cal BC, or are away from the effects of sea-level change, such as the sixth millennium cal BC evidence for clearance from Burnfoothill Moss in Dumfries (Tipping 1995a; 1995b; 1995c). The resolution to this question may lie in the study of foreshore peat deposits and other parts of the drowned landscape.

The presence of Mesolithic coastal activity well-dated to the end of the fifth millennium cal BC in close association with early Neolithic finds means that the wetlands of the West Coast have been relatively well studied in terms of the Mesolithic / Neolithic transition (*eg* Zvelebil and Rowley-Conwy 1986, 73-4). The similarity of pre-Elm Decline clearances, with 'Mesolithic' dates, yet also with signatures associated with Neolithic artefacts, notably polished axes, has been noted. This has given rise to a model which views the change from Mesolithic to Neolithic as essentially one of continuity, where new artefacts are integrated within existing patterns of subsistence and settlement. The scattered palaeoecological record

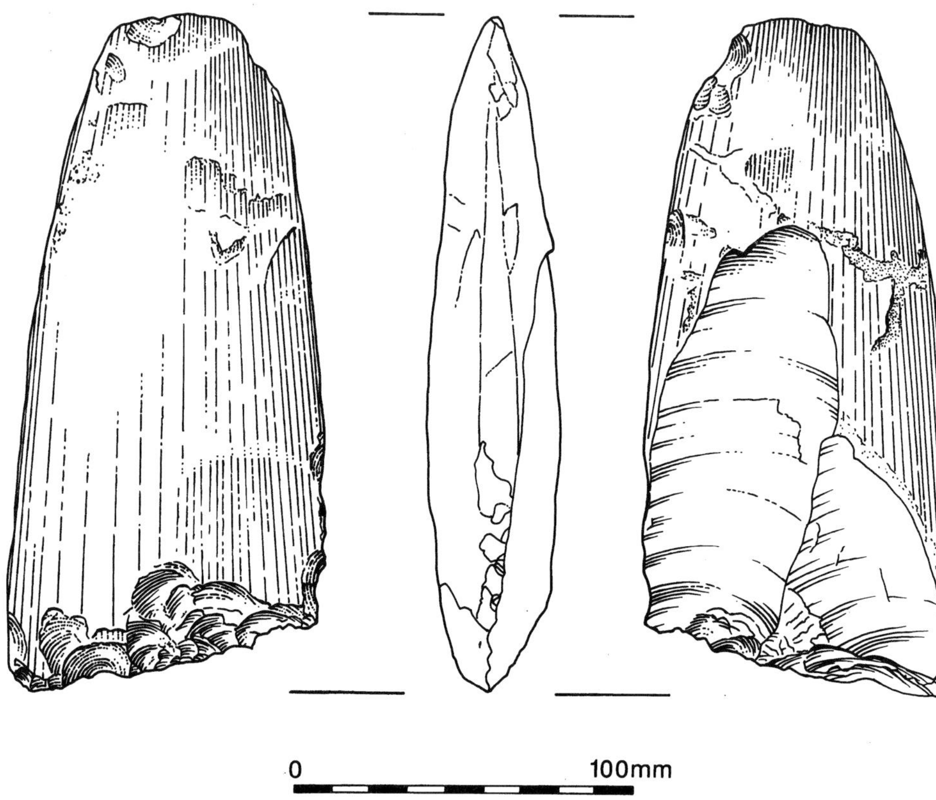

Fig 58 Polished Group VI stone axe from Gilbank Farm (CU33)

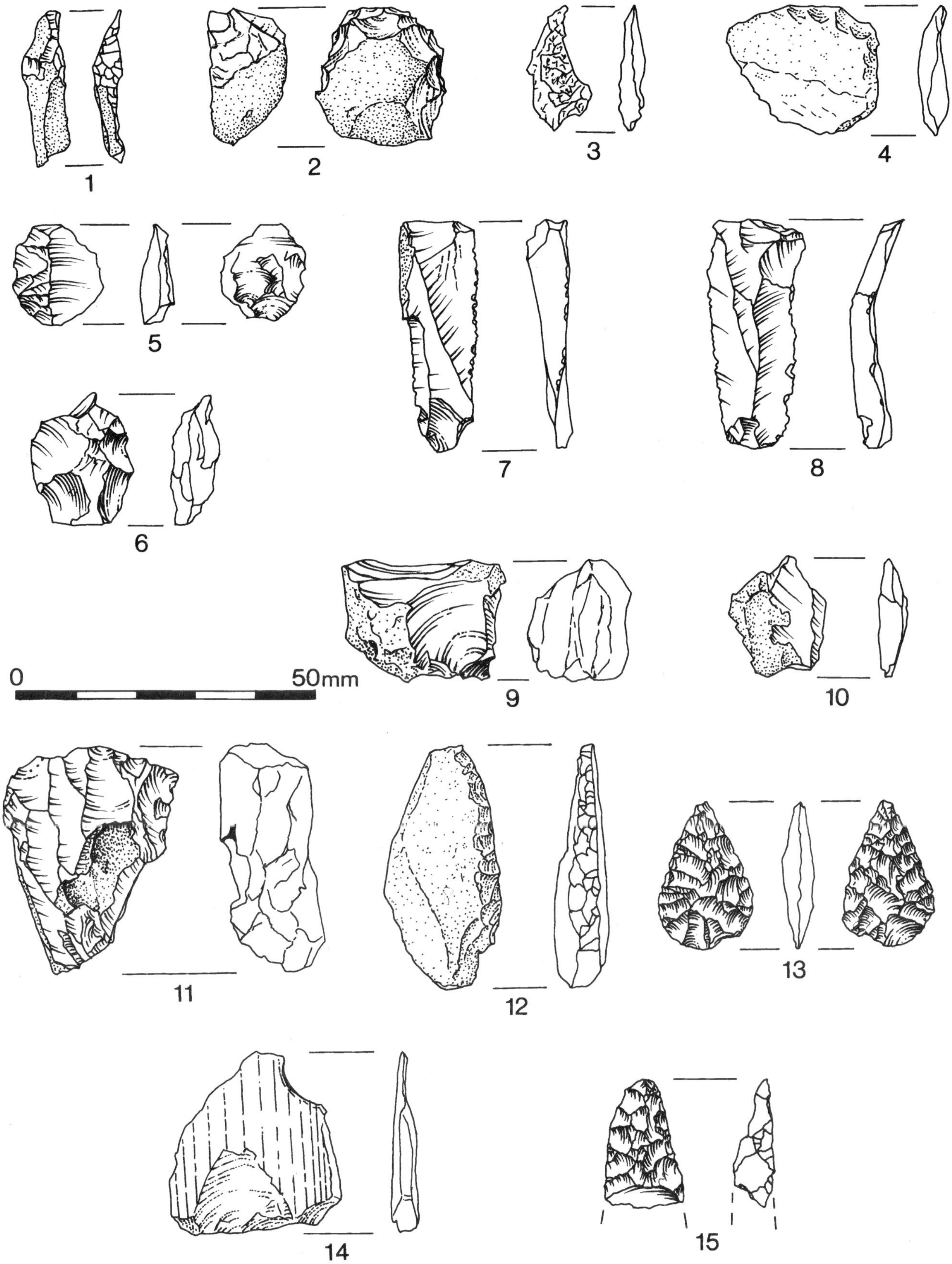

Fig 59 Selection of flint implements collected during the archaeological survey

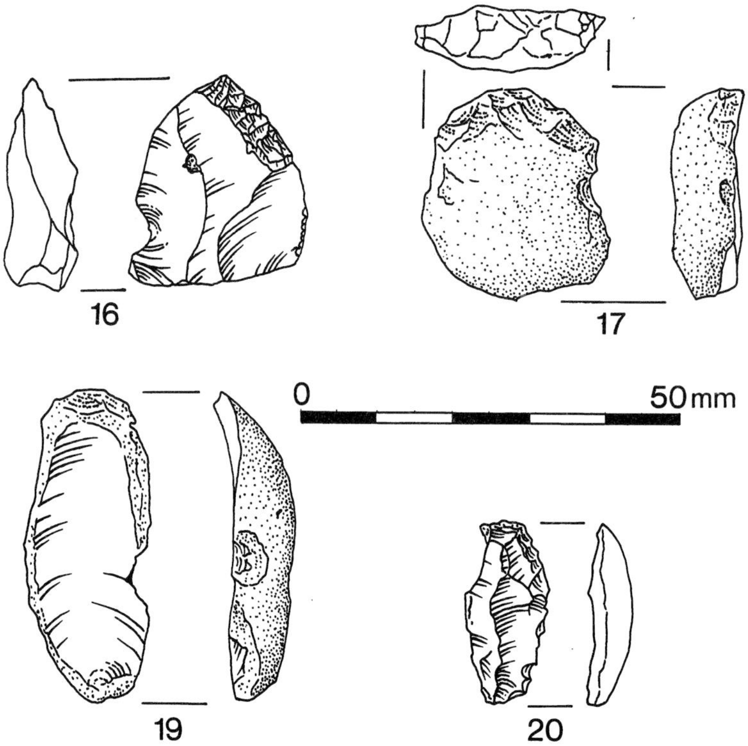

Fig 60 Selection of flint implements collected during the archaeological survey

Plate 49 Giant's Grave Bronze Age standing stone at Kirksanton, West Cumbria

for the early Neolithic in the county suggests continued landscape management by burning, notably in the uplands (Bradley and Edmonds 1993, 138-9), and also of small-scale clearances. Fine-resolution palaeoecological analysis of appropriate peat deposits covering the late fifth and early fourth millennia cal BC may reveal similar patterns.

The greatest change in the artefactual record of the period is the introduction of the polished stone axe, which was produced on an industrial scale at the Group VI sources in Langdale (Fig 58), a fact that has probably skewed the county record to disguise other Neolithic elements. Indeed, it was only the presence of such axes, some apparently being manufactured, that distinguished the Ehenside Tarn site as Neolithic to its excavator (Darbishire 1873). The only indicator for a Neolithic presence over much of lowland Cumbria is the evidence of individual axes which, on their own, are difficult to interpret as a proxy indicator of settlement: there are few settlement traces represented either by physical structures or surface flintwork. Scattered finds of arrowheads, such as those from Walney Island, may (speculatively) represent isolated hunting losses.

The apparent continuity between the late Mesolithic and early Neolithic has been noted, as has the importance of resource-rich wetlands for hunter-gatherers. The role of wetlands for early farmers is, however, less clear. The small amount of early Neolithic flintwork from the Coastal Plain has been taken to suggest a move away from the coast, although the absence of such material from surface surveys is a national phenomenon (Healy 1983). Archaeologists' eyes are also often distracted by the monuments of the Neolithic which tend to be located away from the coast. An alternative view suggests that the same level of coastal settlement and exploitation continued away from the Mesolithic, but there was also an expansion of activity into other parts of the landscape. The differential survival of monuments on now-marginal land is an important factor as well. Work on Neolithic Langdale (Bradley and Edmonds 1993) has suggested an expansive use of the landscape with seasonal grazing, a practice which may, from the Blea Tarn evidence, have had pre-Elm Decline origins. The continuous use of such high land, with its climatic restrictions, may suggest a more intensive use of valley bottoms and coasts, which had more clement weather and better soils. The widespread occurrence of charcoal, allied to small woodland clearances in the available pollen record, may be an indicator of this.

Bewley's (1993) work at Plasketlands on the North Cumbrian Plain may give some clues to the nature of early Neolithic activity and our understanding of the archaeological record where it concerns early farmers. The site, possibly an enclosed settlement, may, from its location between wetlands, have been related to their exploitation. The site was extremely vulnerable to modern agricultural damage and in many other parts of the Coastal Plain, where there is no perched water table, more intensive and regular drainage and ploughing over centuries would have destroyed the remains. The site was also presumed to be much later in date at the outset of fieldwork (*op cit*, 12). These factors of survival, visibility, and archaeologists' assumptions, may have led to the understating of the true extent of Neolithic remains in and around the lowland wetlands. The responses of early farmers to the wetlands themselves is likely to have been variable, dependent upon the available resources and contemporary perceptions of the landscape. There can be little doubt, however, that the results presented in this volume represent only a partial picture of the actual pattern of settlement and exploitation (Figs 59 and 60).

Late Neolithic/Bronze Age

The late Neolithic and Bronze Age are marked by a county-wide change in the archaeological record from a predominantly artefactually-based to a monument-based assemblage (Pl 49): monuments typical of the period, stone circles, henges, and the later settlements and cairnfields of the upland edge, occur widely across the region. In examining this pattern, the factors affecting archaeological visibility and survival are of the first importance; it may be no coincidence that the amount of activity documented on the narrow West Cumbrian Coast, which is bordered by the upland fringe, is greater than that from the broad North Cumbrian Plain.

In many respects the lowland archaeological record for the late Neolithic is very similar to that of the early Neolithic, namely a large number of polished stone axes from insecure contexts with few other finds; the number of sites and chance finds is low until an apparent expansion of activity from the middle Bronze Age onwards. The few known sites of the later Neolithic period, which are restricted to a disparate group of features from the Esk Estuary, are similar to the fragmentary remains of the early Neolithic and equally difficult to detect. The work at the Langdale axe factories established a change in the pattern of axe production in the late Neolithic, although it was probably framed within similar patterns of upland summer grazing and lowland settlement (Bradley and Edmonds 1993, 140-3). The scant lowland palaeoecological evidence suggests a degree of continuity, with small-scale woodland clearances allied to continuing arable cultivation, notably on the West Coast. The use of wetlands is difficult to determine for this period and there is no conclusive evidence that they were preferentially selected (or otherwise) for settlement or exploitation. The important role of the lowlands in the finishing and dispersal of Langdale axes in the late Neolithic is, however, notable, in the light of the late Neolithic radiocarbon dates from Ehenside Tarn (*see Ch 3*).

There are hints of increased landscape exploitation in the early Bronze Age, although the evidence from the lowlands continues to be fragmentary. There are signs that the rate of clearance followed by regeneration increased in South Cumbria, on the West Coast, and on the North Cumbrian Plain where (arguably) systematic cereal cultivation occurs for the first time in this period. In the later Bronze Age there is clear evidence, notably from South Cumbria, for sustained clearance resulting from many individual disturbance episodes, in part for arable cultivation. The palaeoecological data for more intensive exploitation appear to coincide with increased material evidence in the form of stray finds, notably barbed and tanged arrowheads and metalwork. One major contributor to the archaeological record is the finding of axe-hammers, which are concentrated in the Furness Peninsula (Roe 1979, fig 5).

The scattered finds can be linked to the expansion of settlement onto now-marginal land on the upland fringe, evidence for which is preserved as cairnfields, field systems, and settlements. The survival and detection of these widespread sites and finds is closely linked to modern land-use; they are found where there has been no subsequent intensive land-use (such as the uplands), in unstable soils (the Drigg and Walney Island dunes), and through systematic survey of arable fields as they become accessible (*eg* on the West Coast). The latter work has produced many small sites and isolated lithic finds of this date. Surviving monuments are virtually absent from the lowlands, although the concentration of ring ditches, showing as cropmarks on the well-drained soils of the North Cumbrian Plain, and the group of apparent henges in the vicinity of Sellafield, seen on aerial photographs (ACH Olivier pers comm), may indicate the richness of the landscape to be discovered elsewhere in the county's lowlands.

Despite the diversity of the archaeological and palaeoecological records for the Bronze Age, evidence for wetland use is fragmentary. There is little evidence to suggest they were preferentially exploited, although there are no site-specific studies for Cumbria which may illuminate localised moss use, such as the possible evidence for seasonal grazing in the Bronze Age from North Lancashire (Middleton *et al* 1995, 204-6). There are, however, several metalwork finds which may have been intentionally deposited, including finds from South Cumbria and the Salta Moss rapier (*Ch 4*). The find of a stone axe-hammer from Solway Moss itself (*Ch 4*) may suggest that intentional deposition was not solely restricted to metal artefacts.

The sites and finds from the Walney Island and Drigg dunes, and the pits in the Esk Estuary, suggest continued coastal exploitation in the period, although wetland finds are rare. Exceptional in this are Ehenside Tarn (*Ch 3*), which appears to have seen Beaker period settlement, and the trackway from Foulshaw Moss (*Ch 2*), which suggests that the riverine and estuarine wetlands of the Lyth Valley may have been important. The presence of the trackway may also have implications for the exploitation of the surrounding woodlands in the light of the amount of timber required for its construction, although the building of such structures does not necessarily show in pollen records. The (undated) 'palisade' from Bowness Common (*see above p 114*) may have similar implications for the North Cumbrian Plain.

It is likely that the lowlands and their wetlands were an integral part of a widely-settled and exploited landscape by the end of the Bronze Age. The large-scale abandonment of the uplands at the end of the period coincides with apparent climatic deterioration, resulting in marginal lands becoming unviable, and which also may have led to the development of raised mires widely across the landscape. These factors may have restricted the amount of land available for year-round settlement and altered the resources available. Raised mires support very limited flora and fauna and, in their fully developed form, see little human exploitation for subsistence. The presence of such features over large parts of the landscape, particularly in the valleys of South Cumbria and on the North Cumbrian Plain, would have severely constrained populations. Such pressure would have caused particular stress if the mosses were exploited preferentially, for example for grazing (Middleton *et al* 1995). The evidence for this is at present lacking, although future research strategies may seek to elucidate this key period for the use and exploitation of the county's wetlands.

Iron Age and Romano-British period

The apparently clear break present in the archaeological record towards the end of the Bronze Age is reflected in the palaeoecological record for the beginning of the Iron Age. The sustained clearings of the late Bronze Age in South Cumbria subsequently regenerated, although there appears to have been a stable, if low, level of clearance and cereal cultivation on the North Cumbrian Plain. In common with much of North West England, however, the middle and later Iron Age saw the sustained and permanent clearance of woodland. This is visible in pollen diagrams from across Cumbria and appears to represent a real expansion of activity. In South Cumbria much of this appears to be associated with cereal cultivation, and the evidence from the upland fringe and central valleys suggests that the deforestation occurred across the landscape, although the date at which clearance was initiated and the pattern of that clearance probably varied widely owing to local factors. The difference in patterns of regeneration between the Duddon and Lyth Valleys, for example, appears to have been related to localised sea-level change.

The absence of archaeological material associated with this large-scale clearance is a notable feature of the archaeology of the North West. The number of incidental finds and sites declines at the end of the Bronze Age and we lack a basic framework for understanding the pattern of settlement. Once again, we must look to the North Cumbrian Plain for information, where the Abbeytown Ridge has substantial traces of settlements and field systems, for which Iron Age and Romano-British dates have been proposed. The earlier origin of many of these features should not be discounted (*cf* Bewley 1993), however, and the apparently dense Iron Age and Romano-British settlement pattern may represent landscape evolution over a much longer timespan. It is noticeable that our view of settlement is dictated by the presence of well-drained soils which show cropmarks, and the preservation of features on now-marginal land (Higham and Jones 1975). A key question can be posed as to whether there was such intensity of settlement in other parts of the county. Extensive clearance and localised cereal cultivation elsewhere (for instance, on the Pennine scarp, C Skinner pers comm) would suggest permanent settlement and field systems, perhaps associated with extensive grazing over more marginal land, and the evidence from neighbouring Northumberland suggests that cattle may have been particularly important (Higham 1986, 135-8).

The marginality of the mosses to settlement at this time may be suggested by the occurrence of bog bodies from Scaleby and Seascale Mosses, although the fact that these finds were made during nineteenth century peat-cutting leaves their exact context and date difficult to establish. They clearly fit within a pattern of the ritualised disposal of bodies within mosses in the Iron Age and early Roman periods, although the depth at which the Scaleby body was found has led to suggestions of a Bronze Age date (Turner 1988). In terms of the Iron Age of Cumbria, the occurrence of the bodies does suggest that some of the mires were marginal to the main area of settlement and that there had been a move away from wetlands and coasts being an integral part of settlement and subsistence, as they seem to have been in earlier periods. It is likely that this reflects the increased wetness and acidity of the mires. From recent work (Stead *et al* 1986; Turner and Scaife 1995; van der Sanden 1996), it is clear that the apparent distribution of these bodies depends critically on where they were found, when, and by whom. It can be assumed both that many more have been discovered in the past and not reported and that there remain more to be found within undisturbed deposits.

There appears to have been no preferential use of wetlands at this time by the civilian population; there are no stray wetland finds as there were in the late Bronze Age, and settlement is restricted to higher land. The diversity of wetlands, however, means that it is likely that they would have been used for a variety of purposes. The fens south of the Abbeytown Ridge (*eg* Common Moss, Cockley Moss, and Chapel Moss) may have offered seasonal grazing and other opportunities, whereas raised mires may have had few exploitable resources; there is no evidence as yet for pre-medieval peat extraction.

It is clear, however, that during the Roman period the military were well aware of the strategic use of wetlands, overcoming the problems posed by the presence of the coastal mosses in the siting of Hadrian's Wall, to the extent that they surmounted considerable engineering difficulties to construct the Wall along the south shore of the Solway. This contrasts with the positioning of the earlier, late first century, Stanegate, which followed the low Fingland Rigg to the west of Burgh to Kirkbride, thus exploiting the higher, drier land, as well as providing a good view across the Solway to the north.

Medieval and post-medieval

Most recent archaeological work has focused on the prehistoric and Romano-British settlement of the lowlands and there has been little consideration of the later use and exploitation of the peatlands. The NWWS programme sought to overcome this by targeting work on those documentary sources (Pl 50) which were accessible and related directly to wetlands, as well as through fieldwork, such as that on the Solway Cow (*Ch 4*). There remains significant scope for further work, however, particularly for the early medieval period where the archaeological evidence is scant for the region as a whole. There is also considerable potential in terms of early documents and those in collections which have yet to be indexed by the Cumbria Record Office.

Recent work, including the work of this project, has started to identify some of the uses to which the mosses were put in the early medieval period. The importance of this work rests with the fact that archaeological remains are generally the only evidence to come from an extremely poorly documented period. The Solway Cow find and the evidence for hemp retting at Glasson Moss (*Ch 4*) both represent tasks undertaken away from settlement and agricultural activities. The Solway Cow was probably deposited in a large pool on the edge of Solway Moss within an actively growing raised mire. It may be significant that the find was deposited not on the edge of the moss, but *c* 300 m into the mire, although whether this marked simply the 'accessible' limit of the mire, or the fact that the symbolism of place was stronger than elsewhere, is unknown.

Plate 50 Man digging turves, from the Lanercost Cartulary (© The Surtees Society). By kind permission of Cumbria Record Office (Carlisle)

Detailed surface survey of the mosses will undoubtedly reveal further evidence of their incidental uses which have left slight physical traces but no documentary record. Glasson Moss contains some examples of such remains, notably the hemp retting, but also slight traces of peat extraction for the manufacture of naphtha in the early nineteenth century (F Mawby pers comm). Peat was used for a diverse number of industries elsewhere (such as paper, industrial fuel, oils, and chemicals), for example on Dartmoor (Harris 1992, 98-115), for which traces may survive on some of the larger mires in Cumbria.

The available documentary records, the bulk of which date from the post-medieval and modern periods, tend to refer either to disputes or catalogues of land

Plate 51 Peatcutting on Whitbarrow Moss, 1912 (Lancaster University Library)

Plate 52 Evidence for traditional peat cutting (Lyth Valley)

holdings, particularly those with an economic value. What is not represented is the everyday exploitation of undisputed marginal land which covers the majority of Cumbria's wetlands. Thus most small-scale hand extraction of peat remains undocumented, despite the increased importance it had as a fuel during the medieval and post-medieval periods (Winchester 1987, 82, 90). Most of this exploitation would have been small-scale, with each person cutting only sufficient fuel for their own use; an idea of the scale of this can be gleaned from the diary of Isaac Fletcher, who mentions the collection ('leading') of peats regularly through the period 1756–81 (Winchester 1994). Interestingly, Isaac Fletcher's local mire, Mosser Moss, was enclosed and reclaimed for agriculture in 1767, after which there are no records of peat being brought to his house. This may indicate that the moss was the source of the peat, and it suggests that its enclosure and the ending of peat-cutting could be associated with wider economic and social changes.

Most commons and mires were regulated throughout the medieval and post-medieval periods to prevent the wholesale removal of peat for sale. The origins of the use of peat as a fuel evidently predate the earliest documents of the twelfth century, since these clearly indicate established systems of cutting within 'stinted' mires. It was also used extensively in salt production around the county's coast, notably by the monastic houses. The increased use of peat for fuel was related to the lack of available woodland by the post-medieval period (Winchester 1987, 90), and palaeoecological data suggest a largely cleared landscape by the late medieval period, notably in South Cumbria, although some woodland did survive on the North Cumbrian Plain. In some areas near towns, such as in the Lyth Valley near Kendal, there was commercial exploitation on a large scale throughout the post-medieval period which removed most of the area's mires. By the beginning of the twentieth century, peat was regarded as 'poverty fuel' and peat carters and customers were regarded as 'improvident and yet independent' (Palmer 1952, 166-7); domestic-scale peat-cutting died out by the 1950s, usually for want of labour, although there is still some peat cut by traditional methods in South Cumbria (Pls 51 and 52).

A notable feature of the post-medieval period was the commercialisation of peat exploitation, with the shift from domestic to large-scale extraction. In part,

Plate 53 Modern commercial peat cutting, Wedholme Flow

Plate 54 Moss reclaimed for agriculture, Silecroft

this was driven by the establishment of rural industries, notably that of iron processing in South Cumbria, which used a variety of fuel sources, including peat (*eg* Marshall *et al* 1996, 198). In the nineteenth century, however, peat came to be used for a variety of purposes, notably bedding for horses and poultry, and the railways permitted effective transport around the country. In the twentieth century large-scale exploitation of the larger mosses has continued for horticultural peat with a shift from hand-cutting to mechanical production during the 1960s. This continues, somewhat controversially, at three sites in North Cumbria: Wedholme Flow (Pl 53), Bolton Fell Moss, and Solway Moss.

Aside from peat-cutting, evidence for the medieval and post-medieval use of the mosses is somewhat fragmentary. Some documents suggest that the mosses in South Cumbria were used for animal pasturage, notably Angerton and Arnaby Mosses and those in the Lyth Valley. Other such evidence relating specifically to mosses is absent, and it may be that the location of these South Cumbrian mosses immediately adjacent to saltmarsh meant that they were particularly suitable for grazing; sheep are still grazed in such areas. The lack of use of mosses by Holm Cultram Abbey, in the north of the county, is illuminating and contrasts markedly with the evidence for the reclamation of uplands and saltmarsh by Furness Abbey in the south.

The evidence for reclamation of the mosses (Pl 54) prior to the Parliamentary Enclosures is fragmentary and it is often unclear whether drains and dykes mentioned in documents are purely boundaries or whether they were for drainage also, particularly when the term 'dyke' often refers to a mound, rather than a ditch. Both Furness and Holm Cultram Abbeys appear to have delineated at least some of their holdings with banks and drains: for instance, the 'Mossebank' mentioned in the Lyth Valley in 1352 (Farrer 1923–4, II, 124) would appear to suggest a physical structure across the moss. Indeed, the Lyth Valley appears to have been unique in the county, with rights for enclosure being granted from 1315, although no more than 5 % of the mosses had been reclaimed before the late eighteenth century (*Ch 2*): whether this is a further reflection of the close association between the Lyth Mosses and salt-marsh is unknown. There is, however, a clear contrast there with the situation on the North Cumbrian Plain, where the mosses around Holm Cultram Abbey were unreclaimed at the Dissolution and the value of the mosslands appears to have been solely for fuel (*Ch 4*). Piecemeal enclosure, such as that around Bowness Common, appears to have been the small-scale reclamation of worked-out turf dales.

Large-scale reclamation of mosslands was not undertaken until the Parliamentary Enclosures of the late eighteenth to the mid-nineteenth centuries, but, even with this more systematic approach to 'land reform', the effect on the landscape was variable across the county. The Lyth Valley, for example, was wholly enclosed, except for the small remnant areas of moss, whilst elsewhere individual mosses were enclosed, such as Ulverston Moss with many, such as Mosser Moss (in West Cumbria), being enclosed and drained under private schemes. Ehenside Tarn is another example of such small-scale enclosure, with the farmer draining a tarn to reclaim the land (Darbishire 1873), although the area was subsequently permitted to re-flood. On the North Cumbrian Plain, ambitious enclosure schemes for the mosses appear to have been thwarted by a lack of belief that it would result in large amounts of productive land. Whilst some mosses there were simply enclosed for agriculture, most of the areas of surviving moss were enclosed to delimit areas of peat-winning, and in these areas there was only limited drainage.

Palaeoecological survey

The extension of the palaeoecological survey beyond the areas in which archaeological survey was undertaken, encompassing the margins of selected lakes in the Lake District National Park and the basin mires from the fringe of the Central Uplands (Pl 55) and Eastern Cumbria, has provided some indications as to the importance of these sites in our understanding of the palaeoenvironment and archaeology of these areas. Therefore, rapid palaeoecological assessments were made on a number of lake margins (*Appendix 5*). In these areas the rate of change seems to be slow, and dependent upon a complex set of factors, including present management, and wider environmental factors such as the trophic state of the water and sediment inflows.

The preliminary results of the NWWS palaeoecological assessment suggest that most surveyed sites are well-preserved with relatively long sequences of pollen and plant macrofossils. In the absence of radiocarbon dates, the dating of the profiles depends upon pollen assemblages which suggest that sequences back to the beginning of the Holocene may be present. The results have also reinforced the importance of palaeoecological work in areas where conventional archaeological survey is unlikely to be successful. Hints of woodland openings in the Mesolithic period, detected during the assessment of Three Mile Bog (*see Appendix 5*), suggest a promising site for further analysis. The evidence for cereal

Plate 55 Peat on a lake margin, at Littlewater (NY 5090 1705) (© T N Huckerby)

cultivation, for example from Derwentwater, suggests that the valley bottom and lake margins may have been *foci* for settlement throughout the historic period, whilst the identification of suitable sites in the valley bottoms (*eg* White Moss at Rydal) of the Central Fells is likely to prove useful in the study of the prehistory of the Lake District. Three Mile Bog and White Moss alone demonstrate the justification of the palaeoecological strategy in Cumbria. Furthermore, the work of the palaeoecological survey (*see Appendix 7*) from Carlingill in Eastern Cumbria has enhanced that undertaken by Cundill (1976) in understanding the formation of the nexus of deeply gouged gorges in the Lune Valley.

Threats to the resource

The NWWS survey of Cumbria has identified a number of threats to the archaeological and palaeoecological resource which will, over time, lead to the destruction of the wetland cultural landscape by desiccation, physical disturbance, or piecemeal loss. Two points are also apparent from the fieldwork; firstly, the current threats stem directly from past activity, notably from agricultural improvement and drainage, and from the long history of peat-cutting which has left the centres of the old mosses exposed to desiccation. The current expression of historical peat-cutting is commercial peat harvesting, much of which is undertaken under planning permissions granted prior to recent conservation concerns. The second point is that most of the threats are operating over a long time-scale; many of the mire fragments, for example in the Lyth Valley (Pl 56), have been in their current situation for over a century and yet still have potential as palaeoecological archives with pollen and macrofossils (variably) preserved.

It is important to identify the nature of the threats to the wetland archaeological and palaeoecological resource. The Peatlands Campaign Consortium in several publications has identified commercial peat extraction as the main threat to mires (*eg* RSPB 1993) and has based a management strategy on reducing the demand for horticultural peat. Recent evidence, however, suggests that demand for peat has actually increased by a quarter in the decade since the campaign began (DETR 1999). In Cumbria, however, even given the fact that the largest area of cut mire in the North West lies within the county, commercial extraction is arguably not the main threat. Even on sites where peat is still actively being cut, survey has suggested that the rate of attrition is relatively slow (*c* 50–150 mm per year) although this is annually variable. In addition, it has already been noted (*Ch 4 above*) that there is little direct evidence to suggest that raised mires, where all commercial cutting takes place, were the *foci* for settlement and subsistence. Chance finds made in the upper levels of such mires, however, such as the Solway Cow and Solway Sheep, may indicate activity marginal to settlement.

There is one further factor which may suggest that commercial peat extraction may in some circumstances be less of a threat to archaeological remains than other threats, although the damage to the palaeoecological record cannot be understated. Peat is cut by disturbing the upper few centimetres of the peat deposit with a harrow which is then left to dry. The peat beneath this, deeper than *c* 0.25 m from the surface, can be very well preserved, except where immediately adjacent to drains (Pl 57). Thus cut mires have the potential to hold well-preserved archaeological deposits immediately prior to disturbance for cutting.

This means that it is the archaeological importance of the deposits under short/medium-term threat that archaeologists should seek to manage. Thus, for example, at Bolton Fell Moss, most of the peats under immediate threat are of post-medieval date — arguably from the last 200–300 years. Over the medium term, medieval peats may be threatened, although prehistoric peats are unlikely to be affected for perhaps a century. It is in these earlier peats, from the fen stages of the sequence, that organic archaeological remains are most likely to be found (arguably) and so in this instance perhaps the most successful management would be to record the stratigraphy and palaeoecology of the deposits under short-term threat and maintain close links with those working the site to ensure prompt recording of remains. In the longer term, however, we may wish to preserve the prehistoric deposits in their entirety, perhaps as a basis for mire regeneration. This highlights the fact that we should be able to identify the importance of different deposits within different types of mires and communicate this effectively to those managing them.

As with other parts of the North West, the desiccation of peats through drainage, either for agriculture or for peat-cutting, has resulted in significant damage to the resource, and continues to do so. The extent of the damage is, however, difficult to assess. Palaeoecological assessments would appear to suggest very good pollen and macrofossil survival even in mires which have been cut over extensively. This does not mean, however, that survival is guaranteed over the long term; the archaeological

Plate 56 Remnant moss in the Lyth Valley

Plate 57 Cleaning ditches, Solway Moss

and palaeoecological resource within remnant mosses and upstanding hags will continue to decline in quality and extent. Some areas may be more vulnerable to desiccation than others, for instance, fragments of former mires and upstanding peat baulks and mire margins. Equally, fully reclaimed mires, such as Silecroft and Pow Beck, in West Cumbria, will be vulnerable to existing drainage, as was recently demonstrated at Ehenisde Tarn, where worked wood was identified in newly cut drains. This drainage, however, will be maintained as long as the areas are farmed. Where there has been management action, the retention of water within profiles may have reduced the rate of decay.

Recent palaeoecological work has identified the importance of medieval and post-medieval peat deposits and coring has established their scarcity. Past peat-cutting has removed such deposits from most of the mosses in the county and they survive only in large, intact mires, although small fragments may remain elsewhere, such as in the centres of the former mires in the Lyth Valley and under peat baulks in Foulshaw Moss. This survival would, however, need to be established. The deposits that do survive are vulnerable to desiccation, being at the top of the profiles, as well as being exposed to disturbance from surface works, and physical removal where peat-cutting is still practised. The protection of such deposits must be a priority in any scheme of management.

The potential importance of vestigial traces on mire surfaces has already been mentioned (*Ch 4*). One of the most important types of site, and one that is ubiquitous across most mires, is the evidence for peat extraction, both commercial and domestic. In most cases, such remains are the only traces of a practice which has been undertaken for centuries, for which there is little documentary evidence; indeed, the archaeology of peat extraction is very poorly understood and characterised. It seems perhaps surprising that, for a craft which was an integral part of rural life until the middle of the twentieth century, archaeological remains are the only evidence. A great many questions remain relating to the division of the mosses, the operation of common rights, the nature and extent of medieval peat extraction, and the origins of the practice; little is known also of the experimental and short-lived industrial use of peats.

In mires cut by milling, many remains will have been removed, but in less intensively worked areas, mire restoration works pose the greatest threat to archaeological deposits. Unfortunately, the archaeological evidence is under threat from any agency which disturbs the mire surface, and thus also from 'natural' regeneration whereby the baulks and trenches become levelled up as vegetation colonises the lower areas and upper baulks desiccate and fall apart. The time-scale over which the remains deteriorate in such conditions is unknown, and indeed may be variable, although the fact that the Glasson Moss hemp retting pools and drains (*Ch 4*) survived after several centuries of abandonment must be cause for optimism.

There are several wetlands where the threat to the archaeological resource could not be assessed, notably open water areas and lake margins. Most of these are located in the Central Fells, outside the archaeological survey area, but during the survey it was recognised that these were important lowland wetlands, and their management should be considered. How archaeologically-sensitive deposits react in such environments is not fully known, although there may be some damage from the increased growth, in the water margins, of reeds (*Phragmites communis*), the roots of which can penetrate and disturb archaeological deposits (*eg* Leighton Moss in North Lancashire (Middleton *et al* 1995, 134-5)). In many of these situations, however, there is likely to be little or no obvious threat to archaeological remains and management action may remain a low priority. The archaeological and palaeoecological importance of such deposits, however, should be identified and incorporated into land management and water-course plans. The palaeoecological assessments undertaken as a part of this project have made a start in this direction (*see Appendix 5*).

Some of the mosses, notably Foulshaw Moss in South Cumbria, are threatened by existing forestry plantations. This reflects past practice and it is unlikely that new plantings would be undertaken on intact raised mires, now that the importance of lowland raised mire habitats has been recognised (Forestry Commission 1999). New planting may, however, be a continuing threat to vestigial and thin peats in former mire areas which may contain archaeological deposits and palaeoecological archives.

Work to reverse past planting is evident at, for example, Nicols Moss in South Cumbria, where English Nature has cleared plantations and scrub to re-create a raised mire habitat. In terms of the archaeological and palaeoecological archive, existing forestry is, however, very damaging: Wimble (1986), in his work at Foulshaw, identified the damage caused by tree roots and desiccation to pollen and plant macrofossils in the upper stratigraphy. It has been noted in this section that this upper stratigraphy can be the most important element of the mire stratigraphy.

There are thus several agencies which threaten damage to the archaeological resource, including, on occasions, the complete removal of the peat, mostly in the creation of artificial landscaped ponds: it was during one such operation in the Lyth Valley that the skull of a horse was recovered by a JCB driver. The pattern that emerges is one of variability across the county and within mires. Commercial peat-cutting, the most scenically and politically obvious threat, may be less of a danger to the archaeological than to the palaeoecological resource, although machine harvesting does inevitably damage finds, as in the case of the Solway Cow. With the vigilance and interest of the extraction companies, however, it does not necessarily destroy them. Finally, it must always be remembered that most mosses are threatened by desiccation following drainage initiated, in some cases, centuries ago.

Whatever the causes, there is no doubt that the archaeological and palaeoecological resource is threatened and mitigating management action is required. The following section will review current management strategies and identify areas where archaeological management may be appropriate.

Management strategies

Agriculture

The major threat to the smaller mosses, particularly those in enclosed land, and to the margins of the larger, unenclosed mires, is agricultural drainage, much of it instigated in the last century. The scale of the threat is long term although, as repeated work at Ehenside Tarn has shown (*see Ch 3*), it has an insidious effect: desiccating deposits, artefacts, and palaeo-ecological sequences. In addition, there are small, isolated areas of arable which cause physical damage to deposits and the oxidation of surface peat layers. It should be said, however, that this is not a significant problem, when compared to other parts of the North West, such as South West Lancashire (Middleton *et al* forthcoming).

The management of agricultural landscapes has a number of problems; not least is the fact that they are managed by farmers who make a living from the land and, most often, do not have money to protect features of environmental value unless compensated for the income lost in making the land-management changes. For farmers there is little incentive either to spend money or lose land (to, for example, re-wetted mire surfaces) unless compensated. In semi-natural habitats under the management of nature conser-vation bodies, it is arguably easier to incorporate measures that protect archaeology. Schemes can, however, be developed for the protection of archaeological landscapes and features, notably through MAFF's agri-environment programme.

The *Lake District Environmentally Sensitive Area* (ESA) incorporates all of the National Park and its wetlands, including the basin mires, lake margins, upland peats, and mires on the upland fringe. The provisions of the scheme, into which almost any agricultural land-holding can be entered, incorporates far-reaching provisions for the protection of the archaeological and palaeoecological wetland resource, including no new land drainage, protection of known historic features, and no extension of arable by the ploughing up of old pastures. In this case, the farmer receives a series of payments to compensate for the income foregone in the adoption of these and a wide range of other environmental protection measures.

Similar protection can be offered by the Countryside Stewardship programme which targets the environmentally-valuable land outside ESAs. The flexible programme of grants can assist farmers to undertake environmental improvements on their holdings, which include measures to improve the condition of archaeological sites and the restoration and protection of wetlands, which are a nationally-targeted landscape within the scheme. There are also similar cross-compliance measures to those in ESAs related to the extensions of land area under arable and new drainage. On a large, landscape scale, this programme, which is part-funded by the European Union, offers the only practical means by which the wide variety of small wetlands can be managed and their archaeological importance considered.

Nature conservation

The importance of mires for nature conservation has been recognised only relatively recently, although some aspects, such as their value for birds, have been known for significantly longer. The importance of the habitat is matched by its decline: only 4 % of raised mires 'intact' in 1840 survive today (RSNC 1990), and only 6 % of England's raised mires survive (RSPB 1993, 11-12). The accuracy of these figures depends crucially on the reading of the First Series of Ordnance Survey maps of the 1840s, when a significant number of mires were being reclaimed (Middleton *et al* 1995, 208-10). The long history of small-scale peat-cutting will also have damaged 'intact' mires. Whilst the figures may be less dramatic than sometimes presented, there can be no doubt that lowland mire habitat has seen significant losses over the past 150 years. There are also significantly less than 10,000 ha of intact raised mire remaining in Great Britain, which is the minimum threshold for the conservation of an entire habitat by SSSI status.

The nature conservation value of the mires is enshrined in a series of legal designations, of which SSSIs and NNRs have already been mentioned. A significant proportion of raised mire (13,000 ha) has SSSI status, although not all of it is in pristine condition; for instance, the peat workings at Bolton Fell Moss and Wedholme Flow are both in SSSIs (designated long after the workings were established). A significant proportion of the NNRs in Cumbria are lowland wetlands, including Bassenthwaite Lake, Blelham Bog, Bowness Common, Cliburn Moss, Drumburgh Moss, the Duddon Mosses, Finlandrigg Woods, Glasson Moss (South Solway Mosses), Halsenna Moor, North Fen, Roudsea Wood and Mosses, Rusland Moss, Tarn Moss, Thornhill Moss, Walton Moss, and Wedholme Flow.

These national designations have been supplemented by international legislation, notably:

- *EC Council Directive on the conservation of Wild Birds* (79/409/EEC) through which Special Protection Areas (SPAs) are designated. There are none of these in the Cumbrian wetlands;
- *Ramsar Convention on Wetlands of International Importance* (signed in Iran in 1971). Two areas of raised mire, Duddon Mosses and Solway Mosses, are Ramsar sites, mainly for wintering wildfowl populations;
- *EC Directive* on the conservation of *Natural Habitats and of Wild Fauna and Flora* (92/43/EEC), through which Special Areas of Conservation are designated (SACs).

The importance of managing these habitats is also underpinned by Biodiversity, adopted by the UK Government at the 1992 Rio Earth Summit. Several *Habitat Action Plans* have been adopted which recognise the importance of wetlands and the proportion of the world habitat in the UK, each which will result in specific targets and action by named Governmental Departments and Non-Govermental Organisations. Published plans relevant to Cumbria's wetlands include *Reedbeds, Fens, Lowland of Raised Bog, Standing Open Water, and Estuaries* (UK Biodiversity Steering Group 1995), and the draft *Lowland Raised Bog Habitat Action Plan* outlines actions from a wide variety of organisations which can influence the protection and management of the habitat. These include sustainable agriculture, Policy Planning Guidance, Water Level Management Plans, Local, Mineral, and Structure Plans and Common

Land Legislation, and the production and consumption of horticultural peat. Further detail to this complex picture will be added to local management strategies with the future publication of the *Cumbria Biodiversity Action Plan*.

In 1991 English Nature issued a *Position Statement on Lowland Peatland*, subsequently reinforced by Biodiversity targets, which informed much of the ensuing management action in Cumbria under the Lowland Peatland Programme established in 1992. It set out to acquire the most important peatland sites, to conserve existing habitats, and rehabilitate damaged areas (Johnson 1997), the aims of the rehabilitation programme being to:

1 raise captured rainwater to within 100 mm of the mire surface and minimise seasonal fluctuations;

2 develop sample damming techniques;

3 control invasive vegetation to permit peat-forming systems to spread (*op cit*, 324).

In particular, an emphasis was laid on the rehabilitation of mires through positive management works with a focus on mires with significant depths of peat and where there is an 'existing or potential nature conservation value' (English Nature 1991).

The programme of management works, undertaken by both English Nature and the Cumbria Wildlife Trust, has been enhanced since the acquisition of several parcels of land in the north of the county by English Nature from Fisons Horticulutre Ltd (now Levingtons) (Pl 58). Much of this land was formerly commercially worked for peat and mainly comprised trenches and baulks from former hand-block peat-cutting, with vegetation that varied from regenerating *Sphagnum* to dense mature heather. The main areas concerned were at Wedholme Flow, adjacent to Levingtons' worked area, and Whitrigg Common, next to the Glasson Moss NNR. On these sites measures were taken to prevent surface water loss by damming drainage dykes and control of scrub. On both Glasson Moss and Wedholme Flow the results were monitored by a series of dip wells to check the level of the water retained in the mire, initial results suggesting that these relatively straightforward

Plate 58 North West Wetlands Survey staff engaged in field survey

Plate 59 Creating and maintaining micro-habitats to encourage specific species groups to flourish at Scaleby Moss

works have been successful in stemming water loss (Johnson 1997).

Other works have created micro-habitats (Pl 59) to encourage specific species groups, particularly to encourage wet mire species on dry surfaces. This has included the excavation of ponds to attract dragonflies, and the control of invading pines and birch, on Scaleby Moss by the Cumbria Wildlife Trust. On Whitrigg Common, heather is regularly burnt to create stands of differing ages to promote invertebrate diversity.

The effects of these works on the archaeological resource has been limited. Where peat removal has been needed, for instance at Scaleby Moss, exposed sections were examined and their impact on the archaeological resource assessed. There can be little doubt that the work to raise water levels benefits organic archaeological and palaeoecological remains in general. In particular, it will allow the retention of complete sequences, particularly the important upper deposits.

Not all management for nature conservation, however, takes a holistic view and there is little that takes the wetland heritage into account explicitly. At best the works may benefit archaeology incidentally, but at worse real damage can be caused to wetland sequences. The creation of ponds through habitat management for birds and invertebrates may be a significant danger (*eg* Kirby 1992) and there is a pressing need for the impact of such work to be assessed. Many of the problems stem from a lack of appreciation of the potential of wetland archaeology which means that its priorities are not included within management plan objectives. There are cases, however, where archaeology is a 'material consideration', although it is still not central to wetland management (*eg* RSPB 1997, 23).

The conservation works outlined *above* are concerned with management issues for individual sites. A wider approach is taken by Water Level Management Plans, whereby organisations with responsibility for land drainage have an agreement with specific provision for the control of water levels in certain areas (MAFF *et al* 1994). Certain designated areas, such as SSSIs, Scheduled Monuments, SACs, and SPAs, are priorities. In Cumbria, such agreements have been drawn up for Glasson Moss, and Roudsea Wood and Mosses NNRs which will be reviewed prior to further implementation (D Smith pers comm). Catchment Management Plans by the Environment Agency take a wider view by considering water resources within the whole river catchment, a wide approach recommended under the 1996 *EC Water Resources Framework Directive*. Both of these approaches will be of significant value for archaeology and palaeoecology within mires, as long as archaeological values are promoted within the schemes to ensure that the full wetland potential is realised.

Archaeology and palaeoecology

Preservation by record

Most archaeological management within wetlands has come about through the planning process under *Planning Policy Guidance Note 16* (PPG16; DOE 1990) and minerals planning legislation. At the outset of the project, prior to the publication of PPG16, several planning applications on the edge of wetlands were brought to the attention of the project and a watching brief undertaken in two cases, at Kirkbride and Salta. In neither case were archaeological deposits discovered. As a result of PPG16, subsequent planning applications were subject to more formal assessments to identify archaeological and palaeoecological values. An example of such is Kidburngill, near Whitehaven, where an assessment was undertaken of a small peat-filled basin prior to coal extraction. The methodology involved transect coring to examine the stratigraphy, and spot pollen samples to identify preservation and provide pointers to dating (LUAU 1996; 1998). This was undertaken in association with other works to assess and evaluate mining remains in the affected area.

Most of the mosses in Cumbria have the physical evidence for many years of intermittent peat-cutting to fuel the fires of local people. In all but three cases, Bolton Fell Moss, Wedholme Flow, and Solway Moss, such cutting has long ceased. These three sites are still cut commercially for horticultural peat under planning permissions, many of which date back to before the Second World War. Before the mid-1980s, peat extraction was seen as a useful field of employment in relatively remote areas where unemployment could be high. The last planning permission granted for peat extraction was on Solway Moss in 1981, although peat had been taken from the moss commercially for nearly 100 years prior to this. Nature conservation interests had not been recognised prior to this sufficiently to affect the planning process; permission for extraction on Wedholme Flow in 1975, for example, was granted after full consultation with the Nature Conservancy Council. The old Planning Permissions had conditions that limited only the area worked and the anticipated after-use was agriculture or forestry as appropriate. Archaeological and nature conservation priorities were not considered in these permissions and there was little control of extraction methods.

Since the *Minerals Planning Guidance Note 14* (DOE 1995), however, which followed from the Environment Act (1995), all minerals planning permissions were to be reconsidered with conditions on methods of working, environmental impacts, and after-use. As a result, SSSIs were prioritised, first by the Peat Producers Association; Environmental Statements for Bolton Fell Moss and Wedholme Flow were produced in 1999. Solway Moss, which is not an SSSI, will be considered in 2000. These statements have allowed archaeological issues to be addressed fully for the first time; at Wedholme, for example, a programme of mitigative action has been accepted by the site operators, including the regular monitoring of exposed peat faces. A reversion to regenerating mire has been accepted as the after-use, including the retention of a minimum of 1 m of peat at the base of the deposit (B Gorman pers comm).

There can be little doubt that wetland archaeology has suffered within mires which have planning permissions reflecting the priorities of their time. The scattered finds that have been reported from the mires, such as the axe from Solway Moss, may be but a small part of the resource that has been lost, a loss probably considerably enhanced with the change from hand- to machine-cutting since the 1960s. The new legislation, with its full appraisal of environmental impacts of the works, will go some way to redress the balance, particularly with the continuous appraisal of the loss of the resource with periodic rapid survey. There is also the opportunity to reassess the archaeological strategy as the environmental impacts are reviewed every 15 years to bring them up to date. It should be noted, however, that the actual area covered by extraction cannot be restricted, unless compensation can be found to pay the site operators, and so the chances of *in situ* preservation will be negligible, unless through the goodwill of the site owner. The second point is that, whilst the 1 m of peat for regeneration set out by the legislation will address nature conservation concerns, it will not do much for the archaeology, except for anything in that last peat remnant, or protected surfaces underneath it.

Preservation *in situ*

The approaches outlined *above* have depended upon recording archaeology as it is destroyed. In Cumbria, there has, however, been no action specifically directed at the management of wetland archaeology *in situ*. This reflects the lack of systematic wetland data and specific sites to manage. Nationally, all efforts to manage wetland archaeology have taken a site-specific approach (Coles 1995), largely for wholly practical reasons. Management at a landscape scale has rightly tended to rely on initiatives from other bodies, notably English Nature and the Environment Agency (*see above*). This raises two key points; firstly action by other agencies will focus on their priorities, thus English Nature will aim to manage SSSIs and NNRs which are those areas with either intact mire vegetation or the potential for restoration. Mires in poor condition, such as those reclaimed for agriculture and/or severely truncated, will not be restored or managed by English Nature but these may contain important archaeological remains or palaeoecological sequences. Secondly, extensive peat survival under pasture in the county means that it is very difficult to predict exactly where sites and finds will occur and it is thus very difficult to specify site-specific management action. Any proactive action must therefore depend on the archaeological and palaeoecological potential of mires.

The idea of archaeological and palaeoecological potential underpins a great deal of work in wetlands, including the NWWS. The fragmentary record, represented by stray finds and some detailed work, indicated at the outset of the project that the wetlands of the North West had a potential, and it was the aim of the NWWS to further understanding of the record in order to aid management. In certain areas, such as Merseyside (Cowell and Innes 1994) and Lancashire (Middleton *et al* 1995; Middleton *et al* forthcoming), the nature of the record and threat to the wetlands has permitted the potential to be understood, if only to a limited degree (Middleton 1999). Unfortunately, where access to the archaeological record is restricted, the potential is less easy to understand and must rest on an assessment of the existing record and new data from targeted areas which, it is hoped, this volume has provided for Cumbria.

To summarise, several general points have arisen in the course of this volume relating to the potential of the mires, which are presented *below*.

1. All mires are likely to contain evidence for vegetation exploitation by humans as far back as the Neolithic and also possibly the Mesolithic. This may suggest that settlement and other sites are present in the vicinity.

2. In the Bronze Age, mire edge activity was likely to be related to fens and valley mires. Raised mires, due to their acid nature, are perhaps more likely to contain sporadic finds of metalwork and bog bodies, rather than evidence of agricultural exploitation, although indirectly they may provide evidence of landscape changes from the pollen record.

3. Iron Age and Romano-British raised-mire deposits may contain bog bodies. Other mires may have been important within a settled agricultural landscape (*eg* seasonal grazing), and large raised mires in the north may even have had a defensive role. The archaeological traces of these roles, however, are likely to be slight.

4. Medieval use of raised mires was restricted mainly to peat-cutting, traces of which are likely to survive and be recognised once the archaeological signature is understood. Evidence for other sporadic activities may also occur (*eg* hemp retting, and deliberate deposition of objects, animal remains). Other mires may have been used as in the Iron Age and Romano-British periods, although, perhaps in a more intensive manner.

5. Remains of industrial archaeology are of equal importance to those of earlier periods. Mire exploitation included reclamation for agriculture and peat-cutting, including substantial amounts for the iron industry.

6. The palaeoenvironmental record of all mires has both palaeoecological and archaeological value, since fossilised pollen and macrofossils can survive even in remnant and reclaimed mires.

7. The upper sequences of intact mires are particularly valuable, given that, as a result of cutting, they are now rare and can give evidence for medieval and post-medieval environmental change and human activity.

8. For mires in areas where archaeology is difficult to record, owing to land-use and other factors, the palaeoecological record is particularly valuable as a proxy indicator of landscape use.

There is a danger inherent in this sort of exercise of prioritising some issues over others; all mires may produce unexpected evidence, this is one of the exciting facets of wetland archaeology. If mires are to be managed for archaeology, particularly if it is to be enacted and paid for by others, archaeologists have to be able to convey the particularly important elements of any mire which they wish to see protected. There should therefore be an expression of the potential of each mire which moves beyond the blanket statement that all mires are important for archaeology.

Several issues follow from this list in terms of Cumbria, notably the need for further work in certain areas (*see below*) to amplify and add detail to our knowledge, and the need to involve all people who live near to and manage mires. Interesting finds, such as the Solway Cow, no doubt help, although something more is often required to ensure a sustained interest which will influence the way people think about mires. There is a need to overcome the thinking which says that, because mires (particularly raised mires) are not settled now, the same was true in the past. The archaeological remains all speak of human exploitation, from the Mesolithic onwards right into the twentieth century, to produce, through the interaction of human and natural processes, the modern landscape. In many cases, this is classified as important enough to warrant formal nature conservation to ensure biodiversity, and has merit as a cultural landscape within its own right.

Future research

The NWWS work in Cumbria has been a starting point for our understanding of human interaction with wetland environments. There are, however, several areas where further research would substantially enhance our understanding of past human interaction with the landscape, and the significance of the resource itself.

Archaeological and palaeoecological survey in related wetlands

This volume differs from others in the NWWS publication series by having 'lowland' in the title. This was deliberate; of all the NWWS counties, Cumbria contains the greatest diversity of wetlands occupying all topographic zones across the county. In the timescale of the project, it was therefore deemed better to concentrate on limited areas in the lowlands, to provide a comprehensive review of those areas most similar to the wetlands of other NWWS counties. A significant body of past work exists for the wetlands of the uplands, however, in both the Central Fells and increasingly in the Pennines to the east, but the potential of this, whilst significant, is as yet understudied. The project's participation in recent work at Seathwaite, upper Borrowdale (Wild *et al* forthcoming) only serves to reinforce this. It has been proved across the uplands that where there is little peat and archaeological visibility is good, complete relict landscapes can be found (*eg* Cherry and Cherry 1987b; Quartermaine 1989); the potential for buried sites and intact surfaces under upland peat may be considerable.

Similarly, the volume has not discussed in any detail the considerable potential of the valley bottoms in the Central Fells where there are important wetland deposits. Rapid coring has established that long, well-preserved sequences were present, containing a substantial record of human activity and landscape change. All areas of valley-bottom wetland will contain important records of their environs which were intensively used and settled from at least the Neolithic period, and perhaps earlier. In addition, non-intensive land-use in recent centuries may have ensured good survival of deposits. Other wetlands, which, potentially, are of the first importance but were outside the scope of the survey, were foreshores. Most of the Cumbrian coast has a long tidal reach and there are extensive areas of saltmarsh, grazing marsh, and open foreshores of international importance for their nature conservation interest. Recent work by English Heritage and the RCHME has established the archaeological importance of these areas (Fulford *et al* 1997), although the lack of recent detailed survey in the North West has meant that the precise potential of the region is difficult to establish.

A way forward in the interpretation of palaeo-ecological records from small basins (<100 m in diameter) would be provided by a fully multi-disciplinary approach. Pollen analysis from such sites is known to provide a record from the vegetation immediately surrounding the site (Janssen 1966; Jacobson and Bradshaw 1981; Jackson 1990; 1994), and a combination of geochemical, charcoal and pollen analysis at a site in Hungary has provided evidence of soil degradation and human activity related to both farming and mining (Willis *et al* 1998). Cumbria, with its numerous basin mires and extensive archaeological records — including known mining activity — would provide excellent sites for such a study.

The northern limit of the NWWS survey was defined by Cumbria's border with Scotland. As this volume has shown, several very significant mosses lie on the North Cumbrian Plain where survey has been focused. In one case, Solway Moss, survey went up to the Scottish Border. It is clear that there are many landscape similarities between the Scottish and English sides of the Solway Firth and the distribution of mosses is contiguous on both sides. The threats facing the mosses are equal — in Dumfries and Galloway large areas of mossland are either forested or being actively harvested.

There are, however, some interesting differences in the archaeological record between the two areas which require elucidation. The most significant of these is the occurrence of crannogs in Scotland, which appear to be completely absent from England. This may be related to the influence of the Roman presence in England (Barber and Crone 1993, 520–1) and the differences in patterns of settlement on either side of Hadrian's Wall, although it may also reflect a different contemporary view of wetlands. It would be of great use if the results of work on both sides of the Border could be collated to add colour to the picture being revealed.

The archaeology of peat-cutting

It has already been noted (*see above*) that one of the key results from the NWWS work has been the identification of the importance of the remains of peat-cutting. In the North West this has been a relatively understudied area of research, largely because the documentary record is very poor, even though it was a common, everyday experience for most inhabitants in the environs of wetlands. Where work has been undertaken, the results have been illuminating in terms of the manner in which mosses were exploited, for example at Fenn's and Whixall Mosses in Shropshire (Berry *et al* 1996) and on Dartmoor (Harris 1992). The physical traces on mire surfaces are the only evidence for such activity and they are vulnerable to even the smallest changes in mire management. There needs to be a fuller understanding of the archaeological signatures of peat-cutting and peat exploitation, allied to a detailed assessment of those mosses undergoing proactive restoration works.

Liaison and partnership

It is to be hoped that one of the key messages from both this chapter and other recent publications on wetland management is that landscape management on any scale cannot be undertaken by one discipline alone. The North West Wetlands Survey has demonstrated that a good picture of the archaeology of wetlands can only be achieved by the integration of archaeological and palaeoecological techniques. A holistic approach is also required in the wider community, which takes into account all interests that can then be reconciled within an objective-led management strategy. For this to happen there must be liaison and dialogue between partners which means that archaeology, in its widest sense, including palaeoecologists and the Quaternary Science community, has to make its presence felt by bodies representing larger interests which may have conflicting aims. Similarly, there must be liaison with local communities and visitors to present the human dimension of the wetland landscape. There may be opportunities for this, for example, in conjunction with new access planned around the western end of Hadrian's Wall (D McGlade pers comm). To do this, archaeologists must have a clear view of what they are trying to protect, conserve and enhance, and to present it in a manner that can be appreciated by a non-specialist audience (Pl 60). It is to be hoped that this volume is a step in the right direction.

Plate 60 Sunset over Midgeholme Moss, Cumbria

Glossary

Basin mire Mire developed in enclosed waterlogged depressions which have become colonised by peat-forming vegetation. Most often found in areas of glacial deposition where local irregularities of relief such as kames and kettle holes provide the necessary geomorphological environment of hollows with enclosed drainage. Usually characterised by relatively small surface area (*ie* <50 ha) compared to depth (often >5 m).

Biostratigraphy Stratigraphy based on changes in biological assemblages in the sediment (*eg* pollen zones).

Blanket mire Peat developed directly over mineral ground up to a considerable angle of slope (10–25°), normally in an upland environment.

Bog General term for ombrotrophic mires.

Bog Burst Catastrophic failure of structural and hydrological integrity of a peat bog. The cause remains unknown but is always associated with a rapid increase in the amount of water entering the system (usually caused by prolonged heavy rain). Bog bursts are rare events but may have been commoner before modern agriculture improved drainage of mire systems.

BP Before Present. In terms of radiocarbon dates, present is defined as AD 1950.

Bryophyte Member of the major group of the plant kingdom comprising the mosses and liverworts.

Calibration (of radiocarbon dates) See *Radiocarbon dating.*

Cereal type Large grass pollen grain of between *c* 40 μm and 60 μm in size.

Chronozone A chronostratigraphic unit, defined as the time-span of a selected stratigraphic unit of another kind (*eg* a pollen zone) the boundaries of which have been radiocarbon dated (*eg* Flandrian II).

Cropmark Variations in the growth of crops caused by disturbances in the soil due to features completely buried beneath. Cropmarks are most readily identified from the air under favourable lighting conditions.

Dendrochronology Tree ring analysis. In prehistoric examples the dating of some species of tree remains can be achieved by matching variations in ring patterns with examples of known age. Variations in tree ring widths can also be used to shed light on climatic fluctuations in the past.

Devensian The most recent glacial stage, culminating in a brief interstadial (Windermere Interstadial LDeII) followed by a final cold period (The so-called 'Younger Dryas' or Loch Lomond Stadial, LDeIII). The end of the cold conditions at around 10,000 cal BC (*c* 10,000 BP) marked the opening of the Flandrian stage.

Ecosystem An ecological unit comprising distinct plant and animal communities, together with their abiotic environment (*eg* saltmarsh, ombrotrophic mire).

Ecotone Transitional boundary between different types of ecological community.

Elm Decline A fall in elm pollen frequencies as percentages of tree pollen, dated throughout north–west Europe to a period around 4000–3700 cal BC (*c* 5000 BP). It defines the boundary between the Flandrian II and III chronozones (zones VIIa and VIIb in the Godwin system (1975)). Its exact cause is controversial but it may be related to a complex interaction between disease, human activity, and possibly climate.

Environmental archive Term used to describe any discrete body of organic sediment which, by means of palaeoecological analyses, is capable of generating an environmental history for a particular geographical area. In the context of the *North West Wetlands Survey* it is synonymous with the term *peat archive.*

Esker — Long, winding ridges of sand and gravel, unrelated to the surrounding topography, found in previously glaciated regions. They originated beneath the ice, either from continuous deposition at the mouth of a sub-glacial stream as the ice retreated, or from the infilling of the tunnels of these streams before recession.

Fen — General term for minerotrophic mires.

Flandrian I, II and III — Choronzones of the present interglacial stage in Britain, dating from *c* 10,000 cal BC (*c* 10,000 BP) to the present. (See *Holocene*).

Floodplain mire — Minerotrophic mire, characterised by fen communities, frequently developed in lowland alluvial plains traversed by sluggish rivers.

Fluvio-glacial — Descriptive term for deposits or processes connected with the hydrology of glaciers.

GIS — Geographic Information System. A map-based computerised archive and modelling system for geographical and related information.

Grenzhorizont — Literally, 'boundary horizon'. A prominent recurrence surface found in many European ombrotrophic mires. First identified by the German stratigrapher C A Weber and initially thought to be synchronous, reflecting climatic change. Recent research has indicated a wide range of dates for major humification changes but many cluster around the period *c* 600 cal BC (*c* 2500 BP).

Gyttja — Fine detrital mud, usually deposited in open freshwater conditions. Synonym — *Nekron mud.*

Holocene — The present interglacial stage in Britain, dating from *c* 10,000 cal BC (*c* 10,000 BP) to the present.

Humification — Degree of decomposition (of peat).

Hydrosere — A plant succession commencing in waterlogged sites.

Intermediate mire — Mire characterised by a stratigraphic profile indicating the growth of a peat reservoir to cover low-amplitude water partings leading to coalescence with other peat reservoirs in adjacent basins. They are 'intermediate' because they form a type of mire morphology transitional between true raised mires and blanket mires. Synonym: *ridge-raised mire.*

Interstadial — Relatively short-lived warmer period occurring within a full glacial stage (*eg* Windermere Interstadial which lasted perhaps 500-1000 years at the end of the Devensian glacial stage).

Kame — A steep-sided alluvial cone deposited against an ice front.

Kettle hole — Small, steep-sided depression in glacial boulder clays formed by dead ice fragments from a retreated glacier. Kettle holes often become waterfilled to form small lakes or become colonised by peat-forming vegetation and develop into basin mires.

Macrofossil — Usually taken to mean plant macrofossil. A term which covers everything that cannot be considered a microfossil. It effectively includes all plant remains which can be recognized by the naked eye or with the aid of a low power microscope. The semi-decomposed plant remains are not true fossils, in the conventional geological sense of the term (*ie* having petrified tissue), but are sub-fossils. Cell contents are generally replaced by water in the peat sub-fossil material.

Microlith — Generic term for very small worked flint flakes and blades thought to be diagnostic of Mesolithic lithic assemblages.

Minerotrophic — Used to describe a mire whose surface receives water from outside that mire's own limits.

Mire — Peat-producing ecosystem which develops in sites of abundant water supply.

Mire macrotope Mire complex which has been formed by the fusion of isolated mire mesotopes.

Mire mesotope Mire system developed from one original centre of peat formation.

Monocotyledon Abbreviation: *monocot.* Group of flowering plants possessing only one cotyledon. In the context of macrofossil assemblages, often used as a convenient shorthand to encompass undifferentiated remains of grasses, sedges and rushes.

Moss Local (northern England/Scottish lowland) term for mire, usually, but not exclusively, of the acid, ombrotrophic variety.

Ombrotrophic Literally, 'fed by rain'. Used to describe a mire which receives water only directly from the atmosphere in the form of precipitation.

Ontogeny Biological or ecological development through time.

Palaeoecology The study of past interactions between plant and animal communities and their environment.

Palaeoenvironment An ancient environment.

Palynology The science and practice of pollen analysis.

Paludification Initiation of mire formation over mineral ground by the lateral expansion of peat.

Peat Partly-decomposed organic material formed in areas of permanent waterlogging.

Phase-shift Change in macrofossil assemblage indicating a change in surface hydrological conditions (ie from a wetter to a drier situation or *vice versa*).

Pingo Blister-like hill in periglacial terrain resulting from the expansion of sub-surface ground ice embedded within drift. Subsequent melting of the ice core of the pingo can leave a more or less circular depression surrounded by a low ridge.

Pollen zone A body of sediment with a consistent and homogeneous fossil pollen and spore content which differs in type and frequency from that of adjacent bodies of sediment. See *Chronozone.*

Post-glacial Used to denote the temperate period since the end of the Devensian at *c* 10,000 cal BC (*c* 10,000 BP).

Radiocarbon dating A method of dating ancient organic material. The technique involves the measurement of amounts of radiocarbon (which is subject to decay after death) remaining in organic matter. *Calibrated* radiocarbon dates have been converted to calendar years and are distinguished by the use of the suffix 'cal BC' or prefix 'cal AD'.

Raised mire Ombrotrophic mire characterized by a very low amplitude convex profile and usually occupying topographical situations such as level floodplains of river systems and alluvial deposits of estuaries. The ontogeny of raised mires usually indicates that ombrotrophic plant communities have developed over minerotrophic and/or aquatic ones as upward growth of peat has insulated the surface from groundwater influence.

Recurrence surface A stratigraphic feature found in many ombrotrophic mires consisting of a boundary between two peat layers of markedly differing humification, the upper layer being less humified. Thought to be related to climatic changes (particularly increased rainfall) affecting the wetness of the mire surface leading to an increased rate of mire growth and peat accumulation. Recently, this view has been challenged by Chambers (1991) who has suggsted that human disturbance might also play a role. Recurrence surfaces have been classified into a sequence in Europe by the Scandinavian palaeoecologist Granlund and are often given the prefix 'RY' (from *Recurenstyor*), *eg* RYIII being the '*Grenzhorizont*'.

Ridge-raised mire Synonym for intermediate mire.

Sand island	An area of mineral ground exposed by the wasting of overlying peat. Sand islands sometimes reveal evidence for prehistoric human activity pre-dating the paludification of the surface by mire expansion.
Sandur	A large flat area at the margin of an ice sheet. Sediments are deposited on this plain by a few large meltwater streams which then become braided at some distance from the ice and subsequently form deeply cut channels that flood during periods of greatest melting.
Sere	Plant succession.
Sickle gloss	Characteristic wear pattern found on the sharpened edges of lithic implements used for harvesting.
Skirtland	Dark-stained soil derived from decomposed peat. Skirtland represents the edges of a former mire which have retrenched due to disturbance (usually by agriculture).
SMR	Sites and Monuments Record.
Sphagnum	'Bog moss'. Genus of mosses often dominant in ombrotrophic mires which are characterised by their water retentive hyaline cells.
Topmoss	Local term adopted by the *North West Wetlands Survey* to describe untruncated peat deposits.
Type site	In a palynological context, a locality chosen to be a standard for comparison of regional biostratigraphic units.
Valley mire	Mire occurring in small, shallow valleys or channels which are not enclosed so that movement of water along the long axis is possible even with only a slight gradient. Typical of wet, elongated depressions in acidic heathland areas of lowlands.
Wetland	A term which has still to achieve a commonly accepted meaning in common usage. Its ecological meaning remains vague but is usually defined as an area characterised by a water table which, for a significant part of the year, lies close to the substrate in which vegetation is rooted. In an archaeological context it is usually taken to refer to any large area of terrestrial ground containing extensive waterlogged or organically-dominated terrain. For the purposes of the *North West Wetlands Survey*, however, it is restricted to lowland mire systems (raised mires, valley mires, floodplain mires, basin mires) and their mineral ground hinterlands.

Appendix 1: Site gazetteer

Archaeological Survey Sites: Cumbria

This gazatteer contains a complete list of new archaeological sites and findspots discovered by the North West Wetlands Survey in Cumbria. The numbering of the sites represents the order in which they were discovered.

Site CU1
NGR NY 1123 4809
Parish HOLME ST CUTHBERT
Ht (OD) 20 m
Date Mesolithic/Neolithic?
Land-use Arable
Description
Small scatter of mixed flintwork with predominantly Mesolithic/Neolithic traits. A single Bronze Age piece was also present.
Unretouched flake 1, Backed blade 1, Dressing chip 1, Thumbnail scraper 1, Unretouched flake 1.

Site CU2
NGR NY 1125 4790
Parish HOLME ST CUTHBERT
Ht (OD) 22 m
Date Neolithic
Land-use Arable
Description
Small group of Neolithic flints.
Preparation flake 2, Unretouched flake 2, Edge dressed flake 1.

Site CU3
NGR NY 1283 4698
Parish HOLME ABBEY
Ht (OD) 40 m
Date Bronze Age
Land-use Arable
Description
Single piece of irregular waste, very crudely worked. Several impact fractures visible.
Irregular waste 1.

Site CU4
NGR NY 1350 4733
Parish HOLME ABBEY
Ht (OD) 15 m
Date Neolithic
Land-use Arable
Description
Small group of nine flints. All appear to be of early Neolithic type; includes two blades. A small piece of burnt bone was also present.
Unretouched flake 4, Preparation flake 1, Utilised flake 1, Irregular waste 1, Point 1, End scraper 1.

Site CU5
NGR NY 1330 4802
Parish HOLME ABBEY
Ht (OD) 45 m
Date Neolithic/Bronze Age
Land-use Arable
Description
Single piece of worked flint. Undiagnostic.
Irregular waste 1.

Site CU6
NGR Centred on NY 1444 4505
Parish WEST NEWTON
Ht (OD) 15 m
Date Mesolithic/Neolithic
Land-use Arable
Description
Small collection of flints. Two of the pieces exhibit blade characteristics suggesting a date in the early Neolithic or Mesolithic.
Unretouched flake 1, Trimming flake 1, Irregular waste 1.

Site CU7
NGR NY 1283 4855
Parish HOLME ST CUTHBERT
Ht (OD) 35 m
Date Mesolithic/Neolithic
Land-use Arable
Description
Single blade core with a piece of natural flint. Mesolithic or early Neolithic date.
Core 1.

Site CU8
NGR Centred on NY 1150 4695
Parish HOLME ST CUTHBERT
Ht (OD) 45 m
Date Prehistoric
Land-use Arable
Description
Small collection of artefacts. None of the pieces is clearly datable, although the flake knife is very curious — careful flaking on poor quality split pebble. May indicate bi-polar working.
Irregular waste 1, Dressing chip 1, Unretouched flake 1, Flake knife 1.

Site CU9
NGR NY 1079 4715
Parish HOLME ST CUTHBERT
Ht (OD) 34 m
Date Neolithic?
Land-use Arable
Description
Single unretouched flake made of poor quality flint. Possibly Neolithic.
Unretouched flake 1.

Site CU10
NGR NY 1026 4720
Parish HOLME ST CUTHBERT
Ht (OD) 20 m

Date N/A
Land-use Arable
Description
Three pieces of natural flint — no site.

Site CU11
NGR NY 1182 4694
Parish HOLME ST CUTHBERT
Ht (OD) 25 m
Date Neolithic/Bronze Age
Land-use Arable
Description
Collection of three worked pieces. None is diagnostic, although probably of Neolithic/Bronze Age date.
Preparation flake 1, Irregular waste 1, Dressing chip 1.

Site CU12
NGR NY 1625 5040
Parish HOLME ABBEY
Ht (OD) 15 m
Date N/A
Land-use Arable
Description
Scatter of natural flint — no site.

Site CU13
NGR NY 1387 4606
Parish WEST NEWTON
Ht (OD) 22 m
Date N/A
Land-use Arable
Description
Two pieces of natural flint — no site.

Site CU14
NGR NY 2421 5226
Parish ESKRIGG
Ht (OD) 17 m
Date N/A
Land-use Arable
Description
Single piece of natural flint — no site.

Site CU15
NGR NY 1455 4955
Parish HOLME ABBEY
Ht (OD) 35 m
Date Neolithic/Bronze Age
Land-use Arable
Description
Three worked pieces, two of which were preparation flakes. The crude working and poor quality raw material may indicate a Bronze Age date.
Preparation flake 2, Unretouched flake 1.

Site CU16
NGR NY 1468 4931
Parish HOLME ABBEY
Ht (OD) 19 m
Date Neolithic
Land-use Arable
Description
Single, relatively crude blade. Of presumed Neolithic date.
Unretouched flake 1.

Site CU17
NGR NY 1328 5024
Parish HOLME ABBEY
Ht (OD) 21 m
Date Neolithic
Land-use N/A
Description
Axe, in farmer's possession. *See* Middleton (1991b) for further details.

Site CU18
NGR NY 1786 4945
Parish DUNDRAW
Ht (OD) 22 m
Date Neolithic
Land-use Arable
Description
Collection of three Neolithic flints dated by an early Neolithic leaf arrowhead.
Trimming flake 1, Leaf arrowhead 1, Dressing chip 1.

Site CU19
NGR NY 1230 5214
Parish HOLME LOW
Ht (OD) 8 m
Date Bronze Age?
Land-use Arable
Description
Two worked pieces of flint. The crudely-worked core fragment split during the working of a poor-quality flint pebble.
Core fragment 1, Dressing chip 1.

Site CU20
NGR NY 1213 5233
Parish HOLME LOW
Ht (OD) 8 m
Date Mesolithic/Neolithic
Land-use Arable
Description
Single fragment of blade.
Unretouched flake 1.

Site CU21
NGR NY 1202 5227
Parish HOLME LOW
Ht (OD) 8 m
Date Roman
Land-use Arable
Description
Single fragment of samian vessel. Dr 37, central Gaulish manufacture, probably Lezoux, possibly in the style of *Cinnamus*. Mid-late second century.

Site CU22
NGR NY 1190 4751
Parish HOLME ST CUTHBERT
Ht (OD) 39 m
Date Neolithic?
Land-use Arable
Description
Single unretouched flake found to the east of Tarn Dub.
Unretouched flake 1.

Site CU23
NGR NY 1194 4743
Parish HOLME ST CUTHBERT
Ht (OD) 41 m
Date Prehistoric
Land-use Arable
Description
Single unretouched flake made of light grey flint, found near CU22.
Unretouched flake 1.

Site CU24
NGR NY 1303 4868
Parish HOLME ST CUTHBERT
Ht (OD) 35 m
Date N/A
Land-use Arable
Description
Two pieces of natural flint — no site.

Site CU25
NGR NY 1283 4928
Parish HOLME ABBEY
Ht (OD) 23 m
Date Neolithic?
Land-use Arable
Description
Single flint — a point made on a squat flake to make a fine point. Of a type commonly found in earlier Neolithic assemblages.
Point 1.

Site CU26
NGR NY 1385 4968
Parish HOLME ABBEY
Ht (OD) 19 m
Date Medieval
Land-use Pasture
Description
Single body fragment of incompletely reduced, green-glazed medieval pottery. Abraded.

Site CU27
NGR NY 1379 4983
Parish HOLME ABBEY
Ht (OD) 19 m
Date N/A
Land-use Pasture
Description
Single piece of natural flint — no site.

Site CU28
NGR NY 1380 4937
Parish HOLME ABBEY
Ht (OD) 22 m
Date Neolithic
Land-use Arable
Description Broken flake removed from polished stone axe. Visible inspection suggests of Group VI source (Langdale).
Polished stone axe (flake) 1.

Site CU29
NGR NY 1388 4937
Parish HOLME ABBEY
Ht (OD) 22 m
Date Neolithic/Bronze Age
Land-use Arable
Description
Isolated trimming flake. Not closely datable.
Trimming flake 1.

Site CU30
NGR NY 1227 4840
Parish HOLME ST CUTHBERT
Ht (OD) 22 m
Date Bronze Age
Land-use Arable
Description
Scale-flake flint knife associated with a crudely-worked unretouched flake. Both of likely Bronze Age date.
Flake knife 1, Unretouched flake 1.

Site CU31
NGR NY 1194 4798
Parish HOLME ST CUTHBERT
Ht (OD) 40 m
Date Neolithic?
Land-use Arable
Description
Single point. Finely-made on crude flake. Eroded edges, possibly by being in sand.
Point 1.

Site CU32
NGR NY 1114 4587
Parish HOLME ST CUTHBERT
Ht (OD) 21 m
Date Bronze Age
Land-use Arable
Description
Crudely-worked flint pebble with removals off one edge to make small points.
Denticulate 1.

Site CU33
NGR NY 1374 4956
Parish HOLME ABBEY
Ht (OD) 30 m
Date Neolithic/Bronze Age
Land-use Arable
Description
Polished Group VI axe with some plough damage, patinated and worn. Distal end and cutting edge have been reworked. Crudity of flaking and different patination may indicate Bronze Age date for reworking.
Polished stone axe.

Site CU34
NGR NY 0842 4449
Parish ALLONBY
Ht (OD) 7 m
Date Unknown
Land-use Arable
Description
Copper alloy coin. Very worn. Indecipherable.

Site CU35
NGR Centred on NY 0813 4447
Parish ALLONBY
Ht (OD) 7 m
Date Neolithic/Bronze Age; Roman?
Land-use Arable
Description
Isolated unretouched flake associated with several unworked flint nodules, including a large piece which may have derived from chalk. Also, single small fragment Roman? pottery, oxidised white fabric.
Unretouched flake 1.

Site CU36
NGR NY 1162 5270
Parish HOLME LOW
Ht (OD) <10 m
Date Prehistoric
Land-use Arable
Description
Single piece of undatable burnt irregular waste.
Irregular waste 1.

Site CU37
NGR NY 0832 4270
Parish ALLONBY
Ht (OD) 8 m
Date Unknown
Land-use Arable
Description
Ceramic ?net-sinker or small loomweight in hand-fired almost stoneware. Undatable.

Site CU38
NGR NY 3455 7010
Parish KIRKANDREWS
Ht (OD) 25 m
Date Early Medieval
Land-use Peat
Description
Cow bones found on Solway Moss, having been disturbed by peat machine and scattered. An area measuring 7 x 4 m was excavated and the bones proved to be a discrete concentration of 5 x 3 m. Pieces included foot bones and skull fragments and five horn cores: three animals are represented. Radiocarbon-dated to eighth–tenth centuries. *For fuller details see Ch 4.*

Site CU39
NGR NY 3421 6816
Parish KIRKANDREWS
Ht (OD) 25 m
Date Post-medieval
Land-use Peat
Description
Remains of a sheep fleece and some bones found by a peat worker and reported to NWWS staff. An area 5 x 5 m was excavated. Rib, leg bones, some skin and fleece survived. The peat was dated to *c* 2569–2039 cal BC (3840±85 BP) and *c* 3622–3040 cal BC (4595 ± 90 BP), the fleece to cal AD 1440–1955 (265±95 BP) and the bone to cal AD 1470–1955 (230±75 BP), suggesting that the sheep fell into watery peat and died. *For fuller details see Ch 4.*

Site CU40
NGR SD 304 743
Parish BARDSEA
Ht (OD) 0 m
Date Prehistoric
Land-use Peat
Description
Fragments of bone found in a peat-filled channel at the interface of sand and shingle on the beach at Bardsea, south Cumbria. An area 3 x 6 m was excavated, revealing old stream channels running to the sea filled with peat in which were the roughly articulated remains of a leg, possibly red deer. *For fuller details see Ch 2.*

Site CU41
NGR NY 1352 4683
Parish BROMFIELD
Ht (OD) 10 m
Date Unknown
Land-use Pasture
Description
A possible circular structure located in a slightly wet pasture field.

Site CU42
NGR NY 0820 4525
Parish HOLME ST CUTHBERT
Ht (OD) 10 m
Date Neolithic/Bronze Age
Land-use Arable

Description
Isolated flake fragment with some blade-like characteristics.

Unretouched flake 1.

Site	CU43
NGR	SD 5179 7780
Parish	BEETHAM
Ht (OD)	20 m
Date	Medieval
Land-use	Pasture

Description
A small triangular area of often indistinct earthworks on the eastern edge of Holme Moss. Two distinct zones split by former trackway. Cut by dyke which forms the parish boundary at this point. The southern area appears to be ridge and furrow, although it is in an odd location on the floor of a basin. No clear patterns visible, although the earthworks appear to be associated with cultivation.

Site	CU44
NGR	SD 4700 7845
Parish	ARNSIDE
Ht (OD)	6 m
Date	Unknown
Land-use	Pasture

Description
An undisturbed mound with an even profile, 10 m diameter and 0.6 m long, located on a small boulder clay rise on a plateau which marks the shoulder enlargements on the eastern edge of Arnside Moss. No peat remains: it is organic soil on possible marine silts. No stones show through the turf which covers the mound.

Site	CU45
NGR	SD 2250 8310
Parish	MILLOM
Ht (OD)	<10 m
Date	Medieval?
Land-use	Pasture

Description
Ridge and furrow spreading out from Angerton Hall to the floodplain, aligned north–north–west to south–south–east and 3.5 m apart. In the field to the south of the hall is a prominent headland, measuring 2–2.5 m; the headland in the adjacent field is not so obvious. Two fields have been cross-ploughed, obliterating all evidence for ridge and furrow, and two other fields have only faint remains.

Site	CU46
NGR	SD 2200 8340
Parish	MILLOM
Ht (OD)	<10 m
Date	Medieval?
Land-use	Pasture

Description
Ridge and furrow aligned north–north–west to south–south–east and 1.5 m apart, extending 25 m out from the field boundary. Slightly S–shaped. There is no obvious headland.

Site	CU47
NGR	SD 2310 8435
Parish	ANGERTON
Ht (OD)	<10 m
Date	Medieval? / Post-medieval
Land-use	Pasture

Description
Very well-preserved ridge and furrow aligned north–west to south–east and 3.5 m apart. Obvious headland at the north end, headland at the south end not as prominent.

Site	CU48
NGR	SD 2130 8435
Parish	ANGERTON
Ht (OD)	11–15 m
Date	Medieval? / Post-medieval
Land-use	Pasture

Description
Ridge and furrow 3.5–4.0 m wide in places. No headland on the northern edge and a natural headland on the southern tip.

Site	CU49
NGR	SD 2280 8530
Parish	KIRBY IRELETH
Ht (OD)	<10 m
Date	Unknown
Land-use	Pasture

Description
A rectilinear structure, possibly a stock enclosure, 24 x 25 m, with two sides and part of a third surviving to a height of 0.4 m. Leading into the third side is a hollow track, 0.6 m deep, which predates the existing boundaries. The fourth side of the feature may have coincided with the existing field boundary. Within the feature from the faded-out bank lies an anomalous feature attached to the bank itself.

Site	CU50
NGR	SD 2470 8810
Parish	KIRKBY IRELETH
Ht (OD)	10 m
Date	Medieval / Post-medieval?
Land-use	Pasture

Description
A field system and three bloomery mounds, with an associated charcoal-burning area in a clearing in Longfield Wood. The field system contains the well-preserved boundaries of a rectangular enclosure. A nineteenth–century boundary wall overlies the southern edge of the slag heaps. In the woodland there is evidence of ancient coppicing, as well as a charcoal-burning platform. There are three linear clearance cairns at the east end of the clearing. On the far side of the wood is another bloomery, associated with Green Moor Beck.

Site CU51
NGR SD 2330 8840
Parish BROUGHTON WEST
Ht (OD) 7 m
Date Unknown
Land-use Pasture
Description
An old, well-weathered peat-cutting site situated between Thornthwaite Latter Rigg Moss and Kirby Pool, on the edge of the moss. A group of raised banks divide what appears to be peat-cutting sections.

Site CU52
NGR SD 1885 8435
Parish HALLTHWAITES
Ht (OD) 7–8 m
Date Medieval?/Post-medieval
Land-use Pasture
Description
Very faint ridge and furrow aligned north–south and 4.5–5 m apart. Located in a corner of the field. No headland.

Site CU53
NGR NY 1461 5009
Parish HOLME LOW
Ht (OD) 11 m
Date Neolithic
Land-use Pasture
Description
Single core worked to produced blades. Associated with three pieces of burnt stone (59.3 g).
Core 1.

Site CU54
NGR NY 2110 5385
Parish HOLME EAST WAVER
Ht (OD) 15 m
Date Medieval/Post-medieval
Land-use Pasture
Description
Three fields of ridge and furrow to the north of Slightholm. Aligned north–south and surviving to a height of 0.1 m, with 5 m width ridge to ridge in the west fields and 25–30 m in the eastern field. Respects field boundaries of old hedge banks under mature thorn hedges.

Site CU55
NGR NY 0021 0710
Parish LOWSIDE QUARTER
Ht (OD) 25 m
Date Neolithic; Post-medieval
Land-use Arable
Description
Three pieces of worked flint, all made of the same dark grey/brown flint. They may represent debris from the same flaking episode. They can be dated to the earlier part of the Neolithic by a blade. Four pieces of clay pipe were also collected.
Unretouched flake 2, Dressing chip 1.

Site CU56
NGR NY 0717 0018
Parish LOWSIDE QUARTER
Ht (OD) 30 m
Date Neolithic
Land-use Arable
Description
Two flints, both made of poor quality light grey flint. The controled nature of the flaking suggests a possible date in the Neolithic.
Core 1, Preparation flake 1.

Site CU57
NGR NY 0016 0713
Parish LOWSIDE QUARTER
Ht (OD) 30 m
Date Neolithic/Bronze Age
Land-use Arable
Description
Isolated flint find. Difficult to date.
Unretouched flake 1.

Site CU58
NGR NY 0013 0716
Parish LOWSIDE QUARTER
Ht (OD) 30 m
Date Mesolithic/Neolithic
Land-use Arable
Description
Two burnt blade fragments. May be two parts of the same artefact broken in antiquity. Mesolithic or early Neolithic in date.
Unretouched flake 2.

Site CU59
NGR NY 0003 0715
Parish LOWSIDE QUARTER
Ht (OD) 30 m
Date Neolithic
Land-use Arable
Description
Small group of six flints. All displayed similar technological characteristics suggesting that they are all of the same date. All of the pieces could have been made from a very small raw material resource.
Preparation flake 1, Trimming flake 2, Unretouched flake 1, Scraper 1.

Site CU60
NGR NY 0011 0710
Parish LOWSIDE QUARTER
Ht (OD) 30 m
Date Neolithic
Land-use Pasture
Description
Isolated flint find. Unretouched flake of possible Neolithic date.
Unretouched flake 1.

Site CU61
NGR NY 0025 0710
Parish LOWSIDE QUARTER
Ht (OD) 25 m
Date Neolithic
Land-use Pasture
Description
Isolated flint find. Trimming flake removed from blade core, possibly by bi-polar technique.
Trimming flake 1.

Site CU62
NGR SD 063 981
Parish DRIGG AND CARLETON
Ht (OD) 5 m
Date Post-medieval
Land-use Pasture
Description
A system of ridge and furrow earthworks adjacent to the western bank of the River Irt. The earthworks are aligned east–west and measured *c* 2 m from crown to crown.

Site CU63
NGR SD 063 983
Parish DRIGG AND CARLETON
Ht (OD) 5 m
Date Post-medieval
Land-use Pasture
Description
A system of ridge and furrow aligned north–south and situated directly to the north of, and contemporary with, CU62. Separated from CU62 by a small field bank which is also aligned north–south. Like CU62, these measure *c* 2 m crown to crown.

Site CU64
NGR SD 047 986
Parish DRIGG AND CARLETON
Ht (OD) 2 m
Date Bronze Age?
Land-use Scrub
Description
An expanse of peat, first recorded in the 1950s as the site of Mesolithic flint scatters. Investigated in the 1960s by Cherry (1965), who interpreted the horizontal wood on the surface of the peat as the remains of a structure and the burnt stones as a hearth. Flint with Bronze Age affinities was recovered. Re-inspection revealed that the peat had been much eroded and more wood, none of it worked, was exposed. Similar sites in Cheshire and Shropshire are interpreted as Bronze Age burnt mounds (Leah *et al* 1997; 1998). *See Ch 2 for fuller details.*

Site CU65
NGR Centred on NY 3260 6095
Parish BURGH BY SANDS
Ht (OD) 0 m
Date Medieval
Land-use Marsh
Description
An extensive system of ridge and furrow situated on Burgh Marsh. The earthworks measure 8 m from crown to crown and are variously orientated. The presence of medieval earthworks on Burgh Marsh indicates that land reclamation concentrated on the marsh and not the mosses.

Appendix 2
The lithic material collected by the survey

Ninety-five flints were recovered by fieldwalking as part of the NWWS work in Cumbria, between 1991 and 1997. Of these 95, 23 were unworked natural flint pebbles: as in other parts of the NWWS survey, unworked pebbles were collected from the immediate vicinity of artefacts to identify the local raw materials. All of this material was too small and of too poor quality to have been knapped and it is likely that most represent the 'background' naturally-occurring flint. During the analysis, a single piece of burnt bone (CU4/10) was also identified.

The flints are discussed in general terms, since the assemblage cannot be considered discrete, given that it derived from a significant number of sites spread over a large area of, predominantly, northern Cumbria. Whilst some trends are evident in the data, the individual site assemblages were too small, and the date ranges too wide, for detailed inter-site comparisons. The composition of the worked material (72 pieces), which weighed a total of 333.6 g, is presented in Table 7.

The overall typology closely reflected the disparate nature of the lithic material. A broad range of material from different periods was represented by a very small number of diagnostic pieces — a Mesolithic backed blade, and early Neolithic leaf arrowhead, and a Bronze Age thumbnail scraper, for example. The remainder of the assemblage could be dated on the basis of the technology employed in its manufacture. It is to be regretted that North West England shares a common trait with other areas where flint resources were relatively scarce and of poor quality, namely that technological traits for each period cannot be identified with certainty. There were, however, a few indicators that can be used as a guide to the date of the material as a whole.

The earliest component was represented by a series of well-made blades from four sites (CU2, CU6, CU7, CU58), which were of likely Mesolithic or early Neolithic date (these have been marked Mesolithic? in Table 7). The single datable Mesolithic artefact, a backed blade from CU1, was within a mixed assemblage with no apparently contemporary debitage.

A clear early Neolithic element is represented by a leaf arrowhead from CU18, associated with a trimming flake and a dressing chip which may have been of the same date.

Period	Type	No	Totals
Mesolithic	Backed blade	1	1
Mesolithic?	Unretouched flake	7	
	Trimming flake	1	
	Core	1	
	Dressing chip	1	10
Early	Point	1	
Neolithic	Leaf arrowhead	1	2
Neolithic	Unretouched flake	16	
	Trimming flake	6	
	Preparation flake	8	
	Dressing chip	3	
	Irregular waste	2	
	Core	2	
	Utilised flake	1	
	Edge-dressed flake	1	
	Point	2	
	Scraper	3	
	Polished stone axe flake	1	45
Bronze	Unretouched flake	1	
Age	Irregular waste	1	
	Core	1	
	Thumbnail scraper	1	
	Flake knife	2	
	Denticulate	1	7
Undatable	Unretouched flake	1	
	Dressing chip	2	
	Irregular waste	4	7

Table 7 Lithic assemblage from North West Wetlands Survey of Cumbria: overall typology

The bulk of the assemblage was classified as Neolithic or Neolithic/Bronze Age. All of the material of this date was detached from cores using a well-controlled technique, with care in the preparation of platforms and the use of soft and hard hammers. At least some of the material was either blades, or the likely debris from the production of blades. All stages of the production cycle were represented, from cores, through preparation and trimming flakes, to unretouched flakes and dressing chips. A slightly heavier component of the assemblage, where the use of hard hammers was more prevalent, has been classified as Neolithic/Bronze Age in the archive. This appears to represent the heavier component of the Neolithic assemblage — primarily the preparation flakes and the heavier unretouched flakes.

A very few pieces were markedly cruder than others in the assemblage, exhibiting the use of hard hammers with little evidence for core preparation. In this respect, they are of likely Bronze Age date. Some of the material classified as Neolithic or Neolithic/Bronze Age may also be Bronze Age, although individual pieces of debitage can be very difficult to date in isolation. Datable Bronze Age implements included a thumbnail scraper from CU1, a denticulate from CU32, and a flake knife from CU8.

A small number of irregular waste and other crudely-worked pieces could not be closely dated and could have derived from any period represented from the Mesolithic to the Bronze Age. They are predominantly the waste products from the working of poor-quality raw materials.

Overall, the material was generally in good condition, considering that it derived from the ploughsoil. All pieces exhibited some degree of edge damage with occasional rounding of thicker edges, consistent with having been in the ploughsoil for a considerable length of time. Of all the flint, only 21 (29.1%) were broken — a low figure given the amount of movement and abrasion the flints will have received in the ploughsoil. There are, however, some factors which mitigate breakage in this assemblage, notably the generally small size of the material and their thickness in relation to their length. It is also likely that the ploughsoils from which they were recovered may have seen only occasional tillage, certainly with modern equipment. Notably absent from the assemblage were any pieces with very fresh edges indicative of having been in the ploughsoil a short time; this may indicate that the arable cultivation in Cumbria is tending not to damage undisturbed deposits. Little is known, however, of the sub-surface remains of these lithic sites in Cumbria.

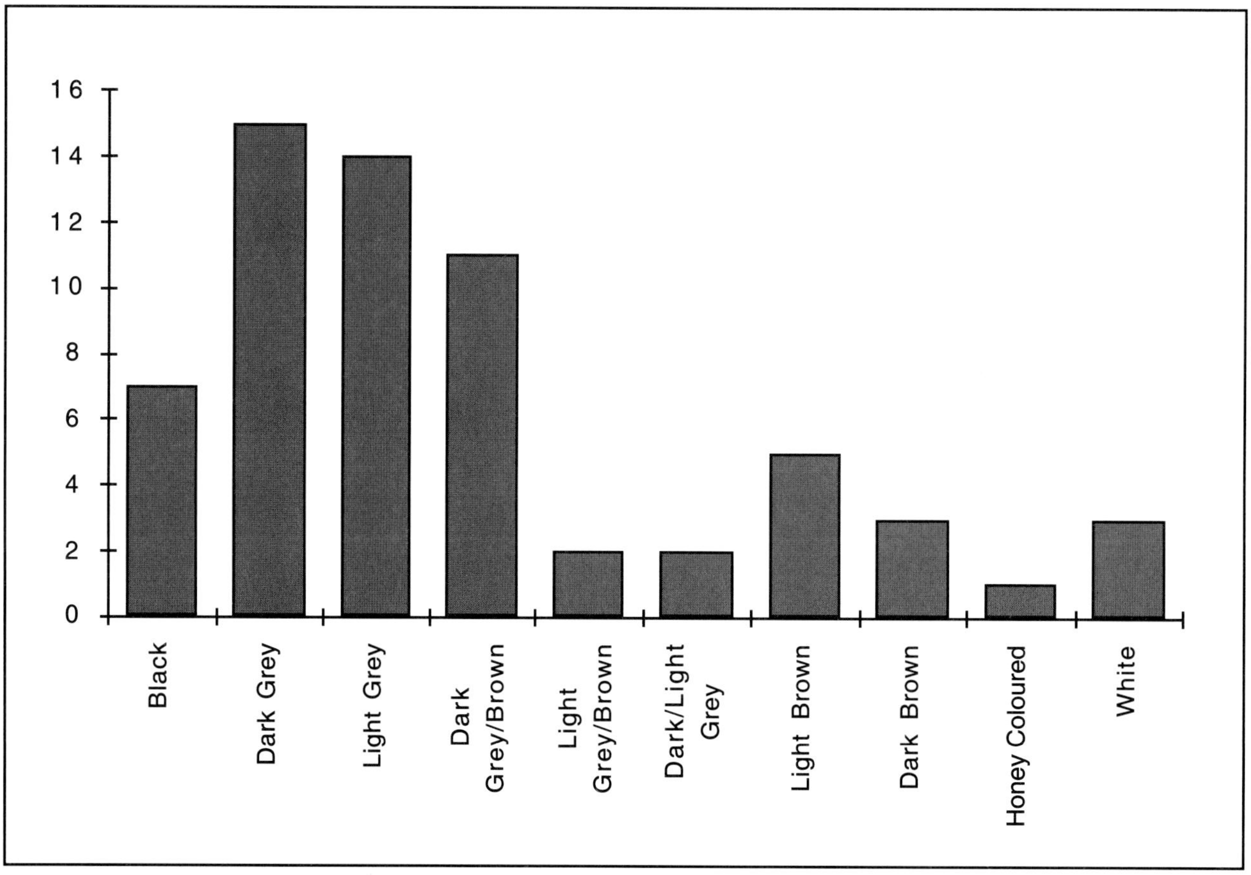

Fig 61 Lithic assemblage collected during archaeological survey: flint colour

A wide range of raw materials was used in the assemblage, perhaps not surprising given the time over which the material was deposited and the large area of collection. The colour of the raw materials is presented in Figure 61.

There was no clear linkage between the date of the material and the raw material used. It is clear from the cortical pieces that the raw material derived from gravels or beach pebble, presumably from the area of the Solway Plain. It is expected that such material would have been widely available although, from the number of flints recovered, not in any great quantity. If the main source was from the coast, as appears likely, then the availability of raw material would have varied through time with changes of sea-level and coastal morphology.

In many respects the assemblage is typical of the material recovered from other parts of the lowland North West, both in terms of overall composition and technology, and in the small number of pieces from each site. Detailed consideration of each site within its archaeological context is best left for discussion elsewhere, although a number of general points can be made.

Firstly, the overall pattern reflects the nature of the survey from which the assemblage derived — single visits to fields usually ploughed every few years. The low yield of finds mirrors the pattern identified by the work of the Cherrys on the West Cumbrian Coast (*see Ch* 3), although their assemblages there were often augmented by repeat visits. This gives a viable assemblage from technological traits from each site to be identified.

The pattern is also very similar to that recorded by Bewley (1994) during limited fieldwalking over a broadly similar area to that covered by the NWWS work in North Cumbria. It is notable that this survey employed many workers walking on closely-spaced transects, yet this recovered the same pattern of finds as two or three workers walking 10 m apart, during the NWWS work. Close scrutiny of the results from each field may illuminate the relative effectiveness of each strategy.

In a similar vein, we know little of the sub-surface archaeology of northern Cumbria — a consideration very relevant in the context of wetland research where the sub-surface potential is all important. Synthesis of the results of recent work on the Solway Plain and West Cumbrian Coast, notably that by Bewley (1994), and Bonsall (Bonsall *et al* 1986) at Eskmeals, would be useful to identify the relationship between sub-surface and surface remains. This has been shown to be crucial in the assessment of the archaeological resource in other parts of the NWWS survey area (for instance, Cowell and Innes 1994; Middleton *et al* 1995).

Appendix 3
Gazetteer of SMR sites on and around Cumbria's wetlands

This gazetteer is of sites of prehistoric to medieval date, in or around wetland areas surveyed archaeologically or palaeoecologically, although the list is not totally exhaustive. It also includes SMR sites within 1 km (approximately) of the major lakes of the Central Fells. The relevant section of the gazetteer from *Peat and the Past* (Howard-Davis *et al* 1988) has also been incorporated. The entry gives a brief summary of the data contained in the Cumbria SMR: in some cases groups of records have been amalgamated together, where there is a high density of similar sites within a small geographical area. The gazetteer has been ordered numerically by chapter, with additional information divided into broad geographical areas appended.

South Cumbria: Chapter 2

Record number 2078
NGR SD 51400 90700
Parish NATLAND
Dating Roman
Description
Fort, damaged by robbing and ploughing.

Record number 2133
NGR SD 22000 82000
Parish KIRKBY IN FURNESS
Dating Neolithic?
Description
Stone celt.

Record number 2151
NGR SD 21000 87000
Parish BROUGTON IN FURNESS
Dating Neolithic
Description
Polished stone axe.

Record number 2162
NGR SD 21610 88300
Parish MILLOM WITHOUT
Dating Bronze Age
Description
Barrow.

Record number 2197
NGR SD 22000 77000
Parish PENNINGTON
Dating Neolithic
Description
Stone axe.

Record number 2226
NGR SD 27420 75090
Parish URSWICK
Dating Bronze Age
Description
Six socketed bronze celts.

Record number 2248
NGR SD 27420 75170
Parish URSWICK
Dating Prehistoric
Description
Palisaded enclosure later modified into a univallate hillfort.

Record number 2265
NGR SD 25000 72800
Parish URSWICK
Dating Neolithic/Bronze Age
Description
Settlement at Stone Close, including finds of saddle querns, a rubber, several celts, animal bones, a palstave and a bronze ring.

Record number 2307
NGR SD 27800 72100
Parish ALDINGHAM
Dating Prehistoric
Description
Cave in a limestone bluff excavated in 1991. Late glacial flints and human remains were found but the burials may be intrusive. A possible late Bronze Age brooch was also found.

Record number 2311
NGR SD 26270 74470
Parish URSWICK
Dating Neolithic
Description
Burial and finds.

Record number 2312
NGR SD 26300 74500
Parish URSWICK
Dating Bronze Age
Description
Stone quern and tanged knife.

Record number 2313
NGR SD 26020 74090
Parish URSWICK
Dating Prehistoric
Description
Farmstead on the summit of Little Urswick Crags, with an oval-shaped enclosure.

Record number 2314
NGR SD 26350 74000
Parish URSWICK
Dating Bronze Age

Description
Hoard of bronze instruments including sword, spear (originally hafted but this crumbled), several brass rings, and axes.

Record number 2329
NGR SD 26100 71600
Parish ALDINGHAM
Dating Bronze Age
Description
Early Bronze Age copper alloy axe from Gleaston Castle.

Record number 2334
NGR SD 27610 72240
Parish ALDINGHAM
Dating Unknown
Description
Two human skulls and other bones found pre-1805 adjacent to cave (Baycliff Haggs). Later sources suggest the caves were too small for burials.

Record number 2341
NGR SD 26000 74000
Parish URSWICK
Dating Neolithic
Description
Stone axes.

Record number 2343
NGR SD 26000 74000
Parish URSWICK
Dating Bronze Age
Description
Axe.

Record number 2351
NGR SD 27500 72310
Parish ALDINGHAM
Dating Unknown
Description
Iron objects said to have been found in Dobson Cave pre-1927.

Record number 2352
NGR SD 27760 72190
Parish ALDINGHAM
Dating Neolithic/Iron Age
Description
Small cave at Bonfire Scar, situated at the foot of a limestone bluff on Scales Haggs. Excavations produced finds of human and varied animal bones and molluscs. Many of the bones had been split longitudinally; some had been fashioned into artefacts although it was suggested that the cave was used for funerary purposes. At least six people were represented, both male and female.

Record number 2366
NGR SD 26800 74800
Parish URSWICK
Dating Bronze Age
Description
Collared urn found in limestone pothole with flint knife and three skeletons, one female and two male.

Record number 2367
NGR SD 23200 74750
Parish DALTON IN FURNESS
Dating Bronze Age
Description
Cist, containing a large quantity of human and animal bones, plus a bronze sword and a spearhead.

Record number 2371
NGR SD 28910 74340
Parish URSWICK
Dating Bronze Age
Description
An oval cairn within which, on the old ground surface, were more than 30 'deposits' of dark earth containing either clay or pottery sherds.

Record number 2374
NGR SD 26700 74200
Parish URSWICK
Dating Prehistoric
Description
Standing stone.

Record number 2379
NGR SD 24120 72500
Parish URSWICK
Dating Bronze Age
Description
An urn containing bones and a bronze 'razor' were discovered during quarrying. Another, containing a smaller urn decorated with criss-crossed diagonal lines, was discovered two years later.

Record number 2386
NGR SD 33340 78120
Parish LOWER ALLITHWAITE
Dating Prehistoric
Description
Large cave (Capeshead) in limestone where the estuary of the River Leven opens upon Ulverston sands, much disturbed by the construction of the railway. The mouth of the cave is *c* 20 ft above the present high water mark and the height from floor to roof is *c* 12 ft. The cave continues for *c* 80 ft. Prior to 1869 some human remains (part of a humerus, and a molar) were found embedded in a stalactite, as well as several pieces of charcoal.

Record number 2397
NGR SD 30500 78300
Parish ULVERSTON
Dating Neolithic

Description
Fragment of polished stone axe.

Record number 2415
NGR SD 39100 75670
Parish LOWER ALLITHWAITE
Dating Neolithic/Bronze/Iron Age
Description
Kirkhead cave, the entrance of which is 2 ft high but the roof rises to 14 ft before descending and forming a series of caves in the limestone. Excavation revealed animal and human bones, charcoal, coin of Domitian, along with bone/stone implements, various pottery sherds, various beads and flints.

Record number 2417
NGR SD 37000 75000
Parish LOWER HOLKER
Dating Neolithic
Description
Several stone axes were found on Winder Moor in 1864-5.

Record number 2423
NGR SD 36500 76500
Parish LOWER ALLITHWAITE
Dating Bronze Age
Description
Perforated axe.

Record number 2427
NGR SD 36000 75000
Parish LOWER HOLKER
Dating Bronze Age
Description
Two or three bronze palstaves ploughed up in a field near Flookburgh before 1886.

Record number 2437
NGR SD 36170 79130
Parish LOWER ALLITHWAITE
Dating Bronze Age
Description
Mound.

Record number 2443
NGR SD 35760 79200
Parish LOWER ALLITHWAITE
Dating Iron Age
Description
Howbarrow hillfort.

Record number 2446
NGR SD 42120 79760
Parish GRANGE
Dating Iron Age/Romano British
Description
A natural stronghold, possibly a hillfort.

Record number 2456
NGR SD 41000 80000
Parish UPPER ALLITHWAITE
Dating Bronze Age
Description
Stone hammer found pre-1936 at Bogrells Farm at *c* 8 ft depth.

Record number 2458
NGR SD 41200 81400
Parish UPPER ALLITHWAITE
Dating Roman
Description
Tombstone.

Record number 2487
NGR SD 46400 85400
Parish CROSTHWAITE AND LYTH
Dating Unknown
Description
Corduroy track brought to light in 1897 or 1898 by peat cutters in Foulshaw Moss, and parts subsequently revealed at later dates. The road consists of cross timbers laid side by side on three lines of supporting logs parallel to the direction of the road. Pointed stakes had been driven at intervals to keep the supports in place. Axe marks were visible on the logs. The road was 15–16 ft wide, and 3–4 ft under the surface of the peat. A wooden 'chariot' wheel or shield was found in 1925.

Record number 2488
NGR SD 48200 80200
Parish MILNTHORPE
Dating Neolithic
Description
Possible long barrow.

Record number 2489
NGR SD 48290 80250
Parish MILNTHORPE
Dating Iron Age/Romano-British
Description
Cave at Haverbrack known as Dog Holes, containing human bones.

Record number 2490
NGR SD 40000 80000
Parish LEVENS
Dating Iron Age/Romano-British
Description
Settlement and finds, including a skeleton, bronze fibula, and ring, and two or three glass beads.

Record number 2492
NGR SD 49400 80800
Parish MILNTHORPE
Dating Bronze Age
Description
Cairn.

Record number 2496
NGR SD 49500 84300
Parish LEVENS
Dating Prehistoric
Description
Mound.

Record number 2504
NGR SD 49750 85870
Parish LEVENS
Dating Bronze Age
Description
Damaged round barrow, described as a 'ring earthwork' by RCHME.

Record number 2506
NGR SD 49750 85870
Parish LEVENS
Dating Bronze Age
Description
Mound.

Record number 2507
NGR SD 49370 86940
Parish LEVENS
Dating Prehistoric
Description
Earthwork.

Record number 2508
NGR SD 49200 86900
Parish LEVENS
Dating Prehistoric
Description
Cairnfield. A number of slight, irregularly-shaped mounds thought by RCHME to be artificial.

Record number 2509
NGR SD 49400 86900
Parish LEVENS
Dating Bronze Age
Description
Tumulus which can no longer be identified. Parts of a type B1 Beaker were found.

Record number 2510
NGR SD 49480 86840
Parish LEVENS
Dating Bronze Age
Description
Burial cairn containing the skeletons of four adults and a child, all placed 'closely doubled up in a sitting posture'. This site may be the same as SMR 2506.

Record number 2545
NGR SD 33000 89000
Parish COLTON
Dating Bronze Age
Description
Perforated stone axe found in 1881–2 during drain digging in low-lying peaty ground.

Record number 2552
NGR SD 37320 88490
Parish COLTON
Dating Medieval / Post-medieval
Description
Bloomery site near Smithy Haw Wood.

Record number 2560
NGR SD 34000 83000
Parish HAVERTHWAITE
Dating Late Neolithic / Early Bronze Age
Description
Perforated stone hammer was found in 1933, 400 yds east of Low Wood Bridge.

Record number 2568
NGR SD 33900 86710
Parish COLTON
Dating Prehistoric
Description
Standing stone.

Record number 2595
NGR SD 23940 67350
Parish BARROW-IN-FURNESS
Dating Early Medieval
Description
Broken ninth century Viking sword.

Record number 2596
NGR SD 23920 67370
Parish BARROW-IN-FURNESS
Dating Early Medieval
Description
Iron dagger, with a straight guard 3 in long and a 'tongue' or handle, but no pommel or knob.

Record number 2603
NGR SD 24000 69200
Parish BARROW-IN-FURNESS
Dating Early Bronze Age
Description
Bronze axe with hammered flanges.

Record number 2605
NGR SD 23920 68660
Parish BARROW-IN-FURNESS
Dating Prehistoric
Description
Whetstone of fine sandstone, square in cross section with the ends broken off, found in Roose Moss.

Record number 2610
NGR SD 24000 69000

Parish ALDINGHAM
Dating Neolithic
Description
Polished stone celt.

Record number 2614
NGR SD 25000 68000
Parish ALDINGHAM
Dating Unknown
Description
Seven human skeletons, buried in two rows, the feet of one row to the heads of the others. A stone battleaxe was discovered nearby.

Record number 2616
NGR SD 20400 62900
Parish DALTON-IN-FURNESS
Dating Mesolithic/Neolithic/Bronze Age
Description
Multi-period site with flints of Mesolithic, Neolithic and Bronze Age date, plus Neolithic and Bronze Age pottery.

Record number 2619
NGR SD 23270 63600
Parish BARROW-IN-FURNESS
Dating Bronze Age
Description
Spearhead found at Piel Castle.

Record number 2621
NGR SD 22650 62320
Parish DALTON-IN-FURNESS
Dating Bronze Age
Description
Perforated adze and an early or middle Bronze Age macehead found in clay during drainage work.

Record number 2623
NGR SD 17800 68500
Parish DALTON-IN-FURNESS
Dating Mesolithic
Description
Microlith in thin soil over boulder clay.

Record number 2624
NGR SD 19000 65100
Parish DALTON-IN-FURNESS
Dating Prehistoric
Description
Chipping floor with anvil stone, blade and scraper found in boulder clay. Fields nearby contain microliths, macehead and Bronze Age-type pottery.

Record number 2625
NGR SD 19400 64700
Parish DALTON-IN-FURNESS
Dating Prehistoric
Description
Stone axe.

Record number 2626
NGR SD 20200 63400
Parish DALTON-IN-FURNESS
Dating Mesolithic/Bronze Age
Description
Blade, scraper, and microlith in boulder clay. A scatter of flint pieces, a fragment of a macehead, and some Bronze Age pottery were found in a field behind the cliff.

Record number 2628
NGR SD 22300 62200
Parish DALTON-IN-FURNESS
Dating Prehistoric?
Description
Settlement site with worked flints and a possible gaming board.

Record number 2641
NGR SD 53070 88730
Parish NATLAND
Dating Iron Age
Description
Castlesteads, small multivallate hillfort located on the summit of the Helm. To the north and south there are two rock-cut basins — possibly wells or dew ponds.

Record number 2645
NGR SD 50400 85900
Parish LEVENS
Dating Roman/Medieval
Description
Sub-rectangular enclosure of medieval date with Roman occupation beneath.

Record number 2724
NGR SD 19980 69990
Parish DALTON-IN-FURNESS
Dating Bronze Age
Description
Copper alloy? axe.

Record number 2725
NGR SD 18330 69460
Parish DALTON-IN-FURNESS
Dating Bronze Age
Description
Stone axe-hammer found in soft ground in 1913.

Record number 2728
NGR SD 18510 68430
Parish DALTON-IN-FURNESS
Dating Mesolithic
Description
Perforated stone axe-hammer.

Record number 2729
NGR SD 18500 68340
Parish DALTON-IN-FURNESS

Dating Mesolithic
Description
Stone axe found in boulder clay in 1903.

Record number 2730
NGR SD 18600 68130
Parish DALTON-IN-FURNESS
Dating Bronze Age
Description
Perforated axe-hammer.

Record number 2731
NGR SD 17640 67720
Parish DALTON-IN-FURNESS
Dating Neolithic
Description
Polished axe.

Record number 2732
NGR SD 18530 66350
Parish DALTON-IN-FURNESS
Dating Prehistoric
Description
Flint scatter.

Record number 2734
NGR SD 19230 66280
Parish DALTON-IN-FURNESS
Dating Prehistoric
Description
Possible chipping floor.

Record number 2735
NGR SD 18950 65130
Parish DALTON-IN-FURNESS
Dating Mesolithic
Description
Flint chipping site.

Record number 2736
NGR SD 18980 65080
Parish DALTON-IN-FURNESS
Dating Mesolithic
Description
Chipping floor.

Record number 2737
NGR SD 10000 60000
Parish DALTON-IN-FURNESS
Dating Prehistoric
Description
Flint flakes.

Record number 2738
NGR SD 17820 68450
Parish DALTON-IN-FURNESS
Dating Mesolithic
Description
Flints.

Record number 2739
NGR SD 18000 68000
Parish DALTON-IN-FURNESS
Dating Unknown
Description
Stone quern.

Record number 2740
NGR SD 19000 66000
Parish DALTON-IN-FURNESS
Dating Bronze Age
Description
Perforated stone axe-hammer.

Record number 2741
NGR SD 19200 66300
Parish DALTON-IN-FURNESS
Dating Bronze Age
Description
Stone celt.

Record number 2742
NGR SD 19200 66300
Parish DALTON-IN-FURNESS
Dating Unknown
Description
Circular perforated stone.

Record number 2743
NGR SD 18400 68200
Parish DALTON-IN-FURNESS
Dating Unknown
Description
Perforated stone.

Record number 2744
NGR SD 19200 66300
Parish DALTON-IN-FURNESS
Dating Unknown
Description
Stone spindle whorl.

Record number 2745
NGR SD 19700 64600
Parish DALTON-IN-FURNESS
Dating Prehistoric
Description
Chipping floor with pestle and a cylinder.

Record number 3086
NGR SD 48400 92500
Parish UNDERBARROW AND BRADLEYFIELD
Dating Unknown
Description
Oval enclosure and settlement on aerial photo-graphs.

Record number 3089
NGR SD 24700 69300

Parish ALDINGHAM
Dating Neolithic
Description
Polished stone axe from Leece.

Record number 3107
NGR SD 49780 59000
Parish LEVENS
Dating Bronze Age
Description
Barrow.

Record number 3114
NGR SD18600 68130
Parish DALTON-IN-FURNESS
Dating Bronze Age
Description
Axe-hammer.

Record number 3161
NGR SD 39780 76540
Parish GRANGE
Dating Prehistoric
Description
Perforated stone axe-hammer.

Record number 3203
NGR SD 27800 74500
Parish DALTON-IN-FURNESS
Dating Mesolithic
Description
Flints.

Record number 3219
NGR SD 27690 96080
Parish CONISTON
Dating Unknown
Description
Three-sided rampart or stone wall forming an enclosure to the south of the pack-horse road, containing a possible hut circle, at Banishead Moor.

Record number 3573
NGR SD 36300 76600
Parish LOWER ALLITHWAITE
Dating Prehistoric
Description
Perforated stone.

Record number 3575
NGR SD 35000 85900
Parish HAVERTHWAITE
Dating Bronze Age
Description
Perforated axe-hammer.

Record number 3578
NGR SD 41000 80000
Parish UPPER ALLITHWAITE
Dating Unknown
Description
Perforated stone implement.

Record number 3585
NGR SD 34000 88000
Parish COLTON
Dating Prehistoric
Description
Axe-hammer?

Record number 3586
NGR SD 21200 88400
Parish BROUGHTON WEST
Dating Prehistoric
Description
Perforated stone axe-hammer, possibly of diorite, found in 1915 at Springs Wood. Hourglass perforation.

Record number 3593
NGR SD 40000 77000
Parish GRANGE
Dating Prehistoric
Description
Stone axe of tuff.

Record number 3596
NGR SD 18550 68440
Parish DALTON-IN-FURNESS
Dating Mesolithic?
Description
Perforated stone.

Record number 3971
NGR SD 18530 66350
Parish DALTON-IN-FURNESS
Dating Prehistoric
Description
Flint chipping floor.

Record number 3973
NGR SD 19200 65100
Parish DALTON-IN-FURNESS
Dating Prehistoric
Description
Flint artefacts, cores and chippings.

Record number 4067
NGR SD 47200 85400
Parish LEVENS
Dating Bronze Age
Description
Flint arrowhead and polished axe.

Record number 4068
NGR SD 49300 86400
Parish LEVENS

Dating Early medieval
Description
Four skeletons oriented in east–west direction, possibly Christian.

Record number 4070
NGR SD 48000 89000
Parish HELSINGTON
Dating Neolithic
Description
Stone adze found at Brigsteer.

Record number 4071
NGR SD 46800 89100
Parish HELSINGTON
Dating Early Bronze Age
Description
Tanged dagger found in peat moss.

Record number 4072
NGR SD 45400 88300
Parish CROSTHWAITE AND LYTH
Dating Neolithic
Description
Polished stone axe of oval section, convex cutting edge.

Record number 4073
NGR SD 44300 87000
Parish CROSTHWAITE AND LYTH
Dating Late Bronze Age
Description
Bronze spearhead with leaf-shaped blade and loops on socket, found in 1920.

Record number 4074
NGR SD 45500 85600
Parish CROSTHWAITE AND LYTH
Dating Neolithic
Description
Broken flint axe found in 1962.

Record number 4080
NGR SD 49600 84700
Parish LEVENS
Dating Unknown
Description
Pebble with countersunk hollows, found in 1959.

Record number 4089
NGR SD 51000 88000
Parish HELSINGTON
Dating Prehistoric
Description
Perforated stone axe-hammer.

Record number 4106
NGR SD 48750 92150
Parish UNDERBARROW AND BRADLEYFIELD
Dating Unknown
Description
Rough-out stone axe found in 1958 among limestone outcrops.

Record number 4108
NGR SD 48300 90800
Parish UNDERBARROW AND BRADLEYFIELD
Dating Bronze Age
Description
Flint arrowhead found at Underbarrow Scar.

Record number 4109
NGR SD 46000 92000
Parish UNDERBARROW AND BRADLEYFIELD
Dating Bronze Age
Description
Arrowhead.

Record number 4132
NGR SD 46300 84700
Parish WITHERSLACK
Dating Bronze Age
Description
Perforated whetstone / hone found 5 ft deep in peat prior to 1903.

Record number 4133
NGR SD 45100 84700
Parish WITHERSLACK
Dating Prehistoric
Description
Flint scraper.

Record number 4140
NGR SD 46000 81000
Parish MEATHOP AND ULPHA
Dating Unknown
Description
Irish elk antlers.

Record number 4146
NGR SD 39000 75000
Parish LOWER ALLITHWAITE
Dating Medieval
Description
Possible chapel at Kirkhead, near the place mentioned in 1199 as Kierkepol.

Record number 4147
NGR SD 37000 75000
Parish LOWER HOLKER
Dating Prehistoric
Description
Several stone hammers were found in a meadow, once part of Winder Moor, below high water mark.

Record number 4149
NGR SD 36700 75700
Parish LOWER HOLKER
Dating Prehistoric
Description
Axe.

Record number 4151
NGR SD 45700 89300
Parish CROSTHWAITE AND LYTH
Dating Prehistoric
Description
Perforated stone implement (macehead?) found in a field prior to 1914.

Record number 4269
NGR SD 49350 93180
Parish UNDERBARROW AND BRADLEYFIELD
Dating Unknown
Description
Settlement with cairns and hut circles, on aerial photographs.

Record number 4366
NGR SD 48400 85100
Parish LEVENS
Dating Unknown
Description
Settlement with platforms around former farmsteads.

Record number 4375
NGR SD 36300 75000
Parish LOWER HOLKER
Dating Prehistoric
Description
Nab Green hoard, comprising many hammers and battle axes of varying sizes and of stone type, as well as axes of brass and copper. *See also* SMR 4147.

Record number 4790
NGR SD 49500 89400
Parish HELSINGTON
Dating Iron Age/Romano-British
Description
Settlement.

Record number 4806
NGR SD 27500 74700
Parish URSWICK
Dating Unknown
Description
Enclosure and settlement on aerial photographs.

Record number 4839
NGR SD 27900 74200
Parish URSWICK
Dating Unknown
Description
Enclosure and dyke.

Record number 4856
NGR SD 24000 74000
Parish DALTON-IN-FURNESS
Dating Unknown
Description
Earthworks on aerial photographs.

Record number 4863
NGR SD 27000 85000
Parish LOWICK
Dating Unknown
Description
Cropmark.

Record number 4940
NGR SD 26100 74200
Parish URSWICK
Dating Unknown
Description
Settlement.

Record number 5152
NGR SD 49300 88500
Parish HELSINGTON
Dating Unknown
Description
Settlement.

Record number 5331
NGR SD 45000 86000
Parish CROSTHWAITE AND LYTH
Dating Unknown
Description
Whitbarrow Bone Cave, from which animal and human bones have been retrieved, possibly redeposited.

Record number 5352
NGR SD 19200 66300
Parish DALTON-IN-FURNESS
Dating Unknown
Description
Stone spindle whorl.

Record number 5544
NGR SD 35000 83000
Parish HAVERTHWAITE
Dating Unknown
Description
Axe.

Record number 5561
NGR SD 30900 79200
Parish EGTON WITH NEWLAND
Dating Bronze Age

Description
Perforated stone axe-hammer.

Record number 5563
NGR SD 30500 78900
Parish EGTON WITH NEWLAND
Dating Neolithic
Description
Butt end of a polished stone axe.

Record number 6506
NGR SD 41700 80700
Parish UPPER ALLITHWAITE
Dating Upper Palaeolithic
Description
Lindale Low Cave, which excavation has revealed as largely undisturbed. Three unassociated flint artefacts were found in separate contexts within the cave, all with slightly abraded edges, all patinated. They include an angle-backed blade, a backed bladelet, and a broken flake or bladelet. The larger blade falls within the range of angle-backed pieces known from Upper Palaeolithic contexts in northern British cave sites. Its presence at this site complements the evidence for Late Pleistocene human occupation of the Morecambe Bay area discovered at Kirkhead Cave, and this extends the known distribution of Upper Palaeolithic sites from North Lancashire into Cumbria.

Record number 13688
NGR SD 13950 80800
Parish MILLOM WITHOUT
Dating Unknown
Description
Double-ditched enclosure shown as cropmark.

Record number 16720
NGR SD 42700 85300
Parish WITHERSLACK
Dating Neolithic
Description
Re-worked fragment of lightweight stone axe, the broken end flaked into a tang.

Record number 17930
NGR SD 24240 69420
Parish ALDINGHAM
Dating Neolithic
Description
Two portions of a rough-out axe, broken in the middle on a hinge fracture.

West Cumbrian Coastal Plain: Chapter 3

Record number 1028
NGR NY 07300 25250
Parish DEAN
Dating Bronze Age
Description
Cup and ring mark on a freestone boulder, found in 1918 in a field ajacent to 'the butts'.

Record number 1038
NGR NY 00340 22510
Parish DISTINGTON
Dating Bronze Age?
Description
Copper alloy spearhead.

Record number 1041
NGR NY 02990 20700
Parish ARLECDON AND FRIZINGTON
Dating Bronze Age
Description
Cremation and urn found in digging a drain: large boot? of baked clay, diamond-shaped ornament and half buried bones, covered by a stone flag.

Record number 1042
NGR NY 00650 21950
Parish DISTINGTON
Dating Prehistoric
Description
Stone hammer found.

Record number 1047
NGR NY 06180 21510
Parish DEAN
Dating Prehistoric
Description
Perforated stone axe-hammer.

Record number 1048
NGR NY 08950 21710
Parish DEAN
Dating Prehistoric
Description
Stone hammer.

Record number 1049
NGR NY 07480 21350
Parish DEAN
Dating Prehistoric
Description
Stone hammer, found in 1856.

Record number 1050
NGR NY 07200 21900
Parish DEAN
Dating Bronze Age
Description
Perforated stone hammer, provenance uncertain.

Record number 1054
NGR NY 02000 26000

Parish WINSCALES
Dating Bronze Age
Description
Large perforated stone axe-hammer.

Record number 1060
NGR NY 09000 28000
Parish DEAN
Dating Unknown
Description
Two unpolished stone axes found west of Eaglesfield.

Record number 1179
NGR NX 95900 12100
Parish ST BEES
Dating Prehistoric
Description
Various flints.

Record number 1180
NGR NX 96200 11800
Parish ST BEES
Dating Prehistoric
Description
Flints.

Record number 1181
NGR NX 96300 11600
Parish ST BEES
Dating Prehistoric
Description
Flint scatter.

Record number 1182
NGR NX 96900 11400
Parish ST BEES
Dating Prehistoric
Description
Flints.

Record number 1183
NGR NX 96800 11000
Parish ST BEES
Dating Prehistoric
Description
Finds.

Record number 1193
NGR NX 96870 12120
Parish ST BEES
Dating Early Medieval
Description
Pre-Norman cross standing in the churchyard.

Record number 1198
NGR NY 01500 10600
Parish ST JOHN BECKERMET
Dating Prehistoric
Description
Destroyed stone circle or standing stones?

Record number 1204
NGR NY 04000 18000
Parish ARLECDON AND FRIZINGTON
Dating Prehistoric
Description
Axe-hammer, now lost.

Record number 1247
NGR NY 98600 07900
Parish LOWSIDE QUARTER
Dating Prehistoric
Description
Flint scatter.

Record number 1248
NGR NY 00270 07050
Parish LOWSIDE QUARTER
Dating Prehistoric/Roman
Description
Neolithic dwelling site, discovered during draining of Ehenside (Gibb) Tarn in 1869; further excavations in 1871. Along the lake edge was a series of hearths with occupation debris. Finds included stone axes of Great Langdale origin, grinding stones, wooden implements, a possible dug-out canoe and paddle. Sandstone rubbers and grinders used in the finishing of the rough-outs from the Langdale axe factories were also found. A large flint arrowhead was found in the last century, together with three others, of a type rare in the district. A scatter of flints, mainly Bronze Age in character, including a side scraper/knife, two flake cores, a single platform blade core and a broken, fire-damaged core, were found south–west of Gibb Tarn, and other flints have been found in the neighbouring area. A saucer quern and a rotary quern of possible Roman date were also found. A quantity of Black Ware pottery of Lyles Hill or Yorkshire Wold class and coarser bowls of Peterborough type were found. The site is radiocarbon-dated to 4500–3040 cal BC (5051±300 BP, C–462; Walker 1966).

Record number 1250
NGR NY 01700 04700
Parish ST BRIDGET BECKERMET
Dating Bronze Age
Description
Settlement site at the south edge of a low-lying damp plain. Normal occupation debris of broken stones and struck flints was found. Flint debris typical of a poverty industry based on beach pebbles, characteristic of west Cumberland coast Bronze Age sites was also found. Artefacts included scrapers, knives, tanged and barbed arrowhead, and a fragment of a small stone axe.

Record number 1256
NGR NY 02700 09400

Parish ST JOHN BECKERMET
Dating Unknown
Description
Enclosure on aerial photographs.

Record number 1258
NGR NY 00790 05280
Parish ST BRIDGET BECKERMET
Dating Prehistoric
Description
Axe.

Record number 1259
NGR NY 01500 06040
Parish ST BRIDGET BECKERMET
Dating Early Medieval
Description
Remains of two crosses, both of sandstone. The earlier, of ninth century date, is round below a square shaft, with a long inscription. The second is of tenth/eleventh century date, with unusual design.

Record number 1260
NGR NY 00410 06090
Parish ST BRIDGET BECKERMET
Dating Prehistoric
Description
Stone slicker.

Record number 1261
NGR NY 00680 06240
Parish LOWSIDE QUARTER
Dating Roman
Description
Roman coin hoard, including at least one coin of Commodus.

Record number 1262
NGR NY 01890 06710
Parish ST JOHN BECKERMET
Dating Early Medieval
Description
Church of pre-Conquest origin. There are fragments of crosses within the church and porch, dated to the tenth/eleventh centuries.

Record number 1263
NGR NY 02000 07200
Parish ST BRIDGET BECKERMET
Dating Prehistoric
Description
Granite querns have been ploughed up.

Record number 1265
NGR NY 04150 01300
Parish SEASCALE
Dating Bronze Age
Description
Flint-chert arrowhead of Bronze Age-type (with barbs as long as tang).

Record number 1266
NGR NY 03800 00800
Parish SEASCALE
Dating Mid-Bronze Age
Description
Ceramic burial urn, containing small bones,found on a shingle terrace at Herding Nab in about 1874.

Record number 1267
NGR NY 04000 02800
Parish PONSONBY
Dating Roman
Description
Coin of Nero, found at Seascale Hall.

Record number 1268
NGR NY 01100 04700
Parish ST BRIDGET BECKERMET
Dating Iron Age/Romano-British
Description
Rotary beehive quern, found in 1896 at Bogholes. A rough-out axe was found two years later.

Record number 1269
NGR NY 04550 01840
Parish SEASCALE
Dating Neolithic
Description
Stone axe of banded tuff, unfinished?

Record number 1270
NGR NY 01000 04800
Parish ST BRIDGET BECKERMET
Dating Neolithic
Description
Stone axe of slate, with brown veins, was found at Boggles (Bogholes) near the mouth of the Ehen.

Record number 1271
NGR NY 09150 00450
Parish IRTON WITH SANTON
Dating Early Medieval
Description
Complete ninth century Northumbrian cross of red sandstone in the churchyard set into a small base.

Record number 1272
NGR NY 07230 03570
Parish GOSFORTH
Dating Early Medieval
Description
At least three pre-Norman wheel-headed crosses; fragments of two are built into the church, the third surviving complete

in the churchyard. The latter stands to 15 ft high and is of red sandstone, characterised by its extreme slenderness, the lower shaft being round and becomes square with height. The iconography links Christian and Norse motifs in a remarkable manner. Other pre-Norman stones are in the church.

Record number 1273
NGR NY 06030 01140
Parish SEASCALE
Dating Prehistoric
Description
Incomplete axe found in 1855.

Record number 1274
NGR NY 05100 01200
Parish GOSFORTH
Dating Early / Middle Bronze Age
Description
Flint knife, found in 1937 in a bog east of Black How Farm.

Record number 1275
NGR NY 06700 01700
Parish GOSFORTH
Dating Roman
Description
Coin of Nerva, found in a field near Hallsenna.

Record number 1276
NGR NY 07940 00390
Parish SEASCALE
Dating Neolithic
Description
Stone axe.

Record number 1277
NGR NY 07900 00600
Parish SEASCALE
Dating Bronze Age
Description
Middle Bronze Age cinerary urn found inverted over a cremated child, young adult and animal bones. About 250 yards south, a stone axe was found 'about two feet six inches under peat beside a very ancient road' (Anon 1944, 161).

Record number 1278
NGR NY 08200 03600
Parish SEASCALE
Dating Prehistoric
Description
Flint arrowhead.

Record number 1279
NGR NY 05700 04000
Parish PONSONBY
Dating Bronze Age
Description
Tanged and barbed, yellowish, flint arrowhead.

Record number 1280
NGR NY 09030 04020
Parish GOSFORTH
Dating Neolithic
Description
Fragment of a stone axe, found in 1886.

Record number 1281
NGR NY 08950 03670
Parish GOSFORTH
Dating Prehistoric
Description
Fragment of an axe, found in 1885.

Record number 1282
NGR NY 05200 01400
Parish SEASCALE
Dating Unknown
Description
Flint knife.

Record number 1283
NGR NY 08900 03300
Parish GOSFORTH
Dating Neolithic
Description
Cutting edge of an axe made of volcanic tuff, found in 1949.

Record number 1284
NGR NY 09100 04200
Parish GOSFORTH
Dating Neolithic
Description
Part of a polished stone axe, found at Rainors farm.

Record number 1285
NGR NY 09100 03900
Parish GOSFORTH
Dating Bronze Age
Description
Copper alloy axe, recorded as found at Bolton wood in 1820, but the actual findspot was near Rainors.

Record number 1286
NGR NY 05470 02560
Parish GOSFORTH
Dating Unknown
Description
'Ancient' paved road.

Record number 1288
NGR NY 03340 02380
Parish SEASCALE
Dating Neolithic

Description
Circle of ten stones with a small central oval cairn at Greycroft, partially excavated and reconstructed in 1956. Bone fragments and organic material were found in the cairn and below. Part of a jet ring, flint flakes, and a scraper were also found. A broken stone axe was near, but not directly associated with, the standing stones. There is also a prone outlier to the north.

Record number 1292
NGR NY 05700 03600
Parish SEASCALE
Dating Bronze Age?
Description
Three stone circles with central cairns (probably Bronze Age, possibly not antiquities). Circle A: 104 ft in diameter; an arc of 90 ft remains. Glacial volcanic tuff and granite, infilled to form a thick stone wall, on a level area on a sloping hillside. Circle B: 72 ft diameter, with 15 or 16 stones of which nine remain, six *in situ*. In the circle is a large area of cobbles, possibly a cairn (C Fell pers comm), of Borrowdale volcanics and granite. This circle has been excavated. Circle C: 24 ft in diameter; there were originally 12 stones, nine of which remain, seven *in situ*. There is a cobbled central area possibly a cairn of granite. The cobbled paving and stone settings have been excavated, producing two small flakes and haematite and possibly burnt bones.

Record number 1293
NGR NY 01500 06000
Parish ST BRIDGET BECKERMET
Dating Medieval
Description
Deserted medieval settlement.

Record number 1297
NGR NY 00850 05110
Parish ST BRIDGET BECKERMET
Dating Iron Age / Roman
Description
Ceramic?

Record number 1298
NGR NY 00540 05520
Parish ST BRIDGET BECKERMET
Dating Prehistoric
Description
Flint scatter.

Record number 1301
NGR NY 00490 07140
Parish ST BRIDGET BECKERMET
Dating Prehistoric
Description
Single standing stone, six to seven feet high, thrown down in 1860.

Record number 1302
NGR NY 03800 02400
Parish SEASCALE
Dating Bronze Age
Description
A 4 ft stone, part of a stone circle.

Record number 1303
NGR NY 04800 00300
Parish SEASCALE
Dating Prehistoric
Description
Flint flakes, scrapers and a knife found in 1965, near the edge of a small hollow which often floods, found within a few square yards of each other.

Record number 1304
NGR NY 04400 00900
Parish SEASCALE
Dating Prehistoric
Description
Flint scatter from the edge of a boggy hollow, south of Seascale, known as Bailey Ground.

Record number 1308
NGR NY 07530 03150
Parish GOSFORTH
Dating Neolithic
Description
Stone axe with three rubbing stones, found 2 ft 6 in underground.

Record number 1309
NGR NY 05900 01100
Parish SEASCALE
Dating Prehistoric
Description
Several flint scrapers, a knife, and a core and a struck flake of volcanic tuff found in1965.

Record number 1310
NGR NY 05300 00560
Parish SEASCALE
Dating Prehistoric
Description
Flint flake, end scraper and knife.

Record number 1311
NGR NY 05000 00200
Parish SEASCALE
Dating Neolithic
Description
Three flint-chert scrapers and two blades found as well as a fragment of a pointed butt of a stone axe.

Record number 1312
NGR NY 05000 00800
Parish SEASCALE
Dating Mesolithic
Description
Small flint scatter.

Record number 1313
NGR NY 09400 03700
Parish GOSFORTH
Dating Bronze Age
Description
Flint artefacts: utilised blade (1.4 in), small scraper, and barbed and tanged arrowheads.

Record number 1338
NGR SD 06700 99100
Parish DRIGG AND CARLETON
Dating Neolithic
Description
Stone polished axe of Cumbrian-type.

Record number 1376
NGR SD 09600 97000
Parish DRIGG AND CARLETON
Dating Unknown
Description
Settlement seen on aerial photographs.

Record number 1377
NGR SD 07800 96900
Parish DRIGG AND CARLETON
Dating Unknown
Description
Boat.

Record number 1378
NGR SD 08810 95810
Parish MUNCASTER
Dating Roman
Description
Fort at Ravenglass on the site of an earlier fortlet . An extensive cilivian settlement extended to the north and east and there is a well-preserved bath house.

Record number 1381
NGR SD 07900 93500
Parish BOOTLE
Dating Bronze Age
Description
Occupation site discovered west of Broom Bank on the sand dunes. Flint scatter.

Record number 1382
NGR SD 08900 91900
Parish BOOTLE
Dating Unknown
Description
Finds, *see* Cherry (1969).

Record number 1383
NGR SD 08300 91800
Parish BOOTLE
Dating Prehistoric
Description
Flint and pottery found on raised beach with peaty deposits near Williamson's Moss.

Record number 1384
NGR SD 08590 91600
Parish BOOTLE
Dating Prehistoric
Description
Flint-chert, *see* Cherry (1969).

Record number 1385
NGR SD 08400 90150
Parish BOOTLE
Dating Prehistoric
Description
Flint-chert, *see* Cherry (1969).

Record number 1386
NGR SD 08100 91100
Parish BOOTLE
Dating Mesolithic
Description
Flint scatter on Skelda Hill. *See also* SMR 1382-1387, 4222.

Record number 1387
NGR SD 08900 91600
Parish BOOTLE
Dating Mesolithic
Description
Flint scatter found on a low, boulder clay ridge. *See also* SMR 1382-1386, 4222.

Record number 1388
NGR SD 06700 99100
Parish DRIGG AND CARLETON
Dating Neolithic
Description
Polished stone axe found in pasture near 'White Garth'.

Record number 1389
NGR SD 06820 98950
Parish DRIGG AND CARLETON
Dating Bronze Age
Description
Dark red perforated stone hammer or mace found in 1960 or 1962 at Drigg.

Record number 1390
NGR SD 06900 96400
Parish DRIGG AND CARLETON
Dating Prehistoric
Description
Alleged tumulus reported to be situated on the sand dunes. *See also* SMR 4430.

Record number 1391
NGR SD 08930 92530
Parish BOOTLE

Dating Mesolithic
Description
Large flint scatter, an extensive settlement site at Monk Moors, found by J Cherry on a terrace at *c* 15 m above sea level, 1.6 km from the present coast. An area of the site, excavated in 1974, produced three hearths, stakeholes, and numerous flints. Charcoal from a hearth provided a radiocarbon date of *c* 5970–5350 cal BC (6750±155 BP, BM–1216).

Record number 1392
NGR SD 08800 93700
Parish WABERTHWAITE
Dating Iron Age/Romano-British
Description
Camp known from aerial photographs but with corroborating finds.

Record number 1393
NGR SD 08850 93720
Parish BOOTLE
Dating Roman
Description
Coins found in 1837.

Record number 1394
NGR SD 04700 98500
Parish DRIGG AND CARLETON
Dating Mesolithic
Description
Sauveterrian flints.

Record number 1395
NGR SD 07900 93400
Parish BOOTLE
Dating Prehistoric/medieval
Description
Occupation site to the west of a ruined house on Eskmeals sand dunes. This is Cherry's site II (Cherry 1963). Finds include scrapers, cores, knives, an arrow-tip, a fabricator, iron slag, medieval pottery fragments, and an Iron Age segmented blue glass bead.

Record number 1396 - 1398
NGR SD 05 97, SD 05 96, SD 06 96
Parish DRIGG AND CARLETON
Dating Late Mesolithic–Bronze Age
Description
Extensive flint chipping sites discovered *c*1958 along the coast between the Irt and the shore at Drigg.

Record number 1399
NGR SD 08000 94200
Parish BOOTLE
Dating Prehistoric (mainly Bronze Age)/ medieval
Description
Settlement site at Eskmeals discovered in 1935 with a flint-working area, hearths, iron-working evidence and cairns. Finds covered a wide range of dates and included barbed and tanged transverse arrowheads, a petit tranchet derivative, flakes, scrapers, knives, bronze fragments, and second century melon-shaped blue glazed beads. At SD 081 950, fragments of a saddle quern and rubber were found.

Record number 1400
NGR SD 08100 93000
Parish BOOTLE
Dating Prehistoric
Description
Bronze Age flint-working site found on Eskmeals Sands, west of the look-out tower.

Record number 1445
NGR SD 10700 97500
Parish MUNCASTER
Dating Unknown
Description
Cairn.

Record number 1448
NGR SD 12800 96500
Parish MUNCASTER
Dating Unknown
Description
Cairn.

Record number 1460
NGR SD 09900 85300
Parish WHICHAM
Dating Neolithic/Bronze Age
Description
Annaside stone circle.

Record number 1461
NGR SD 04900 98300
Parish DRIGG AND CARLETON
Dating Unknown
Description
Flint flakes and quern fragment found in 1932 in a hollow of sandhills. *See also* SMR 1394.

Record number 1471
NGR SD 15130 81270
Parish MILLOM WITHOUT
Dating Bronze Age
Description
Two circles at Lacra, one containing middle Bronze Age overhanging rim urn.

Record number 1472
NGR SD 13610 81100
Parish MILLOM WITHOUT
Dating Prehistoric
Description
The Giant's Grave, two standing stones of unhewn granite located a short distance from Kirksanton village, with

evidence of prehistoric rock art: cup marks. Hutchinson (1794) describes it as being a small tumulus or burial mound upon which the two stones were erected.

Record number 1474
NGR SD 11200 85700
Parish WHICHAM
Dating Neolithic/Bronze Age
Description
Stone circle with eight large stones.

Record number 1476
NGR SD 11200 87100
Parish WHICHAM
Dating Neolithic/Bronze Age
Description
Hall Force stone circle.

Record number 1478, 1493
NGR SD 10 84
Parish WHICHAM
Dating Neolithic
Description
Site of two concentric stone circles at Gutterby. A cairn stood 185 m to the south.

Record number 1480
NGR SD 14920 80990
Parish MILLOM WITHOUT
Dating Bronze Age
Description
Stone circle at Lacra. Burnt bone, soil and charcoal found but no primary burial or pottery.

Record number 3556
NGR NY 05400 00800
Parish SEASCALE
Dating Mesolithic
Description
Microlith.

Record number 3557
NGR SD 08500 91000
Parish WABERTHWAITE
Dating Mesolithic
Description
Microliths.

Record number 3558
NGR SD 08600 95500
Parish MUNCASTER
Dating Mesolithic
Description
Flint scatter.

Record number 3559
NGR NY 04200 00900
Parish SEASCALE
Dating Mesolithic
Description
Microliths.

Record number 3561
NGR NX 96400 12500
Parish ST BEES
Dating Mesolithic
Description
Flint debris, and microliths, leaf arrowheads and barbed arrowheads found on the beach.

Record number 3562
NGR NX 96400 12500
Parish ST BEES
Dating Mesolithic
Description
Flints.

Record number 3676
NGR NX 98600 08000
Parish LOWSIDE QUARTER
Dating Mesolithic
Description
Flint scatter.

Record number 3677
NGR NX 99780 06900
Parish LOWSIDE QUARTER
Dating Neolithic
Description
Polished axe found in a semi-swamp on the west side of Silver Tarn.

Record number 3709
NGR NY 12500 02000
Parish IRTON WITH SANTON
Dating Bronze Age
Description
Extensive cairnfield with a series of complex cairns, possibly hut circles and standing stones, cut by several sunken tracks.

Record number 3984
NGR SD 10030 95100
Parish WABERTHWAITE
Dating Early Medieval
Description
Cross shaft and fragment, both of red sandstone. The cross is rectangular, tapering slightly towards the top and stands 2 m high set into a sandstone socket. It is decorated on all four sides. The fragment is somewhat earlier (ninth century), again decorated on all four sides.

Record number 4155
NGR NY 00000 07000
Parish LOWSIDE QUARTER
Dating Prehistoric

Description
Flint scatter south-west of Ehenside.

Record number 4156
NGR NX 99000 07000
Parish LOWSIDE QUARTER
Dating Mesolithic / Neolithic
Description
Flint scatter.

Record number 4157
NGR NX 99000 07000
Parish LOWSIDE QUARTER
Dating Prehistoric
Description
Flint scatter north of Nethertown.

Record number 4158
NGR NX 99000 07000
Parish LOWSIDE QUARTER
Dating Prehistoric
Description
Flint scatter south of Nethertown.

Record number 4222
NGR SD 08100 94980
Parish MUNCASTER
Dating Neolithic
Description
Polished stone axe, hollow-based arrowhead, flint scatter and pottery found in wet peaty hollow on Williamson's Moss. Pollen analysis by W Pennington suggested forest clearance *c* 3000 BC (5000 BP).

Record number 4300
NGR SD 045000 98600
Parish DRIGG AND CARLETON
Dating Mesolithic
Description
Hearth.

Record number 4424
NGR SD 08100 94980
Parish BOOTLE
Dating Prehistoric
Description
Broken saddle quern and a rubber, found on the raised beach at Eskmeals in 1960.

Record number 4425
NGR SD 08850 93720
Parish BOOTLE
Dating Roman
Description
Coins found in 1837.

Record number 4427
NGR SD 09390 97120
Parish DRIGG AND CARLETON
Dating Unknown
Description
Enclosure on aerial photographs.

Record number 4428
NGR SD 06790 96600
Parish DRIGG AND CARLETON
Dating Romano-British
Description
Bloomery site with iron slag scattered over a wide area.

Record number 4430
NGR SD 06800 96400
Parish DRIGG AND CARLETON
Dating Romano-British
Description
Spindle whorl, fragments of a Roman glass bangle, Romano-British pot sherds and flints found.

Record number 4431
NGR SD 06900 99200
Parish DRIGG AND CARLETON
Dating Prehistoric
Description
Stone axe of Cumbrian type found at Drigg.

Record number 4609
NGR SD 07300 99400
Parish DRIGG AND CARTLETON
Dating Unknown
Description
Stone quern.

Record number 4612
NGR NY 03600 09600
Parish ST JOHN BECKERMET
Dating Prehistoric
Description
Stone quern.

Record number 4615
NGR NY 00300 09300
Parish HAILE
Dating Prehistoric
Description
Stone quern.

Record number 4616
NGR NX 98300 09300
Parish LOWSIDE QUARTER
Dating Prehistoric
Description
Stone quern.

Record number 4670
NGR NX 96200 10800
Parish ST BEES

Dating Prehistoric
Description
Exposed forest beds on the beach.

Record number 4718
NGR SD 12500 95800
Parish MUNCASTER
Dating Unknown
Description
Settlement with cairns and hut circles on aerial photographs.

Record number 4723
NGR SD 10399 96550
Parish MUNCASTER
Dating Early Medieval
Description
Cross shaft of red sandstone, tapering towards the top, with interlace on sides, lozenge pattern at front and back.

Record number 4758
NGR NY 03030 08800
Parish HAILE
Dating Early Medieval
Description
Fragment of a spiral cross-shaft built into Haile Church.

Record number 4845
NGR SD 14970 81330
Parish MILLOM WITHOUT
Dating Bronze Age
Description
Stone circle on aerial photographs.

Record number 4906
NGR SD 09000 93000
Parish BOOTLE
Dating Roman
Description
Brass pot found prior to 1806 at Eskmeals near the site of a Roman encampment. *See* SMR 1393.

Record number 5115
NGR SD 09200 86100
Parish BOOTLE
Dating Unknown
Description
Enclosure on aerial photographs.

Record number 5116
NGR SD 09500 85300
Parish WHICHAM
Dating Unknown
Description
Earthwork on aerial photographs.

Record number 5330
NGR NY 04000 02000
Parish GOSFORTH
Dating Prehistoric
Description
Bog body found during peat-cutting *c* 1 ft under the surface. Part of the body had been destroyed, but the hands, left ear and feet were intact. A hazel 'walking stick' was found by his side (Turner 1989).

Record number 5347
NGR SD 09000 96000
Parish MUNCASTER
Dating Unknown
Description
Two ceramic urns, together with other finds, found in peaty material.

Record number 5351
NGR SD 15020 80960
Parish MILLOM WITHOUT
Dating Bronze Age
Description
Lacra stone circle with packing in two of the four standing stones and some charcoal.

Record number 5368
NGR NY 15800 43800
Parish WESTNEWTON
Dating Unknown
Description
Circular enclosure appearing as cropmark.

Record number 6155
NGR NY 13000 02000
Parish IRTON WITH SANTON
Dating Prehistoric
Description
Flints found at Mecklin Park.

Record number 6363
NGR SD 07000 97000
Parish DRIGG AND CARLETON
Dating Early Medieval
Description
Coin of Canute, found on the Drigg sand dunes.

Record number 6442
NGR NY 00200 06500
Parish LOWSIDE QUARTER
Dating Prehistoric
Description
Small, steeply worked scraper.

Record number 6443
NGR NY 00100 06700
Parish LOWSIDE QUARTER
Dating Prehistoric
Description
Flint scatter.

Record number 6447
NGR NY 01700 04400
Parish ST BRIDGET BECKERMET
Dating Bronze Age
Description
Concentration of flints.

Record number 6449
NGR NY 01800 04400
Parish ST BRIDGET BECKERMET
Dating Prehistoric
Description
Small concentration of flints associated with SMR 6447.

Record number 6450
NGR NY 01600 04200
Parish ST BRIDGET BECKERMET
Dating Prehistoric
Description
Flint scatter.

Record number 6451
NGR NY 01700 04100
Parish ST BRIDGET BECKERMET
Dating Prehistoric
Description
Flint scatter.

Record number 6460
NGR NY 05400 02700
Parish GOSFORTH
Dating Neolithic
Description
Small rough-out axe of volcanic tuff, found below a disused sand quarry.

Record number 6462
NGR NY 05400 00600
Parish DRIGG AND CARLETON
Dating Neolithic
Description
Rough-out of Cumbrian-type axe.

Record number 6464
NGR NY 07000 00000
Parish DRIGG AND CARLETON
Dating Prehistoric
Description
Two flakes of slightly patinated flint.

Record number 6465
NGR NY 05800 00200
Parish DRIGG AND CARLETON
Dating Prehistoric
Description
Flint scatter near Moorside Farm.

Record number 6475
NGR NX 98500 08200
Parish LOWSIDE QUARTER
Dating Mesolithic
Description
Large concentration of over 1200 flints.

Record number 6476
NGR NX 98800 08400
Parish LOWSIDE QUARTER
Dating Mesolithic
Description
Flints including patinated bladed cores, blades, an end scraper, and two knives.

Record number 6478
NGR NX 99400 07400
Parish LOWSIDE QUARTER
Dating Bronze Age
Description
Few scattered flints.

Record number 6479
NGR NX 99700 06900
Parish LOWSIDE QUARTER
Dating Prehistoric
Description
Flint scatter in a field above the northern end of Silver Tarn.

Record number 6480
NGR NX 99500 06900
Parish LOWSIDE QUARTER
Dating Prehistoric
Description
Flint scatter on the edge of Silver Tarn.

Record number 6481
NGR NX 98700 07700
Parish LOWSIDE QUARTER
Dating Mesolithic
Description
Concentration of flints.

Record number 6498
NGR SD 06000 98000
Parish DRIGG AND CARLETON
Dating Unknown
Description
Small pebble marked with human features, of uncertain authenticity, found at Drigg.

Record number 13546
NGR NY 02400 10700
Parish ST JOHN BECKERMET
Dating Unknown
Description
Earthwork enclosure.

Record number 13688
NGR SD 13950 80800
Parish MILLOM WITHOUT

Dating Unknown
Description
Double-ditched enclosure.

Record number 13927
NGR SD 07900 93800
Parish BOOTLE
Dating Prehistoric/Iron Age
Description
Hearth, anvil, Iron Age bead and flint scatter.

Record number 13928
NGR SD 08500 91900
Parish BOOTLE
Dating Neolithic/Early Bronze Age
Description
Flint scatter.

Record number 13929
NGR SD 08400 91900
Parish BOOTLE
Dating Mesolithic
Description
Concentration of heavily-patinated flints.

Record number 13930
NGR SD 08400 91800
Parish BOOTLE
Dating Neolithic
Description
Dense scatter of mainly lightly-patinated flints, including leaf-shaped arrowhead from Williamson's Moss.

Record number 13931
NGR SD 08300 91700
Parish BOOTLE
Dating Neolithic/Bronze Age
Description
Flint scatter, leaf-shaped arrowhead, petit tranchet, and pottery, from area fringing Williamson's Moss.

Record number 13932
NGR SD 08300 91600
Parish BOOTLE
Dating Mesolithic
Description
Few, lightly-patinated microliths.

Record number 13933
NGR SD 08200 91700
Parish BOOTLE
Dating Bronze Age
Description
Burnt tanged and barbed arrowheads and flint scatter.

Record number 13935
NGR SD 08800 91900
Parish BOOTLE
Dating Mesolithic
Description
Flint scatter on the former shoreline.

Record number 13936
NGR SD 08700 91900
Parish BOOTLE
Dating Post-Mesolithic
Description
Flint scatter.

Record number 13937 – 13942
NGR SD 08 91
Parish BOOTLE
Dating Mesolithic
Description
An area with a concentration of flint scatters. SMR 13940 is near edge of Williamson's Moss (Bonsall *et al* 1994).

Record number 13943
NGR SD 08700 91600
Parish BOOTLE
Dating Prehistoric
Description
Few flints.

Record number 13944
NGR SD 08800 91400
Parish BOOTLE
Dating Neolithic
Description
Flint scatter.

Record number 13946 – 13949
NGR SD 08 90, SD 08 91, SD 08 92
Parish BOOTLE
Dating Mesolithic/Late Mesolithic
Description
Skelda Hill, concentration of flint scatters.

Record number 13950 – 13953
NGR SD 08 90, SD 08 91, SD 08 92
Parish BOOTLE
Dating Mesolithic/Late Mesolithic
Description
Monk Moors, concentration of flint scatters.

Record number 13954 – 13955
NGR SD 08 90, SD 08 91, SD 08 92
Parish BOOTLE
Dating Mesolithic/Late Mesolithic
Description
Eskmeals Pool, concentration of flint scatters.

Record number 13956
NGR SD 09300 92800
Parish WABERTHWAITE
Dating Mesolithic

Description
Heavy surface concentration of flints on the probable continuation of an eroded shore line.

Record number 13957
NGR SD 09400 92900
Parish WABERTHWAITE
Dating Mesolithic
Description
Flint scatter at Eskmeals Pool IV.

Record number 16924
NGR SD 06000 97000
Parish DRIGG AND CARLETON
Dating Neolithic
Description
Axe rough-out on the beach at Drigg and a small barbed and tanged arrowhead.

Record number 17803
NGR NX 96500 11500
Parish ST BEES
Dating Prehistoric
Description
St Bees golf course, flint scatter (Cherry 1983).

Record number 17804
NGR NX 96500 11700
Parish ST BEES
Dating Mesolithic/Bronze Age
Description
St Bees golf course (Cherry and Cherry 1983), flint scatter including a tanged and barbed arrowhead, microliths, back blade and cores.

Record number 17827
NGR NX 95000 13400
Parish ST BEES
Dating Prehistoric
Description
Large sub-triangular grey flint and a small scraper of yellow flint.

Record number 17914
NGR SD 08600 96800
Parish MUNCASTER
Dating Prehistoric
Description
Two flakes and a bifacially worked point.

Record number 18908
NGR SD 11700 82200
Parish WHICHAM
Dating Prehistoric
Description
1065 flints discovered, including barbed and tanged arrowhead, 11 scrapers and 10 blades.

North Cumbrian Plain: Chapter 4

Record number 42
NGR NY 48890 71120
Parish STAPLETON
Dating Medieval/Post-medieval
Description
Map (British Museum) 1590 marks 'Sowtermore' with a tower symbol.

Record number 47
NGR NY 52100 78100
Parish STAPLETON
Dating Iron Age
Description
Lynchets, earthworks.

Record number 50
NGR NY 54600 79800
Parish STAPLETON
Dating Unknown
Description
Cairns and field system.

Record number 51
NGR NY 53100 79900
Parish STAPLETON
Dating Bronze Age?
Description
Cairns.

Record number 81
NGR NY 52000 78300
Parish STAPLETON
Dating Unknown
Description
Enclosure and ditch known from documentary evidence.

Record number 82
NGR NY 52000 78600 NY 51900 78200
Parish STAPLETON
Dating Unknown
Description
Settlement, one mound with 'flagging'.

Record number 88
NGR NY 54850 79950
Parish STAPLETON
Dating Early Bronze Age
Description
Food vessel found in 1972 while quarrying for sand.

Record number 94
NGR NY 58200 76100
Parish BEWCASTLE
Dating Iron Age
Description
Bronze cauldron found in Black Moss, High Grains.

Record number 144
NGR 37900 67700
Parish ARTHURET
Dating Unknown
Description
Unspecified burial.

Record number 148
NGR NY 19770 61480
Parish BOWNESS
Dating Roman
Description
Estimated position of milefort 2 of the Cumberland Coast series.

Record number 149
NGR NY 19470 61160
Parish BOWNESS
Dating Roman
Description
Possible site of turret 2A of the Cumberland Coast series.

Record number 150, 5738
NGR NY 19200 60800
Parish BOWNESS
Dating Unknown
Description
Folk tradition of great army at 'Campfield'. Later aerial photographs have shown cropmarks of enclosures, possibly fields. Two parallel ridges are visible from the ground (150). Bracelet of gold found in vicinity in 1809 (5738).

Record number 151
NGR NY 19030 60780
Parish BOWNESS
Dating Roman
Description
Turret 2B of the Cumberland Coast series.

Record number 152
NGR NY 18620 60520
Parish BOWNESS
Dating Roman
Description
Milefortlet 3 of Cumberland Coast series.

Record number 153
NGR NY 18160 60410
Parish BOWNESS
Dating Roman
Description
Turret 3A of the Cumberland Coast series.

Record number 154
NGR NY 17640 60310
Parish BOWNESS
Dating Roman
Description
Turret 3B of the Cumberland Coast series.

Record number 155
NGR NY 21200 61300
Parish BOWNESS
Dating Prehistoric
Description
Palisade, probably prehistoric, found by peat diggers in 1903.

Record number 156
NGR NY 23010 62750
Parish BOWNESS
Dating Prehistoric
Description
Knockcross mound, possibly barrow. Sword hilt, part of blade of unknown date and origin found there.

Record number 157
NGR NY 24100 62300
Parish BOWNESS
Dating Prehistoric/medieval
Description
Tradition of upright stone on Fishers Cross Mound (perhaps a cross shaft), and a stone coffin, skeleton and coins found there. Calculated position of Turret 78B of Hadrian's Wall.

Record number 158
NGR NY 23020 62720
Parish BOWNESS
Dating Roman
Description
Knockcross Roman camp, plotted from aerial photographs. *See also* SMR 156.

Record number 159
NGR NY 20920 61970
Parish BOWNESS
Dating Roman
Description
Site of milefortlet 1 of the Cumberland Coast series.

Record number 160
NGR NY 21860 62670
Parish BOWNESS
Dating Roman
Description
End of Hadrian's Wall discernible in 1707 *c* 1/4 of a mile west of Bowness. Likely existence of Turret 80A here.

Record number 161
NGR NY 22830 62560
Parish BOWNESS
Dating Roman
Description
Turret 79B of Hadrian's Wall.

Record number 163
NGR NY 22410 62570
Parish BOWNESS
Dating Medieval
Description
Bowness Tower.

Record number 165
NGR NY 23590 62240
Parish BOWNESS
Dating Roman
Description
Milecastle 79 of Hadrian's Wall, built on razed remains of earlier Turf Wall milecastle. Both are built on artificial platform as area is subject to flooding.

Record number 167
NGR NY 22350 62570
Parish BOWNESS
DatingRoman
Description
Extramural settlement to south–east of Bowness fort.

Record number 168
NGR NY 24540 61350
Parish BOWNESS
Dating Roman
Description
Milecastle 78 of Hadrian's Wall.

Record number 169
NGR NY 24310 61750
Parish BOWNESS
Dating Roman
Description
Turret 78A of Hadrian's Wall.

Record number 172
NGR NY 22000 61000
Parish BOWNESS
Dating Bronze Age
Description
Millstone grit axe-hammer.

Record number 173
NGR NY 22000 62000
Parish BOWNESS
Dating Bronze Age
Description
Polished black stone adze, probably from Bowness.

Record number 174
NGR NY 23000 61000
Parish BOWNESS
Dating Bronze Age
Description
Axe-hammer, flat iron in shape.

Record number 175
NGR NY 22000 62000
Parish BOWNESS
Dating Neolithic?
Description
Polished stone axe, rectangular in shape.

Record number 176
NGR NY 22000 62000
Parish BOWNESS
Dating Bronze Age
Description
Axe-hammer with rounded butt.

Record number 177
NGR NY 22200 62590
Parish BOWNESS
Dating Roman
Description
Inscribed slab of red sandstone, dedicated to Matras. Probable temple site.

Record number 188
NGR NY 34000 69000
Parish KIRKANDREWS
Dating Prehistoric
Description
Report of various stray finds, some prehistoric, found during peat-cutting, which may suggest a settlement.

Record number 189
NGR NY 33340 68310
Parish KIRKANDREWS
Dating Unknown
Description
Aerial photograph shows a circular enclosure/ring ditch.

Record number 191
NGR NY 33090 66760
Parish KIRKANDREWS
Dating Medieval/Post-medieval
Description
Map (British Museum) 1590 has marked 'Ye Gardes' with a tower symbol.

Record number 196
NGR NY 35000 64000
Parish ROCKCLIFFE
Dating Medieval/Post-medieval
Description
Map (British Museum) 1590 marks 'Thomwhathill' with tower symbol to north of Rockelis Cast (*sic*).

Record number 198
NGR NY 34000 69000
Parish KIRKANDREWS
Dating Neolithic

Description
Double-expanded perforated stone axe-hammer, found during peat digging in Solway Moss.

Record number 199
NGR NY 37000 68000
Parish ARTHURET
Dating Neolithic
Description
Hafted stone axe.

Record number 200
NGR NY 37000 68700
Parish ARTHURET
Dating Bronze Age
Description
Copper alloy socketed spearhead.

Record number 204
NGR NY 39000 62000
Parish WEST LINTON
Dating Neolithic
Description
Polished flint axe.

Record number 227
NGR NY 43000 63000
Parish SCALEBY
Dating Neolithic
Description
Polished stone axe found at Horsegills, Scaleby Moss.

Record number 260
NGR NY 53800 65400
Parish BURTHOLME
Dating Bronze Age
Description
Stone hammer of 'pestle-type'.

Record number 275
NGR NY 56000 63000
Parish NETHER DENTON
Dating Bronze Age
Description
Copper alloy spear.

Record number 277
NGR NY 56000 63000
Parish NETHER DENTON
Dating Bronze Age
Description
Stone axe-hammer.

Record number 287
NGR NY 53180 64940
Parish BURTHOLME
Dating Mesolithic
Description
Perforated stone axe.

Record number 322
NGR NY 61570 66150
Parish WATERHEAD
Dating Prehistoric
Description
Possible promontory fort underlying Birdoswald Roman fort.

Record number 329
NGR NY 62030 66410
Parish WATERHEAD
Dating Roman
Description
Milecastle 49 of Hadrian's Wall.

Record number 330
NGR NY 61540 66280
Parish WATERHEAD
Dating Roman
Description
Turret 49A of Hadrian's Wall, found in 1945 within Birdoswald fort.

Record number 331
NGR NY 61800 66400
Parish WATERHEAD
Dating Early Medieval
Description
Anglo-Saxon disc-headed gilt bronze pin, probably eighth century, found during clearance of Hadrian's Wall east of Birdoswald.

Record number 333
NGR NY 61630 66310
Parish WATERHEAD
Dating Roman
Description
Probable site of *vicus* adjoining Birdoswald Fort.

Record number 335
NGR NY 61550 66120
Parish WATERHEAD
Dating Roman
Description
Rectangular Roman enclosure, 62 x 112 ft, associated with second century pottery, within native promontory fort (SMR 322), which was revealed by excavation in 1931–33. May have been connected with the Stanegate signalling system.

Record number 336
NGR NY 61130 66200
Parish WATERHEAD
Dating Roman
Description
Turret 49B of Hadrian's Wall.

Record number 337
NGR NY 60670 66010

Parish WATERHEAD
Dating Roman
Description
Milecastle 50 of Hadrian's Wall.

Record number 339
NGR NY 61190 66120
Parish WATERHEAD
Dating Roman
Description
Turret 49B (Turf Wall) of Hadrian's Wall.

Record number 343
NGR NY 61500 66300
Parish WATERHEAD
Dating Roman
Description
Birdoswald Fort.

Record number 344
NGR NY 61200 66200
Parish WATERHEAD
Dating Roman
Description
Cremation cemetery found in 1959. Contained several pots which belong to first half of fourth century.

Record number 369
NGR NY 17700 50600
Parish STAPLETON
Dating Medieval
Description
St Thomas Chapel or Bedehouse, an almshouse founded for resident Brothers in the sixteenth century, dissolved in 1538. Dependent on Holm Cultram Abbey.

Record number 370
NGR NY 17730 50980
Parish HOLME ABBEY
Dating Medieval
Description
Boundary ditch for Holm Cultram Abbey. Earlier thought to be a motte and moat.

Record number 375
NGR NY 13050 50100
Parish HOLME ABBEY
Dating Prehistoric
Description
Stone celt found in 1890.

Record number 376
NGR NY 15800 51980
Parish HOLME LOW
Dating Neolithic
Description
Stone axe found in 1898 200 yds west–south–west of Mireside.

Record number 377
NGR NY 19000 58000
Parish BOWNESS
Dating Neolithic
Description
Unpolished stone axe.

Record number 378
NGR NY 19000 58000
Parish BOWNESS
Dating Neolithic
Description
Polished stone axe, pointed butt type, found at Anthorn, among debris brought from Longtown.

Record number 382
NGR NY 28300 52540
Parish AIKTON
Dating Medieval
Description
Down Hall, manorial seat of John de Morville.

Record number 393
NGR NY 29900 50200
Parish THURSBY
Dating Unknown
Description
Series of conjoined rectangular enclosures with at least one linear ditch, appearing as cropmarks.

Record number 395
NGR NY 23000 56000
Parish KIRKBRIDE
Dating Bronze Age
Description
Perforated stone axe-hammer found in Kirkbride parish in 1858.

Record number 396
NGR NY 23000 56000
Parish KIRKBRIDE
Dating Neolithic
Description
Unpolished stone axe.

Record number 397
NGR NY 26000 60000
Parish BOWNESS
Dating Iron Age
Description
Macedonian silver coin found at Drumburgh.

Record number 398
NGR NY 25800 52400
Parish AIKTON
Dating Medieval
Description
Medieval bronze cauldron found in the River Wampool near Gamelsby in 1962.

Record number 404
NGR NY 32740 54730
Parish ORTON
Dating Medieval
Description
Great Orton once enclosed by ditch and rampart called the 'Ring Fence'.

Record number 406
NGR NY 34500 54800
Parish ORTON
Dating Prehistoric/Roman
Description
Cropmarks on aerial photographs show enclosures.

Record number 412
NGR NY 33090 59090
Parish BURGH BY SANDS
Dating Medieval
Description
Probably Norman motte and bailey. A later ditch may have been moat of a grange. Mention is also made of ancient fishponds, but no traces remain.

Record number 413
NGR NY 33370 59080
Parish BURGH BY SANDS
Dating Roman
Description
Roman culvert or small bridge carrying the Military Way across east branch of Powburgh Beck south of crossing point of Hadrian's Wall.

Record number 415
NGR NY 32800 59100
Parish BURGH BY SANDS
Dating Roman
Description
Site of Roman fort of *Aballava* (Burgh 2). Position of Turret 71B of Hadrian's Wall should underlie fort.

Record number 416
NGR NY 32420 59100
Parish BURGH BY SANDS
Dating Roman
Description
Remains of Milecastle 72 of Hadrian's Wall.

Record number 418
NGR NY 33790 59200
Parish BEAUMONT
Dating Prehistoric/Roman
Description
Milecastle 71 of Hadrian's Wall. Two worked flints were found, stratified near footings of the walls, also two small fragments of Roman pottery.

Record number 421
NGR NY 31440 59220
Parish BURGH BY SANDS
Dating Roman
Description
Turret 72B of Hadrian's Wall.

Record number 424
NGR NY 34900 57100
Parish BEAUMONT
Dating Unknown
Description
Three ring ditches on highest part of Common, reported by Whellan (1850). He also mentions earlier find of urn.

Record number 425
NGR NY 34000 59000
Parish BEAUMONT
Dating Roman
Description
Temporary camp recorded from aerial photographs east of Burgh by Sands, between Hadrian's Wall and Vallum, immediately south of Milecastle 71.

Record number 426
NGR NY 34300 58300
Parish BEAUMONT
Dating Roman
Description
Temporary Roman camp, half a mile south of Hadrian's Wall.

Record number 428
NGR NY 33000 59000
Parish BURGH BY SANDS
Dating Roman
Description
Buildings of a civil settlement east of fort at *Aballava* on aerial photograph.

Record number 429
NGR NY 32000 59000
Parish BURGH BY SANDS
Dating Iron Age/Romano-British
Description
Naked, horned, armed god, holding a sphere and a spheroid object, from Burgh by Sands.

Record number 444
NGR NY 32800 54200
Parish ORTON
Dating Medieval
Description
Shaft of former market cross of Great Orton.

Record number 480
NGR NY 33350 53460
Parish ORTON
Dating Neolithic
Description
Stone axe.

Record number 515, 596–599
NGR NY 10 43, NY 11 44, NY 11 43
Parish WESTNEWTON
Dating Iron Age / Romano-British
Description
Five sites, known as The Batteries, lie near Aiglegill farm north–west of Aspatria, four of them on Pow Rigg Hill spur. A stone hammer, an adze, and pointed stone were found.

Record number 584
NGR NY 11460 46170
Parish HOLME ST CUTHBERT
Dating Iron Age / Romano-British
Description
Settlement.

Record number 590
NGR NY 08700 48500
Parish HOLME ST CUTHBERT
Dating Roman
Description
Roman fortlet south–south–west of Castle Fields, from aerial photographs.

Record number 591
NGR NY 08650 48500
Parish HOLME ST CUTHBERT
Dating Roman
Description
Roman cemetery and pyre site, a short distance south of Beckfoot Fort, revealed by coastal erosion.

Record number 592
NGR NY 08800 47800
Parish HOLME ST CUTHBERT
Dating Romano-British
Description
Cropmarks of a possible settlement site.

Record number 603
NGR NY 09800 45200
Parish HOLME ST CUTHBERT
Dating Unknown
Description
Rectangular ditched enclosure, on aerial photographs, 125 m south–east of Old Mawbray.

Record number 604
NGR NY 08990 47900
Parish HOLME ST CUTHBERT
Dating Roman
Description
Cropmark of a rectangular ditched enclosure.

Record number 605
NGR NY 09670 45760
Parish HOLME ST CUTHBERT
Dating Romano-British
Description
Probable native settlement (Edderside) appearing as cropmark. An unusual feature of the site was a series of pits forming a rectangle at the south–west end with a central posthole.

Record number 606
NGR NY 09900 45000
Parish HOLME ST CUTHBERT
Dating Unknown
Description
Oval enclosure south of Edderside.

Record number 607
NGR NY 09700 46200
Parish HOLME ST CUTHBERT
Dating Roman
Description
Oval ditched enclosure (Plasketlands) appearing as cropmark.

Record number 608
NGR NY 10200 49900
Parish HOLME ST CUTHBERT
Dating Iron Age / Romano-British
Description
Aerial photograph of settlement site.

Record number 609
NGR NY 06900 40200
Parish CROSSCANONBY
Dating Iron Age
Description
Small multivallate hillfort located on the summit and south–east slope of Swarthy Hill. On aerial photographs.

Record number 611
NGR NY 11100 47400
Parish HOLME ST CUTHBERT
Dating Medieval
Description
Deserted medieval settlement. *See also* SMR 593.

Record number 612
NGR NY 10500 47100
Parish HOLME ST CUTHBERT
Dating Medieval
Description
Shrunken medieval village.

Record number 613
NGR NY 07830 46100
Parish HOLME ST CUTHBERT
Dating Roman
Description
Tower 16B of Cumberland Coast series.

Record number 617
NGR NY 11000 47000
Parish HOLME ABBEY
Dating Unknown
Description
Cropmark sites.

Record number 619
NGR NY 08300 47020
Parish HOLME ST CUTHBERT
Dating Roman
Description
Site of milefortlet 16 of Cumberland Coast series.

Record number 620
NGR NY 08124 46550
Parish HOLME ST CUTHBERT
Dating Roman
Description
Site of Cote How tower 16A of Cumberland Coast series.

Record number 621
NGR NY 07800 45200
Parish HOLME ST CUTHBERT
Dating Roman
Description
Possible site of signal station.

Record number 622
NGR NY 07710 45640
Parish HOLME ST CUTHBERT
Dating Roman
Description
Milefortlet 17 of Cumberland Coast series, on aerial photographs on the tip of Dubmill Point.

Record number 623
NGR NY 08370 47510
Parish HOLME ST CUTHBERT
Dating Roman
Description
Site of tower 15B of Cumberland Coast series.

Record number 624
NGR NY 08500 47940
Parish HOLME ST CUTHBERT
Dating Roman
Description
The buried remains of Bank Mill Tower 15A of Cumberland Coast series.

Record number 625
NGR NY 08960 48840
Parish HOLME ST CUTHBERT
Dating Roman
Description
Beckfoot Fort.

Record number 626
NGR NY 09040 48890
Parish HOLME ST CUTHBERT
Dating Roman
Description
Vicus to east and north of Beckfoot Fort.

Record number 634
NGR NY 11070 47340
Parish HOLME ST CUTHBERT
Dating Unknown
Description
Stone hammer found in 1873.

Record number 635
NGR NY 14040 47970
Parish HOLME ABBEY
Dating Unknown
Description
Flint arrowhead found in 1898.

Record number 636
NGR NY 14270 48010
Parish HOLME ABBEY
Dating Unknown
Description
Flint spearhead found in 1898.

Record number 637
NGR NY 11580 45320
Parish HOLME ST CUTHBERT
Dating Neolithic
Description
Fragment of polished stone axe found during drainage work on Chapel Moss.

Record number 638
NGR NY 13400 49400
Parish HOLME ABBEY
Dating Neolithic
Description
Polished stone axe found at Foulsyke.

Record number 639
NGR NY 19300 47500
Parish BROMFIELD
Dating Neolithic
Description
Polished stone axe of crystalline rock.

Record number 640
NGR NY 11950 43000
Parish WESTNEWTON
Dating Unknown
Description
Stone adze found in 1877.

Record number 643
NGR NY 14000 42000
Parish ASPATRIA
Dating Prehistoric
Description
Perforated axe-hammer of yellow sandstone.

Record number 644
NGR NY 14000 42000
Parish ASPATRIA
Dating Bronze Age
Description
Bronze palstave found before 1860. It has a stop ridge, but no side loop.

Record number 645
NGR NY 14000 42000
Parish ASPATRIA
Dating Late Bronze Age
Description
Gold pennanular armlet or bracelet, with broadened ends.

Record number 650
NGR NY 09000 49000
Parish HOLME ST CUTHBERT
Dating Neolithic?
Description
Quern found at Beckfoot.

Record number 651
NGR NY 08888 43240
Parish ALLONBY
Dating Neolithic
Description
Two unpolished stone axes found during ploughing in 1927.

Record number 652
NGR NY 09190 44440
Parish HOLME ST CUTHBERT
Dating Neolithic
Description
Polished celt found in 1914 while cutting a new course for Dub Beck.

Record number 683
NGR NY 31500 49800
Parish THURSBY
Dating Bronze Age
Description
Complete enlarged food vessel and the rim of a second.

Record number 684
NGR NY 32540 48750
Parish THURSBY
Dating Unknown
Description
Cropmark.

Record number 685
NGR NY 32040 49480
Parish THURSBY
Dating Unknown
Description
Rectangular cropmark.

Record number 686
NGR NY 32090 49060
Parish THURSBY
Dating Unknown
Description
Rectangular cropmark.

Record number 698
NGR NY 39240 45060
Parish DALSTON
Dating Prehistoric
Description
Two rectangular enclosures.

Record number 791
NGR NY 03530 35210
Parish MARYPORT
Dating Romano-British
Description
Two ditches converging at a probable entrance to a building, with a cobbled foundation and a road. Finds include a granite quernstone and a Crambeck rim sherd. Also a Romano-British settlement and trackway visible as cropmarks, including infilled ditches, a sub-circular enclosure within a larger oval enclosure. The trackway runs in a south–eastern direction for *c* 110 m.

Record number 901
NGR NY 36040 34230
Parish CALDBECK
Dating Bronze Age
Description
Bronze spearhead with loops on socket and leaf-shaped blade.

Record number 2629
NGR NY 35000 50000
Parish DALSTON
Dating Roman
Description
Tombstone.

Record number 2968
NGR NY 33500 58600
Parish BURGH BY SANDS
Dating Iron Age/Romano-British
Description
Settlement on aerial photographs.

Record number 2973
NGR NY 24250 33640

Parish CALDBECK
Dating Iron Age
Description
Large univallate hillfort, two round cairns, and a medieval shieling. The enclosure wall is surrounded by a drystone rampart.

Record number 2991
NGR NY 27600 57900
Parish KIRKBAMPTON
Dating Unknown
Description
Possible settlement appearing on aerial photographs. *See also* SMR 3330.

Record number 2993
NGR NY 25700 50400
Parish WOODSIDE
Dating Unknown
Description
Polygonal enclosure, appearing on aerial photographs.

Record number 3022
NGR NY 18500 52200
Parish HOLME EAST WAVER
Dating Roman
Description
Possible remains of a Roman road.

Record number 3026
NGR NY 10000 48000
Parish HOLME ST CUTHBERT
Dating Unknown
Description
Cropmark.

Record number 3182
NGR NY 08900 42900
Parish ALLONBY
Dating Iron Age / Romano-British
Description
Settlement on aerial photographs.

Record number 3183 – 3186
NGR NY 09 47, NY 09 48
Parish HOLME ST CUTHBERT
Dating Unknown
Description
Cropmarks on aerial photographs of settlement nucleus (3183), field systems (3184), and enclosures (3185, 3186).

Record number 3188
NGR NY 10100 45200
Parish HOLME ST CUTHBERT
Dating Unknown
Description
Cropmarks of field system.

Record number 3189
NGR NY 10100 45100
Parish HOLME ST CUTHBERT
Dating Unknown
Description
Circular enclosure, on aerial photographs.

Record number 3190
NGR NY 10300 46200
Parish HOLME ST CUTHBERT
Dating Unknown
Description
Settlements on aerial photographs.

Record number 3191
NGR NY 10800 43500
Parish WESTNEWTON
Dating Unknown
Description
Enclosure appearing as cropmark.

Record number 3192
NGR NY 10870 47490
Parish HOLME ST CUTHBERT
Dating Unknown
Description
Curving semi-circular cropmark cut by linear cropmark.

Record number 3193
NGR NY 10900 46700
Parish HOLME ST CUTHBERT
Dating Unknown
Description
Enclosure appearing as cropmark.

Record number 3194
NGR NY 11000 46300
Parish HOLME ST CUTHBERT
Dating Unknown
Description
Possible rectangular enclosure appearing as cropmark.

Record number 3195
NGR NY 11000 43100
Parish BRIGHAM
Dating Unknown
Description
Enclosure showing as cropmark.

Record number 3196
NGR NY 11800 45800
Parish HOLME ST CUTHBERT
Dating Unknown
Description
Rectangular settlement site and field system appearing as cropmark.

Record number 3222
NGR NY 26800 52200
Parish AIKTON
Dating Unknown
Description
Probable field system appearing as cropmark.

Record number 3237
NGR NY 11900 46300
Parish HOLME ST CUTHBERT
Dating Unknown
Description
Curvilinear field system appearing on aerial photographs.

Record number 3238
NGR NY 12100 47800
Parish HOLME ST CUTHBERT
Dating Unknown
Description
Curvilinear enclosure, hut circles and dykes recorded by aerial photographs.

Record number 3239
NGR NY 12200 43800
Parish WESTNEWTON
Dating Unknown
Description
Probable site nucleus appearing on aerial photographs.

Record number 3240
NGR NY 12400 47900
Parish HOLME ST CUTHBERT
Dating Unknown
Description
Subrectangular fields with enclosed settlement appearing as cropmarks.

Record number 3241
NGR NY 12400 42100
Parish ASPATRIA
Dating Unknown
Description
Oval enclosure appearing as cropmark.

Record number 3242–3251
NGR NY 12 48, NY 12 47, NY 13 48, NY 13 47
Parish HOLME ST CUTHBERT
Dating Unknown
Description
Concentration of cropmarks on aerial photographs, including settlement sites (3243, 3245, 3248), field systems (3242, 3243, 3244), enclosures and ring ditches (3247).

Record number 3252, 3253
NGR NY 13 44
Parish WESTNEWTON
Dating Unknown
Description
Two ditched circular enclosures appearing as cropmarks.

Record number 3254
NGR NY 13900 43600
Parish WESTNEWTON
Dating Unknown
Description
Settlement at Warwick Hall on aerial photographs.

Record number 3255
NGR NY 14100 45400
Parish BROMFIELD
Dating Unknown
Description
Subrectangular sites appearing as cropmarks.

Record number 3256
NGR NY 14000 46500
Parish HOLME ABBEY
Dating Unknown
Description
Possible settlement/hut circle appearing as cropmark.

Record number 3257
NGR NY 13500 41400
Parish ASPATRIA
Dating Unknown
Description
Circular enclosure appearing as cropmark.

Record number 3258
NGR NY 14300 44700
Parish WESTNEWTON
Dating Unknown
Description
Enclosure on aerial photographs.

Record number 3259
NGR NY 14500 48300
Parish HOLME ABBEY
Dating Unknown
Description
Hut circles and field system appearing as cropmarks.

Record number 3260 – 3262
NGR NY 14 48
Parish HOLME ABBEY
Dating Unknown
Description
Two ditched compounds and site with hut circles and field system (3262) appearing on aerial photographs.

Record number 3264
NGR NY 15900 43900
Parish WESTNEWTON
Dating Unknown
Description
Circular enclosure on aerial photographs.

Record number 3265 – 3269
NGR NY 16 50, NY 16 51, NY 16 52

Parish HOLME ABBEY
Dating Unknown
Description
Area of cropmarks (1 x 3 km) of enclosures and settlement (3268).

Record number 3270
NGR NY 17300 49800
Parish HOLME ABBEY
Dating Unknown
Description
Sub-rectangular nucleus appearing as cropmark.

Record number 3272
NGR NY 18000 45500
Parish BROMFIELD
Dating Unknown
Description
Settlement on aerial photograph.

Record number 3273
NGR NY 18100 45700
Parish BROMFIELD
Dating Unknown
Description
Cropmark.

Record number 3274, 3276, 3277
NGR NY 18 48
Parish BROMFIELD
Dating Unknown
Description
Square kilometre with cropmarks of a settlement (3274) and two enclosures.

Record number 3278
NGR NY 19100 49100
Parish DUNDRAW
Dating Unknown
Description
Oval enclosure appearing as cropmark.

Record number 3280
NGR NY 19700 51300
Parish HOLME EAST WAVER
Dating Unknown
Description
Oval ditched enclosure appearing as cropmark.

Record number 3281
NGR NY 20800 49900
Parish DUNDRAW
Dating Unknown
Description
Field system appearing as cropmark.

Record number 3282
NGR NY 20900 61700
Parish BOWNESS
Dating Unknown
Description
Single ditch appearing as cropmark. *See also* SMR 159 which may be same site.

Record number 3283
NGR NY 21200 49400
Parish DUNDRAW
Dating Unknown
Description
Possible settlement nucleus discovered as cropmark, but traces of earthworks also on the ground as rectangular platform.

Record number 3284
NGR NY 21300 62300
Parish BOWNESS
Dating Unknown
Description
Possible settlement appearing as cropmark.

Record number 3286
NGR NY 21700 49700
Parish DUNDRAW
Dating Unknown
Description
Spur site with single ditch appearing as cropmark.

Record number 3288
NGR NY 22700 61000
Parish BOWNESS
Dating Roman
Description
Possible rubbish pits, and Roman road, appearing as cropmark.

Record number 3293
NGR NY 23100 62600
Parish BOWNESS
Dating Unknown
Description
Settlement appearing as cropmark.

Record number 3297
NGR NY 23433 61400
Parish BOWNESS
Dating Unknown
Description
Possible settlement with pits, appearing as cropmark.

Record number 3298
NGR NY 23600 50700
Parish WOODSIDE
Dating Unknown
Description
Pits, ditch, and occupation spread appearing as cropmarks. SMR 9655 may be the same site.

Record number 3300
NGR NY 24000 51100
Parish WOODSIDE
Dating Unknown
Description
Linear ditch, possible settlement site appearing as cropmark.

Record number 3302
NGR NY 24900 56800
Parish BOWNESS
Dating Unknown
Description
Possible settlement and dykes appearing as cropmarks.

Record number 3303
NGR NY 25110 57250
Parish BOWNESS
Dating Prehistoric/Romano-British
Description
Prehistoric enclosure and associated trackway, together with a later Romano-British farmstead, on very slightly elevated ground to the west–north–west of Fingland.

Record number 3304
NGR NY 25300 56900
Parish BOWNESS
Dating Unknown
Description
Cropmark.

Record number 3305, 3307, 3308
NGR NY 25 55
Parish AIKTON
Dating Unknown
Description
Cropmarks of trackway and two settlement sites.

Record number 3306
NGR NY 25300 57400
Parish BOWNESS
Dating Unknown
Description
Cropmark, also flint scatter and sherd of late medieval reduced ware.

Record number 3310
NGR NY 25400 57500
Parish BOWNESS
Dating Unknown
Description
Sub-rectangular nucleus and field system appearing as cropmarks.

Record number 3311
NGR NY 25700 54700
Parish KIRKBAMPTON
Dating Iron Age/Romano-British?
Description
Two sub-rectangular enclosures.

Record number 3312 – 3315
NGR NY 25 54, NY 25 55, NY 26 54
Parish KIRKBAMPTON
Dating Unknown
Description
Area containing a number of cropmarks of enclosures and a possible settlement site (3315).

Record number 3318 – 3324
NGR NY 26 52, NY 27 52, NY 26 53
Parish AIKTON
Dating Unknown
Description
Area containing a number of cropmarks of possible settlement sites (3319, 3320, 3324), enclosures (3318, 3321, 3323), and field systems (3322).

Record number 3325
NGR NY 27100 57500
Parish KIRKBAMPTON
Dating Unknown
Description
Road/trackway appearing as cropmark.

Record number 3326
NGR NY 27100 50500
Parish WOODSIDE
Dating Unknown
Description
Possible settlement site appearing as cropmark.

Record number 3330
NGR NY 27800 57600
Parish BOWNESS
Dating Unknown
Description
Ditched oval enclosure appearing as cropmark. *See also* SMR 2991.

Record number 3332
NGR NY 28100 56900
Parish KIRBAMPTON
Dating Unknown
Description
Site on aerial photographs.

Record number 3379
NGR NY 29400 58300
Parish BURGH BY SANDS
Dating Prehistoric/Romano-British
Description
Group of enclosures, two rectangular and one oval. This site is one of a series of similar settlements possibly associated with a Roman trackway/road between Kirkbride and Burgh Roman forts.

Record number 3386
NGR NY 30200 58100
Parish BURGH BY SANDS
Dating Iron Age/Romano-British
Description
Oval enclosure at Boustead Hill.

Record number 3388
NGR NY 30900 56400
Parish KIRKBAMPTON
Dating Unknown
Description
Possible hut circles appearing as cropmarks.

Record number 3390
NGR NY 31400 59700
Parish BURGH BY SANDS
Dating Unknown
Description
Site on aerial photographs.

Record number 3391
NGR NY 31900 57600
Parish BURGH BY SANDS
Dating Roman
Description
Cropmarks of sub-rectangular enclosure: auxiliary fort.

Record number 3394
NGR NY 33300 58300
Parish BURGH BY SANDS
Dating Romano-British
Description
Triple-ditched enclosure. Similarities with Northumbrian native settlements but may be Roman fortlet.

Record number 3397
NGR NY 24500 56100
Parish AIKTON
Dating Unknown
Description
Cropmark.

Record number 3398
NGR NY 35200 60800
Parish BEAUMONT
Dating Unknown
Description
Site on aerial photographs.

Record number 3405
NGR NY 41100 61400
Parish STANWIX RURAL
Dating Unknown
Description
Settlement and field system.

Record number 3428
NGR NY 55200 33700
Parish LANGWATHBY
Dating Iron Age and Romano-British
Description
Enclosure.

Record number 3435
NGR NY 43000 63500
Parish SCALEBY
Dating Prehistoric
Description
Remains of a human skeleton buried *c* 8–9 ft below the surface, closely embedded in the lowest stratum of black peat, discovered in 1845 in Scaleby Moss. It was wrapped in a 'hide garment' of sewn pieces bound together by strong leather thongs. Parts of the skull with black hair, many teeth and a large proportion of the brain, was discovered later. A stick, 3 ft long, 12 inches in circumference, was grasped in one hand. *See* Turner 1988.

Record number 3604
NGR NY 17700 50800
Parish HOLME ABBEY
Dating Medieval
Description
Cistercian house of St Mary at Holm Cultram; largest monastic house in Cumberland, founded in 1150, dissolved in 1538.

Record number 3692
NGR NY 03530 35210
Parish MARYPORT
Dating Bronze Age
Description
Ewanrigg burial containing nine cremations, and a tumulus with a robbed cist.

Record number 3697
NGR NY 09700 48700
Parish HOLME ST CUTHBERT
Dating Unknown
Description
Cropmark.

Record number 3699
NGR NY 08900 48400
Parish HOLME ST CUTHBERT
Dating Iron Age/Romano-British
Description
Settlement appearing as cropmark.

Record number 3700
NGR NY 09000 46900
Parish HOLME ST CUTHBERT
Dating Unknown
Description
Site on aerial photographs.

Record number 3731
NGR NY 18510 51970

Parish HOLME EAST WAVER
Dating Medieval
Description
Cruck-framed barn at Raby Grange may be part of the grange belonging to Holm Cultram Abbey.

Record number 3751
NGR NY 23700 61800
Parish BOWNESS
Dating Unknown
Description
Earthworks on aerial photographs.

Record number 3752
NGR NY 20200 61500
Parish BOWNESS
Dating Unknown
Description
Site on aerial photographs.

Record number 3769
NGR NY 32860 59100
Parish BURGH BY SANDS
Dating Roman / medieval
Description
St Michael's Church, mentioned prior to 1202, has late Norman doorway and 1350 pele tower. Inscribed Roman altar found in walled-up doorway in 1881 may suggest temple site.

Record number 3770
NGR NY 34400 57100
Parish BEAUMONT
Dating Unknown
Description
Ring ditch on aerial photographs.

Record number 3771
NGR NY 33200 57600
Parish BURGH BY SANDS
Dating Roman
Description
Cropmark complex – pits and small enclosures in north end of area and several other features including a droveway between two fields to the south–east, where a substantial ditch defines the probable limits of settlement.

Record number 3772
NGR NY 33000 59700
Parish BURGH BY SANDS
Dating Unknown
Description
Site on aerial photographs.

Record number 3849
NGR NY 50600 56300
Parish HAYTON
Dating Unknown
Description
Tumulus or motte from aerial photographs.

Record number 3875
NGR NY 50070 71590
Parish STAPLETON
Dating Medieval
Description
Tower at Harperhill marked on map dated 1590.

Record number 3876
NGR NY 50000 71700
Parish STAPLETON
Dating Unknown
Description
Earthworks.

Record number 3877
NGR NY 26700 57000
Parish KIRKBAMPTON
Dating Romano-British
Description
Cropmark, part of settlement.

Record number 4195
NGR NY 13200 51300
Parish HOLME LOW
Dating Unknown
Description
Settlement site.

Record number 4239
NGR NY 02600 35000
Parish MARYPORT
Dating Prehistoric / Roman / medieval
Description
Rise How tower, together with the buried remains of a pre-Roman burial mound / barrow, and a medieval grain-drying kiln.

Record number 4395
NGR NY 32350 58250
Parish BURGH BY SANDS
Dating Roman
Description
Rectangular enclosure; when excavated revealed Roman fort (Burgh 1). Parts dated by pottery to early second century.

Record number 4405
NGR NY 15800 42100
Parish ASPATRIA
Dating Unknown
Description
Oval enclosure appearing as cropmark on aerial photographs.

Record number 4406
NGR NY 14800 42700
Parish ASPATRIA
Dating Unknown
Description
Enclosure, possible ring ditch, appearing as cropmark on aerial photographs.

Record number 4521
NGR NY 09000 45000
Parish HOLME ST CUTHBERT
Dating Middle Bronze Age
Description
Salta Moss bronze rapier.

Record number 4522
NGR NY 08000 46000
Parish HOLME ST CUTHBERT
Dating Neolithic
Description
Two perforated axe-hammers and a polished axe found at Mawbray.

Record number 4523
NGR NY 08000 46000
Parish HOLME ST CUTHBERT
Dating Late Bronze Age
Description
Socketed and looped spearhead found at Dubmill Farm, Mawbray.

Record number 4524
NGR NY 09000 48000
Parish HOLME ST CUTHBERT
Dating Neolithic
Description
Four polished and two unpolished stone axes found in Newton, Mawbray.

Record number 4525
NGR NY 08800 46500
Parish HOLME ST CUTHBERT
Dating Unknown
Description
Mound.

Record number 4538
NGR NY 46100 58600
Parish STANWIX RURAL
Dating Prehistoric
Description
Two bone harpoon heads.

Record number 4568
NGR NY 51300 60800
Parish BRAMPTON
Dating Bronze Age
Description
Two supposed barrows, one natural.

Record number 4584
NGR NY 22370 62640
Parish BOWNESS
Dating Roman / Medieval / Post-medieval
Description
Late Norman church with with reused stones from Hadrian's Wall.

Record number 4597
NGR NY 51600 63300
Parish BRAMPTON
Dating Unknown
Description
Cropmark at Castlesteads.

Record number 4607
NGR NY 04800 19300
Parish ARLECDON AND FRIZINGTON
Dating Prehistoric?
Description
Stone quern.

Record number 4627, 6283
NGR NY 32000 59000
Parish BURGH BY SANDS
Dating Iron Age / Romano-British
Description
Female head, possibly wearing a diadem, found at Burgh by Sands (4627). It appears to be part of a ritual vessel (spout and handle?). Also fragment of pot decorated with three human heads (6283).

Record number 4639
NGR NY 17000 50000
Parish HOLME ABBEY
Dating Medieval
Description
Possibly a separate hospital for women existed outside the walls of Holm Cultram Abbey.

Record number 4677
NGR NY 26400 57200
Parish BOWNESS
Dating Roman
Description
The Stanegate Roman road.

Record number 4689
NGR NY 23200 61400
Parish BOWNESS
Dating Medieval
Description
Bastle, used as a barn in 1977.

Record number 4700
NGR NY 15300 41100
Parish ASPATRIA
Dating Prehistoric?
Description
Two possible barrows at river's edge, appearing as cropmarks.

Record number 4880–4885
NGR NY 08 46
Parish HOLME ST CUTHBERT
Dating Unknown
Description
Three stone axes (4880, 4882, 4883), two stone axe-hammers (4884, 4885) and one stone adze (4881) from Mawbray.

Record number 4908
NGR NX 98800 20000
Parish PARTON
Dating Prehistoric
Description
Tumulus.

Record number 4955
NGR NY 51800 78200
Parish STAPLETON
Dating Bronze Age
Description
Cairn with cist and urns.

Record number 5103
NGR NY 45400 60000
Parish STANWIX RURAL
Dating Unknown
Description
Enclosure.

Record number 5118
NGR NY 31000 57000
Parish BURGH BY SANDS
Dating Unknown
Description
Square enclosure appearing as cropmark.

Record number 5119
NGR NY 10800 50800
Parish HOLME ST CUTHBERT
Dating Unknown
Description
Enclosure on aerial photographs.

Record number 5124
NGR NY 47100 41600
Parish HESKET
Dating Unknown
Description
Enclosure / cropmarks.

Record number 5146
NGR NY 43400 64500
Parish SCALEBY
Dating Unknown
Description
Standing stone.

Record number 5182
NGR NY 43300 64300
Parish SCALEBY
Dating Unknown
Description
Enclosure.

Record number 5183
NGR NY 43100 65600
Parish KIRKLINTON MIDDLE
Dating Unknown
Description
Enclosure.

Record number 5184
NGR NY 43600 66100
Parish KIRKLINTON MIDDLE
Dating Unknown
Description
Earthworks.

Record number 5185
NGR NY 43000 67800
Parish HETHERSGILL
Dating Unknown
Description
Cropmark.

Record number 5203
NGR NY 43470 65870
Parish KIRKLINTON MIDDLE
Dating Unknown
Description
Earthworks.

Record number 5214
NGR NY 55000 62000
Parish BRAMPTON
Dating Unknown
Description
Mound on aerial photographs.

Record number 5281
NGR NY 51200 17500
Parish BAMPTON
Dating Unknown
Description
Settlement site on aerial photographs.

Record number 5267
NGR NY 45000 78700
Parish NICHOLFOREST
Dating Unknown
Description
Dykes and earthworks.

Record number 5376
NGR NY 19400 46200
Parish BROMFIELD
Dating Unknown
Description
Cropmark.

Record number 5649
NGR NY 34000 59000
Parish BEAUMONT
Dating Medieval
Description
Motte of the Le Brun family, said to have been situated on the site of a milecastle of Hadrian's Wall.

Record number 5710
NGR NY 47900 79300
Parish NICHOLFOREST
Dating Prehistoric?
Description
Cist.

Record number 5783
NGR NY 60700 66000
Parish WATERHEAD
Dating Roman
Description
Course of Roman Military Way associated with Hadrian's Wall.

Record number 5784
NGR NY 60600 66000
Parish WATERHEAD
Dating Roman
Description
Road north from High House. *See also* SMR 5783.

Record number 5786
NGR NY 61500 66100
Parish WATERHEAD
Dating Roman
Description
Crossing over Vallum system.

Record number 5796
NGR NY 61800 66100
Parish WATERHEAD
Dating Unknown
Description
Field system at Underheugh Farm.

Record number 5815
NGR NY 61950 66550
Parish WATERHEAD
Dating Unknown
Description
Possible earthworks but may also be natural feature.

Record number 5837
NGR NY 53900 55900
Parish CASTLE CARROCK
Dating Bronze Age
Description
Barrow with flint-chert find.

Record number 5844
NGR NY 25700 51700
Parish WOODSIDE
Dating Medieval
Description
Tarn Castle.

Record number 5918
NGR NY 16900 42270
Parish ALLHALLOWS
Dating Unknown
Description
Earthwork.

Record number 5981
NGR NY 55300 41900
Parish KIRKOSWALD
Dating Unknown
Description
Enclosures and dykes appearing as cropmarks.

Record number 6052
NGR NY 19300 32000
Parish HOLME EAST WAVER
Dating Unknown
Description
Cropmarks.

Record number 6057
NGR NY 08700 43320
Parish ALLONBY
Dating Unknown
Description
Three enclosures, two circular, one subcircular.

Record number 6111
NGR NY 32000 57000
Parish BURGH BY SANDS
Dating Unknown
Description
Series of enclosures, indicating a settlement site, appearing as cropmarks.

Record number 6123
NGR NY 25400 52100
Parish WOODSIDE
Dating Medieval
Description
Traces of ridge and furrow in several fields near Gamelsby.

Record number 6169
NGR NY 24100 54150
Parish WOODSIDE
Dating Medieval
Description
Area of earthworks west of River Wampool. Includes ridge and furrow, headlands and flood bank for the river.

Record number 6299
NGR NY 23340 61500
Parish BOWNESS
Dating Roman
Description
Probably two temporary camps superimposed rather than one site with alterations. Sherds of second century pottery were found.

Record number 6304
NGR NY 61510 66320
Parish WATERHEAD
Dating Medieval / Post-medieval
Description
Possible site of farmstead / tenement at Birdoswald in Tudor times or earlier.

Record number 6474
NGR NY 08000 46000
Parish HOLME ST CUTHBERT
Dating Prehistoric
Description
Stone axe from Mawbray.

Record number 6485
NGR NY 27300 57700
Parish BOWNESS ON WINDERMERE
Dating Roman
Description
Roman palisade and defensive ditch on Fingland Rigg containing early second-century pottery.

Record number 6486
NGR NY 31700 58800
Parish BURGH BY SANDS
Dating Roman
Description
Fort (Burgh 3) with an annexe on the south–west side. Second-century and medieval pottery found, with two undiagnostic flints.

Record number 6692
NGR NY 32220 59300
Parish BURGH BY SANDS
Dating Unknown
Description
Three cropmark enclosures.

Record number 6697
NGR NY 08700 47300
Parish HOLME ST CUTHBERT
Dating Unknown
Description
Rectilinear cropmarks.

Record number 6698
NGR NY 08200 44400
Parish ALLONBY
Dating Unknown
Description
Ridge and furrow.

Record number 6866
NGR NY 10800 44300
Parish WESTNEWTON
Dating Unknown
Description
Cropmarks which may represent an oval enclosure.

Record number 6869
NGR NY 14400 48100
Parish HOLME ABBEY
Dating Unknown
Description
Possible settlement site from aerial photographs.

Record number 6882
NGR NY 21100 62100
Parish BOWNESS
Dating Unknown
Description
Cropmarks.

Record number 6883
NGR NY 18900 60500
Parish BOWNESS
Dating Unknown
Description
Cropmarks including possible sub-oval enclosure.

Record number 6887
NGR NY 13300 48400
Parish HOLME ABBEY
Dating Unknown
Description
Cropmark complex.

Record number 6889
NGR NY 26900 52200
Parish AIKTON
Dating Roman
Description
Square-ditched enclosure with linear feature.

Record number 6891
NGR NY 32100 59700
Parish BURGH BY SANDS
Dating Unknown
Description
Linear feature in several fields.

Record number 6892
NGR NY 31400 58500
Parish BURGH BY SANDS
Dating Unknown
Description
Cropmarks of apparently adjoining enclosures.

Record number 6893
NGR NY 32600 57700
Parish BURGH BY SANDS
Dating Unknown
Description
Linear feature and possible earthwork.

Record number 6894
NGR NY 33900 59500
Parish BEAUMONT
Dating Unknown
Description
Possible square enclosure and other cropmarks.

Record number 6895
NGR NY 23400 61400
Parish BOWNESS
Dating Unknown
Description
Cropmarks and linear features.

Record number 6896
NGR NY 20600 61500
Parish BOWNESS
Dating Unknown
Description
Cropmarks.

Record number 6928
NGR NY 22600 62000
Parish BOWNESS
Dating Unknown
Description
Diverse cropmarks, mainly linear, some of which relate to earlier field boundaries.

Record number 6929
NGR NY 21400 55700
Parish HOLME EAST WAVER
Dating Unknown
Description
Linear cropmarks, some of which appear to relate to earlier field boundaries.

Record number 6930
NGR NY 21700 55700
Parish HOLME EAST WAVER
Dating Unknown
Description
Cropmarks and earthworks of rectangular enclosures/ field boundaries.

Record Number 7611
NGR NY 16450 07370
Parish IRTON WITH SANTON
Dating Prehistoric?
Description
Settlement site identified as hut circle.

Record number 9528
NGR NY 08050 44710
Parish HOLME ST CUTHBERT
Dating Roman?
Description
Faint cropmark corner may represent a Roman installation on the supposed site of tower 17B of Cumberland Coast series.

Record number 9531
NGR NY 08910 44290
Parish ALLONBY
Dating Unknown
Description
Visible remains of straight ridge and furrow and possible remains of an old field system.

Record number 9564
NGR NY 12130 44130
Parish WESTNEWTON
Dating Unknown
Description
Enclosure shown as sub-oval cropmark.

Record number 9569 – 9571
NGR NY 12 48, NY 12 49
Parish HOLME ST CUTHBERT
Dating Unknown
Description
Area with a number of cropmarks of possible enclosures.

Record number 9614
NGR NY 19880 51650
Parish HOLME EAST WAVER
Dating Unknown
Description
Faint cropmark suggests a possible sub-oval enclosure.

Record number 9655
NGR NY 23690 50850
Parish WOODSIDE
Dating Unknown

Description
Roughly rectangular cropmark enclosure with some possible pits and a rather curvilinear length of ditch.

Record number 9656
NGR NY 23960 51520
Parish WOODSIDE
Dating Unknown
Description
Cropmark.

Record number 9659
NGR NY 24750 54520
Parish AIKTON
Dating Unknown
Description
Cropmark.

Record number 9676
NGR NY 27960 52840
Parish AIKTON
Dating Unknown
Description
Part of a roughly rectangular cropmark, may be enclosure.

Record number 9687
NGR NY 22590 61250
Parish BOWNESS
Dating Unknown
Description
Small semi-circular cropmark.

Record number 9718
NGR NY 30530 58720
Parish BURGH BY SANDS
Dating Unknown
Description
Possible rectangular or squarish cropmark.

Record number 9723
NGR NY 31590 59550
Parish BURGH BY SANDS
Dating Unknown
Description
Two cropmarks, known as Hill Farm.

Record number 10333
NGR NY 23490 60930
Parish BOWNESS
Dating Post-medieval?
Description
Charcoal works, noted as being already disused on First Series OS map (1860).

Record number 10370
NGR NY 35300 70800
Parish KIRKANDREWS
Dating Medieval? / Post-medieval
Description
Building complex with courtyard, called Nineveh.

Record number 10525
NGR NY 43390 64010
Parish KIRKLINTON MIDDLE
Dating Medieval?
Description
The name 'Moat Lodge' suggests the possible site of a moat.

Record number 10638
NGR NY 33200 68030
Parish KIRKANDREWS
Dating Roman
Description
Length of Roman road.

Record number 10919
NGR NY 51650 37740
Parish LAZONBY
Dating Medieval / Post-medieval
Description
Watermill.

Record number 11338
NGR NY 51670 70620
Parish STAPLETON
Dating Medieval? / Post-medieval
Description
Bogs (Ruins) as marked on the 1866 OS Map. There seems to be two small, ruined, unroofed buildings with a small enclosure.

Record number 13506
NGR NY 39300 53300
Parish CUMMERSDALE
Dating Neolithic
Description
Broken leaf-shaped arrowhead.

Record number 13553
NGR NY 11000 29900
Parish BRIGHAM
Dating Unknown
Description
Possible trackway.

Record number 13660
NGR NY 32920 59170
Parish BURGH BY SANDS
Dating Roman
Description
Hadrian's Wall, and free-standing statuette of a genius.

Record number 13661
NGR NY 32400 59000
Parish BURGH BY SANDS

Dating Roman
Description
Section of Vallum.

Record number 13672
NGR NY 32740 59070
Parish BURGH BY SANDS
Dating Roman
Description
Section of Vallum.

Record number 13925
NGR NY 52500 68000
Parish WALTON
Dating Unknown
Description
Large areas of ridge and furrow, dykes and earthworks in the vicinity of Solmain Farm.

Record number 15198
NGR NY 32730 59130
Parish BURGH BY SANDS
Dating Roman
Description
Roman altar.

Record number 15237
NGR NY 34400 58400
Parish KIRKANDREWS
Dating Roman
Description
Circular watch towers from aerial photographs (GDB Jones).

Record number 16567
NGR NY 13750 48860
Parish HOLME ABBEY
Dating Unknown
Description
Two rectangular enclosures shown as cropmarks on aerial photograph.

Record Number 16701
NGR NY 16700 16500
Parish HOLME ABBEY
Dating Unknown
Description
Linear and other earthworks on aerial photographs.

Record number 16707
NGR NY 26700 51800
Parish AIKTON
Dating Unknown
Description
Cropmarks and lynchets on aerial photographs.

Record number 16710
NGR NY 34500 66300
Parish BEAUMONT
Dating Unknown
Description
Cropmarks.

Record number 16788
NGR NY 34500 69500
Parish KIRKANDREWS
Dating Early Medieval
Description
Remains of cattle from Solway Moss.

Record number 16789
NGR NY 34500 69500
Parish KIRKANDREWS
Dating Post-medieval
Description
Remains of sheep from Solway Moss.

Record number 16932
NGR NY 17000 50000
Parish HOLME ABBEY
Dating Bronze Age
Description
Collared urn from Aldoth.

West Cumbrian upland fringe: Appendix 5

Record number 108
NGR SD 26400 84780
Parish KIRKBY IRELETH
Dating Neolithic
Description
Polished stone axe.

Record number 1405
NGR SD 18300 90400
Parish MILLOM WITHOUT
Dating Unknown
Description
Enclosure.

Record number 1432
NGR SD 19000 92100
Parish ULPHA
Dating Prehistoric?
Description
Settlement and cairns.

Record number 2110
NGR SD 25510 86900
Parish SUBBERTHWAITE
Dating Bronze Age
Description
Cairn.

Record number 2117
NGR SD 27000 85000
Parish SUBBERTHWAITE

Dating Bronze Age
Description
Axe or celt.

Record number 2118
NGR SD 27000 85000
Parish SUBBERTHWAITE
Dating Bronze Age
Description
Perforated axe-hammer.

Record number 2119
NGR SD 27000 85000
Parish SUBBERTHWAITE
Dating Mesolithic?
Description
Pebble.

Record number 2120
NGR SD 27000 85000
Parish SUBBERTHWAITE
Dating Bronze Age
Description
Copper alloy armlet of Hallstatt type.

Record number 2133
NGR SD 22000 82000
Parish ANGERTON
Dating Neolithic
Description
Stone celt found at Kirkby-in-Furness in 1896.

Record number 2141
NGR SD 25370 86170
Parish KIRKBY IRELETH
Dating Prehistoric / Medieval
Description
Enclosure, situated on an elevated plateau although lower ground to south–west and south–east is boggy. There are cairns and possible settlement traces *c* 350 yds north–east of the site. *See also* SMR 3211.

Record number 2146
NGR SD 24550 88250
Parish KIRKBY IRELETH
Dating Medieval / Post-medieval
Description
Two heaps of cinder and scoriae on the east bank of Moor Beck indicate a bloomery site. *See also* SMR 2147.

Record number 2147
NGR SD 24700 88020
Parish KIRKBY IRELETH
Dating Medieval / Post-medieval
Description
In a field called Cinder Field are two heaps of cinder and scoriae. *See also* SMR 2146.

Record number 2148
NGR SD 24670 85510
Parish KIRKBY IRELETH
Dating Medieval / Post-medieval
Description
Ashlack Hall, late sixteenth – seventeenth century, also said to mark the site of an extensive bloomery.

Record number 2156
NGR SD 26500 89500
Parish KIRKBY IRELETH
Dating Prehistoric
Description
Small hut?

Record number 2157
NGR SD 26600 89060
Parish KIRKBY IRELETH
Dating Unknown
Description
Mound.

Record number 2158
NGR SD 26940 89050
Parish KIRKBY IRELETH
Dating Unknown
Description
Mound.

Record number 2159
NGR SD 26500 89000
Parish KIRKBY IRELETH
Dating Prehistoric?
Description
Cairn.

Record number 2160, 2161
NGR SD 25 88
Parish KIRKBY IRELETH
Dating Prehistoric
Description
Clearance cairns.

Record number 2163
NGR SD 25100 88300
Parish KIRKBY IRELETH
Dating Bronze Age
Description
Ring cairns and other cairns at Giant's Grave excavated in 1842.

Record number 2164
NGR SD 25860 88100
Parish KIRKBY IRELETH
Dating Prehistoric?
Description
Cairn.

Record number 2165–2168
NGR SD 25 87
Parish KIRKBY IRELETH
Dating Prehistoric
Description
Clearance cairns, field systems, ridge and furrow. Some other cairns appear to be burial cairns.

Record number 2170
NGR SD 26440 85760
Parish LOWICK/KIRKBY IRELETH
Dating Prehistoric
Description
Cairn.

Record number 2171
NGR SD 26400 85540
Parish KIRKBY IRELETH
Dating Prehistoric
Description
Ill-defined 'ring cairn'.

Record number 2172
NGR SD 26600 85400
Parish SUBBERTHWAITE
Dating Unknown
Description
Mound.

Record number 2173–2175
NGR SD 26 85
Parish KIRKBY IRELETH
Dating Unknown
Description
Mounds.

Record number 2176
NGR SD 26450 84900
Parish KIRKBY IRELETH
Dating Bronze Age
Description
Cairn.

Record number 2177
NGR SD 26510 84800
Parish BROUGTON–IN–FURNESS
Dating Unknown
Description
Mound.

Record number 2178
NGR SD 26400 84700
Parish SUBBERTHWAITE
Dating Prehistoric?
Description
Cairn.

Record number 2179
NGR SD 28000 84230
Parish SUBBERTHWAITE
Dating Bronze Age
Description
Ring cairn.

Record number 2181
NGR SD 25130 82990
Parish KIRKBY IRELETH
Dating Bronze Age
Description
Ring cairn with internal cist.

Record number 2184
NGR SD 26000 82000
Parish KIRKBY IRELETH
Dating Bronze Age
Description
Circular enclosure and stone circle.

Record number 2186
NGR SD 24300 84900
Parish KIRKBY IRELETH
Dating Unknown
Description
Settlement with circular platform.

Record number 2187
NGR SD 25470 87340
Parish KIRKBY IRELETH
Dating Prehistoric
Description
Complex stone-walled upland settlement.

Record number 3206
NGR SD 25000 87910
Parish KIRKBY IRELETH
Dating Unknown
Description
Clearance cairns; together with natural terracing they form definite plots which are probably an extension of the field system to the south–east. *See also* SMR 2160.

Record number 3207
NGR SD 24870 87350
Parish KIRKBY IRELETH
Dating Prehistoric?
Description
Cairn.

Record number 3210
NGR SD 25470 87340
Parish KIRKBY IRELETH
Dating Unknown
Description
Settlement.

Record number 3211
NGR SD 25640 86360
Parish KIRKBY IRELETH
Dating Prehistoric
Description
Group of four cairns. Eastern cairn includes a large stone, probably once erect but now leaning at an angle of *c* 45°. *See also* SMR 2160.

Record number 3212
NGR SD 25450 88460
Parish KIRKBY IRELETH
Dating Bronze Age
Description
Cairn.

Record number 4396
NGR SD 28000 84000
Parish SUBBERTHWAITE
Dating Unknown
Description
Settlements.

Record number 4407
NGR SD 26230 86100
Parish KIRKBY IRELETH
Dating Prehistoric
Description
Two possible cairns.

Record number 4408
NGR SD 25980 86720
Parish KIRKBY IRELETH
Dating Prehistoric?
Description
Cairn on Mawthwaite Moss.

Record number 4409
NGR SD 25130 88040
Parish KIRKBY IRELETH
Dating Prehistoric
Description
Cairnfield: almost circular cairn and an elongated cairn centred at SD 2504 8824. Clearance cairns centred at SD 2513 8804 connected with a field system, possibly of prehistoric date. *See also* SMR 2160.

Record number 4822
NGR SD 24100 85000
Parish KIRKBY IRELETH
Dating Unknown
Description
Rectilinear platform.

Record number 4827
NGR SD 24200 84600
Parish KIRKBY IRELETH
Dating Unknown
Description
Enclosure.

Record number 5572
NGR SD 23700 83900
Parish KIRKBY IRELETH
Dating Neolithic
Description
Polished stone axe.

Record number 6823
NGR NY 09000 28000
Parish DEAN
Dating Medieval
Description
Robert de Eaglesfield was given licence to impark at Eaglesfield in 1338.

Record number 9394, 9396, 9398 – 9418, 9420, 9421
NGR SD 25 88, SD 24 88, SD 25 87
Parish KIRKBY IRELETH
Dating Prehistoric/Medieval
Description
Area rich in clearance cairns, with some settlement evidence, possible hut platforms (9405), funerary monuments including 'Giant's Grave' (9407), cairns (9408), and possible medieval shielings (9414, 9417).

Record number 9422
NGR SD 25600 87700
Parish KIRKBY IRELETH
Dating Medieval
Description
Boundary banks/walls. Possible limits of medieval farmstead. Also clearance cairns.

Record number 9423
NGR SD 25450 87550
Parish KIRKBY IRELETH
Dating Prehistoric?
Description
Ill-defined cairnfield.

Central Cumbria: Appendix 5

Record number 855
NGR NY 18900 30000
Parish WYTHOP
Dating Medieval
Description
Deserted medieval settlement.

Record number 879
NGR NY 25320 34130
Parish IREBY
Dating Unknown
Description
Earthworks of a small rectilinear, non-defensive enclosure.

Record number 885
NGR NY 24750 34450
Parish IREBY
Dating Medieval
Description
Moated site located in a valley bottom 450 m south–east of Overwater Hall.

Record number 886
NGR NY 20170 30820
Parish WYTHOP
Dating Iron Age
Description
Castle How Hillfort is on the summit of Castle How, a rocky hillock rising steeply above the western shore of Bassenthwaite Lake.

Record Number 1065
NGR NY 13900 20300
Parish LOWESWATER
Dating Unknown
Description
Three-sided earthwork with a substantial bank enclosing a space 1.5 miles west of Lanthwaite Green adjoining Bargate Farm. Outside the south–west corner is a further smaller enclosure (*See also* SMR 1083).

Record Number 1071
NGR NY 15060 20230
Parish LOWESWATER
Dating Medieval
Description
Loweswater Peel is an unusual example of a moated manor on the west shore of Crummock Water. A natural feature – in this case a low hillock protruding into the lake – was modified by the cutting of ditches, remaining predominantly waterlogged today, on its landward side, to create a moated site defended by marshy ground to the north.

Record number 1073
NGR NY 10000 20000
Parish LAMPLUGH
Dating Bronze Age
Description
Copper alloy looped and socketed axe, found in a moss around 1872. It has a broad, kite-shaped blade.

Record number 1076
NGR NY 10830 28640
Parish DEAN
Dating Neolithic
Description
Stone axe-hammer of local quartzite found at Moorland Close.

Record number 1078
NGR NY 11800 28800
Parish COCKERMOUTH
Dating Bronze Age
Description
Bronze flanged axe.

Record Number 1091
NGR NY 15970 21010
Parish BUTTERMERE
Dating Romano-British
Description
Enclosed hut circle and associated annexe on relatively flat land on Lanthwaite Green.

Record Number 1092
NGR NY 15940 20510
Parish BUTTERMERE
Dating Unknown
Description
Enclosure.

Record Number 1093
NGR NY 15400 20800
Parish BUTTERMERE
Dating Prehistoric?
Description
Enclosure, alleged hillfort.

Record number 1095
NGR NY 15360 23500
Parish BLIND BOOTHEL
Dating Unknown
Description
Possible settlement site.

Record Number 1097
NGR NY 20700 28900
Parish WYTHOP
Dating Unknown
Description
Possible house platform.

Record Number 1101
NGR NY 27620 20350
Parish ST JOHN'S CASTLERIGG AND WYTHBURN
Dating Medieval
Description
Ancient huts recorded east of Falcon Crag – possible shielings.

Record Number 1102
NGR NY 25940 21320
Parish ABOVE DERWENT
Dating Medieval/Post-medieval
Description
Ruins on St Herbert's Island may be connected with a summer house mentioned by Sandford *c* 1675, or a chapel built there in 1375, or a commonhouse.

Record number 1103
NGR NY 23500 24200
Parish ABOVE DERWENT
Dating Roman
Description
Remains of road in the field by Braithwaite Station.

Record Number 1104
NGR NY 25160 23050
Parish ABOVE DERWENT
Dating Unknown
Description
Four unpolished greenstone celts found in 1901 while digging a fishpond at Moss-Garth and 30 yds away a further half-polished celt found in 1902.

Record Number 1105
NGR NY 26490 21890
Parish ST JOHN'S CASTLERIGG AND WYTHBURN
Dating Medieval
Description
Earthworks on slightly sloping ground on west side of Lords Island. They may have been used for archery butts.

Record Number 1107
NGR NY 26560 21910
Parish ST JOHN'S CASTLERIGG AND WYTHBURN
Dating Medieval/Post-medieval
Description
Remains of manor house *c* 1460, destroyed in 1709.

Record Number 1108
NGR NY 26380 21370
Parish ST JOHN'S CASTLERIGG AND WYTHBURN
Dating Medieval
Description
Possible bloomery on the highest part of St Herbert's Island: traces of iron slag and black earth.

Record Number 1110
NGR NY 26000 23000
Parish KESWICK
Dating Prehistoric?
Description
Stone axe from Keswick. Possibly a miner's tool.

Record Number 1111
NGR NY 25400 22800
Parish ABOVE DERWENT
Dating Prehistoric
Description
Flint arrowhead found in 1901 near Portinscale.

Record Number 1117
NGR NY 26000 23000
Parish KESWICK
Dating Prehistoric
Description
Stone adze found in or near Keswick.

Record Number 1118
NGR NY 26000 23000
Parish KESWICK
Dating Prehistoric
Description
Part of polished stone axe found at Keswick.

Record Number 1119
NGR NY 26420 23710
Parish KESWICK
Dating Prehistoric?
Description
Grooved stone axe found at Keswick.

Record Number 1121
NGR NY 26000 23000
Parish KESWICK
Dating Prehistoric
Description
Stone celt found near Keswick.

Record Number 1123
NGR 326000 523000
Parish KESWICK
Dating Middle Bronze Age
Description
Looped palstave from Keswick.

Record Number 1161
NGR NY 44200 21200
Parish MATTERDALE
Dating Medieval
Description
Possible deserted medieval village.

Record Number 1164
NGR NY 46770 24620
Parish DACRE
Dating Prehistoric
Description
Small multivallate hillfort on the summit of Dunmallard Hill overlooking the foot of Ullswater

Record Number 1165
NGR NY 45130 24350
Parish MATTERDALE
Dating Prehistoric
Description
Maiden Castle is a rare example of a univallate defended enclosure, on a plateau on the south–east-facing slopes of Soulby Fell and includes a roughly circular enclosure defended by a rampart, ditch and counterscarp bank. There are internal earthworks including possible house platforms.

Record Number 1196
NGR NY 09000 14800
Parish ENNERDALE AND KINNISIDE
Dating Medieval
Description
Shieling of almost square type, perhaps associated with the nearby sheepfold. *See also* SMR 3689.

Record Number 1219
NGR NY 15600 17700
Parish BORROWDALE
Dating Medieval
Description
Series of huts representing permanent occupation rather than shielings.

Record Number 1220
NGR NY 15590 17580
Parish LOWESWATER
Dating Medieval
Description
Remains of shieling settlement of at least 10 rectangular buildings and other associated garths and fields on the fellside close to the mouth of Scale Beck to the west of Crummock Water. A circular structure probably represents a hearth and finds of slag to the north–west may suggest the existence of a bloomery. May be related to SMR 1219.

Record Number 1221
NGR NY 19090 15580
Parish BUTTERMERE
Dating Romano-British
Description
Farmstead, including a sub-circular enclosure containing two hut circles.

Record Number 1223
NGR NY 14690 19430
Parish LOWESWATER
Dating Prehistoric
Description
Stone hammer found in 1868.

Record Number 1225
NGR NY 12000 14700
Parish ENNERDALE AND KINNISIDE
Dating Prehistoric
Description
Scattered cairnfield of 15–20 small clearance cairns. *See also* SMR 3714, a bloomery site.

Record number 1230
NGR NY 24750 10640
Parish BORROWDALE
Dating Neolithic
Description
Stone axe factory on Glaramara.

Record number 1234
NGR NY 25100 19200
Parish ABOVE DERWENT
Dating Neolithic
Description
Stone axe rough-out, found at south–west corner of Derwentwater in 1960.

Record Number 1243
NGR NY 38600 19400
Parish MATTERDALE
Dating Unknown
Description
Four early settlement sites in Glencoyne Park beside Ullswater.

Record Number 1244
NGR NY 31350 15420
Parish ST JOHN'S CASTLERIGG AND WYTHBURN
Dating Iron Age
Description
Two bronze penannular armlets with expanded ends.

Record Number 1245
NGR NY 31000 17000
Parish ST JOHN'S CASTLERIGG AND WYTHBURN
Dating Prehistoric
Description
Chipped axe of volcanic ash with lateral facets.

Record Number 1246
NGR NY 32000 13000
Parish ST JOHN'S CASTLERIGG AND WYTHBURN
Dating Prehistoric
Description
Polished stone axe found at Wythburn.

Record Number 1322
NGR NY 14540 03870
Parish ESKDALE
Dating Medieval
Description
Bloomery site visible as a rectangular earthwork on aerial photographs. There is a turf-covered mound of iron slag.

Record number 1339
NGR NY 37300 03500
Parish LAKES
Dating Roman
Description
Civil settlement north of the Roman fort (1877) at Ambleside.

Record Number 1506
NGR SD 29500 91900

Parish COLTON
Dating Medieval
Description
Remains perhaps representing a fortalice and ironworks.

Record Number 1513
NGR NY 42000 19000
Parish MARTINDALE
Dating Prehistoric
Description
Possible stone circle and cairn. A maze of hollow-ways/ sledgates is said to be nearby.

Record Number 1514
NGR NY 44300 19300
Parish MARTINDALE
Dating Prehistoric?
Description
Three cairns and other possible cairns, plus an enclosure, and other unidentified stone structures on either side of the track east from Lanty Tarn towards Howtown. Another, more well-defined rectangular enclosure with internal stone feature has been identified south of Lanty Tarn.

Record number 1518
NGR NY 49980 16030
Parish BAMPTON
Dating Iron Age/Romano-British
Description
Remains of a small farmstead on a narrow plateau on the hillside near Haweswater, consisting of two hut circles and part of the surrounding enclosure wall. Also, quarry/mine remains *c* 50 m above the site.

Record number 1542
NGR NY 51000 18000
Parish BAMPTON
Dating Neolithic
Description
Stone axe, found in a wall at Bampton.

Record number 1563
NGR NY 50170 16260
Parish BAMPTON
Dating Bronze Age
Description
Round cairn on the south-facing hillside of Burn Banks overlooking Haweswater.

Record Number 1597 – 1599
NGR NY 43870 19180
Parish MARTINDALE
Dating Prehistoric?
Description
Three groups of small mounds and grassed heaps on the north side of a col 200-250 m OD.

Record number 1633
NGR SD 27300 92500
Parish BLAWITH AND SUBBERTHWAITE
Dating Prehistoric?
Description
Cairn on Torver Low Common.

Record number 1634
NGR SD 27700 92200
Parish BLAWITH AND SUBBERTHWAITE
Dating Bronze Age
Description
Round cairns.

Record number 1636
NGR SD 27140 91280
Parish BLAWITH AND SUBBERTHWAITE
Dating Prehistoric
Description
Cairn on Stable Harvey Moss.

Record number 1637
NGR SD 26510 90950
Parish BLAWITH AND SUBBERTHWAITE
Dating Prehistoric
Description
Cairn.

Record number 1638
NGR SD 27220 90440
Parish BLAWITH AND SUBBERTHWAITE
Dating Prehistoric
Description
Oval cairn.

Record number 1850
NGR NY 37250 03650
Parish LAKES
Dating Roman
Description
Cropmarks suggest a temporary camp roughly parallel to the Roman fort at Ambleside (SMR 1877).

Record number 1859
NGR NY 34000 07000
Parish LAKES
Dating Neolithic/Bronze Age
Description
Stone spearhead or sharpener, and an axe or chisel, found in Dunnerbeck, Rydal.

Record number 1860
NGR NY 33840 05270
Parish LAKES
Dating Unknown
Description
A quern found prior to 1877.

Record number 1861
NGR NY 34100 05300
Parish LAKES

Dating Neolithic?
Description
Rough-out axe of unspecified type.

Record number 1862
NGR NY 34400 06700
Parish LAKES
Dating Neolithic
Description
Polished stone axe.

Record number 1865
NGR NY 33000 07000
Parish LAKES
Dating Neolithic
Description
Small stone axe.

Record number 1867
NGR NY 36400 06320
Parish LAKES
Dating Unknown
Description
The Mount, an unclassified terraced mound 9 ft high.

Record number 1868
NGR NY 36200 06200
Parish LAKES
Dating Roman
Description
Urn containing ashes with Roman finds from Rydal.

Record number 1869
NGR NY 36930 05510
Parish LAKES
Dating Medieval
Description
Foundations of Rydal Old Hall.

Record number 1870
NGR NY 35220 06490
Parish LAKES
Dating Medieval
Description
Probable course of medieval boundary fence of Rydal, emparked in 1277.

Record number 1877
NGR NY 37250 03400
Parish LAKES
Dating Roman
Description
Ambleside (*Galava*) Fort.

Record number 1880
NGR NY 36800 02000
Parish LAKES
Dating Neolithic
Description
Polished stone axe found at a depth of 5–6 ft near Low Wray.

Record number 1881
NGR NY 37200 03400
Parish LAKES
Dating Unknown
Description
Three net sinkers found at Borrans, Ambleside.

Record number 1885
NGR NY 37700 03180
Parish LAKES
Dating Roman
Description
Jug found at Ambleside.

Record number 1889
NGR NY 31700 03900
Parish LAKES
Dating Unknown
Description
Enclosure on aerial photographs.

Record number 1892
NGR NY 36700 05600
Parish LAKES
Dating Unknown
Description
Field system on aerial photographs.

Record number 1894
NGR NY 37900 03000
Parish LAKES
Dating Unknown
Description
Possible site (circular) on aerial photographs.

Record Number 1895
NGR NY 38800 02000
Parish LAKES
Dating Unknown
Description
Foundations of a rectangular building appearing as cropmark.

Record number 1896
NGR NY 38600 02100
Parish LAKES
Dating Unknown
Description
Possible rectangular ditch and cockpit on aerial photographs.

Record Number 1897
NGR NY 38900 01700
Parish LAKES

Dating Unknown
Description
Circular structure on aerial photographs.

Record Number 1898
NGR NY 39200 01100
Parish LAKES
Dating Prehistoric
Description
Possible ring ditch on aerial photographs.

Record Number 1905
NGR NY 40000 00000
Parish LAKES
Dating Prehistoric
Description
Two stone axes from near Windermere.

Record number 1941
NGR NY 54710 00680
Parish WHITWELL AND SELSIDE
Dating Medieval
Description
Moated site forming a rectangular enclosure.

Record Number 2029
NGR SD 30330 95650
Parish CONISTON
Dating Medieval / Post-medieval
Description
Bloomery site at Water Park, perhaps sixteenth century. *See also* SMR 2030.

Record Number 2030
NGR SD 30290 95400
Parish CONISTON
Dating Medieval / Post-medieval
Description
Large, seventeenth century? bloomery excavated in 1897. *See also* SMR 2029.

Record Number 2047
NGR SD 39280 96470
Parish WINDERMERE
Dating Roman / Post-medieval
Description
Tessellated pavement found in an area north of Belle Isle Lodge? during building work in 1774. Possible Roman well in the basement.

Record Number 2048
NGR SD 39900 99700
Parish WINDERMERE
Dating Neolithic
Description
Polished stone axe found at Calgarth.

Record Number 2049
NGR SD 39900 99400
Parish WINDERMERE
Dating Bronze Age
Description
Perforated stone axe-hammer found at Calgarth.

Record Number 2051
NGR SD 37640 93570
Parish SATTERTHWAITE
Dating Medieval / Post-medieval
Description
Three bloom smithies, in existence at the Dissolution of Furness Abbey.

Record Number 2052
NGR SD 37690 93610
Parish SATTERTHWAITE
Dating Medieval / Post-medieval
Description
Remains of a forge ceasing work in 1715. *See also* SMR 2687.

Record Number 2055
NGR SD 37050 91250
Parish SATTERTHWAITE
Dating Medieval / Post-medieval
Description
Graythwaite Hall, possibly sixteenth / seventeenth century, but with medieval plan of central hall and flanking wings.

Record Number 2059
NGR SD 37000 91000
Parish SATTERTHWAITE
Dating Early Medieval?
Description
A stone with a small channel and a semi-circular recess, similar to a basin, perhaps part of a mill.

Record number 2154, 2155
NGR SD 27 89
Parish BLAWITH AND SUBBERTHWAITE
Dating Prehistoric
Description
Possible burial mounds and cairns.

Record number 2156
NGR SD 26500 89500
Parish SUBBERTHWAITE
Dating Prehistoric?
Description
Cairns on Woodland Fell.

Record number 2491
NGR SD 26680 89030
Parish SUBBERTHWAITE
Dating Bronze Age
Description
Cairn.

Record Number 2682
NGR SD 37500 94200

Parish CLAIFE
Dating Medieval/Post-medieval
Description
Charcoal pit.

Record Number 2685
NGR SD 38400 93900
Parish CLAIFE
Dating Unknown
Description
Possible quay, seen in Windermere.

Record Number 2687
NGR SD 3770093600
Parish SATTERTHWAITE
Dating Medieval/Post-medieval
Description
Farm building, once the office for a forge at Cunsey. The site was in use before 1584, known as 'Les Smithies'.

Record Number 2689
NGR SD 38800 98900
Parish CLAIFE
Dating Medieval/Post-medieval
Description
Barn originally used for storage of grain and sheep products between Kendal and Hawkshead.

Record Number 2934
NGR NY 47000 25000
Parish DACRE
Dating Prehistoric
Description
Antiquarian tradition of four concentric circles in marshy ground adjacent to the River Eamont.

Record number 2959
NGR NY 46400 23310
Parish BARTON
Dating Medieval
Description
Moated site and assoicated annexe close to Ullswater.

Record number 2973
NGR NY 34250 33640
Parish CALDBECK
Dating Iron Age
Description
Carrock Fell hillfort.

Record Number 2981
NGR NY 38800 15100
Parish PATTERDALE
Dating Medieval?
Description
Possible village, described by Hay (1936). May really be up to 12 charcoal burners' platforms.

Record number 2987
NGR NY 30000 02900
Parish SKELWITH
Dating Unknown
Description Settlements on aerial photo-graphs.

Record number 3143
NGR SD 29700 90900
Parish COLTON
Dating Medieval/Post-medieval
Description
According to Marshall and Davies-Shiel (1977), there was an extensive bloomery site at Selside Beck.

Record Number 3144
NGR NY 44300 20600
Parish MATTERDALE
Dating Medieval/Post-medieval?
Description
Bloomery.

Record Number 3147
NGR SD 29180 92590
Parish CONISTON
Dating Medieval/Post-medieval
Description
Bloomery.

Record Number 3224
NGR SD 29720 93860
Parish TORVER
Dating Medieval
Description
Site of bloomery recorded on the OS second edition map as Roman.

Record Number 3227
NGR SD 29970 94370
Parish TORVER
Dating Medieval
Description
Bloomery on edge of Coniston.

Record number 3232
NGR SD 27320 92050
Parish TORVER
Dating Medieval?
Description
Remains of bloomery mound.

Record Number 3233
NGR SD 29060 91500
Parish TORVER
Dating Medieval
Description
Site of two bloomeries, with much waste, on shore of Coniston.

Record Number 3234
NGR SD 28520 91760
Parish TORVER
Dating Medieval
Description
Bloomery.

Record number 3235
NGR SD 27400 91800
Parish TORVER
Dating Prehistoric
Description
Cairn.

Record Number 3236
NGR SD 29780 91060
Parish COLTON
Dating Medieval / Post-medieval
Description
Possible bloomery said to date between the enclosure of the park in 1339 and the Dissolution.

Record number 3356
NGR SD 30370 96450
Parish CONISTON
Dating Medieval / Post-medieval
Description
Fish pond and remains of a rectangular building at Coniston Hall.

Record Number 3531
NGR NY 16610 17700
Parish BUTTERMERE
Dating Early medieval
Description
Iron spearhead found in Crummock Water.

Record number 3535
NGR NY 35010 01700
Parish SKELWITH
Dating Unknown
Description
Enclosure on aerial photographs.

Record number 3542
NGR NY 36950 02700
Parish SKELWITH
Dating Medieval / Post-medieval?
Description
Platforms on the shore at Brathay Bay on Windermere may have been associated with charcoal burning or possibly supported permanent dwellings.

Record Number 3605
NGR 312700 514900
Parish ENNERDALE AND KINNISIDE
Dating Medieval
Description
Smithy Beck shielings / settlement site. *See* SMR 3712, 3716.

Record Number 3634
NGR SD 31000 95200
Parish CONISTON
Dating Medieval / Post-medieval
Description
Bloomery partly excavated in 1984.

Record Number 3689
NGR NY 08880 14680
Parish ENNERDALE AND KINNISIDE
Dating Medieval
Description
Shieling.

Record Number 3712, 3716
NGR NY 12670 15010, NY 15280 14970
Parish ENNERDALE AND KINNISIDE
Dating Medieval
Description
Twelve shielings or miners' huts are strung out along the north side of Smithy Beck. *See also* SMR 3714. They differ from normal shielings in that they are surrounded by outer walls which leaves little space at the front and sides.

Record Number 3714
NGR NY 12000 14700
Parish ENNERDALE AND KINNISIDE
Dating Medieval
Description
Bloomery at Smithy Beck. *See also* SMR 1225, a cairnfield.

Record Number 3724
NGR NY 18800 33700
Parish BLINDCRAKE
Dating Medieval
Description
Deserted village.

Record Number 3737
NGR NY 28000 22000
Parish BORROWDALE
Dating Mesolithic
Description
Flints, possibly a working site.

Record Number 3738
NGR NY 27800 22500
Parish BORROWDALE
Dating Mesolithic
Description
Flints, possibly a working site.

Record number 3753
NGR NY 33000 07000

Parish LAKES
Dating Medieval
Description
Three fulling mills mentioned at Grasmere in the thirteenth and fourteenth centuries.

Record number 3755
NGR NY 36550 06350
Parish LAKES
Dating Medieval / Post-medieval
Description
Rydal Hall.

Record number 4064
NGR NY 35000 05000
Parish LAKES
Dating Prehistoric / Medieval?
Description
Dug-out canoe found on Loughrigg Fell.

Record Number 4226
NGR NY 30300 18800
Parish ST JOHN'S CASTLERIGG AND WYTHBURN
Dating Medieval
Description
Possible shieling.

Record Number 4232
NGR NY 22400 29100
Parish BASSENTHWAITE
Dating Roman
Description
Suggested site of signal station.

Record Number 4235
NGR NY 30000 19000
Parish ST JOHN'S CASTLERIGG AND WYTHBURN
Dating Bronze Age?
Description
Bronze axe found at Shoulthwaite.

Record number 4281
NGR NY 56300 12300
Parish SHAP
Dating Prehistoric
Description
Cairnfield and settlement site.

Record number 4435
NGR NY 37300 03400
Parish LAKES
Dating Mesolithic
Description
Core-trimming flake from Borrans Road, Ambleside.

Record Number 4671
NGR NY 30000 17000
Parish ST JOHN'S CASTLERIGG AND WYTHBURN
Dating Unknown
Description
Crude stone enclosures on top of Cockrigg Bank.

Record Number 4675
NGR NY 40000 20000
Parish MATTERDALE
Dating Roman
Description
Possible Roman road at Gowbarrow Park Head.

Record number 4731
NGR NY 38800 02100
Parish LAKES
Dating Medieval
Description
Alleged deserted village with supposed sunken way, a cockpit and diverse earthworks.

Record number 4734
NGR NY 34400 06000
Parish LAKES
Dating Medieval
Description
Coin hoard reported at Banerigg.

Record Number 4760
NGR NY 12800 14000
Parish ENNERDALE AND KINNISIDE
Dating Prehistoric
Description
Possible clearance mounds and short stretches of walling.

Record Number 4904
NGR NY 26100 22300
Parish KESWICK
Dating Bronze Age?
Description
Stone circle.

Record number 4930
NGR SD 25000 87100
Parish KIRKBY IRELETH
Dating Unknown
Description
Settlement with hut platforms.

Record number 4931
NGR SD 25500 87000
Parish KIRKBY IRELETH
Dating Unknown
Description
Hut platforms.

Record number 4932
NGR SD 25400 86800
Parish KIRKBY IRELETH
Dating Unknown
Description
Settlement with hut platforms.

Record Number 5235
NGR NY 31070 16320
Parish ST JOHN'S CASTLERIGG AND WYTHBURN
Dating Prehistoric
Description
Cairns.

Record number 5281
NGR NY 51200 17500
Parish BAMPTON
Dating Unknown
Description
Settlement on aerial photographs.

Record number 5318
NGR NY 37300 03700
Parish LAKES
Dating Roman
Description
Several traces of extramural settlement to the east and north of Ambleside Roman fort, including corduroy road.

Record Number 5472
NGR NY 45700 24800
Parish DACRE
Dating Romano-British
Description
Possible settlement (heavily disturbed) on Soulby Fell.

Record Number 5497
NGR NY 29990 518850
Parish ST JOHN'S CASTLERIGG AND WYTHBURN
Dating Iron Age/Early Medieval
Description
Shoulthwaite Gill bivallate hillfort, possibly post-Roman.

Record number 5504
NGR NY 34710 01890
Parish LAKES
Dating Unknown
Description
Mound.

Record Number 5536
NGR SD 31720 95100
Parish CONISTON
Dating Medieval/Post-medieval
Description
Lawson Park, once a grange of Furness Abbey. Present house is seventeenth century.

Record number 5537
NGR SD 31300 97000
Parish CONISTON
Dating Mesolithic
Description
Mace head found about 6 ft deep in peat, resting on the natural soil.

Record number 5538
NGR SD 30000 97000
Parish CONISTON
Dating Unknown
Description
Quern found at Coniston.

Record number 5574
NGR SD 29400 89800
Parish COLTON
Dating Neolithic
Description
Polished stone axe.

Record number 5575
NGR SD 29200 88900
Parish COLTON
Dating Roman
Description
Broken beehive rotary quern from Nibthwaite.

Record Number 5648
NGR NY 20300 28400
Parish WYTHOP
Dating Medieval
Description
Reported site of moat. Hugh de Louthre had a park at Wythop as early as 1367; exact location unknown.

Record Number 6122
NGR SD 30000 92100
Parish COLTON
Dating Medieval/Post-medieval
Description
Cairns and an enclosure wall.

Record Number 6129
NGR SD 30280 95280
Parish TORVER
Dating Neolithic?
Description
Polished stone axe.

Record Number 6322
NGR NY 26700 23500
Parish KESWICK
Dating Medieval
Description
Hazard area for Keswick medieval town.

Record number 6338
NGR NY 24200 10200
Parish BORROWDALE
Dating Neolithic
Description
Stone axe factory at Looking Steads.

Record Number 6354
NGR SD 29900 92500
Parish COLTON
Dating Medieval / Post-medieval
Description
Bloomery site.

Record number 6355
NGR SD 30500 96200
Parish CONISTON
Dating Medieval / Post-medieval
Description
Bloomery mound at Lands Point.

Record Number 6362
NGR SD 40000 00000
Parish LAKES
Dating Medieval
Description
Medieval village site.

Record Number 6419
NGR SD 31000 95000
Parish CONISTON
Dating Medieval / Post-medieval
Description
Area containing 74 charcoal pitsteads, charcoal burners' and bark peelers' huts, sammel holes, potash kilns and bloomeries.

Record Number 6588
NGR NY 40500 18400
Parish MARTINDALE
Dating Medieval / Post-medieval?
Description
Twin drystone hut bases, sited on a grassy rake above Ullswater. They are probably bothies guarding the lakeward route for summering cattle on Place Fell.

Record Number 6664
NGR NY 31300 14700
Parish ST JOHN'S CASTLERIGG AND WYTHBURN
Dating Unknown
Description
Three hut bases adjoining an ancient trackway.

Record Number 6665
NGR NY 31200 14800
Parish ST JOHN'S CASTLERIGG AND WYTHBURN
Dating Unknown
Description
Possible turf hut with crude double base.

Record number 6717
NGR NY 57000 14000
Parish SHAP
Dating Medieval
Description
Hazard area for Shap medieval village.

Record Number 6802
NGR NY 46800 22400
Parish BARTON
Dating Medieval
Description
Documentary evidence of a park at Barton in 1279, owned by Roger de Lancaster.

Record number 6804
NGR NY 33000 07000
Parish LAKES
Dating Medieval
Description
Documentary evidence for medieval park at Grasmere as early as 1297.

Record Number 7618
NGR NY 16750 07330
Parish NETHER WASDALE
Dating Unknown
Description
Enclosure walls identified from aerial photographs and in the field.

Record Number 7619
NGR NY 16520 07810
Parish NETHER WASDALE
Dating Unknown
Description
Enclosures situated on the south face of Great Door identified from aerial photographs and in the field.

Record Number 7620
NGR NY 16600 07100
Parish NETHER WASDALE
Dating Unknown
Description
Large D-shaped enclosure with no visible entrance.

Record Number 7664
NGR NY 14930 05460
Parish NETHER WASDALE
Dating Prehistoric?
Description
Cairnfield identified from aerial photographs and in the field.

Record Number 7694
NGR NY 16700 06900
Parish NETHER WASDALE
Dating Unknown
Description
Stone enclosure with a curvilinear wall to the north–west.

Record Number 7709
NGR NY 15280 05830
Parish ESKDALE
Dating Unknown
Description
Small rectangular enclosure identified from aerial photographs and in the field. Possibly associated with bloomery (SMR 1328).

Record Number 7841
NGR NY 14330 04940
Parish NETHER WASDALE
Dating Unknown
Description
Field enclosures identified from aerial photographs and field observations.

Record Number 7847
NGR NY 14520 03870
Parish NETHER WASDALE
Dating Unknown
Description
Earthworks identified from aerial photographs and in the field.

Record Number 7849
NGR NY 14220 03640
Parish NETHER WASDALE
Dating Unknown
Description
Field systems identified from aerial photographs and in the field.

Record Number 7850
NGR NY 14000 04320
Parish NETHER WASDALE
Dating Unknown
Description
Linear field boundary identified from aerial photographs and in the field.

Record Number 7855
NGR NY 14280 03320
Parish ESKDALE
Dating Unknown
Description
Enclosures identified from aerial photographs and in the field.

Record Number 7889
NGR NY 17100 07200
Parish NETHER WASDALE
Dating Unknown
Description
Sub-circular enclosure identified from aerial photographs and in the field.

Record Number 7890
NGR NY 18280 06370
Parish ESKDALE
Dating Unknown
Description
Nine buildings or enclosures identified from aerial photographs and in the field.

Record Number 7891
NGR NY 18250 08100
Parish NETHER WASDALE
Dating Unknown
Description
Enclosures or linear features identified from aerial photographs and in the field.

Record Number 7892
NGR NY 17900 07610
Parish NETHER WASDALE
Dating Unknown
Description
Field boundaries identified from aerial photographs and in the field.

Record Number 7893
NGR NY 17300 06300
Parish ESKDALE
Dating Unknown
Description
Buildings or enclosures identified from aerial photographs and in the field.

Record Number 7896
NGR NY 17980 05990
Parish ESKDALE
Dating Unknown
Description
A long sub-rectangular building identified from aerial photographs and in the field.

Record number 8323
NGR NY 50240 16650
Parish BAMPTON
Dating Unknown
Description
Two banks seen as earthworks.

Record number 8324
NGR NY 50300 16500
Parish BAMPTON
Dating Unknown
Description
Cairns and banks of possible field system.

Record number 8325
NGR NY 5010 16300
Parish SHAP RURAL
Dating Unknown
Description
Seven cairns, three of which are larger than normal clearance cairns. There are three banks (possible field boundaries) and the remains of a small building.

Record Number 8726
NGR NY 47900 23500
Parish BARTON
Dating Prehistoric
Description
Two flints, found on Askham Fell; one is possibly from eastern England, the other from Yorkshire or Durham.

Record Number 9553
NGR NY 18760 34820
Parish BLINDCRAKE
Dating Unknown
Description
Cropmarks.

Record number 11034
NGR NY 25650 34170
Parish IREBY
Dating Unknown
Description
Possible site of a castle, indicated by the name Castle How.

Record Number 11537
NGR NY 14200 03500
Parish NETHER WASDALE
Dating Unknown
Description
Enclosure walls and cairns on aerial photographs.

Record Number 11538
NGR NY 13800 03400
Parish NETHER WASDALE
Dating Unknown
Description
Enclosure walls and cairns on aerial photographs.

Record Number 11813
NGR NY 22450 25950
Parish ABOVE DERWENT
Dating Medieval/Post-medieval?
Description
Thornthwaite lead mine.

Record number 11814
NGR NY 22280 25620
Parish ABOVE DERWENT
Dating Medieval/Post-medieval?
Description
Site of a lead mine, with a disused level.

Record Number 11862
NGR NY 14460 21440
Parish LOWESWATER
Dating Unknown
Description
Pond, a marsh is marked on the 1974 OS Map.

Record Number 11863
NGR NY 14260 21360
Parish LOWESWATER
Dating Unknown
Description
Fish pond.

Record Number 12039
NGR NY 25100 19500
Parish ABOVE DERWENT
Dating Medieval/Post-medieval
Description
Lead mine on the south–west bank of Derwent Water.

Record Number 12061
NGR NY 39720 16820
Parish PATTERDALE
Dating Medieval/Post-medieval
Description
Site of Patterdale smithy.

Record Number 12063
NGR NY 38270 17100
Parish PATTERDALE
Dating Medieval/Post-medieval
Description
Site of forge at Rowhead.

Record Number 12064
NGR NY 08170 15300
Parish ENNERDALE AND KINNISIDE
Dating Medieval/Post-medieval
Description
Ennerdale corn mill.

Record Number 12072
NGR NY 24100 19300
Parish ABOVE DERWENT
Dating Medieval/Post-medieval?
Description
Workings and levels around Brunt Crag.

Record Number 12094
NGR NY 31930 19720
Parish ST JOHN'S CASTLERIGG AND WYTHBURN
Dating Medieval/Post-medieval
Description
Legburthwaite corn mill.

Record Number 12135
NGR NY 26400 18210
Parish BORROWDALE
Dating Unknown
Description
Cairn.

Record Number 12137
NGR NY 25300 18460
Parish BORROWDALE
Dating Medieval / Post-medieval
Description
Lead mine.

Record Number 12418
NGR NY 22450 26000
Parish ABOVE DERWENT
Dating Medieval / Post-medieval
Description
Thornthwaite mine.

Record Number 12684
NGR NY 38040 15570
Parish PATTERDALE
Dating Unknown
Description
Braesteads Stone.

Record Number 12685
NGR NY 38390 16090
Parish PATTERDALE
Dating Unknown
Description
Grassthwaitehow Old Castle.

Record Number 13530
NGR NY 48000 23000
Parish BARTON
Dating Unknown
Description
Trackways and dykes on aerial photographs.

Record number 13561
NGR NY 56500 12700
Parish SHAP
Dating Unknown
Description
Settlement, earthworks, dykes and hollow-ways on aerial photographs.

Record number 13562
NGR NY 56500 12500
Parish SHAP
Dating Prehistoric
Description
Hut circle / ring bank on aerial photographs.

Record Number 13566
NGR NY 20000 28100
Parish WYTHOP
Dating Unknown
Description
Dykes on aerial photographs.

Record number 13567
NGR NY 25500 25200
Parish UNDER SKIDDAW
Dating Unknown
Description
Parchmarks.

Record number 13772
NGR NY 33700 05300
Parish LAKES
Dating Prehistoric
Description
Possible burial mound, although it may be natural.

Record number 13784
NGR NY 51910 16800
Parish BAMPTON
Dating Unknown
Description
Area of earthworks on aerial photographs.

Record number 13815
NGR SD 37700 86600
Parish STAVELEY
Dating Roman
Description
Two possible Roman forts, not visible on the ground.

Record Number 13867
NGR NY 42500 19900
Parish MARTINDALE
Dating Prehistoric
Description
Worked flint (scraper?) found between Howtown and Sandwick.

Record number 13986
NGR NY 36800 02700
Parish LAKES
Dating Early medieval
Description
Possible runes incised on a small flat pebble.

Record number 15110
NGR SD 29000 96000
Parish CONISTON
Dating Unknown
Description
Dykes and enclosures on an aerial photograph.

Record Number 15202
NGR NY 40300 20300
Parish MATTERDALE
Dating Medieval

Description
Ridge and furrow.

Record Number 15900
NGR NY 15670 20600
Parish BUTTERMERE
Dating Medieval
Description
Bloomery.

Record Number 15909
NGR NY 15080 19340
Parish BUTTERMERE
Dating Late Medieval
Description
Potash kiln with remains of charcoal.

Record Number 15910
NGR NY 16150 19500
Parish BUTTERMERE
Dating Medieval
Description
Bloomery on the shore of Buttermere.

Record Number 15911
NGR NY 16280 19300
Parish BUTTERMERE
Dating Medieval
Description
Bloomery.

Record number 15921
NGR NY 37050 03930
Parish LAKES
Dating Medieval?
Description
Bloomery.

Record number 15929
NGR SD 27260 92040
Parish TORVER
Dating Medieval?
Description
Bloomery.

Record Number 15931
NGR SD 29100 91500
Parish BLAWITH AND SUBBERTHWAITE
Dating Medieval
Description
Bloomery.

Record Number 15932
NGR SD 29080 91280
Parish BLAWITH AND SUBBERTHWAITE
Dating Medieval
Description
Bloomery, on shore of Coniston.

Record Number 15933
NGR SD 29120 91030
Parish BLAWITH AND SUBBERTHWAITE
Dating Medieval
Description
Bloomery near SMR 15932.

Record number 15934
NGR SD 28880 90100
Parish BLAWITH AND SUBBERTHWAITE
Dating Medieval?
Description
Bloomery.

Record number 15936
NGR SD 30810 96950
Parish CONISTON
Dating Medieval?
Description
Bloomery.

Record number 15937
NGR SD 31220 96200
Parish CONISTON
Dating Medieval?
Description
Bloomery.

Record number 15938
NGR SD 31190 96000
Parish CONISTON
Dating Medieval?
Description
Bloomery.

Record Number 15947
NGR SD 40340 98790
Parish WINDERMERE
Dating Medieval
Description
Small bloomery.

Record Number 15965
NGR SD 39000 88700
Parish STAVELEY IN CARTMEL
Dating Medieval/Post-medieval
Description
Stone-fronted charcoal pitstead.

Record number 16433
NGR SD 27240 91350
Parish BLAWITH AND SUBBERTHWAITE
Dating Prehistoric?
Description
Site of Fisher High cairn.

Record Number 16679
NGR SD 30300 93220

Parish SATTERTHWAITE
Dating Medieval/Post-medieval
Description
Charcoal pitstead.

Record Number 16680
NGR SD 30490 93610
Parish SATTERTHWAITE
Dating Medieval/Post-medieval
Description
Charcoal pitstead and hut.

Record Number 16724
NGR NY 39600 19800
Parish MATTERDALE
Dating Prehistoric/Roman
Description
Roman fort/fortlet or possible settlement site on aerial photographs.

Record Number 16725
NGR NY 44900 21000
Parish MARTINDALE
Dating Middle Bronze Age
Description
Bronze spearhead found near Ullswater.

Record Number 16743
NGR SD 38000 92000
Parish WINDERMERE
Dating Prehistoric
Description
Grooved 'net sinker' found in Windermere.

Record number 16850
NGR NY 56800 13300
Parish SHAP
Dating Prehistoric
Description
Shap Stone Circle.

Record Number 16922
NGR NY 16400 16900
Parish LOWESWATER
Dating Medieval/Post-medieval
Description
Charcoal pitsteads.

Record Number 18209
NGR SD 38420 90130
Parish CARTMEL FELL
Dating Medieval/Post-medieval
Description
Forge suggested by place-name evidence (Smithy Wood).

Record Number 18422
NGR NY 37160 00730
Parish CLAIFE
Dating Medieval/Post-medieval
Description
Building used as hunting lodge.

Record Number 30009
NGR NY 15080 19340
Parish LOWESWATER
Dating Late medieval
Description
Potash kiln, charcoal.

Record Number 30010
NGR NY 18810 15780
Parish BUTTERMERE
Dating Medieval
Description
Bloomery.

Record Number 30014
NGR SD 29990 94420
Parish TORVER
Dating Medieval
Description
Bloomery.

Record Number 30015
NGR SD 29080 91770
Parish BLAWITH AND SUBBERTHWAITE
Dating Medieval
Description
Bloomery.

Record Number 30016
NGR SD 31150 95810
Parish CONISTON
Dating Medieval
Description
Bloomery.

Record Number 30017
NGR SD 31070 95570
Parish CONISTON
Dating Medieval
Description
Bloomery.

Record Number 30018
NGR SD 39200 92800
Parish CARTMEL FELL
Dating Medieval
Description
Small bloomery.

Record Number 30114
NGR SD 29100 94760
Parish TORVER
Dating Medieval/Post-medieval
Description
Potash kiln.

Record Number 30116
NGR SD 28600 91580
Parish BLAWITH AND SUBBERTHWAITE
Dating Medieval/Post-medieval
Description
Potash kiln.

Record Number 30117
NGR SD 29980 92070
Parish COLTON
Dating Medieval/Post-medieval
Description
Chopwood kiln.

Record Number 30118
NGR SD 29500 91900
Parish COLTON
Dating Medieval/Post-medieval
Description
Bloomery.

Record Number 30119
NGR SD 29760 91620
Parish COLTON
Dating Medieval/Post-medieval
Description
Potash kiln.

Record Number 30133
NGR SD 38000 89000
Parish CARTMEL FELL
Dating Medieval/Post-medieval
Description
At least 11 charcoal pitsteads and two barkpeelers' huts.

Record Number 30134, 30135
NGR SD 38930 88680
Parish STAVELEY IN CARTMEL
Dating Medieval/Post-medieval
Description
Charcoal pitstead.

Record Number 30138
NGR SD 37700 93650
Parish SATTERTHWAITE
Dating Medieval/Post-medieval
Description
Bloomsmithy.

Record Number 30140
NGR SD 37750 94680
Parish CLAIFE
Dating Medieval/Post-medieval
Description
Potash kiln.

Record Number 30143
NGR SD 38070 92450
Parish SATTERTHWAITE
Dating Medieval/Post-medieval
Description
Chopwood kiln.

Record Number 30144
NGR SD 37270 93650
Parish SATTERTHWAITE
Dating Medieval/Post-medieval
Description
Chopwood kiln.

Record Number 30149
NGR SD 37300 99900
Parish CLAIFE
Dating Medieval/Post-medieval
Description
Potash kiln.

Record Number 30150 – 30157
NGR SD3195
Parish CONISTON
Dating Medieval/Post-medieval
Description
Potash kilns.

Record Number 30159
NGR SD 30100 94800
Parish TORVER
Dating Medieval/Post-medieval
Description
Potash kiln.

Record Number 30161
NGR SD 31020 95260
Parish CONISTON
Dating Medieval/Post-medieval
Description
Stringhearth dated to 1331–1537.

Record Number 30169
NGR SD 30420 93460
Parish COLTON
Dating Medieval/Post-medieval
Description
At least four barkpeelers' huts.

Record Number 30185
NGR SD 38570 96380
Parish CLAIFE
Dating Medieval/Post-medieval
Description
Potash kiln.

Record Number 30186, 30187
NGR SD 38580 96280, SD 38670 96130
Parish CLAIFE
Dating Medieval/Post-medieval
Description
Potash kilns.

Record Number 30201, 30202
NGR SD 29990 94600, SD 29930 94650
Parish TORVER
Dating Medieval/Post-medieval
Description
Potash kiln.

Record Number 30206
NGR SD 28100 93020
Parish TORVER
Dating Prehistoric?
Description
Burnt mound.

Record Number 30232
NGR SD 36960 88600
Parish COLTON
Dating Medieval/Post-medieval
Description
Potash kiln.

Record Number 30233
NGR SD 37080 88650
Parish COLTON
Dating Medieval/Post-medieval
Description
Chopwood kiln.

Record Number 30234
NGR SD 36950 88570
Parish COLTON
Dating Medieval/Post-medieval
Description
Barkpeeler's hut.

Record Number 30249–30255
NGR SD3888
Parish STAVELEY-IN-CARTMEL
Dating Medieval/Post-medieval
Description
Potash kiln, barkpeelers' huts, and charcoal pitsteads.

Record Number 30256
NGR SD 38460 89120
Parish STAVELEY-IN-CARTMEL
Dating Medieval/Post-medieval
Description
Potash kiln.

Record Number 30257
NGR SD 38460 89090
Parish STAVELEY-IN-CARTMEL
Dating Medieval/Post-medieval
Description
Two barkpeelers' huts.

Record Number 30342
NGR SD 40750 96750
Parish WINDERMERE
Dating Medieval/Post-medieval
Description
Potash kiln.

Record Number 30352
NGR SD 40480 97800
Parish WINDERMERE
Dating Medieval/Post-medieval
Description
Potash kiln.

Record Number 30360
NGR SD 40220 98660
Parish WINDERMERE
Dating Medieval/Post-medieval
Description
Potash kiln.

Record Number 30368
NGR SD 40510 98840
Parish WINDERMERE
Dating Medieval/Post-medieval
Description
Fulling and corn mills.

Record Number 30369
NGR SD 40590 99030
Parish WINDERMERE
Dating Medieval/Post-medieval
Description
Fulling mill.

Record Number 30370
NGR SD 40410 97560
Parish WINDERMERE
Dating Medieval/Post-medieval
Description
Corn and fulling mills.

Record Number 30371
NGR SD 40300 96500
Parish WINDERMERE
Dating Medieval/Post-medieval
Description
Fulling mill.

Record Number 30461
NGR NY 08220 14980
Parish ENNERDALE AND KINNISIDE
Dating Medieval/Post-medieval
Description
Potash kiln.

Record Number 30462
NGR NY 09930 14900
Parish ENNERDALE AND KINNISIDE
Dating Medieval/Post-medieval
Description
Potash kiln.

Record Number 30466
NGR NY 09300 16500
Parish ENNERDALE AND KINNISIDE
Dating Medieval/Post-medieval
Description
A logwood dye mill.

Record Number 30502
NGR NY 12930 03790
Parish NETHER WASDALE
Dating Medieval/Post-medieval
Description
Corn and saw mills.

Record Number 30505
NGR NY 14400 03800
Parish NETHER WASDALE
Dating Medieval/Post-medieval
Description
Pot kiln.

Record Number 30520
NGR NY 11250 14240
Parish ENNERDALE AND KINNISIDE
Dating Medieval/Post-medieval
Description
Potash kiln.

Record Number 30521
NGR NY 16670 18760
Parish BUTTERMERE
Dating Medieval/Post-medieval
Description
Chopwood kiln.

Record Number 30522
NGR NY 19400 15020
Parish BUTTERMERE
Dating Medieval/Post-medieval
Description
Mill.

Record Number 30523
NGR NY 18680 15720
Parish BUTTERMERE
Dating Medieval/Post-medieval
Description
Bloomery.

Record Number 30524
NGR NY 17450 16980
Parish BUTTERMERE
Dating Medieval/Post-medieval
Description
Mill.

Record Number 30529
NGR NY 10980 15180
Parish ENNERDALE AND KINNISIDE
Dating Prehistoric?
Description
Hut circles.

Record Number 30530
NGR NY 12360 14880
Parish ENNERDALE AND KINNISIDE
Dating Medieval/Post-medieval
Description
Barkpeeler's hut.

Record Number 30542
NGR NY 13120 21650
Parish LOWESWATER
Dating Medieval/Post-medieval
Description
Potash kiln.

Record Number 30544
NGR NY 12190 22800
Parish LOWESWATER
Dating Medieval/Post-medieval
Description
Potash kiln.

Record Number 30545
NGR NY 14650 21800
Parish LOWESWATER
Dating Medieval/Post-medieval
Description
Potash kiln.

Record Number 30548
NGR NY 13020 21570
Parish LOWESWATER
Dating Medieval/Post-medieval
Description
Fulling mill.

Record Number 30549
NGR NY 13810 20370
Parish LOWESWATER
Dating Medieval/Post-medieval
Description
Fulling mill.

Record Number 30550
NGR NY 13960 20860
Parish LOWESWATER
Dating Medieval/Post-medieval
Description
Fulling mill.

Record Number 30551
NGR NY 14660 21610
Parish LOWESWATER
Dating Medieval/Post-medieval
Description
Fulling mill.

Record Number 30582
NGR NY 24160 19150
Parish ABOVE DERWENT
Dating Medieval/Post-medieval
Description
Chopwood kiln.

Record Number 30583
NGR NY 24940 19760
Parish ABOVE DERWENT
Dating Medieval/Post-medieval
Description
Potash kiln.

Record Number 30588
NGR NY 26520 18730
Parish BORROWDALE
Dating Medieval/Post-medieval
Description
Mill.

Record Number 30589
NGR NY 27000 19700
Parish BORROWDALE
Dating Medieval/Post-medieval
Description
Bloomsmithy.

Record Number 30592
NGR NY 23430 27450
Parish UNDER SKIDDAW
Dating Medieval/Post-medieval
Description
Potash kiln.

Record Number 30595, 30597 – 30601
NGR NY20 28, NY20 29
Parish WYTHOP
Dating Medieval/Post-medieval
Description
Ridge and furrow with associated potash kiln (30595), ruined building (30599) and charcoal pitstead (30601).

Record Number 30617
NGR NY 26360 23860
Parish KESWICK
Dating Medieval/Post-medieval
Description
Woollen mill.

East Cumbria: Appendix 5

Record number 558
NGR NY 54550 55130
Parish CASTLE CARROCK
Dating Iron Age
Description
Hallsteads, triangular settlement with ditch and banks separating it from higher ground, nearly surrounded by bog or marsh, which was destroyed early in the century; the settlement was disturbed in 1904 by waterworks.

Record number 559
NGR NY 53900 55950
Parish CASTLE CARROCK
Dating Bronze Age
Description
Barrow known from documentary evidence.

Record number 566
NGR NY 54120 50290
Parish CUMREW
Dating Prehistoric?
Description
Cumrew III, a mound on aerial photographs.

Record number 568
NGR NY 54000 57700
Parish CASTLE CARROCK
Dating Prehistoric?
Description
Burial cairn with skeleton excavated in 1775.

Record number 575
NGR NY 50300 54200
Parish CUMREW
Dating Unknown
Description
Cairn on OS map.

Record number 579
NGR NY 53390 51010
Parish CUMWHITTON
Dating Bronze Age
Description
Six standing stones.

Record number 581
NGR NY 54000 57700
Parish HAYTON
Dating Prehistoric
Description
Bronze celt.

Record number 582
NGR NY 54000 55000
Parish CASTLE CARROCK
Dating Unknown
Description
Stone querns found near Castle Carrock.

Record number 583
NGR NY 54000 55000
Parish CASTLE CARROCK
Dating Bronze Age
Description
Three gold penannular armlets found in a possible tumulus in 1796.

Record number 585
NGR NY 53300 51990
Parish CARLATTON
Dating Bronze Age
Description
Arrowheads.

Record number 731
NGR NY 55000 48200
Parish AINSTABLE
Dating Bronze Age?
Description
Stone circle known from documentary evidence.

Record number 744
NGR NY 55900 49120
Parish AINSTABLE
Dating Neolithic/Bronze Age
Description
Axe-hammer.

Record number 746
NGR NY 52000 40000
Parish LAZONBY
Dating Neolithic
Description
Polished axe.

Record number 764
NGR NY 50660 40400
Parish LAZONBY
Dating Iron Age
Description
Settlement and enclosure.

Record number 765
NGR NY 50770 40310
Parish LAZONBY
Dating Iron Age/Medieval
Description
Settlement and enclosure.

Record number 935
NGR NY 53000 37400
Parish GREAT SALKELD
Dating Early Bronze Age
Description
Food vessel found in 1972 while quarrying for sand.

Record number 1003
NGR NY 66440 31690
Parish CULGAITH
Dating Prehistoric
Description
A 'native' settlement site in Littledale Beck.

Record number 1716
NGR NY 63700 13000
Parish CROSBY RAVENSWORTH
Dating Unknown
Description
Possible entrenchments of a camp found on the highest part of Bank Moor. Faint traces of a boundary bank near Raise Howe cairn (SMR 1721).

Record number 1718
NGR NY 64890 13190
Parish ASBY
Dating Bronze Age/Early Medieval
Description
Bowl barrow with four bodies, including one skeleton laid full length with an iron knife at the waist, and one cremation.

Record number 1719
NGR NY 64900 13200
Parish ASBY
Dating Neolithic/Bronze Age
Description
One rough-out axe and two polished axes found near Gaythorne Hall. Also a stone axe-hammer.

Record number 1720
NGR NY 63110 13120
Parish CROSBY RAVENSWORTH
Dating Prehistoric
Description
Small cairn containing four small skeletons in cists, a gold-plated bronze buckle, and animal bones, probably those of a horse.

Record number 1721
NGR NY 63710 13030
Parish CROSBY RAVENSWORTH
Dating Bronze Age?
Description
Raise Howe cairn, located on summit.

Record number 1755
NGR NY 68000 10000
Parish ASBY
Dating Unknown
Description
Quern in good condition.

Record number 1954
NGR NY 68400 07300
Parish CROSBY GARRETT
Dating Prehistoric
Description
Two burial cairns at Rayseat Pike with possible mortuary structure. Possibly six unaccompanied cremated remains apparently burnt *in situ*; also secondary burials.

Record number 2069
NGR SD 48700 93800
Parish UNDERBARROW AND BRADLEYFIELD
Dating Medieval?

Description
Settlement; five mounds, possibly pillow mounds, at Cunswick Hall.

Record number 2075
NGR SD 53620 90800
Parish KENDAL
Dating Medieval
Description
Earthworks, possible parish boundary.

Record number 2084
NGR SD 56600 97000
Parish PATTON
Dating Unknown
Description
Settlement and field systems.

Record number 2086
NGR SD 48600 93800
Parish UNDERBARROW AND BRADLEYFIELD
Dating Unknown
Description
Settlement north of Cunswick Hall on aerial photographs.

Record number 2104
NGR SD 62430 99410
Parish SEDBERGH
Dating Prehistoric / Post-medieval?
Description
Gibbet Hill cairn to the west of the Roman road, later used as gallows.

Record number 2460
NGR SD 55900 98100
Parish PATTON
Dating Bronze Age
Description
Middle Bronze Age looped and socketed axe found at edge of Whinfell Tarn.

Record number 2519
NGR SD 50000 78900
Parish BEETHAM
Dating Unknown
Description
Settlement and cairn.

Record number 2641
NGR SD 53070 88730
Parish NATLAND
Dating Iron Age / Romano-British
Description
Castlesteads, a small multivallate hillfort on the summit of the Helm.

Record number 2663
NGR SD 62200 99700
Parish TEBAY
Dating Iron Age / Romano-British
Description
Probable settlement near Roman road.

Record number 2795
NGR NY 68700 31800
Parish MILBURN
Dating Unknown
Description
Possible settlement site on aerial photographs.

Record number 2801
NGR NY 62000 26400
Parish KIRKBY THORE
Dating Roman
Description
Milestone apparently *in situ* on the east side of road north of Kirkby Thore.

Record number 2855
NGR NY 58000 25000
Parish CLIBURN
Dating Prehistoric
Description
Eight flint flakes.

Record number 2861
NGR NY 56290 26290
Parish BROUGHAM
Dating Bronze age
Description
Ceramic urns, cropmarks, stone circle.

Record number 3091
NGR SD 62800 99300
Parish SEDBERGH
Dating Unknown
Description
Remains of enclosures, cairns and hut circles may indicate a settlement site.

Record number 3117
NGR SD 46800 98700
Parish NETHER STAVELEY
Dating Prehistoric
Description
Scraper.

Record number 3338
NGR SD 39000 89000
Parish STAVELEY
Dating Medieval / Post-medieval
Description
Bloomery.

Record number 3374
NGR SD 48600 93700

Parish UNDERBARROW AND BRADLEYFIELD
Dating Unknown
Description
Settlement on aerial photographs.

Record number 3375
NGR SD 45700 94440
Parish CROOK
Dating Prehistoric
Description
Possible cist with a shallow inhumation.

Record number 3555
NGR NY 58000 25000
Parish CLIBURN
Dating Mesolithic
Description
Flint/chert.

Record number 3579
NGR SD 37000 86200
Parish STAVELEY
Dating Bronze Age
Description
Perforated stone axe-hammer found at Newby Bridge.

Record number 3624
NGR SD 54700 96000
Parish WHINFELL (585)
Dating Medieval
Description
Patton, a deserted village with class D earthworks.

Record number 3901
NGR NY 63200 13000
Parish CROSBY RAVENSWORTH
Dating Unknown
Description
Earthwork on St Joseph aerial photographs but nothing on RAF aerial photographs of 1954.

Record number 3906
NGR NY 62810 28680
Parish NEWBIGGIN
Dating Medieval
Description
Newbiggin Hall, founded *c*1332.

Record number 3911
NGR NY 66520 29320
Parish MILBURN
Dating Medieval
Description
Howgill Castle, probably fourteenth century.

Record number 3922
NGR NY 66500 31900
Parish CULGAITH
Dating Iron Age/Romano-British
Description
Possible settlement on aerial photographs.

Record number 4018
NGR SD 55000 89000
Parish OLD HUTTON AND HOLMESCALES
Dating Medieval
Description
According to Machell (CRO (C)/D&C/Machell Mss) there was a chapel at St Sundays Beck, '...known only by tradition'.

Record number 4019
NGR SD 56000 89000
Parish OLD HUTTON AND HOLMESCALES
Dating Medieval
Description
According to Machell (CRO (C)/D&C/Machell Mss) there was a chapel at Bridge End.

Record number 4085
NGR SD 54900 89200
Parish OLD HUTTON AND HOLMESCALES
Dating Prehistoric
Description
Stone axe, hanging from the roof rafters at Blease Hall.

Record number 4086
NGR SD 56000 88000
Parish OLD HUTTON AND HOLMESCALES
Dating Prehistoric
Description
Perforated stone hammer.

Record number 4092
NGR SD 56000 98000
Parish PATTON EAST
Dating Pehistoric
Description
Perforated stone.

Record number 4113
NGR SD 53000 97000
Parish SKELSMERGH
Dating Unknown
Description
Stone hammer.

Record number 4115
NGR SD 52400 99300
Parish STRICKLAND ROGER
Dating Bronze Age?
Description
Perforated stone axe-hammer from Garnett Bridge.

Record number 4116
NGR SD 54900 98300
Parish PATTON
Dating Prehistoric
Description
Axe.

Record number 4117
NGR SD 55000 98000
Parish PATTON
Dating Unknown
Description
Stone quern.

Record number 4285
NGR SD 59600 85000
Parish MANSERGH
Dating Unknown
Description
Small circular enclosure.

Record number 4309
NGR NY 63300 13300
Parish CROSBY RAVENSWORTH
Dating Prehistoric
Description
Cairnfield with perhaps one burial cairn.

Record number 4328
NGR NY 68000 29000
Parish MILBURN
Dating Roman
Description
Altar, found in Burney Beck.

Record number 4368
NGR SD 60970 84860
Parish MANSERGH
Dating Unknown
Description
Table-tomb-like object of stones at Rigmarden Park.

Record number 4507
NGR NY 53010 41450
Parish LAZONBY
Dating Unknown
Description
Hut circles on aerial photographs.

Record number 4594
NGR SD 45400 95250
Parish CROOK
Dating Unknown
Description
Flints found in the same field as two lead spindle whorls.

Record number 4622
NGR SD 48300 95100
Parish STRICKLAND KETEL
Dating Unknown
Description
Settlement.

Record number 4623
NGR SD 48900 94900
Parish STRICKLAND KETEL
Dating Unknown
Description
Field system on aerial photographs.

Record number 4625
NGR SD 45100 95000
Parish CROOK
Dating Unknown
Description
A ?bronze vessel with iron parts, lying in a wet place south of Crook Church.

Record number 4626
NGR SD 45000 97400
Parish NETHER STAVELEY
Dating Unknown
Description
Settlement with cairn on aerial photographs.

Record number 4717
NGR NY 61900 26760
Parish TEMPLE SOWERBY
Dating Unknown
Description
Circular enclosure, known from documentary source and aerial photograph.

Record number 4869
NGR NY 54600 00900
Parish FAWCETT FOREST
Dating Medieval/Post-medieval
Description
Possible charcoal burner's platform.

Record number 4871
NGR NY 55200 00150
Parish WHITWELL AND SELSIDE
Dating Medieval/Post-medieval
Description
Possible charcoal burners' platforms.

Record number 4872
NGR NY 55400 00800
Parish WHINFELL
Dating Medieval/Post-medieval
Description
Rectangular enclosure, perhaps the remains of a farmstead.

Record number 4928
NGR NY 64960 13280
Parish ASBY
Dating Medieval

Description
Various earthworks near Scale Beck linked to a twelfth century grant of land to the Hospital of St Peter at York.

Record number 4939
NGR NY 60300 26300
Parish BROUGHAM
Dating Medieval
Description
Earthworks next to the present Julian Bower farmhouse may be buildings standing in the fourteenth century.

Record number 4986
NGR SD 53800 95900
Parish SKELSMERGH
Dating Unknown
Description
Settlement.

Record number 4990
NGR NY 54580 55240
Parish CASTLE CARROCK
Dating Unknown
Description
House platform.

Record number 5129
NGR NY 58300 25000
Parish CLIBURN
Dating Unknown
Description
Cropmark.

Record number 5130
NGR NY 62100 26100
Parish TEMPLE SOWERBY
Dating Unknown
Description
Rectangular enclosure appearing as cropmark.

Record number 5131
NGR NY 61300 28100
Parish TEMPLE SOWERBY
Dating Unknown
Description
Sinuous cropmark and part of small circular feature.

Record number 5150
NGR SD 53900 90500
Parish NEW HUTTON
Dating Unknown
Description
Earthworks on aerial photographs.

Record number 5157
NGR SD 53270 87720
Parish STAINTON
Dating Unknown
Description
Earthworks.

Record number 5161
NGR SD 53200 88400
Parish STAINTON
Dating Unknown
Description
Enclosure on aerial photographs.

Record number 5162
NGR SD 53300 88600
Parish STAINTON
Dating Unknown
Description
Possible enclosure.

Record number 5163
NGR SD 54100 90100
Parish NEW HUTTON
Dating Prehistoric?
Description
Scooped enclosure.

Record number 5164
NGR SD 54200 90200
Parish NEW HUTTON
Dating Unknown
Description
Rectangular structure on aerial photographs.

Record number 5166
NGR SD 53900 96500
Parish SKELSMERGH
Dating Medieval/Post-medieval?
Description
Remains of farmstead on the site of an earlier one.

Record number 5167
NGR SD 53950 97000
Parish SKELSMERGH
Dating Unknown
Description
Small square earthwork.

Record number 5168
NGR SD 54400 99900
Parish WHITWELL AND SELSIDE
Dating Unknown
Description
Settlement site.

Record number 5216
NGR NY 54000 51000
Parish CUMREW
Dating Bronze Age
Description
Cumrew II cairn known from doumentary evidence.

Record number 5217
NGR NY 54000 51000
Parish CUMREW
Dating Neolithic?
Description
Cumrew II megalithic tomb and cairn.

Record number 5218
NGR NY 54000 55100
Parish CASTLE CARROCK
Dating Bronze Age
Description
Cumrew IV, cist and ceramic urn known from documentary evidence.

Record number 5490
NGR NY 58000 25000
Parish CLIBURN
Dating Prehistoric
Description
Flint implements.

Record number 5999
NGR NY 62400 27100
Parish TEMPLE SOWERBY
Dating Unknown
Description
Cropmarks.

Record number 6099
NGR NY 63900 13000
Parish CROSBY RAVENSWORTH
Dating Prehistoric?
Description
Possible mound.

Record number 6131
NGR SD 62300 99600
Parish TEBAY
Dating Unknown
Description
Enclosure.

Record number 6178
NGR NY 61700 26500
Parish TEMPLE SOWERBY
Dating Unknown
Description
Cropmark.

Record number 6197
NGR NY 63100 15900
Parish CROSBY RAVENSWORTH
Dating Medieval
Description
Earthworks, mostly ridge and furrow and associated banks and trackways, plus disused quarries near Maulds Meaburn.

Record number 6372
NGR SD 53800 96100
Parish SKELSMERGH
Dating Unknown
Description
Possible settlement in small valley beside a beck.

Record number 6680
NGR SD 53300 96300
Parish SKELSMERGH
Dating Medieval
Description
Possible double oval settlement site.

Record number 6721
NGR NY 66000 15000
Parish BANDLEYSIDE
Dating Medieval
Description
Extensive earthworks and ridge and furrow of Drybeck village.

Record number 6777
NGR NY 62000 28000
Parish NEWBIGGIN
Dating Medieval
Description
Earthworks and ridge and furrow of Newbiggin village.

Record number 6778
NGR NY 61000 27000
Parish TEMPLE SOWERBY
Dating Medieval
Description
Earthworks and strip fields of Temple Sowerby village.

Record number 6779
NGR NY 61000 28000
Parish TEMPLE SOWERBY
Dating Medieval
Description
Earthworks and ridge and furrow of Acorn Bank village.

Record number 6784
NGR NY 51000 37000
Parish LAZONBY
Dating Medieval
Description
Possible evidence for village at Scarfoot Tarn.

Record number 8037
NGR NY 61980 28660
Parish CUMREW
Dating Unknown
Description
Small earthwork on aerial photographs.

Record number 8138 – 8141, 8283
NGR NY 73 30, NY 74 30, NY 74 29
Parish DUFTON
Dating Unknown
Description
Mining areas and structures on aerial photographs and maps.

Record number 8150
NGR NY 74650 29550
Parish DUFTON
Dating Unknown
Description
Mining area with enclosures on aerial photographs and maps: possible lead mine.

Record number 8151
NGR NY 74600 29930
Parish DUFTON
Dating Unknown
Description
Enclosures, defined by low banks, on aerial photographs.

Record number 8262
NGR NY 67670 29040
Parish MILBURN
Dating Unknown
Description
Earthworks on aerial photographs, with field observation.

Record number 8267
NGR NY 66000 31000
Parish MILBURN
Dating Unknown
Description
Earthworks on aerial photographs, also ridge and furrow.

Record number 8268
NGR NY 66480 30940
Parish MILBURN
Dating Unknown
Description
Quarry pits on aerial photographs.

Record number 8272
NGR NY 68020 31570
Parish MILBURN
Dating Unknown
Description
Curvilinear enclosure on aerial photographs.

Record number 9372
NGR NY 56400 30900
Parish LANGWATHBY
Dating Unknown
Description
Cropmark.

Record number 9940
NGR NY 67320 09460
Parish ORTON
Dating Prehistoric?
Description
Possible field boundaries (stone wall remains) associated with a hut circle with enclosure and a field / enclosure.

Record number 9999
NGR NY 66290 31970
Parish CULGAITH
Dating Prehistoric
Description
Sub-circular enclosure.

Record number 10072
NGR NY 32220 54570
Parish ORTON
Dating Prehistoric
Description
Single standing stone, on first edition OS map (1860).

Record number 13765
NGR NY 51200 37500
Parish HESKET
Dating Neolithic?
Description
Possible long cairn.

Record number 13766
NGR SD 58500 88600
Parish OLD HUTTON AND HOLMESCALES
Dating Unknown
Description
Field boundaries and enclosures on aerial photographs.

Record number 13767
NGR SD 58600 88500
Parish OLD HUTTON AND HOLMESCALES
Dating Neolithic?
Description
Possible long cairn.

Record number 13768
NGR SD 58300 88800
Parish OLD HUTTON AND HOLMESCALES
Dating Unknown
Description
Enclosure and boundary banks on aerial photographs.

Record number 13774
NGR NY 65200 14000
Parish CROSBY RAVENSWORTH
Dating Medieval?

Description
Lynchets and possible farmstead site, with extensive area of well-defined ridge and furrow.

Record number 13861
NGR NY 68300 09500
Parish ASBY
Dating Medieval?
Description
Possible farmstead.

Record number 13862
NGR NY 67900 10000
Parish ASBY
Dating Unknown
Description
Rectilinear building platform with associated square structure *c* 200 ft west of SMR 13861.

Record number 13869
NGR NY 54000 36000
Parish GREAT SALKELD
Dating Unknown
Description
Place-name evidence for dyke system.

Record number 14264, 14266
NGR SD 53090 87230
Parish STAINTON
Dating Medieval/Post-medieval
Description
Mill Bridge Mill, with mill race.

Record number 14743
NGR SD 53500 89400
Parish STAINTON
Dating Unknown
Description
Rectilinear field system, dykes, and ridge and furrow on aerial photographs.

Record number 14846
NGR SD 53100 89000
Parish NATLAND
Dating Unknown
Description
Site of a rectilinear field system and possibly a cairnfield, on aerial photographs.

Record number 15147
NGR NY 68050 09300
Parish ORTON
Dating Unknown
Description
Enclosure in the form of a rough arc.

Record number 15148
NGR NY 68850 09800
Parish ORTON
Dating Unknown
Description
Traces of an enclosure and walls.

Record number 15150
NGR NY 68000 08200
Parish ASBY
Dating Bronze Age?
Description
Stone circle, possibly the remains of a kerbed cairn.

Record number 15151
NGR NY 68500 08500
Parish ASBY
Dating Prehistoric
Description
Several flints found at a small disused quarry.

Record number 15152
NGR NY 67200 08000
Parish ORTON
Dating Unknown
Description
Sub-rectangular enclosure.

Record number 15153
NGR NY 67200 07900
Parish ORTON
Dating Unknown
Description
Small enclosure with turf-covered walls.

Record number 15154
NGR NY 67200 07800
Parish ORTON
Dating Unknown
Description
Sub-rectangular enclosure.

Record number 15186
NGR NY 64970 13050
Parish CROSBY RAVENSWORTH
Dating Prehistoric/Romano-British?
Description
Large D-shaped enclosure — possibly defended.

Record number 15698
NGR NY 72125 28965
Parish DUFTON
Dating Prehistoric?
Description
High Scald Fell Cairns.

Record number 15699
NGR NY 72405 27730
Parish DUFTON
Dating Bronze Age?

Description
Seamore cairn stands north–west of Seamore Tarn.

Record number	15718
NGR	NY 73455 29920*
Parish	DUFTON
Dating	Prehistoric?

Description
Woolbeck mound lies west of Knock Coal Shop. Another mound lies about half a mile north–east.

Record number	16536
NGR	SD 54700 89600
Parish	NEW HUTTON
Dating	Holocene

Description
Wetland / former tarn.

Record number	16581
NGR	SD 53660 87720
Parish	STAINTON
Dating	Unknown

Description
Earthwork.

Record number	16582
NGR	SD 54300 87900
Parish	STAINTON
Dating	Unknown

Description
Possible earthwork.

Record number	16583
NGR	SD 54400 88000
Parish	OLD HUTTON AND HOLMESCALES
Dating	Unknown

Description
Possible earthwork.

Record number	16584
NGR	SD 54300 88300
Parish	STAINTON
Dating	Unknown

Description
Possible earthworks.

Record number	16585
NGR	SD 54900 87700
Parish	PRESTON RICHARD
Dating	Unknown

Description
Possible earthwork, traces of trackway.

Record number	16586
NGR	SD 54700 87700
Parish	PRESTON RICHARD
Dating	Unknown

Description
Possible earthwork.

Record number	16653
NGR	SD 55550 75800
Parish	BURTON IN KENDAL (511)
Dating	Unknown / Medieval?

Description
Circular enclosure, ridge and furrow, hollow-ways, and other earthworks, on aerial photographs.

Record number	16663
NGR	SD 55600 85600
Parish	PRESTON PATRICK
Dating	Medieval?

Description
Traces of ridge and furrow, possible lynchets, and grubbed-out field boundaries.

Record number	16726
NGR	SD 58900 88000
Parish	KILLINGTON
Dating	Bronze Age

Description
Small angle-flanged axe of bronze found on a heap of peat / soil dredged from Tarneybank Tarn in 1998.

Record number	16993
NGR	NY 63000 25850
Parish	KIRKBY THORE
Dating	Medieval

Description
Strip fields and ridge and furrow on aerial photograph.

Appendix 4
Cumbria's wetlands: a list of extant and improved landscapes identified from cartographic sources and field observatiom

The following mosses, situated at 200 m OD or below, are divided into four main sections, the first three reflecting the descriptive text in Chapters 2 to 4; the last section briefly lists other mosses in areas subject to palaeoecological assessment only. Table 8 lists all raised mires in Cumbria with their area, vegetation type, and designations (Lindsay and Immirzi 1996).

South Cumbria: Chapter 2
Lyth and Winster Valleys

Site Name	Bellart How Moss (4 m OD)
NGR	SD 452 831
Parish	MEATHOP AND ULPHA
Source	Ordnance Survey 1862 6" first edition Westmorland Sheet 42

Decription
This lies to the south-west of the Lyth Valley, on the western side of the River Kent, merging into Foulshaw Moss to the south. Most of its surface was planted with commercially-grown conifers, although certain parts did support small areas of mire vegetation. It was assumed that the surfaces were largely intact and untruncated by cutting, although there was extensive evidence for earlier extraction around the periphery. On the north–western edge there were former turf dales which had been reclaimed and enclosed into fields; the peat over most of this area was less than 1 m thick. To the south, where it bordered the forestry, a cleaned dyke revealed at least 3–4 m of peat under the trees. The depth of peat in the central area may suggest that, despite the desiccating action of the trees, the peat in lower levels may be in good condition and it may contain high quality archaeological remains, possibly including the remains of trackways, as well as palaeo-ecological archives. The site is, however, under long-term threat from the desiccating action of the trees and the drainage of the surrounding agricultural land. It should be noted that hand peat-cutting continues at this site, although it was difficult at the time of survey to make an accurate assessment of the threat this posed to the peat deposits.

Site Name	Blake Bank Moss (20 m OD)
NGR	SD 459 905
Parish	UNDERBARROW AND BRADLEYFIELD
Source	Ordnance Survey 1863 6" first edition Westmorland Sheet 38

Decription
This is the most northerly of the mosses in the main part of the Lyth Valley, occupying the whole of the upper tip of the valley. To the south it was in part defined by a tongue of dry ground extending from the western moss edge, separating Blakebank Moss from Ulpha Moss to the south. The moss survives as a *c* 2 ha area of (possibly) intact peat under Blakebank Wood, surrounded by pasture land with thin peat less than 1 m deep, covering *c* 25 ha. Surrounding this is a ring, varying in thickness, of remnant organic soils which quickly gave way to dryland soils. From the modern surface topography it would appear that the moss formed a discrete hydrological unit within a small embayment at the end of the valley. The most valuable element is the centre which contains several metres of peat and, possibly, intact surfaces. The limited nature of the remains means that they are under severe threat from desiccation around their margins and there must be a question as to how long they survive before becoming too dry for the survival of pollen and plant remains.

Site Name	Castle Moss (10 m OD)
NGR	SD 418 797
Parish	GRANGE OVER SANDS
Source	Ordnance Survey 1850 6" first edition Lancashire Sheet 13

Decription
This is a small moss south of the B5277 (the former A590) east of Lindale. It is defined by artificial boundaries on three sides (moss tracks and drains) and by the rising ground of Low Meathop to the south. It may originally have formed part of Meathop Moss (Fig 14) to the north–east, although this was difficult to establish at the time of survey. Cushatt Wood probably represents the original centre of the moss with *c* 10 ha of peat up to 3 m deep under the trees and the adjacent track to the south. This was surrounded by a ring of thinner peats (< 1 m deep), of variable width. Organic soils were present in the remainder of the area, giving way to former salt-marsh deposits to the south–west. This moss has some palaeoecological potential in the relatively deep peats in its centre, although these were being dessicated and disturbed by the tree cover and the drainage of surrounding land. The archaeological potential has been shown by the discovery of a possible trackway in the silts underlying the peats near Castle Head Cottage, suggesting that it is not just the mire deposits which are important in this area.

Site Name	Catcrag Moss (10 m OD)
NGR	SD 448 823
Parish	MEATHOP AND ULPHA
Source	Ordnance Survey 1862 6" first edition Westmorland Sheet 42

Description
No field data.

Site Name	Cock Moss (10 m OD)
NGR	SD 465 890
Parish	UNDERBARROW AND BRADLEYFIELD
Source	Ordnance Survey 1863 6" first edition Westmorland Sheet 38

Decription
Cock Moss formed the southern part of Underbarrow Moss.

Site Name	Cowmire Moss (3 m OD)
NGR	SD 462 875
Parish	CROSTHWAITE AND LYTH
Source	Ordnance Survey 1:25,000 Sheet 627

Decription
This small area of peat may originally have been part of the larger Lyth Moss to the south–east, and possibly extended to the valley edge at Dale Cottage and northwards to fill the small embayment near Flodder Hall. At the time of survey the moss survived as a small dome of peat covering *c* 1 ha, standing *c* 3 m proud of the surrounding land. It was surrounded by a small fragment of thin peat and organic soils extending north, south, and east to the River Gilpin. It is assumed that the peat was a small remnant in the centre of the moss, which had subsequently been enclosed. The place-name 'Dale' Cottage suggests earlier peat-cutting on the site. The poor survival of peats means that the moss has little archaeological or palaeoecological potential.

Site Name	Cunswick Tarn (140 m OD)
NGR	SD 489 938
Parish	UNDERBARROW AND BRADLEYFIELD
Source	Ordnance Survey 1:25,000 Outdoor Leisure sheet 7

Decription
This was referred to in Howard-Davis *et al* (1988).

Site Name	Foulshaw Moss (3 m OD)
NGR	SD 465 836
Parish	MEATHOP AND ULPHA
Source	Ordnance Survey 1862 6" first edition Westmorland Sheet 42

Decription
This forms a large area (400 ha) of peat in the south-west of the Lyth Valley to the west of the River Kent (Figs 11 and 14), contiguous with Bellart How Moss and Stakes Moss to the north. It also incorporates Ulpha Moss to the south. To the south–east, the moss borders Ulpha Fell and the eastern edge of the moss was strongly defined by a line of fields to the west of Main Drain, which separated the peat from the reclaimed coastal silts. The sharp edge between the peats and the coastal silts is clearly a product of recent reclamation and peat-cutting, and thus the true relationship between the different wetland deposits was difficult to determine from surface survey. The centre of the moss was planted with a large commercial conifer plantation which was present on peats, up to 6 m deep, and some thinner peats, particularly on the southern edge. The upper stratigraphy has been removed (Burton and Hodgson 1987, 43), although some *Sphagnum* peats remain in places overlying fen carr and *Phragmites* peats over marine alluvium. Whilst the deposits within the moss may be relatively well-preserved, as indicated by Wimble's core (1986), the whole deposit is under threat from long-term desiccation and damage to the upper layers from transpiration and root damage. Clear felling and re-planting will only exacerbate the problem in the future.

Site Name	Hell Moss (90 m OD)
NGR	SD 436 854
Parish	WITHERSLACK
Source	Ordnance Survey 1:25,000 sheet 627

Decription
No field data.

Site Name	Helsington Moss (10 m OD)
NGR	SD 470 890
Parish	UNDERBARROW AND BRADLEYFIELD
Source	Ordnance Survey 1863 6" first edition Westmorland Sheet 38

Decription
This formerly extensive mire in the centre of the Lyth Valley (Fig 11) is now represented by remnant peats surrounded by thinner deposits under agriculture. It is likely that the moss covered a substantial portion of the valley, defined by the River Gilpin to the east and the valley edge to the west. It was contiguous with Underbarrow Moss to the north and incorporated Savinhill Moss, which was indistinguishable on the ground. The existing remnant in the centre comprised an irregular area of *c* 15 ha of woodland, deeply indented on the western side with narrow turf dales from which peat had previously been cut. The peat in these areas was largely less than 1 m thick, in contrast to the at least 3–4 m in the central woodland. The condition of the peat is variable in the central area, although that on the western side appeared to be more intact, perhaps because of less intensive exploitation on the side away from Kendal – the main market for Lyth Valley peat. An irregular band of thinner remnant peats surrounded the centre, notably in enclosed peat dales on the north–east side. Thick peats were also present under the track on the southern side of the moss. Recent shrinkage was evidenced north of the centre, with the exposure of tree roots in field boundaries. The moss has the potential to offer a valuable opportunity for an understanding of landscape history whilst also being under threat from desiccation over the long term from peripheral drainage and tree transpiration. The peats around the core are likely to be too disturbed to be of significant value.

Site Name	Helton Moss (10 m OD)
NGR	SD 419 849
Parish	UPPER ALLITHWAITE
Source	Ordnance Survey 1:25,000 sheet 627

Decription
This was referred to in Howard–Davis *et al* (1988).

Site Name	High Moss (14 m OD)
NGR	SD 422 865
Parish	CARTMEL FELL

Source	Ordnance Survey 1851 6" first edition Lancashire Sheet 9

Decription

Located north of Helton Moss, High Moss incorporated Low Moss, both being parts of the same mire system which appeared originally to have occupied a long, thin depression in the Winster Valley. It was separated from Helton Moss and Stock Moss by small areas of dry ground. To the west the peats were bounded by the River Winster, although this may not have been the river's original course; there was some evidence for a palaeo-channel, perhaps of the main river, north of Stang Hill. Full assessment of peat survival was limited by a lack of access to the northern part of the moss. However, over much of the area, little peat appeared to survive, with thin peats being confined to the centre, including Cow Hill Wood and the area to the north, including the alluviated area around the river. The poor state of the peat suggests that this moss may be of limited palaeoecological and archaeological potential, although the drainage of the moss was poor and so waterlogged peats may survive.

Site Name	Key Moss (10 m OD)
NGR	SD 429 843
Parish	WITHERSLACK
Source	Ordnance Survey 1:25,000 sheet 627

Decription

The name survives as that of a farm on the eastern edge of the Winster Valley, although the survey did not reveal any surviving deposits. The topography did not demonstrate ill-drained basins, where peat may originally have formed. It is likely that the name refers to the northern section of Nichols Moss, *c* 0.5 m to the south.

Site Name	Levens Moss (5 m OD)
NGR	SD 479 858
Parish	UNDERBARROW AND BRADLEYFIELD
Source	Ordnance Survey 1862 6" first edition Westmorland Sheet 42

Decription

This extensive area of valley moss, totalling approximately 7 km^2 encompasses a large part of the Lyth Valley that has evidently been extensively enclosed and drained. The current land-use is dominated by pasture with one small area of mixed woodland. The fieldwork did not reveal any evidence of peat deposits on the site and it is considered to have little or no significant archaeological or palaeo-ecological value.

Site Name	Low Green (5 m OD)
NGR	SD 420 826
Parish	UPPER ALLITHWAITE
Source	Ordnance Survey 1850 6" first edition Lancashire Sheet 13

Decription

This moss lay on the western side of the Winster Valley, and filled a small embayment created by Back Holme, a crag which projects across the valley floor. It is likely that it originally formed part of the larger Nichols Moss on the eastern bank of the River Winster, to which it was joined by a small band of organic soils north of Nichols Wood. Thin peats were present over most of the moss intercut with, and bounded by, organic soils. The original moss would have filled virtually all of the embayment, although divided in the north by Back Holme. The field pattern comprised reclaimed turf dales which extended into the centre of the moss, and fields which appear to derive from amalgamated dales, a pattern unusual for South Cumbria, given that in most mosses a small area of intact, or near intact, peat was left in the centre. It is, however, similar to the Cheshire mosses where intensive exploitation, particularly in the nineteenth century, meant that common rights of turbary were exploited to the full (Leah *et al* 1997, 215). The archaeological and palaeoecological potential of the moss is likely to be relatively low, given the poor peat survival.

Site Name	Low Moss (10 m OD)
NGR	SD 422 862
Parish	CARTMEL FELL
Source	Ordnance Survey 1851 6" first edition Lancashire Sheet 9

Decription

No field data.

Site Name	Lyth Moss (5 m OD)
NGR	SD 467 872
Parish	UNDERBARROW AND BRADLEYFIELD
Source	Ordnance Survey 1862 6" first edition Westmorland Sheet 42

Decription

Lyth Moss lay on the western side of the valley bounded by the River Gilpin to the east, and was contiguous with Cowmire Moss. The distribution of peat at the time of survey was patchy, with only small fragments of thin peat within a wider area of organic soils. The greatest survival was between the two upstanding crags within the former moss, High Heads and Low Heads. It is, however, likely that peat survives under the recent alluvium around the River Gilpin. Given the state of peat preservation, this moss is unlikely to retain a high archaeological value, although the juxtaposition of the crags, peat, and the river may offer some interesting opportunities.

Site Name	Meathop Moss (3 m OD)
NGR	SD 443 814
Parish	MEATHOP AND ULPHA
Source	Ordnance Survey 1868 6" first edition Westmorland Sheet 46

Decription

A large area of surviving mossland lies on the southern edge of the Winster Valley (Figs 11 and 14), to the west of Foulshaw Moss and separated from it by a canalised drain within alluvium, suggesting that the two systems were originally independent. Meathop Moss also merges into

Nichols Moss to the north–west, where peat deposits join the two, and it incorporates Catcrag Moss to the north. The moss comprises a central area of possibly intact mossland covering *c* 50 ha, which at the time of survey was part wooded, although a substantial area in the centre had been cleared to permit regeneration of moss vegetation. Deep sections also survived under the tracks into the moss, around which the peat had been cut for fuel and had wasted as a result of drainage. Outside this was a narrow ring of reclaimed deep peats which formed the edge of the cut-over area, distinguished by a distinct break of slope. These merged into organic soils which, in turn, gave way to alluvium to the south and east. To the south–east the moss merged with Castle Moss. Most of the central area of the moss, and part of Catcrag Moss, have been designated as a SSSI and are managed by Cumbria Trust for Nature Conservation as a nature reserve. The main interest of the site is for its mire vegetation and invertebrate fauna. Peat survival was clearly extremely good in the centre of the moss, which represents an extremely important palaeo-ecological and archaeological resource. The boreholes along the line of the 1982/3 pipe trench (*Ch 2*) also indicate the potential of underlying marine and freshwater alluvial deposits.

Site Name	Nichols Moss (10 m OD)
NGR	SD 432 822
Parish	WITHERSLACK
Source	Ordnance Survey 1862 6" first edition Westmorland Sheet 41

Decription
This large moss is contiguous with Meathop Moss and lies to the north–west of it on the north side of the A590. The centre of deep peat, *c* 93 ha in extent, is designated as a SSSI and is under the management of English Nature. The SSSI supports important raised mire communities, surrounded by pine and birch woodlands. It also has important invertebrate populations, including butterflies and moths, and on its eastern edge there is a vegetational sequence from open mire to acidic oak woodland. The centre of the moss appeared, from surface survey, to be untruncated, although there is evidence for peat-cutting on the edge, including the northern part of the SSSI, where the peats thin to less than 1 m in depth and give way to enclosed pasture fields. To the west, the peats were of variable depth, perhaps indicating the former presence of turf dales, particularly around The Grove. The moss is bounded by organic soils and alluvium around the River Winster. It is clearly of the greatest significance for archaeology and palaeoecology, particularly in the light of its proximity to Meathop and Foulshaw Mosses. There is every likelihood of well-preserved organic remains preserved within the peats in both the centre and the periphery.

Site Name	Outley Moss (100 m OD)
NGR	SD 362 818
Parish	LOWER ALLITHWAITE
Source	Ordnance Survey 1:25,000 sheet 626

Decription
No field data.

Site Name	Park Moss (3 m OD)
NGR	SD 483 871
Parish	UNDERBARROW AND BRADLEYFIELD
Source	Ordnance Survey 1862 6" first edition Westmorland Sheet 42

Decription
A small area of mossland is adjacent to the eastern edge of the Lyth Valley where Brigsteer Park adjoins the wetlands. Although contiguous with Quags Moss and Levens Moss, the field pattern in this area suggests that it may have been exploited differently in the past: the regular strip-like fields appear to be enclosed turf dales. These, however, contrast with the more recent strips in Quags Moss immediately to the west. Only residue organic soils and thin peats survive in this area and so its archaeological and palaeoecological value is likely to be limited.

Site Name	Quags Moss (3 m OD)
NGR	SD 482 874
Parish	UNDERBARROW AND BRADLEYFIELD
Source	Ordnance Survey 1862 6" first edition Westmorland Sheet 42

Decription
This is an area of relatively good peat survival to the west of Cotes, north of Levens village. It lay in the centre of the Lyth Valley and was probably contiguous with Levens Moss and Park Moss, although these two now survive only as remnants. The moss itself has a small central area of deep peats, perhaps with intact surfaces, under woodland (*c* 3 ha), surrounded by a significant expanse of thin peats under pasture, much of which was in former turf dales which were aligned into the centre of the moss. The main area was enclosed by roads (such as Quags Causeway), which presumably skirted the original moss. This moss offers real potential as a palaeoecological archive, although its potential for well-preserved organic remains may be limited. There were several palaeo-channels in the area, the study of which may offer insights into the human use of the pre-moss landscape.

Site Name	Rawson Moss (7 m OD)
NGR	SD 465 854
Parish	UNDERBARROW AND BRADLEYFIELD
Source	Ordnance Survey 1862 6" first edition Westmorland Sheet 42

Decription
A significant area of former moss lies to the north of Stakes Moss (Fig 11), defined by the A590 and A5074 roads. It lay on the west side of the Lyth Valley and took its name from Rawsons (*sic*), a farm on the moss edge, perhaps the name of the first tenant after enclosure. The peat distribution was variable, with an outer ring of organic soils surrounding a

smaller area of thin peat and small areas of thick (< *c* 3 m OD) peat. The largest of these deposits lay under *c* 10 ha of woodland south–west of Gilpin Farm, which may include uncut peat; Gilpin Farm itself is a former peat-cutter's house. Despite the generally poor peat survival over most of the area, the presence of relatively thick peats, including potentially untruncated sequences, means that its palaeoecological potential is likely to be high. The presence of thick peats adjacent to Stakes Moss, where at least one wooden trackway has been discovered, may also prove significant.

Site Name Roudsea Wood (20 m OD)
NGR SD 331 821
Parish HAVERTHWAITE
Source Ordnance Survey 1:25,000 sheet 626
Decription
This was referred to in Howard-Davis *et al* (1988).

Site Name Savin Hill Moss (10 m OD)
NGR SD 466 890
Parish UNDERBARROW AND BRADLEYFIELD
Source Ordnance Survey 1863 6" first edition Westmorland Sheet 38
Decription
No field data.

Site Name Stakes Moss (3 m OD)
NGR SD 465 848
Parish UNDERBARROW AND BRADLEYFIELD
Source Ordnance Survey 1862 6" first edition Westmorland Sheet 42
Decription
Stakes Moss (Fig 11) lies north of Foulshaw Moss and south of Rawson Moss, and was probably originally contiguous with both. It had two distinct areas: to the east, the thick peat is covered with commercial forestry plantation, a continuation of the larger plantings to the south. Adjacent to this was a series of long, thin fields which led into the moss from the east, clearly former turf dales and peat cuttings, as is evidenced by the 'stepped' edge to the deep peat where it adjoins the fields; however, it did not prove possible to examine these fields during the survey. The peat in Stakes Moss presents an important archaeological and palaeoecological resource, because of the deep, potentially untruncated sequences and the presence of a trackway discovered at the turn of the twentieth century (*Ch 2*). This was discovered as the peat was cut away and significant sections may still remain in the undisturbed peat. The afforestation of the moss will, however, continue to affect the peat detrimentally by desiccating and disturbing the upper surfaces and will damage a potentially important archaeological resource.

Site Name Stock Moss (15 m OD)
NGR SD 423 870
Parish CARTMEL FELL
Source Ordnance Survey 1851 6" first edition Lancashire Sheet 9
Decription
A small, isolated moss lies immediately to the north of High Moss, and is separated from it by a thin strip of dry ground. The whole moss extends to *c* 6 ha, most of which survives as an unenclosed centre of apparently deep peat, although there was some evidence to suggest that this may have been heavily truncated by peat-cutting; it did not appear to stand very high compared to the thin peats surrounding it. The peat was bounded by crags to the east and west, and alluvium around the River Winster to the north. The potential of this moss was difficult to establish by surface survey, although if the peats are untruncated, it would represent a useful palaeoecological resource and would provide an interesting contrast to the larger Nichols and Meathop Mosses to the east.

Site Name Ulpha Moss (5 m OD)
NGR SD 460 817
Parish MEATHOP AND ULPHA
Source Ordnance Survey 1868 6" first edition Westmorland Sheet 46
Decription
No field data.

Site Name Underbarrow Moss (5 m OD)
NGR SD 469 904
Parish UNDERBARROW AND BRADLEYFIELD
Source Ordnance Survey 1863 6" first edition Westmorland Sheet 38
Decription
This is the central moss in the 'spine' of peat that runs down the centre of the upper part of the Lyth Valley, with Blakebank Moss to the north and Helsington/Savinhill Moss to the south. It was similar in form to both of these, having a centre of deep, potentially undisturbed peat under woodland, surrounded by deep, truncated peats, thin peats, and organic soils in succession. In the past all the main mosses were undoubtedly contiguous and would have been indistinguishable on the ground. In the centre of the moss, the deepest peat was at least 5 m thick and stood proud of the surrounding peat deposits under pasture. The area of thick peat is probably the largest in the upper valley, at *c* 25 ha, and represents a significant resource. This area is confined to the west by the rising dry ground at Holmes Wood and it has been truncated to the east by peat-cutting evidenced by enclosed former peat dales. The surviving peats, in conjunction with the mosses to the north and south, offer a valuable opportunity to study the vegetation history of the upper valley and provide a useful contrast to the basin mires north of the valley head.

Site Name Witherslack Hall (40 m OD)
NGR SD 439 855
Parish WITHERSLACK

Source Ordnance Survey 1:25,000 sheet 627
Decription
This was referred to in Howard-Davis *et al* (1988).

Rusland Pool **(Fig 15)**

Site Name Back Moss (5 m OD)
NGR SD 311 796
Parish EGTON WITH NEWLAND
Source Ordnance Survey 1850 6" first edition Lancashire Sheet 10
Decription
This was probably part of Newland Moss, one of the coastal mosses north of Ulverston. It is now extensively drained.

Site Name Barker Moss (8 m OD)
NGR SD 345 785
Parish LOWER HOLKER
Source Ordnance Survey 1851 6" first edition Lancashire Sheet 17
Decription
This is no longer marked on maps, but it seems to have been a continuation of the coastal peats of Reake and White Mosses. It is now extensively drained, and down to pasture, with no remaining peat.

Site Name Border Moss (10 m OD)
NGR SD 345 871
Parish COLTON
Source Ordnance Survey 1851 6" first edition Lancashire Sheet 8
Decription
This is part of the Rusland Pool mosses, the area now being used for pasture or woodland, with very little peat remaining.

Site Name Burn Barrow Moss (4 m OD)
NGR SD 343 826
Parish HAVERTHWAITE
Source Ordnance Survey 1851 6" first edition Lancashire Sheet 12
Decription
This is a continuation to the east of Fish House Moss. Several areas remain uncut, and as a whole it supports a typical mire vegetation.

Site Name Crook Moss (50 m OD)
NGR SD 373 857
Parish COLTON
Source Ordnance Survey 1851 6" first edition Lancashire Sheet 8
Decription
No field data.

Site Name Deanholme Moss (10 m OD)
NGR SD 342 813
Parish LOW ALLITHWAITE
Source Ordnance Survey 1851 6" first edition Lancashire Sheet 12
Decription
This is part of the same coastal valley mire complex as Stribers, Fish House, and Deer Dike Mosses. It has been drained, but large areas of uncut peat remain under pasture.

Site Name Deer Dyke Moss (5 m OD)
NGR SD 338 821
Parish HAVERTHWAITE
Source Ordnance Survey 1851 6" first edition Lancashire Sheet 12
Decription
This is part of the same mire complex as Deanholme, Stribers and Fish House Mosses, and contains areas of uncut peat, although it has been largely drained.

Site Name Doup Moss (25 m OD)
NGR SD 333 889
Parish COLTON
Source Ordnance Survey 1851 6" first edition Lancashire Sheet 8
Decription
This moss, no longer named on current maps, formed the northern end of the valley mire of Rusland Pool. There are extensive peat remains and the area is now under woodland.

Site Name Ellerside Moss (7 m OD)
NGR SD 351 801
Parish LOW ALLITHWAITE
Source Ordnance Survey 1851 6" first edition Lancashire Sheet 12
Decription
This is a continuation to the south of the valley mire of Stribers Moss. Parts remain uncut, and support a typical mire vegetation.

Site Name Elmlath Moss (7 m OD)
NGR SD 345 858
Parish COLTON
Source Ordnance Survey 1851 6" first edition Lancashire Sheet 8
Decription
This forms the southern part of Rusland Pool, with Border Moss to the north. It is now drained, although shallow peats remain under woodland or pasture.

Site Name Fish House Moss (5 m OD)
NGR SD 338 827
Parish HAVERTHWAITE
Source Ordnance Survey 1851 6" first edition Lancashire Sheet 12
Decription
This is part of a large complex of coastal valley mires south of Rusland, which include Stribers and Deer Dike Mosses. It is drained, but there are still some areas that are uncut.

Site Name Haybridge Moss (7 m OD)
NGR SD 338 878

Parish COLTON
Source Ordnance Survey 1851 6" first edition Lancashire Sheet 8

Decription

This is the southern part of Rusland Pool, forming an estuarine raised mire. Little intact peat remains, and the moss has been colonised by birch and pine.

Site Name Hilleter Moss (8 m OD)
NGR SD 335 880
Parish COLTON
Source Ordnance Survey 1851 6" first edition Lancashire Sheet 8

Description

This forms another section of Rusland Pool.

Site Name Ireland Moss (8 m OD)
NGR SD 335 845
Parish HAVERTHWAITE
Source Ordnance Survey 1851 6" first edition Lancashire Sheet 12

Decription

This moss is situated on the Leven Estuary, to the south of Haybridge Moss. It is now colonised by pine, or is under more formal plantations, following drainage and, presumably, also extensive cutting. There are, however, some areas of deep peat remaining.

Site Name Mean House Moss (5 m OD)
NGR SD 324 824
Parish COLTON
Source Ordnance Survey 1851 6" first edition Lancashire Sheet 12

Decription

This is an estuarine coastal mire on the Leven, west of Roudsea Wood. Today, no peat remains and the area is under pasture.

Site Name Moss Wood (70 m OD)
NGR SD 324 863
Parish COLTON
Source Ordnance Survey 1:25,000 sheet 626

Decription

There is place-name evidence for a moss in this area, but no data exists to confirm this.

Site Name Newland Moss (5 m OD)
NGR SD 305 795
Parish EGTON WITH NEWLAND
Source Ordnance Survey 1850 6" first edition Lancashire Sheet 10

Decription

This is an extensively drained coastal moss.

Site Name Oxenpark Moss (7 m OD)
NGR SD 334 885
Parish COLTON
Source Ordnance Survey 1851 6" first edition Lancashire Sheet 8

Decription

This is part of Rusland Pool.

Site Name Ready Mire (5 m OD)
NGR SD 323 846
Parish COLTON
Source Ordnance Survey 1851 6" first edition Lancashire Sheet 12

Decription

This coastal mire forms the northern part of Roam Moss. Peat of more than 1 m depth remains under woodland.

Site Name Reake Moss (10 m OD)
NGR SD 345 794
Parish LOW ALLITHWAITE
Source Ordnance Survey 1851 6" first edition Lancashire Sheet 12

Decription

This forms the southern section of the large coastal estuarine moss which includes White, Stribers, Fish House and Deer Dike Mosses. It is now drained and cut, although some peat deeper than 1 m still exists.

Site Name Roam Moss (5 m OD)
NGR SD 325 840
Parish COLTON
Source Ordnance Survey 1851 6" first edition Lancashire Sheet 12

Decription

This is a coastal estuarine moss of a type characteristic of Morecambe Bay. It has been drained and extensively cut, with very little peat of more than 1 m remaining. It is now under pasture.

Site Name Rough Moss (6 m OD)
NGR SD 339 855
Parish COLTON
Source Ordnance Survey 1851 6" first edition Lancashire Sheet 8

Decription

This is a small remnant of moss within Rusland Pool. Shallow peat is recorded beneath pasture.

Site Name Rusland Moss (6 m OD)
NGR SD 336 885
Parish COLTON
Source Ordnance Survey 1851 6" first edition Lancashire Sheet 8

Decription

This forms the major component of Rusland Pool, largely under pasture or woodland, with little peat now remaining.

Site Name Sandy Moss (10 m OD)
NGR SD 310 802
Parish EGTON WITH NEWLAND
Source Ordnance Survey 1850 6" first edition Lancashire Sheet 11

Decription

No field data.

Site Name Stribers Moss (7 m OD)
NGR SD 347 816
Parish LOW ALLITHWAITE
Source Ordnance Survey 1851 6" first edition Lancashire Sheet 12

Decription

This forms the major component of a large coastal mire complex including Fish House, White, and Deer Dike Mosses. The moss has been drained and probably cut, although it was possible to walk only the boundaries of the moss.

Site Name Thwaite Moss (55 m OD)
NGR SD 340 899
Parish COLTON
Source Ordnance Survey 1851 6" first edition Lancashire Sheet 8; Ordnance Survey 1:25,000 sheet 626

Decription

No field data.

Site Name Waitham Moss (5 m OD)
NGR SD 353 781
Parish LOWER HOLKER
Source Ordnance Survey 1850 6" first edition Lancashire Sheet 16

Decription

This was formerly the southern part of the Ellerside Moss complex, now under pasture. Peat is no longer recorded.

Site Name Waste Mires (20 m OD)
NGR SD 335 859
Parish COLTON
Source Ordnance Survey 1851 6" first edition Lancashire Sheet 8

Decription

This place-name in the Rusland Valley suggests that a moss formerly existed.

Site Name White Moss (6 m OD)
NGR SD 346 802
Parish LOW ALLITHWAITE
Source Ordnance Survey 1851 6" first edition Lancashire Sheet 12

Decription

This formed a component of the large coastal mire complex centring on Stribers Moss. Some deep peat remains, possibly even including some top moss, but there is also evidence of cutting and drainage.

Site Name White Stock Moss (10 m OD)
NGR SD 332 889
Parish COLTON
Source Ordnance Survey 1851 6" first edition Lancashire Sheet 8

Decription

This is probably the old name for the north–western tip of Rusland Moss. The area is now under pasture, with peats less than 1 m deep remaining.

Duddon Mosses (including Low Furness, **Fig 16***)*

Site Name Angerton Moss (8 m OD)
NGR SD 217 849
Parish ANGERTON
Source Ordnance Survey 1850 6" first edition Lancashire Sheet 10

Decription

This moss, together with Herd House Moss and Waitham Common, forms a large area of coastal moss north of Kirkby-in-Furness. A small area of peat with a depth of greater than 1 m remains under pasture, although most is now degraded.

Site Name Bank End Moss (10 m OD)
NGR SD 225 852
Parish ANGERTON
Source Ordnance Survey 1850 6" first edition Lancashire Sheet 10

Decription

This joins the south–eastern corner of White Moss. It shows clear evidence of drainage, as well as containing some baulks from former peat-cutting, under scrub. Quite large areas of peat greater than 1 m in depth survive.

Site Name Black Moss (7 m OD)
NGR SD 228 883
Parish ANGERTON
Source Ordnance Survey 1851 6" first edition Lancashire Sheet 7

Decription

This forms the remnants of a valley mire north–east of Broughton-in-Furness. It has mainly been cut, although some areas of peat with a depth greater than 1 m remain, under scrub or woodland. To the south, at Middlescough, the land was inaccessible as the vegetation was impenetrable.

Site Name Far Side Moss (20 m OD)
NGR SD 240 880
Parish KIRKBY IRELETH
Source Ordnance Survey 1851 6" first edition Lancashire Sheet 7

Decription

This was originally part of Heathwaite Moss, but has been completely destroyed, with no peat remaining.

Site Name Fox Moss (10 m OD)
NGR SD 215 853
Parish ANGERTON
Source Broughton Enclosure Award 1847 (LRO/AE)

Decription

This moss was at the southern end of White Moss, and has now been extensively drained, with peat baulks and old field boundaries surviving. Some peat with a depth greater than 1 m also exists, under pasture.

Site Name Heathwaite Moss (10 m OD)
NGR SD 240 875
Parish KIRKBY IRELETH
Source Ordnance Survey 1851 6" first edition Lancashire Sheet 7
Decription
This moss forms a discrete area just to the north of White Moss. It has clearly been cut, although peat survives to a depth of greater than 1 m under scrub, with occasional pockets of trees. Only the marginal fields were examined.

Site Name Herd House Moss (5 m OD)
NGR SD 220 842
Parish ANGERTON
Source Ordnance Survey 1850 6" first edition Lancashire Sheet 10
Decription
This forms an extensive area of coastal moss north of Kirkby-in-Furness and south of Angerton Moss. It has been drained, although some areas of peat greater than 1 m in depth remain under scrub or pasture.

Site Name Leece Mosses (10 m OD)
NGR SD 238 689
Parish ALDINGHAM
Source Ordnance Survey 1:25,000 sheet 647
Decription
These form part of a former valley mire, which includes Roosecote and Stank Mosses. Only organic soils under pasture remain.

Site Name Low Moss (10 m OD)
NGR SD 207 884
Parish BROUGHTON WEST
Source Ordnance Survey 1850 6" first edition Lancashire Sheet 6
Decription
This is a valley mire beside the River Lickle, north–west of Broughton-in-Furness. A small remnant of peat greater than 1 m in depth survives but otherwise only mineral soils are present either under pasture or scrub. The place-name, however, suggests that there was once a more extensive moss.

Site Name Mere Tarn (20 m OD)
NGR SD 266 718
Parish ALDINGHAM
Source Ordnance Survey 1:25,000 sheet 635
Decription
A small tarn measures approximately 0.2 km^2, and is surrounded by marshland that contains no artificial drainage.

Site Name Park Moss (170 m OD)
NGR SD 185 909
Parish ULPHA
Source Ordnance Survey 1:25,000 Outdoor Leisure Sheet 6
Decription
A possible raised bog was located directly to the west of the upper reaches of the Duddon Valley, at a height of between 130 m and 170 m. Fieldwork showed this to be open heathland with no evident walls or drains.

Site Name Roosebeck Moss (30 m OD)
NGR SD 257 687
Parish ALDINGHAM
Source Ordnance Survey 1:25,000 sheet 647
Decription
This former coastal raised mire has only organic soils remaining under pasture.

Site Name Roosecote Moss (30 m OD)
NGR SD 236 686
Parish ALDINGHAM
Source Ordnance Survey 1:25,000 sheet 647
Decription
This is part of a former valley mire which includes Leece and Stank Mosses. Only organic soils now under pasture remain.

Site Name Rosthwaite Moss(6 m OD)
NGR SD 233 892
Parish ANGERTON
Source Ordnance Survey 1851 6" first edition Lancashire Sheet 7
Decription
This is no longer marked on maps, but was probably part of a valley mire alongside Kirkby Pool; no peat survives.

Site Name Row Ridding Latter Rigg Moss (8m OD)
NGR SD 235 886
Parish KIRKBY IRELETH
Source Ordnance Survey 1851 6" first edition Lancashire Sheet 7
Decription
This forms a valley mire alongside Kirkby Pool.

Site Name Stank Moss (10 m OD)
NGR SD 237 688
Parish ALDINGHAM
Source Ordnance Survey 1:25,000 sheet 647
Decription
This is part of a former valley mire, which includes Roosecote and Leece Mosses. Only organic soils under pasture remain.

Site Name Star Mire (10 m OD)
NGR SD 235 868
Parish KIRKBY IRELETH
Source Ordnance Survey 1851 6" first edition Lancashire Sheet 7
Decription
This is no longer in existence, but was probably originally a valley mire alongside Kirkby Pool, close to White and Heaththwaite Mosses. No peat survives.

Site Name Thornthwaite Latter Rigg Moss (7 m OD)
NGR SD 235 882
Parish KIRKBY IRELETH
Source Ordnance Survey 1851 6" first edition Lancashire Sheet 7
Decription
This is a valley mire alongside Kirkby Pool.

Site Name Tuto Moss (70 m OD)
NGR SD 283 869
Parish SUBBERTHWAITE
Source Ordnance Survey 1:25,000 sheet 626
Decription
No field data.

Site Name Ulverston Moss (10 m OD)
NGR SD 305 784
Parish ULVERSTON
Source Ordnance Survey 1850 6" first edition Lancashire Sheet 16
Decription
No field data.

Site Name Urswick Tarn
NGR SD 270 745
Parish URSWICK
Source Ordnance Survey 1:25,000 sheet 635
Decription
This was referred to in Howard-Davis *et al* (1988).

Site Name Waitham Common (8 m OD)
NGR SD 222 848
Parish ANGERTON
Source Ordnance Survey 1850 6" first edition Lancashire Sheet 10
Decription
This forms a continuation to the coastal mosses of Angerton and Herd House. The moss has been drained, but some areas of peat greater than 1 m deep remain, covered by scrub.

Site Name White Moss (10m OD)
NGR SD 228 868
Parish ANGERTON
Source Ordnance Survey 1851 6" first edition Lancashire Sheet 7
Decription
This is the northern portion of White Moss, which is covered by vegetation typical of bog communities, with some wet heath, scrub and mixed woodland. It has been extensively drained but only marginal peat-cutting has been undertaken. This was extensively sampled by Wimble (1986).

Site Name White Moss (10 m OD)
NGR SD 217 855
Parish ANGERTON
Source Ordnance Survey 1850 6" first edition Lancashire Sheet 10
Decription
This forms the southern portion of White Moss, which has been extensively drained. The central part is covered with scrub and pasture, whilst gardens now cover the edge. This was extensively sampled by Wimble (1986).

Site Name Wreaks Moss (10 m OD)
NGR SD 226 862
Parish ANGERTON
Source Ordnance Survey 1851 6" first edition Lancashire Sheet 7
Decription
This forms a discrete area in the central part of White Moss. It has been extensively drained and there is now a causeway through it, but nevertheless peat survives to a depth greater than 1 m. It is covered by scrub and woodland, with some pasture to the north–west.

The West Cumbrian Coastal Plain: Chapter 3
***The Esk Estuary* (Figs 23–26)**

Site Name Black Moss (90 m OD)
NGR NY 029 108
Parish ST JOHN BECKERMET
Source Egremont Enclosure Award 1783 CRO(C)/QRE/1/30, Ordnance Survey 1865 first edition 72
Decription
A small raised bog occupies a natural depression in the boulder clay. Situated 2 km to the east of Egremont, Black Moss is the most westerly example of a lowland raised bog in England, and the only remaining habitat of its kind in the locality. The bog remains relatively intact and unmodified, particularly when compared with the surrounding farmland, which is dissected by numerous field walls and drains. The low domed structure of the peat body is clearly visible, and has the potential for a considerable depth of peat. Where peripheral watercourses do not bound the margins of the bog, there is a natural transition from acid mire to the vegetation of the surrounding mineral soils. Typical transitional communities at Black Moss include marginal fen and birch carr. Across the main body of the site, a moderately wet bog community exists which is dominated by cross-leaved heath and ling heather. The western portion of the bog is generally the wettest area, and *Sphagnum* lawns with common cottongrass, bog asphodel, and hair-moss dominate the flat, quaking surface. Towards the south, the open fen becomes patchy but maintains a good *Sphagnum* cover. The potential is good for the preservation of both archaeological and palaeo-ecological material.

Site Name Bogholes (5 m OD)
NGR NY 010 047
Parish ST BRIDGET BECKERMET
Source Ordnance Survey 1:25,000 sheet 593
Decription
No field data.

Site Name Brownbank Moss (34 m OD)
NGR NY 053 024
Parish GOSFORTH
Source Ordnance Survey 1863 first edition 78
Decription
A small moss, 2 km south–west of Gosforth on the coastal plain, is situated just below the 35 m contour. The northern edge is truncated by the B5344, the construction of which undoubtedly had some negative impact on the site. Part is currently under deciduous woodland.

Site Name Church Moss (10 m OD)
NGR NY 016 057
Parish ST BRIDGET BECKERMET
Source Ordnance Survey 1:25,000 sheet 593
Decription
A low-lying area, measuring approximately 600 sq m, of open water and marshland occupies a drumlin hollow. The western edge of the site is bounded by a disused railway embankment, which partially dams the moss, and the southern side is covered in scrub. A small amount of peat was recorded as remaining, although it is unlikely that the moss has a history of being either extensively cut or drained.

Site Name Distington Moss (160 m OD)
NGR NY 007 208
Parish MORESBY
Source Ordnance Survey 1:25,000 sheet 593
Decription
This is situated on the north–western edge of, and contiguous with, Moresby Moss and represents the most extensive peat in the area. The moss is surrounded by walled farmland, and is evidently undergoing drainage. Vegetation includes royal fern and bog rosemary, the latter at its only site in West Cumbria. Coniferous trees form the vegetative cover of the northern edge of the moss, with scrub dominating the eastern edge.

Site Name Eskmeals Pool (10 m OD)
NGR SD 099 893
Parish BOOTLE
Source Ordnance Survey 1:25,000 sheet 625
Decription
This comprises a low-lying area of coastal heathland with improved and re-seeded grassland and pasture, although the margins have not been ploughed. Mesolithic deposits and artefacts have been identified in the near vicinity (Bonsall *et al* 1991).

Site Name Hallsenna Moor (30 m OD)
NGR NY 065 005
Parish SEASCALE
Source Ordnance Survey 1:25,000 Outdoor Leisure Sheet 6
Decription
Situated on the coastal plain approximately 1.5 km north of Drigg, and 3 km south–east of Seascale, this small basin mire is one of the few lowland heath and peatland complexes remaining in the country, and is the largest in West Cumbria. It contains a wide range of habitats, which form a mosaic including wet and dry heath, nutrient-poor fen, basin mire and woodland. The dry heath is characterised by abundant heather (*Calluna vulgaris*) and a variety of associated species. The margins of the mire are dominated by birch and willow scrub, which give way to improved grassland. Farm walls on Hallsenna Moor are conspicuous by their absence, although the moss has evidently been cut over and partially drained with at least one extant drain. The site is designated as a SSSI and a National Nature Reserve.

Site Name Lantern Moss Tarn (20 m OD)
NGR NY 006 055
Parish ST BRIDGET BECKERMET
Source Ordnance Survey 1:25,000 sheet 593
Decription
A water-filled tarn, adjacent to a caravan park, is partially filled and overgrown with rushes and other aquatic plants. The fields in the vicinity were not systematically walked due to the pastoral cover.

Site Name Moresby Moss (150 m OD)
NGR NY 010 202
Parish MORESBY
Source Ordnance Survey 1:25,000 sheet 593
Decription
This is a former large basin mire, contiguous with Distington Moss to the south–east. The moss comprised an extensive peat deposit, which is being walled and drained, and extensive damage has been inflicted by opencast coal mining. The vegetation is predominantly heathland and scrub, with agricultural activity being pastoral, mostly rough grazing.

Site Name Moss Plantation (10 m OD)
NGR NY 119 978
Parish MUNCASTER
Source Ordnance Survey 1:25,000 Outdoor Leisure Sheet 6
Decription
A moss, totalling 0.2 km^2, is currently planted with coniferous woodland within enclosed farmland.

Site Name Nor Moss (40 m OD)
NGR SD 109 928
Parish WABERTHWAITE
Source Ordnance Survey 1:25,000 Outdoor Leisure Sheet 6
Decription
No field data.

Site Name Parkhouse Moss (10 m OD)
NGR NY 128 984
Parish MUNCASTER
Source Ordnance Survey 1:25,000 Outdoor Leisure Sheet 6

Decription
No field data.

Site Name Shoemire Wood (30 m OD)
NGR NY 102 926
Parish WABERTHWAITE
Source Ordnance Survey 1:25,000 Outdoor Leisure Sheet 6
Decription
No field data.

Site Name Silver Tarn (10 m OD)
NGR NX 998 069
Parish LOWSIDE QUARTER
Source Ordnance Survey 1:25,000 sheet 593
Decription
This sinuous wetland site lies within 0.5 km of the coastline, midway between the villages of Nethertown and Braystones, within the drumlin fields of West Cumbria. It comprises a suite of three separate but related features originating as post-glacial hollows in boulder clay and later forming kettlehole tarns, one of only two examples in the country of intact formations of this type. The site exhibits typical stages in the development of kettlehole vegetation from open water, represented by Harnsey Moss, through the acid poor-fen of Silver Tarn, to a transitional basin fen stage reflected in the Hollas Moss communities. Silver Tarn is the largest mire of this kettlehole complex, lying across two hollows forming linked units referred to as Silver Tarn east and west. The permanently high water-table is maintained by ground water springs, surface seepage, and rainfall, with a movement of water from east to west and beyond. The eastern moss is particularly wet, comprising only a thin mat of vegetation overlying open water or semi-liquid peat. A mosaic of poor-fen communities is supported, with *Sphagnum* spp dominating, forming mossy lawns across the central areas. Whilst remaining as a distinct pocket of peat, there is evidence of some extraction and drainage.

Site Name Snelling's Mire (30 m OD)
NGR NX 999 082
Parish ST JOHN'S BECKERMET
Source Ordnance Survey 1:25,000 sheet 593
Decription
No field data.

Other Coastal Mosses

Site Name Baldmire (20 m OD)
NGR SD 136 819
Parish WHICHAM
Source Ordnance Survey 1:25,000 sheet 625
Decription
A linear valley mire comprises unwalled marshland and heath with springs, in an area totalling 1.2 km^2. The surrounding area has yielded a wealth of archaeological material and has excellent palaeoenvironmental potential. The pasture is dominated by *Juncus conglomeratus*.

Site Name Barfield Tarn (30 m OD)
NGR SD 106 869
Parish WHICHAM
Source Ordnance Survey 1:25,000 sheet
Decription
A lowland tarn, 0.2 km in extent, is surrounded by reedswamp and marshland. It was referred to in Howard-Davis *et al* (1988).

Site Name Bigert Mire (200 m OD)
NGR NY 178 927
Parish MUNCASTER
Source Ordnance Survey 1:25,000 Outdoor Leisure Sheet 6
Decription
No field data.

Site Name Bleamoss Pool (40 m OD)
NGR SD 839 169
Parish MILLOM WITHOUT
Source Ordnance Survey 1:25,000 sheet 625
Decription
No field data.

Site Name Bog Hole (40m OD)
NGR SD 098 848
Parish WHICHAM
Source Ordnance Survey 1:25,000 sheet 625
Decription
A coastal area of 0.4 km^2 comprises enclosed farmland. Although there are no details pertaining to the site, the name does imply wetland conditions.

Site Name Ehenside Tarn (20 m OD)
NGR NY 002 071
Parish LOWSIDE QUARTER
Source Ordnance Survey 1:25,000 sheet 593
Decription
This reduced but still water-filled tarn, also known as Gibb Tarn, covers an area measuring approximately 100 m by 20 m. Following intensive drainage during the 1860/70s, during which significant archaeological material was excavated (Darbishire 1873), the former moss now comprises two water-filled depressions colonised by reeds. These form part of the group of mires in the drumlin fields of Beckermet. The fields immediately to the north of Ehenside Tarn have been ploughed although organic soils remain along the western edge. It was referred to in Howard-Davis *et al* (1988).

Site Name Kirksanton Moss (10 m OD)
NGR SD 134 806
Parish MILLOM WITHOUT
Source Ordnance Survey 1:25,000 sheet 625

Decription
This typical northern moss comprises walled farmland with marsh and heathland. The dominant vegetative species include invading willow and an area of pure alder in the centre of the moss, with mainly tall marsh species, such as meadowsweet, common valerian, yellow iris and marsh woundwort, and water forget-me-not and celery-leaved buttercup occurring in the drainage ditches. The geographical extent is approximately 1 km^2. At the northern end are standing stones (Pl 49), and stone circles are sited 1 km to the north–east.

Site Name The Mosses (18 m OD)
NGR SD 112 842
Parish WHICHAM
Source Ordnance Survey 1:25,000 sheet 625
Decription
This area is enclosed and drained, with heath and woodland growing. It is situated to the west of a railway line. The site covers an area of approximately 0.25 km^2.

Site Name Toppingmoss Plantation (30 m OD)
NGR SD 147 833
Parish WHICHAM
Source Ordnance Survey 1:25,000 sheet 625
Decription
No field data.

The North Cumbrian Plain: Chapter 4
***The Abbeytown Ridge* (Figs 33–6)**

Site Name Bellwater Moss (10 m OD)
NGR NY 156 478
Parish HOLME ABBEY
Source Ordnance Survey 1865 first edition 22
Decription
This area of intensively drained mossland with heath vegetation and extensive peat forms part of a former floodplain mire. The sandy soils of the adjacent moraine deposits situated on the edge of the floodplain are currently under pasture and offer the potential for archaeological remains.

Site Name Bowness Common (15 m OD)
NGR NY 205 602
Parish BOWNESS
Source Ordnance Survey 1:25,000 sheet 544
Decription
This large raised mire along the coastal plain to the south of the Solway Firth extends southwards to within 1 km of the River Wampool. The exact limits of the peat are undetermined since it is contiguous with Glasson Moss and Whitrigg Common to the east. Small pockets of woodland exist, both in the central area and around the fringes of the moss, with both coniferous and deciduous species. It has been little damaged by peat-cutting and contains cereal pollen from the Early Bronze Age (Walker 1966). A wealth of archaeological material is known from the vicinity.

Site Name Bromfield Mire
NGR NY 165 477
Parish BROMFIELD
Source Ordnance Survey 1865 first edition 22
Decription
No field data.

Site Name Chapel Moss (10 m OD)
NGR NY 115 452
Parish HOLME ST CUTHBERT
Source Ordnance Survey 1:25,000 sheet 566
Decription
This is an area of intensively drained moss, with heath and mossland on peat, covering a total area of approximately 0.7 km^2.

Site Name Cockley Moss (14 m OD)
NGR NY 124 457
Parish HOLME ST CUTHBERT
Source Ordnance Survey 1:25,000 sheet 566
Decription
This is an area of intensively drained moss, with heathland and some scrub or peat, covering a total area of approximately 0.25 km^2.

Site Name Colmire (10 m OD)
NGR NY 230 505
Parish WAVERTON
Source Ordnance Survey 1865 first edition 23
Decription
No field data.

Site Name Common Moss (10 m OD)
NGR NY 150 472
Parish HOLME CULTRAM
Source Ordnance Survey 1:25,000 sheet 566
Decription
The moss comprises an area of enclosed, drained farmland containing heathland and peat deposits, covering a total area of approximately 0.5 km^2.

Site Name Cowfold Moss (15 m OD)
NGR NY 143 470
Parish HOLME ABBEY
Source Ordnance Survey 1:25,000 sheet 566
Decription
The moss comprises an area of enclosed, drained farmland containing heathland and peat deposits, covering a total area of approximately 0.3 km^2.

Site Name Cowper Bog (15 m OD)
NGR NY 124 445
Parish HOLME ST CUTHBERT
Source Ordnance Survey 1:25,000 sheet 566

Decription
The moss comprises an area of enclosed, drained bog with heath and scrub vegetation and some deposits of peat, covering a total area of approximately 0.25 km^2.

Site Name Drumburgh Moss (10 m OD)
NGR NY 255 585
Parish DRUMBURGH
Source Ordnance Survey 1:25,000 sheet 557
Decription
This is a nature reserve managed by the Cumbria Trust for Nature Conservation, incorporating approximately 1.3 km^2. It comprises an area of intact, open moss, which is completely vegetated. There has been some peat-cutting around the edges, particularly in the northern portion.

Site Name Fingland Moss (10 m OD)
NGR NY 255 578
Parish BOWNESS
Source Ordnance Survey 1:25,000 sheet 557
Decription
The site comprises an enclosed, drained peat moss with heath vegetation and a little mixed woodland, covering a total area of approximately 0.5 km^2.

Site Name Glasson Moss (15 m OD)
NGR NY 236 605
Parish BOWNESS
Source Ordnance Survey 1:25,000 sheet 544
Decription
This is a large raised mire, contiguous with Bowness Common to the west and Whitrigg Common to the south, covering an area of approximately 2 km^2. The large and ombrotrophic nature of the moss permitted a regional pollen record to be established (Jacobson and Bradshaw 1981). The moss has a documented history of being cut for peat extraction, and this is particularly prevalent on the margins, associated with drains and walling. Fires have also caused extensive damage, firstly in 1968 and latterly in 1976, when the moss was virtually destroyed. *See Ch 4* for further details.

Site Name Hangingshaw Moss (20 m OD)
NGR NY 113 463
Parish HOLME ST CUTHBERT
Source Ordnance Survey 1865 first edition 27
Decription
This is a small raised mire on the Solway Plain, encompassing an area of approximately 0.25 km^2. The vegetative cover comprises drained heath scrub and marshland, dissected by farm walls.

Site Name Kelsick Mire (10–15 m OD)
NGR NY 195 495
Parish DUNDRAW
Source Ordnance Survey 1865 first edition 72
Decription
No field data.

Site Name Kelsick Moss (10 m OD)
NGR NY 180 492
Parish DUNDRAW
SourceOrdnance Survey 1865 first edition 72
Decription
No field data.

Site Name Little Bampton Moss (10 m OD)
NGR NY 274 564
Parish KIRKBAMPTON
Source Ordnance Survey 1:25,000 sheet 557
Decription
This is a small area of enclosed, drained peat with heathland and mixed woodland vegetation. To the south, pastoral agriculture dominates the landscape, with some peripheral rough grazing. The moss covers an area of approximately 0.15 km^2.

Site Name Orton Moss (50 m OD)
NGR NY 340 545
Parish ORTON
Source Ordnance Survey 1:25,000 sheet 557
Decription
A small basin mire, measuring approximately 0.5 km^2, is located within a designated SSSI. The area is completely wooded with both deciduous and coniferous species, hampering identification of the precise edges of the moss. The site is fairly dry and no peat is exposed, thus it is thought to contain only a small deposit of peat. The area is largely walled and contains some artificial drainage.

Site Name Oulton Moss (15–20 m OD)
NGR NY 253 512
Parish WOODSIDE
Source Ordnance Survey 1:25,000 sheet 557
Decription
An enclosed and drained peat moss with heath and mixed woodland vegetation. The current land-use includes pastoral agriculture and wet meadow. It covers an area of approximately 0.25 km^2 and was referred to in Howard-Davis *et al* (1988).

Site Name Salta Moss (15–20 m OD)
NGR NY 083 451
Parish HOLME ST CUTHBERT
Source Ordnance Survey 1865 first edition 22
Decription
A small raised mire, located to the north of Allonby, forms part of the Solway Plain mosses. The surface cover comprises drained peat with marshland, ponds, and heath vegetation.

Site Name Tarns (15 m OD)
NGR NY 115 474
Parish HOLME ST CUTHBERT
Source Ordnance Survey 1865 first edition 27
Decription
This area of enclosed, drained farmland contains scrub, heath, marshland and mires. The land, which totals 0.25

km^2, has been improved and is currently entirely dedicated to pasture.

Site Name Tewfit Moss (20 m OD)
NGR NY 193 467
Parish BROMFIELD
Source Ordnance Survey 1865 first edition 23
Decription
This measures approximately 0.5 km^2 and comprises enclosed and drained farmland with some heath vegetation.

Site Name Thurston Lough (20 m OD)
NGR NY 320 564
Parish ORTON
Source Ordnance Survey 1:25,000 sheet 557
Decription
No field data.

Site Name Wedholme Flow (10–15 m OD)
NGR NY 220 530
Parish WOODSIDE; KIRKBRIDE AND HOLME EAST WAVER
Source Ordnance Survey 1865 first edition 22
Decription
This former raised mire is located on the Solway Plain, covering an area of *c* 615 ha between the Rivers Waver and Wampool. The moss is drained by several small streams and overlies both boulder clay and, in the river floodplains, marine alluvium. The modern limits are defined by agricultural land, some of which comprises reclaimed organic soils, and consequently the margins have been intensively drained. Whilst there is a history of peat-cutting, the eastern and central parts continue to support an intact peatland vegetation with no evidence of peat removal or truncation. In these areas, peat survives to a depth of 7 m, whilst in other parts the depth of survival is somewhat less. The peat can roughly be divided into two units: up to 4 m of *Sphagnum* peat overlie thinner sedge and woody peats. The uncut, deep centre is the most valuable element of the moss, and is designated as a SSSI. The surface presents a mosaic of vegetation and levels which bear witness to the long history of peat-cutting on the site. The fringes have all been truncated by hand cutting for domestic use, with commercial extraction particularly prevalent on the eastern side. Peat harvesting continues in this area, although some parts are now being actively regenerated by English Nature.

Site Name Whitrigg Common (15 m OD)
NGR NY 237 601
Parish BOWNESS
Source Ordnance Survey 1:25,000 sheet 544
Decription
This area of raised mire is contiguous with Bowness Common to the west and Glasson Moss to the north, forming part of the Solway Plain mosses. The eastern edge is bounded by a small copse of deciduous trees, separating it from Bowness Common. The Cumberland Moss Litter Company cut it for peat extraction until 1948. Currently it is completely overgrown with *Calluna.*

Site Name Whitrigglees Moss (10 m OD)
NGR NY 248 581
Parish BOWNESS
Source Ordnance Survey 1:25,000 sheet 557
Decription
No field data.

Upland/Lowland Fringe of North Cumbrian Plain

Site Name Archy Moss (200 m OD)
NGR NY 037 137
Parish CLEATOR MOOR
Source Ordnance Survey 1:25,000 Outdoor Leisure Sheet 4
Decription
No field data.

Site Name Crosshill Moss (150 m OD)
NGR NY 187 346
Parish BLINDCRAKE
Source Ordnance Survey 1:25,000 sheet 575
Decription
No field data.

Site Name Dubbs Moss (90 m OD)
NGR NY 103 293
Parish GREYSOUTHERN
Source Ordnance Survey 1:25,000 Outdoor Leisure Sheet 4
Decription
This formed an NWWS palaeoecological sampling site. *See Appendix 5.*

Site Name Leady Moss (150 m OD)
NGR NY 103 237
Parish BLINDBOTHEL
Source Ordnance Survey 1:25,000 Outdoor Leisure Sheet 4
Decription
No field data.

Site Name Little Tarn (190 m OD)
NGR NY 2495 3385
Parish IREBY AND ULDALE
Source Ordnance Survey 1:25,000 Sheet 576
Decription
This formed an NWWS palaeoecological sampling site. *See Appendix 5.*

Site Name Mockerkin Tarn (120 m OD)
NGR NY 083 232
Parish DEAN
Source Ordnance Survey 1:25,000 Outdoor Leisure sheet 4
Decription
This was referred to in Howard-Davis *et al* (1988).

Site Name North Mosses (120 m OD)
NGR NY 061 212
Parish DISTINGTON
Source Ordnance Survey 1:25,000 Outdoor Leisure Sheet 4
Decription
No field data.

Site Name Scawthwaite Mire (200 m OD)
NGR NY 247 363
Parish IREBY AND ULDALE
Source Ordnance Survey 1:25,000 sheet 597
Decription
No field data.

Site Name South Mosses (120 m OD)
NGR NY 061 210
Parish DISTINGTON
Source Ordnance Survey 1:25,000 Outdoor Leisure Sheet 4
Decription
No field data.

East of Eden

Site Name Ash Moss (186 m OD)
NGR NY 598 699
Parish KINGWATER
Source Ordnance Survey 1:25,000 sheet 545
Decription
No field data.

Site Name Ash Moss Plantation (200 m OD)
NGR NY 585 702
Parish KINGWATER
Source Ordnance Survey 1:25,000 sheet 532
Decription
No field data.

Site Name Black Moss (120 m OD)
NGR NY 483 478
Parish WETHERAL
Source Ordnance Survey 1:25,000 sheet 568
Decription
No field data.

Site Name Black Moss Plantation (25 m OD)
NGR NY 423 601
Parish STANWIX RURAL
Source Ordnance Survey 1:25,000 sheet 545
Decription
No field data.

Site Name Bogside (125 m OD)
NGR NY 553 743
Parish ASKERTON
Source Ordnance Survey 1:25,000 sheet 532
Decription
No field data.

Site Name Bolton Fell Moss (95–110 m OD)
NGR NY 490 690
Parish STAPLETON
Source Ordnance Survey 1865 first edition 11
Decription
This is a large raised mire north of Walton Moss, which has been cut commercially, although areas of intact peat still remain. NWWS monitored the drainage ditches, but little was revealed. *See Chapter 4* and Barber (1981) for further details.

Site Name Breaks Moss (80–90 m OD)
NGR NY 502 658
Parish WALTON
Source Ordnance Survey 1865 first edition 11
Decription
This is a part of Walton Moss, which remains intact in some areas. NWWS monitored the edge of the bog.

Site Name Broomhill Moss (95–100 m OD)
NGR NY 495 665
Parish IRTHINGTON
Source Ordnance Survey 1865 first edition 11
Decription
This forms part of Walton and Breaks Mosses.

Site Name Dhu Bog (150 m OD)
NGR NY 535 756
Parish BEWCASTLE
Source Ordnance Survey 1:25,000 sheet 532
Decription
No field data.

Site Name Haithwaite Moss (100 m OD)
NGR NY 444 764
Parish NICHOLFOREST
Source Ordnance Survey 1:25,000 sheet 532
Decription
No field data.

Site Name Jock's Moss (90 m OD)
NGR NY 444 760
Parish NICHOLFOREST
Source Ordnance Survey 1:25,000 sheet 532
Decription
No field data.

Site Name Little Moss (85–90 m OD)
NGR NY 512 661
Parish WALTON
Source Ordnance Survey 1865 first edition 11
Decription
No field data.

Site Name Longrigg Moss (110 m OD)
NGR NY 507 718
Parish STAPLETON
Source Ordnance Survey 1:25,000 sheet 532

Decription
No field data.

Site Name Malls Flow (160 m OD)
NGR NY478 742
Parish SOLPORT
Source Ordnance Survey 1:25,000 sheet 532
Decription
No field data.

Site Name Midgeholme Moss (140 m OD)
NGR NY 610 665
Parish WATERHEAD
Source Ordnance Survey 1:25,000 sheet 546
Decription
No NWWS field data, but *see* Wilmott 1997 for further comment.

Site Name Mossend (155 m OD)
NGR NY 520 758
Parish BEWCASTLE
Source Ordnance Survey 1:25,000 sheet 532
Decription
No field data.

Site Name Mossfoot (130 m OD)
NGR NY 533 726
Parish ASKERTON
Source Ordnance Survey 1:25,000 sheet 532
Decription
No field data.

Site Name Parkriggmoss Wood (115 m OD)
NGR NY 460 724
Parish SOLPORT
Source Ordnance Survey 1:25,000 sheet 532
Decription
No field data.

Site Name Peppermoss (200 m OD)
NGR NY 471 778
Parish NICHOLFOREST
Source Ordnance Survey 1:25,000 sheet 532
Decription
No field data.

Site Name Rockcliffe Moss (12 m OD)
NGR NY 373 626
Parish WESTLINTON
Source Ordnance Survey 1:25,000 sheet 544
Decription
No field data.

Site Name Rosetrees Moss (8 m OD)
NGR NY 345 662
Parish KIRKANDREWS
Source Ordnance Survey 1:25,000 sheet 544
Decription
No field data.

Site Name Scaleby Moss (36 m OD)
NGR NY 430 635
Parish SCALEBY
Source Ordnance Survey 1:25,000 sheet 545
Decription
This is a former raised mire, which has produced a bog body (Turner (1988) and was the site of palaeoecological work by Walker (1966).

Site Name Solway Moss (20 m OD)
NGR NY 349 690
Parish KIRKANDREWS
Source Ordnance Survey 1:25,000 sheet 544
Decription
This large raised mire is being cut commercially but deep deposits of peat still remain, supporting a typical mire vegetation. NWWS monitored drainage ditches, and it also produced the remains of cattle and a sheep (*Ch 4*). Its potential for both archaeological and palaeoecological data remains great.

Site Name The Boggs (13 m OD)
NGR NY 390 635
Parish WESTLINTON
Source Ordnance Survey 1:25,000 sheet 544
Decription
No field data.

Site Name The Flosh (95 m OD)
NGR NY 502 674
Parish WALTON
Source Ordnance Survey 1865 first edition 11
Decription
No field data.

Site Name Walton Moss (90–10 0m OD)
NGR NY 510 655
Parish WALTON
Source Ordnance Survey 1865 first edition 11
Decription
This extensive raised mire has been the site of extensive palaeoecological work (Hughes 1997).

Site Name Whitehill Moss (15 m OD)
NGR NY 382 630
Parish WESTLINTON
Source Ordnance Survey 1:25,000 sheet 544
Decription
No field data.

Site Name White Moss (30 m OD)
NGR NY 462 606
Parish STANWIX/ SCALEBY
Source Ordnance Survey 1:25,000 sheet 545

Decription
The moss covers 39 ha and has been modified by past drainage and peat-cutting. Wet heath vegetation covers the major part of the mire with heather, hare's-tail cottongrass, and *Sphagnum recurvum*, with several deeper cuttings in this area containing open water with *Sphagnum cuspidatum* and *Sphagnum recurvum* and raised mire plants such as cranberry, bog asphodel and common cottongrass. Birch and scots pine are now invading the drier wet heath areas. The mire is fringed in places by a number of plant communities, *eg* Scot's pine woodland, alder carr, acidic heathland and wet fen meadow.

Other Cumbrian Mosses: Appendix 5
Central Cumbria

Site Name Beacon Tarn (170 m OD)
NGR SD 273 902
Parish BLAWITH
Source Ordnance Survey 1:25,000 Outdoor Leisure 1:25,000 sheet 6
Decription
This formed an NWWS palaeoecological sampling site. *See Appendix 5.*

Site Name Black Dyke Moss (90 m OD)
NGR NY 330 989
Parish HAWKSHEAD
Source Ordnance Survey 1:25,000 Outdoor Leisure Sheet 7
Decription
No field data.

Site Name Black Moss (70 m OD)
NGR NY 230 302
Parish BASSENTHWAITE
Source Ordnance Survey 1:25,000 sheet 597
Decription
No field data.

Site Name Black Moss (<50 m OD)
NGR SD 354 997
Parish HAWKSHEAD
Source Ordnance Survey 1851 6" first edition Lancashire Sheet 2
Decription
No field data.

Site Name Black Moss (150 m OD)
NGR SD 430 981
Parish WINDERMERE
Source Ordnance Survey 1:25,000 Outdoor Leisure Sheet 7
Decription
No field data.

Site Name Blea Moss (150 m OD)
NGR NY 295 035
Parish LAKES
Source Ordnance Survey 1:25,000 Outdoor Leisure Sheet 6
Decription
No field data.

Site Name Blelham Bog (70–80 m OD)
NGR NY 366 006
Parish LAKES
Source Ordnance Survey Outdoor Leisure Sheet 7
Decription
No field data.

Site Name Braithwaite Moss (74 m OD)
NGR NY 234 249
Parish ABOVE DERWENT
Source Ordnance Survey 1:25,000 Outdoor Leisure Sheet 4
Decription
This formed an NWWS palaeoecological sampling site. *See Appendix 5.*

Site Name Brantrake Moss (190 m OD)
NGR NY 154 980
Parish MUNCASTER
Source Ordnance Survey 1:25,000 Outdoor Leisure Sheet 6.
Decription
No field data.

Site Name Burney Tarn (161 m OD)
NGR SD 255 859
Parish KIRKBY IRELETH
Source Ordnance Survey 1:25,000 sheet 626
Decription
This formed an NWWS palaeoecological sampling site. *See Appendix 5.*

Site Name Calvert Trust Land/Throstle Shaw, Bassenthwaite (90 m OD)
NGR NY 2360 2720
Parish UNDERSKIDDAW
Source Ordnance Survey 1:25,000 Outdoor Leisure sheet 4
Decription
This formed an NWWS palaeoecological sampling site. *See Appendix 5.*

Site Name Cannon Dub, Derwent Water (85 m OD)
NGR NY 262 189
Parish ST JOHN'S CASTLERIGG AND WYTHBURN
Source Ordnance Survey 1:25,000 Outdoor Leisure sheet 4
Decription
This formed an NWWS palaeoecological sampling site. *See Appendix 5.*

Site Name Castle How, Bassenthwaite (90 m OD)
NGR NY 202 308
Parish WYTHOP
Source Ordnance Survey 1:25,000 Outdoor Leisure sheet 4
Decription
This formed an NWWS palaeoecological sampling site. *See Appendix 5.*

Site Name Coniston (49 m OD)
NGR SD 288 905
Parish BLAWITH
Source Ordnance Survey 1:25,000 Outdoor Leisure sheet 6
Decription
This formed an NWWS palaeoecological sampling site. *See Appendix 5.*

Site Name Gawthwaite Moss (194 m OD)
NGR SD 265 848
Parish KIRKBY IRELETH
Source Ordnance Survey 1:25,000 sheet 626
Decription
No field data.

Site Name Graythwaite (130 m OD)
NGR SD 1135 2300
Parish LOWESWATER
Source Ordnance Survey 1:25,000 Outdoor Leisure Sheet 4
Decription
This formed an NWWS palaeoecological sampling site. *See Appendix 5.*

Site Name Grey Stone Moss (200 m OD)
NGR SD 262 864
Parish SUBBERTHWAITE
Source Ordnance Survey 1:25,000 sheet 626
Decription
No field data.

Site Name Hollow Moss Beck (200 m OD)
NGR NY 206 930
Parish DUNNERDALE WITH SEATHWAITE
Source Ordnance Survey 1:25,000 Outdoor Leisure Sheet 6
Decription
No field data.

Site Name Mawthwaite Moss (140 m OD)
NGR SD 256 867
Parish SUBBERTHWAITE
Source Ordnance Survey 1:25,000 sheet 626
Decription
No field data.

Site Name Mere Moss (200 m OD)
NGR NY 271 929
Parish TORVER
Source Ordnance Survey 1:25,000 Outdoor Leisure Sheet 6
Decription
No field data.

Site Name Mire House (130 m OD)
NGR NY 315 239
Parish ST JOHN'S CASTLERIGG AND WYTHBURN
Source Ordnance Survey 1:25,000 Outdoor Leisure Sheet 5
Decription
No field data.

Site Name Mosedale (possibly Bramely) (170 m OD)
NGR NY 109 234
Parish LOWESWATER
Source Ordnance Survey Landranger sheet 89
Decription
This formed an NWWS palaeoecological sampling site. *See Appendix 5.*

Site Name Moss Dikes (150 m OD)
NGR NY 311 240
Parish ST JOHN'S CASTLERIGG AND WYTHBURN
Source Ordnance Survey 1:25,000 Outdoor Leisure Sheet 5
Decription
No field data.

Site Name Moss Dub (130 m OD)
NGR NY 145 136
Parish ENNERDALE AND KINNISIDE
Source Ordnance Survey 1:25,000 Outdoor Leisure Sheet 4
Decription
No field data.

Site Name Moss Eccles Tarn (180 m OD)
NGR SD 372 969
Parish CLAIFE
Source Ordnance Survey 1:25,000 Outdoor Leisure Sheet 7
Decription
No field data.

Site Name Mosser (200 m OD)
NGR NY 115 248
Parish BLINDBOTHEL
Source Ordnance Survey 1:25,000 Outdoor Leisure Sheet 4
Decription
No field data.

Site Name Moss Side (100 m OD)
NGR SD 471 976
Parish STAVELEY

Source Ordnance Survey 1:25,000 Outdoor Leisure Sheet 7

Decription
No field data.

Site Name Newbiggin Moss (190 m OD)
NGR NY 476 308
Parish DACRE
Source Ordnance Survey 1:25,000 sheet 577
Decription
No field data.

Site Name Newton Moss (190 m OD)
NGR NY 478 309
Parish DACRE
Source Ordnance Survey 1:25,000 sheet 577
Decription
No field data.

Site Name Nor Moss (190 m OD)
NGR SD 375 992
Parish CLAIFE
Source Ordnance Survey 1:25,000 Outdoor Leisure Sheet 7
Decription
No field data.

Site Name Out Dubs Tarn (70 m OD)
NGR SD 367 948
Parish HAWKSHEAD
Source Ordnance Survey 1851 6" first edition Lancashire Sheet 2
Decription
This was referred to in Howard-Davis *et al* (1988).

Site Name Peat Howe (100 m OD)
NGR NY 268 144
Parish BORROWDALE
Source Ordnance Survey 1:25,000 Outdoor Leisure Sheet 4
Decription
No field data.

Site Name Priest Pot/Esthwaite North Fen (50 m OD)
NGR SD 357 979
Parish CLAIFE
Source Ordnance Survey 1851 6" first edition Lancashire Sheet 5
Decription
This was referred to in Howard-Davis *et al* (1988).

Site Name Red Moss Beck (200 m OD)
NGR NY 215 915
Parish DUNNERDALE WITH SEATHWAITE
Source Ordnance Survey 1:25,000 Outdoor Leisure Sheet 6
Decription
No field data.

Site Name Rough Mire (69 m OD)
NGR NY 232 261
Parish ABOVE DERWENT
Source Ordnance Survey 1:25,000 Outdoor Leisure Sheet 4
Decription
No field data.

Site Name Scab Moss (50 m OD)
NGR SD 377 987
Parish CLAIFE
Source Ordnance Survey 1:25,000 Outdoor Leisure Sheet 7
Decription
No field data.

Site Name Shoulthwaite Moss (172 m OD)
NGR NY 307 201
Parish ST JOHN'S CASTLERIGG AND WYTHBURN
Source Ordnance Survey 1:25,000 Outdoor Leisure Sheet 4
Decription
No field data.

Site Name Sineytarn Tarn (200 m OD)
NGR NY 160 010
Parish ESKDALE
Source Ordnance Survey 1:25,000 Outdoor Leisure Sheet 6
Decription
No field data.

Site Name Stable Harvey Moss (200 m OD)
NGR NY 278 918
Parish BLAWITH
Source Ordnance Survey 1:25,000 Outdoor Leisure Sheet 6
Decription
This formed an NWWS palaeoecological sampling site. *See Appendix 5.*

Site Name Subberthwaite Common (Robbed cairn) (150 m OD)
NGR SD 257 881
Parish SUBBERTHWAITE
Source Ordnance Survey 1:25,000 sheet 626
Decription
This formed an NWWS palaeoecological sampling site, with sampling occurring near a cairn. *See Appendix 5.*

Site Name Ullock Moss (77 m OD)
NGR NY 249 230
Parish ABOVE DERWENT
Source Ordnance Survey 1:25,000 Outdoor Leisure Sheet 4

Decription
No field data.

Site Name Waterson Moss (90 m OD)
NGR SD 352 997
Parish HAWKSHEAD
Source Ordnance Survey 1:25,000 Outdoor Leisure Sheet 7
Decription
No field data.

Site Name White Moss Common (142 m OD)
NGR NY 345 078
Parish LAKES
Source Ordnance Survey 1:25,000 Outdoor Leisure Sheet 7
Decription
No field data.

Site Name White Moss (<70 m OD)
NGR NY 3510 0650
Parish LAKES
Source Ordnance Survey 1:25,000 Outdoor Leisure sheet 7
Decription
This formed an NWWS palaeoecological sampling site. *See Appendix 5.*

Site Name Windermere (40 m OD)
NGR NY 370 010
Parish SKELWITH
Source Ordnance Survey 1:25,000 Outdoor Leisure sheet 7
Decription
This was referred to in Howard-Davis *et al* (1988).

Eastern Cumbria

Site Name Abbot Moss (100 m OD)
NGR NY 510 425
Parish LAZONBY
Source Ordnance Survey 1:25,000 sheet 568
Decription
This was referred to in Howard-Davis *et al* (1988).

Site Name Archer Moss (440 m OD)
NGR NY 633 006
Parish TEBAY
Source Ordnance Survey 1:25,000 sheet 607
Decription
This formed an NWWS palaeoecological sampling site. *See Appendix 7.*

Site Name Birk Moss (190 m OD)
NGR NY 439 939
Parish CROOK
Source Ordnance Survey 1:25,000 Outdoor Leisure Sheet 7
Decription
No field data.

Site Name Birkrigg Bog (90 m OD)
NGR SD 537 866
Parish PRESTON RICHARD
Source Ordnance Survey 1:25,000 sheet 627
Decription
This formed an NWWS palaeoecological sampling site. *See Appendix 5.*

Site Name Black Dub (135 m OD)
NGR NY 527 528
Parish CARLATTON
Source Ordnance Survey 1:25,000 sheet 558
Decription
No field data.

Site Name Black Moss (125 m OD)
NGR NY 535 558
Parish CASTLE CARROCK
Source Ordnance Survey 1:25,000 sheet 558
Decription
No field data.

Site Name Black Moss Tarn (160 m OD)
NGR SD 548 971
Parish PATTON
Source Ordnance Survey 1:25,000 Outdoor Leisure Sheet 7
Decription
This formed an NWWS palaeoecological sampling site. *See Appendix 5.*

Site Name Brownrigg Moss / Long Moss (150 m OD)
NGR NY 520 367
Parish GREAT SALKELD
Source Ordnance Survey 1:25,000 sheet 577
Decription
This formed an NWWS palaeoecological sampling site. *See Appendix 5.*

Site Name Brundrigg Moss (191 m OD)
NGR SD 566 923
Parish NEW HUTTON
Source Ordnance Survey 1:25,000 Outdoor Leisure Sheet 7
Decription
No field data.

Site Name Crook Hall (130 m OD)
NGR SD 448 944
Parish CROOK
Source Ordnance Survey 1:25,000 Outdoor Leisure sheet 7
Decription
This formed an NWWS palaeoecological sampling site. *See Appendix 5.*

Site Name Cumwhitton Moss (120 m OD)
NGR NY 515 520
Parish CUMWHITTON
Source Ordnance Survey 1:25,000 sheet 558
Decription
No field data.

Site Name Egholme Peat Moss / Heights (220 m OD)
NGR SD 603 861
Parish MANSERGH
Source Ordnance Survey 1:25,000 sheet 628
Decription
This formed an NWWS palaeoecological sampling site. *See Appendix 5.*

Site Name Eskrigg (190 m OD)
NGR SD 576 889
Parish OLD HUTTON AND HOLMESCALES
Source Ordnance Survey 1:25,000 Sheet 627
Decription
This formed an NWWS palaeoecological sampling site. *See Appendix 5.*

Site Name Garnet Farm Moss (140 m OD)
NGR SD 537 967
Parish SKELMSERGH
Source Ordnance Survey 1:25,000 Outdoor Leisure Sheet 7
Decription
This formed an NWWS palaeoecological sampling site, but deposits did not prove suitable for assessment.

Site Name Gatebeck (70 m OD)
NGR SD 549 859
Parish PRESTON PATRICK
Source Ordnance Survey 1:25,000 Sheet 627
Decription
This formed an NWWS palaeoecological sampling site. *See Appendix 5.*

Site Name Helms Moss (150 m OD)
NGR NY 684 899
Parish DENT
Source Ordnance Survey 1:25,000 sheet 628
Decription
No field data.

Site Name Hood Tarn (160 m OD)
NGR SD 571 876
Parish OLD HUTTON AND HOLMESCALES
Source Ordnance Survey 1:25,000 Sheet 627
Decription
This formed an NWWS palaeoecological sampling site. *See Appendix 5.*

Site Name Kirkby Thore (110 m OD)
NGR NY 645 252
Parish KIRKBY THORE
Source Ordnance Survey 1:25,000 sheet 578
Decription
This was referred to in Howard-Davis *et al* (1988).

Site Name Littlewater (240 m OD)
NGR NY 5090 1705
Parish BAMPTON
Source Ordnance Survey 1:25,000 Outdoor Leisure Sheet 5
Decription
This formed an NWWS palaeoecological sampling site. *See Appendix 5.*

Site Name Long Moss (120 m OD)
NGR NY 514 540
Parish HAYTON
Source Ordnance Survey 1:25,000 sheet 558
Decription
No field data.

Site Name Melmerby Mire (140 m OD)
NGR NY 596 384
Parish GLASSONBY
Source Ordnance Survey 1:25,000 sheet 577
Decription
No field data.

Site Name Middleham Moss (145 m OD)
NGR NY 520 415
Parish LAZONBY
Source Ordnance Survey 1:25,000 sheet 568
Decription
No field data.

Site Name Moresby Tarn (190 m OD)
NGR NY 691 213
Parish MURTON
Source Ordnance Survey 1:25,000 sheet 578
Decription
No field data.

Site Name Moorthwaite Moss (120 m OD)
NGR NY 512 511
Parish CUMWHITTON
Source Ordnance Survey 1:25,000 sheet 558
Decription
No field data.

Site Name Moss Side (45 m OD)
NGR SD 539 821
Parish LUPTON
Source Ordnance Survey 1:25,000 sheet 627
Decription
No field data.

Site Name Moss Wood (105 m OD)
NGR NY 526 328
Parish LANGWATHBY
Source Ordnance Survey 1:25,000 sheet 577
Decription
No field data.

Site Name St Sunday's Beck Bog (100 m OD)
NGR SD 553 895
Parish OLD HUTTON AND HOLMESCALES
Source Ordnance Survey 1:25,000 sheet 627
Decription
This formed an NWWS palaeoecological sampling site. *See Appendix 5.*

Site Name Skelsmergh Tarn (110 m OD)
NGR SD 534 967
Parish SKELSMERGH
Source Ordnance Survey 1:25,000 Outdoor Leisure sheet 7
Decription
This was referred to in Howard-Davis *et al* (1988).

Site Name Tarnybank/Three Mile Bog (180 m OD)
NGR SD 594 875
Parish KILLINGTON
Source Ordnance Survey 1:25,000 sheet 627
Decription
This formed an NWWS palaeoecological sampling site. *See Appendix 5.*

Site Name Tewfit Mires (180 m OD)
NGR NY 817 135
Parish WINTON
Source Ordnance Survey 1:25,000 sheet 598
Decription
No field data.

Site Name Thackmire (158 m OD)
NGR NY 358 397
Parish SEBERGHAM
Source Ordnance Survey 1:25,000 sheet 576
Decription
No field data.

Site Name The Helm: Oxenholme (100 m OD)
NGR SD 5290 8910
Parish NATLAND
Source Ordnance Survey 1:25,000 sheet 627
Decription
This formed an NWWS palaeoecological sampling site. *See Appendix 5.*

Site Name The Moss: Temple Sowerby (100 m OD)
NGR NY 615 271
Parish NEWBIGGIN
Source Ordnance Survey 1:25,000 sheet 578
Decription
No field data.

Site Name White Mires (135 m OD)
NGR NY 599 371
Parish GLASSONBY
Source Ordnance Survey 1:25,000 sheet 577
Decription
No field data.

Primary (P)
P1* **Natural or near-natural vegetation**
P2* **Degraded vegeatation (usually burnt or dry)**
P3* **Drained**
P4 **Open-canopy scrub or woodland**
P5 **Closed-canopy woodland**
U **Condition not determined yet**

Secondary (S)
S1* **Revegetated or regenerating cutover**
S2 **Commercial or domestic workings**

Archaic (A)
A1 **Bog soils in agricultural use**
A2 **Built development**

* **Classes of raised bogs considered to be 'active' under Annex 1 of the EC Habitats Directive**

NGR	Site Name	Area (ha)	Major Land –Cover	'Best' Land –Cover	Primary Natural Area P1	Primary Degraded Area P2, P3	SSSI Area	NNR	Planning Consent Area	Comments
SD 191 848	Arnaby Moss	58	S1	P4	–	–	Y	–	–	Duddon Mosses SSSI
SD 223 852	Bank End Moss	214	S1	P1	48	56	356	–	–	Duddon Mosses SSSI
NY 029 108	Black Moss	13	P2	P2	–	–	16	–	–	–
NY 422 676	Black Snib	67	P4	P3	–	3	–	–	–	–
SD 458 906	Blakebank Moss	10	P4	P4	–	–	–	–	–	–
NY 492 691	Bolton Fell	401	S2	P1	12	–	316	–	284	–
NY 202 595	Bowness Common	1,106	P2	P1	100	580	759	–	–	Bowness Common SSSI
NY 306 574	Broughmoor Wood	38	P5	P5	–	–	–	–	–	–
SD 462 900	Cock Moss	23	P4	P4	–	–	–	–	–	–
NY 363 719	Dicks Moss	12	A1	A1	–	–	–	–	–	–
NY 344 713	Drownedcow Moss	6	A1	A1	–	–	–	–	–	–
NY 254 584	Drumburgh Moss	205	P2	P1	20	100	189	–	–	CWT Reserve
SD 348 802	Ellerside Moss	246	P3	P1	20	150	Y	Y	–	Roudsea Wood and Mosses SSSI
SD 345 857	Elmlath Moss	10	P5	P5	–	–	–	–	–	–
SD 338 825	Fish House/Deer Dyke	143	P3	P1	15	100	476	Y	–	Roudsea Wood and Mosses SSSI
SD 458 824	Foulshaw Moss	287	P5	P2	–	5	–	–	–	–
NY 238 604	Glasson Moss	258	P2	P1	38	75	225	Y	–	Glasson Moss SSSI
NY 334 720	Greenwrae Flow	1	P2	P2	–	1	–	–	–	–
NY 115 468	Hangingshaw Moss	1	P3	P2	–	–	–	–	–	–
NY 380 615	Harker Moss	12	P4	P3	–	3	–	–	–	–
SD 240 875	Heathwaite Moss	40	S1	P1	10	–	Y	–	–	Duddon Mosses SSSI
SD 421 866	High Moss	7	A1	A1	–	–	–	–	–	–
SD 421 860	High Moss	4	A1	A1	–	–	–	–	–	–
NY 454 651	Hunley Moss11	P2	P2	–	5	–	–	–	–	
SD 333 844	Ireland Moss	25	P5	P4	–	–	–	–	–	–
NY 048 045	Lingbank Moor	10	P5	P5	–	–	–	–	–	–

Table 8 Raised mires in Cumbria (from Lindsay and Immirzi 1996)

NGR	Site Name	Area (ha)	Major Land –Cover	'Best' Land –Cover	Primary Natural Area P1	Primary Degraded Area P2, P3	SSSI Area	NNR	Planning Consent Area	Comments
SD 432 823	Meathop/Nichols	148	P4	P1	30	–	160	–	–	Two SSSIs (Meathop and Nichols Mosses)
SD 463 873	Lyth Moss	2	A1	A1	–	–	–	–	–	–
NY 413 620	Mampus Wood	5	P5	P4	–	–	–	–	–	–
SD 308 794	Newland Moss	11	P3	P3	–	8	–	–	–	–
SD 311 796	Newland Moss	8	P3	P3	–	–	–	–	–	–
SD 303 793	Newland Moss	4	P3	P3	–	–	–	–	–	–
SD 305 794	Newland Moss	1	P3	P3	–	–	–	–	–	–
NY 332 695	Newton Flow	16	P5	P5	–	–	–	–	–	–
NY 252 512	Oulton Moss	53	P3	P2	–	20	24	–	–	–
SD 480 876	Park Moss	33	P4	P3	–	5	–	–	–	–
SD 465 855	Rawsons Moss	10	A1	A1	–	–	–	–	–	–
SD 324 839	Roam Moss	32	P3	P3	–	15	–	–	–	–
NY 345 662	Rosetrees Moss	23	A1	A1	–	–	–	–	–	–
SD 335 883	Rusland Moss	63	P5	P4	–	–	96	Y	–	Rusland Valley Mosses SSSI
NY 085 450	Salta Moss	125	S1	S1	–	–	45	–	–	Salta Moss SSSI
SD 466 889	Savinhill Moss	24	P5	P2–	–	5	–	–	–	–
NY 430 636	Scaleby Moss	99	S1	P2	–	–	10	69	–	Scaleby Moss SSSI
NY 420 628	Scaleby Moss	61	S1	S1	–	–	–	–	–	–
SD 189 856	Shaw Moss	9	S1	P2	–	2	Y	–	–	Duddon Mosses SSSI
NY 345 692	Solway Moss	436	S2	S2	–	–	–	–	364	–
SD 422 870	Stock Moss	6	A1	A1	–	–	–	–	–	–
SD 343 859	The Grams	18	P5	P5	–	–	–	–	–	–
NY 221 531	Wedholme Flow	831	S2	P1	125	–	780	Y	230	–
SD 231 874	White Moss	118	A1	P1	30	25	Y	–	–	Duddon Mosses SSSI
NY 461 606	White Moss – Crosbymoor	31	S1	P2	–	–	39	–	–	White Moss – Crosbymoor
SD 326 823	–	19	P3	P3	–	10	–	–	–	–
NY 443 706	–	13	P5	P5	–	–	–	–	–	–
SD 431 805	–	12	P5	P5	–	–	–	–	–	–
NY 354 642	–	8	P4	P4	–	–	–	–	–	–
NY 431 624	–	8	P4	P4	–	–	–	–	–	–

Table 8 Raised mires in Cumbria cont'd (from Lindsay and Immirzi 1996)

NGR	Site Name	Area (ha)	Major Land –Cover	'Best' Land –Cover	Primary Natural Area P1	Primary Degraded Area P2, P3	SSSI Area	NNR	Planning Consent Area	Comments
SD 477 850	–	7	P4	P4	–	–	–	–	–	–
NY 567 617	–	5	P4	P4	–	–	–	–	–	–
NY 516 609	–	4	A1	A1	–	–	–	–	–	–
SD 345 870	–	4	A1	A1	–	–	–	–	–	–
SD 419 826	–	4	A1	A1	–	–	–	–	–	–
NY 566 618	–	3	P4	P4	–	–	–	–	–	–
SD 207 883	–	3	P5	P5	–	–	–	–	–	–
NY 033 063	–	2	A1	A1	–	–	–	–	–	–
N 097 001	–	2	P5	P5	–	–	–	–	–	–
NY 567 618	–	2	P4	P4	–	–	–	–	–	–
NY 033 061	–	1	A1	A1	–	–	–	–	–	–
SD 420 825	–	1	A1	A1	–	–	–	–	–	–
NY 020 060	–	1	A1	A1	–	–	–	–	–	–
NX 992 124	–	1	A1	A1	–	–	–	–	–	–
NY 029 062	–	1	P5	P5	–	–	–	–	–	–
NY 514 608	–	1	A1	A1	–	–	–	–	–	–
NX 990 127		1	A1	A1	–	–	–	–	–	–

Definitions for headings of columns in table 8:

Major Land Cover is the main form of land cover recorded for the site.
Best Land Cover records the land class closest to the near natural conditions found on the site. In sites subject to extensive agriculture land claim, these may be revegetating cuttings rather than near-natural bog.
Primary Natural Area is estimated from aerial photographs if an area of primary natural bog was recorded (P1).
Primary Degraded Area is estimated from aerial photographs if an area of primary/degraded or drained bog was recorded (P2 or P3).
NNR = Natural Nature Reserve.
Planning Consent Area is one with current planning consents for peat extraction as of December 1994.
Comments notes if the site is a County Trust Reserve or if SSSI name differs from the site name.

Table 8 Raised mires in Cumbria cont'd (from Lindsay and Immirzi 1996)

Appendix 5
Gazetteer of Cumbria environmental archives assessed

Palynological and stratigraphical methods are outlined in Chapter 1. The data from these assessments are summarised in Table 9, which gives the site names, depth of deposits, possible age, and potential of each. A palaeoecological and archaeological chronological framework is also provided in Table 10, so that the reader can relate the various periods mentioned in the text, both to each other and to a generalised timescale derived from radiocarbon dates. Throughout the text, the term 'Holocene', the present post-glacial period, has been used to denote a general timespan, alongside the more specific palaeoecological chronozones mentioned in Table 10. An asterisk (*) denotes where a spot sample has been taken.

Beacon Tarn
NGR SD 273 902
Height OD <170 m

The site lies 1–2 km west of the southern tip of Coniston Water. The stratigraphy, from an area west of the present tarn, began developing in the early Holocene when the tarn was more extensive than it is today. The organic gyttja with bands of clay indicates either fluctuating water levels or the inwash of mineral material as the result of unstable soil conditions surrounding the site. Terrestrialisation occurred with the formation of an acid mire community of *Eriophorum* and *Calluna*, which in turn developed into a *Sphagnum* moss. The area around the tarn is rich in burial and clearance cairns (eg SMR 1636, SMR 1638 and SMR 2154–2156); today the site is rough hill pasture.

Depth (m)	stratigraphy
0–0.50	unsampled
0.50–1.00	H9 *Sphagnum* (*0.60 m)
1.00–1.25	H7 ditto + *Eriophorum* / *Calluna*
1.25–1.75	organic gyttja + occasional clay
1.75+	light grey silty clay + stones (*1.75 m)

Palynological assessment

Two pollen samples were examined. The pollen spectrum at 1.75 m suggests that the sediment was initiated early in Flandrian I with birch as the dominant pollen type but with some hazel, willow and juniper recorded. Herbaceous pollen was approximately 7 % of the total. At 0.60 m there is considerable evidence of burning and the pollen spectrum is dominated by hazel and heather with some alder, but very low values of other tree taxa, and 6 % herbaceous pollen.

Potential

The site would be of very limited use as it is almost certainly truncated. Pollen preservation was very good at 1.75 m but poor at 0.60 m.

Birkrigg Bog (Fig 62)
NGR SD 537 866
Height OD 90 m

Birkrigg Bog is located east of Levens in the south of the county; it is probably a basin mire. Peats of more than 2 m developed on mineral soil when organic deposition was initiated at some time in Flandrian II with the formation of a fen which later developed into a fen carr and then an acid mire with *Eriophorum* and *Calluna*. Today the site is under pasture.

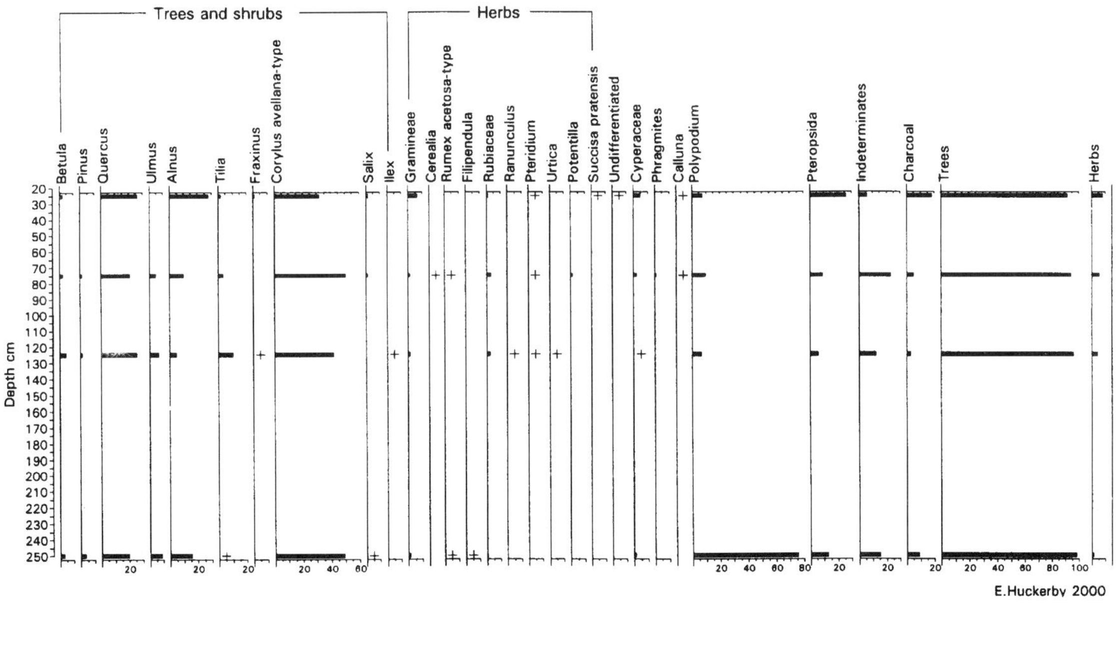

Fig 62 Birkrigg Bog: simplified pollen diagram of palynological assessment

Site	NGR	Altitude (approx mOD)	Depth of Sequence (m)	Possible age range assessed	Samples assessed	Potential
Beacon Tarn	SD 273 902	<170	1.75	Fl I — unknown	2	Limited
Birkrigg Bog	SD 537 866	90	2.60	Fl II	4	Poor
Black Moss Tarn	SD 548 971	160	4.00	Fl I and II	4	Limited to Mesolithic
Braithwaite Moss, Bassenthwaite	NY 2365 2490	74	1.55	Fl III	2	Limited
Brownrigg Moss (Long Moss)	NY 520 367	150	1.90	Fl II and III	3	Good
Burney Tarn	SD 255 859	161	6.20	Fl I, II and part of III	8	Good
Calvert Trust Land Bassenthwaite	NY 2360 2720	90	7.09	Fl I, II and III	5	Good
Castle How, Bassenthwaite	NY 202 308	90	1.30	Fl I and II	5	Poor
Coniston	SD 288 905	49	3.70	L De and Fl I	9	Limited but good
Crook Hall	SD 448 944	130	5.25	L De, Fl I, II and part of III	9	Good
Derwent Water (Cannon Dub)	NY 262 189	85	3.90	Fl II and III	7	Moderate–Good
Dubbs Moss	NY 103 293	90	3.70	Fl I, II and part of III	7	Good
Eskrigg	SD 576 889	190	5.50	L De? Fl I, II and III	6	Good
Gatebeck	SD 549 859	70	3.00	Fl I and late Fl III	6	Limited
Glebe Farm	NY 185 346	130–140	7.00	Fl I, II and III?	6	Good
Graythwaite	NY 1135 2300	130	2.10	Unknown	4	Poor
Heathwaite	SD 255 869	120	3.95	Unknown	4	Poor
Heights (Egholme Peat Moss)	SD 603 861	220	4.20	Fl I, II and part of III	8	Good
Hood Tarn	SD 571 876	160	5.50	Fl II and III	8	Good
Little Tarn	NY 2495 3385	190	6.70	Fl I, II and III	7	Good
Littlewater	NY 5090 1705	240	8.50+	Probably Fl I, II and III	10	Good
Mosedale (possibly Bramely)	NY 109 234	170	2.90	Fl I, II and part of III	4	Good
Oxenholme (The Helm)	SD 5290 8910	100	4.50	L De, Fl I, II and III	8	Good
Robbed Cairn (Subberthwaite Common)	SD 257 881	150	5.10	Fl I, II and part of III	5	Medium
St Sundays Beck Bog	SD 553 895	100	5.00	Fl I and II	7	Medium
Stable Harvey Moss	SD 278 917	200	4.70	Fl I, II and part of III	6	Medium
Three Mile Bog (Tarneybank)	SD 594 875	180	7.00+	Fl I, II and ?L H'De	8	Good
White Moss	NY 3510 0650	<70	8.50+	?Fl I, II and III	10	Very Good

Fl = Flandrian
L De = Late Devensian

Table 9 Summary of palaeoecological sites

	Chronozones	Approx Dates Calendar years	Archaeological Periods		Approx Dates radiocarbon years BP	Blytt – Sernander Periods	
HOLOCENE	Flandrian III		Post-medieval		0	Sub Atlantic	Post-glacial
		1500			450		
			Medieval				
		1066			1000		
			Early Medieval				
		410					
			Romano-British				
		AD43			2000		
			Iron Age				
		700BC			2500		
			Bronze Age	Late		Sub-Boreal	
					3000		
				Early			
		2500			4000		
			Neolithic	Late			
		3200					
				Early			
	Elm Decline				5000	Elm Decline	
	Flandrian II	4000	Mesolithic	Late		Atlantic	
					6000		
		5900			7000		
	Flandrian I: Id					Boreal	
					8000		
	Ic	7089					
	Ib	7791			9000		
	Ia			Early		Pre-Boreal	
		10,000			10,000		
LATE DEVENSIAN	Late Devensian III		Late Upper Palaeolithic			Upper Dryas	Late-glacial
					11,000		
	Late Devensian II					Allerød	
					12,000		
	Late Devensian I					Lower Dryas	
					13,000		

Table 10 Palaeoecological and archaeological chronological framework

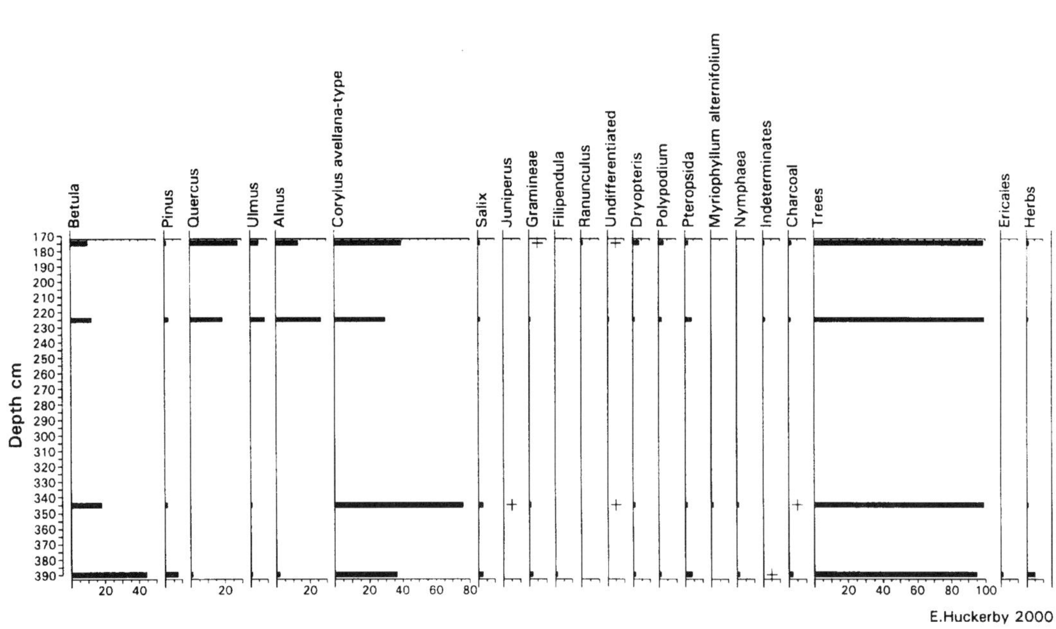

Fig 63 Black Moss Tarn: simplified pollen diagram of palynological assessment

Depth (m)	**stratigraphy**
0–0.25	H10 amorphous peat
0.25–0.50	H9 ditto + *Calluna* (*0.25 m)
0.50–1.00	H7 wood peat + charcoal (*0.75 m)
1.00–1.50	H6 ditto (*1.25 m)
1.50–2.00	H6 wood peat
2.00–2.30	H8/9 sedge (*2.30 m)
2.30–2.50	organic silty clay + *Cladium* (*2.60 m)
2.60+	light grey silty clay

Palynological assessment

The sample at 2.50 m suggests a mixed woodland of oak, elm, alder and hazel with very high values of *Polypodium* spores indicating some disturbance. Charcoal was recorded at 9 %. At 1.25 m lime (*Tilia*) pollen was recorded at values >10 %, and this probably represents a significant number of lime trees in the woodland; alternatively, a few lime trees were growing in close proximity to the site. High values of lime pollen are unusual as Cumbria is on the northern limits of the taxon and values of more than 10 % may suggest alarge presence around the site (Huntley and Birks 1983, 391–410). However, some caution is necessary because pollen preservation was poor and lime is easily recognised because of the grain size, distinctive morphology, and also it is resistance to damage, thereby making it more easily identifiable. Therefore these values may be an artefact caused by preservation rather than the number of lime trees in the area. There is some evidence of woodland disturbance at 0.75–0.25 m with herbaceous pollen reaching nearly 10 % at 0.25 m. The woodland composition at 0.25 m is changed, with elm and lime declining along with hazel and the subsequent rise of oak and alder.

Potential

The site's potential is extremely limited as the site has been badly truncated, most sediments relevant to periods later than the Neolithic may have been removed. In addition, deposition of organic deposits was only initiated some time in Flandrian II. Ecologically it may be of interest if the high lime values are genuine, and for this reason it would warrant additional palynological analysis. Pollen preservation was poor in all samples.

Black Moss Tarn (Fig 63)
NGR SD 548 971
Height OD 160 m

The site is 11 km north–east of Kendal and is a small tarn with organic deposits surrounding it. There are 4 m of well-preserved organic deposits, which began forming in Flandrian I on silty clay surrounding the tarn, with the formation of a fen carr for the period covered by the stratigraphy. A layer of *Polytrichum* moss at 2.25 m suggests a drier phase during Flandrian II. Today the site is under rough pasture. There is some evidence of prehistoric activity in the area with the recovery of a perforated stone axe (SMR 4115), and a possible settlement site (SMR 5168) recorded by Tom Clare.

Depth (m)	**stratigraphy**
0–1.75	sloppy amorphous peat
1.75–2.00	H7 wood peat (*1.75 m)
2.00–230	H7 *Polytrichum* (*2.25 m)
2.30–3.50	ditto + wood fragments (*3.45 m)
3.50–4.00	H9 ditto
4.00+	light grey silty clay

Palynological assessment

Four samples were analysed palynologically from this site, the two lower suggesting a Flandrian I date, with birch and hazel scrub being replaced by a hazel-dominated community. The pollen spectra from the samples at 2.25 m and 1.75 m suggest a mixed woodland characteristic of Flandrian II, with oak, alder, elm, birch and hazel. Charcoal and herb pollen are extremely low.

Potential

The site is probably truncated in Flandrian III and therefore would provide a record only of the Mesolithic. Pollen preservation was basically good.

Braithwaite Moss, Bassenthwaite

NGR NY 2365 2490

Height OD 74 m

Braithwaite Moss developed in the flood plain of the Newlands Beck, which flows from the Derwent Fells north-east of Buttermere into Bassenthwaite Lake. The 1.55 m of peats indicate that a fen carr developed on the silty clay of the flood plain, whilst the upper 0.50 m of silty peat suggest a fluctuating water table. There are archaeological records in the area for the Roman (road, SMR 1103) and later periods (parchmarks, SMR 13567, and lead mines, SMR 11814).

Depth (m)	stratigraphy
0–0.50	H9 silty peat
0.50–1.55	H8 wood peat + monocot (*0.75 m, 1.55 m)
1.55+	light grey silty clay

Palynological assessment

Two samples from this site were analysed palynologically but proved inconclusive. At 1.55 m there is 82 % tree pollen, which is dominated by 65 % alder, and 18 % herbaceous pollen, mainly grass. This may represent an alder carr in Flandrian III. By 0.75 m there is 77 % tree pollen, including birch (19 %) and hazel (47 %), with 22 % herbaceous pollen. The latter is more varied than in the previous sample and the main components of it are grass, *Filipendula*, *Melampyrum*, and *Potentilla*. The two samples may represent local carr vegetation or a more regional picture of clearance.

Potential

Further assessment is necessary to ascertain the potential of the archive. Some caution must be exercised in the palynological interpretation of this site because of its position. During flooding episodes this could seriously affect the pollen rain entering the sediments as some could be of fluvial origin from the uplands and may represent contemporary or reworked pollen from older deposits eroded by the rivers. However, pollen was abundant and well-preserved.

Brownrigg Moss (Long Moss)

NGR NY 520 367

Height OD 150 m

Brownrigg Moss is a valley mire, which developed in a col between two low-lying hills 13 km north of Penrith. Well-preserved stratigraphy of 1.9 m of peat was initiated in the early Holocene when an acid mire of *Eriophorum* and *Calluna* and then *Sphagnum* moss developed on the underlying sand. Today the site is under rough pasture. The archaeological record includes evidence of the Early Bronze Age (a food vessel (SMR 935) and Hesket Long Cairn (SMR 13765)) and also place-name evidence for a dyke system (SMR 13869).

Depth (m)	stratigraphy
0–0.50	H3 *Sphagnum* (*0.50 m)
0.50–1.00	H7 ditto (*0.80 m)
1.00–1.50	H7 *Eriophorum* / *Calluna* (*1.50 m)
1.50–1.90	H9 ditto
1.90+	sand (*1.90m)

Palynological assessment

Three samples were analysed palynologically. The basal one at 1.90 m has 38 % tree and shrub pollen (birch 23 %, juniper 2 %, and willow 5 %), nearly 6 % Crowberry (*Empetrum*), 55 % herb (grass, buttercup (*Ranunculus*) and others) and sedge pollen is at very high values. This suggests an open community characteristic of the early Holocene. At 1.50 m the pollen is suggestive of Flandrian III, with tree pollen 77 %, herbs 5 %, and heather 18 %. The tree taxa recorded are alder (11 %), birch (15 %), oak (23 %), hazel (35 %), and elm (2 %). However, at 0.50 m, the landscape is basically cleared of all trees with a *Calluna* (43 %) heath and grassland, with 45 % herbaceous pollen, mainly of pastoral affinity but with some evidence of cereal cultivation.

Potential

The site may potentially represent the entire Holocene sequence. Although the sediment accumulation is relatively shallow, other similar sites, such as Wet Sleddale (Chinn and Innes 1995), have been shown to record the entire Holocene sequence in less than 2.00 m of peat. Pollen preservation was good and further work to ascertain if there is an hiatus could be advisable.

Burney Tarn (Fig 64)

NGR SD 255 859

Height OD 161 m

The site is located south-west of Coniston Water near Kirkby Moor, adjacent to a small tarn in a hollow surrounded by low-lying fells rising to 200–250 m OD. The 6.2 m of well-preserved stratigraphy indicates that the tarn was once more extensive than it is today, with gyttja recorded from 4.0 m to 6.2 m. Above this, the water level fell and a fen carr formed, later developing into an acid mire at 1.50 m, with *Eriophorum*, *Calluna* and *Sphagnum*,

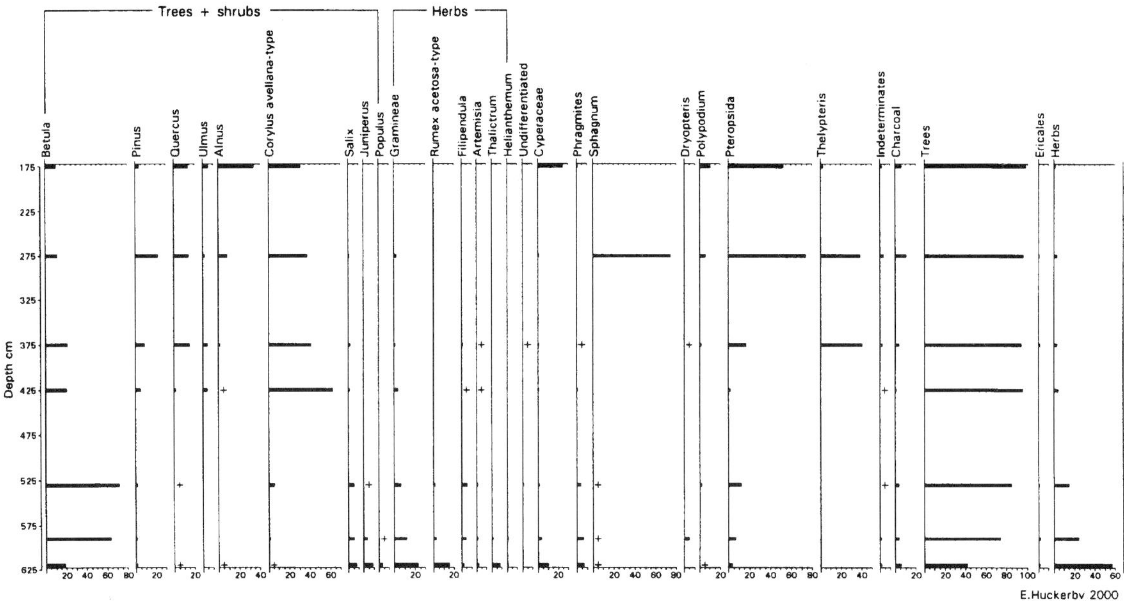

Fig 64 Burney Tarn: simplified pollen diagram of palynological assessment

before becoming wetter at 0.50 m when *Sphagnum* dominates. The tarn apparently formed as the ice melted after Late Devensian III. The hills around the site are rich in archaeological remains, with possible Bronze Age ring cairns (SMR 2170–2171), and clearance cairns (SMR 2160, SMR 4407), enclosures (SMR 2141), and a medieval hall (SMR 2148).

Depth (m)	**stratigraphy**
0–0.50	H6 *Sphagnum*
0.50–150	H4 *Eriophorum / Calluna / Sphagnum* (*0.75 m,1.25 m)
1.50–3.50	H8 wood peat + charcoal (*1.75 m, 2.75 m, 3.25 m)
3.50–4.00	H9 *Polytrichum* (*3.75 m)
4.00–5.30	gyttja + monocot (*4.25 m, 4.80 m, 5.30 m)
5.30–6.20	silty gyttja (*5.90 m, 6.20 m)
6.20+	light grey silty clay

Palynological assessment

Organic deposition was initiated early in Flandrian I when the vegetation was an open scrub with birch, willow, juniper and poplar. A wide variety of herbaceous pollen was recorded, *eg* rockrose (*Helianthemum*) and marsh gentian (*Gentiana pneumonanthe*-type). The vegetation surrounding the site became dominated by birch and was probably less open in character, as fewer herb taxa are recorded. The Holocene hazel maximum developed between 5.25 m and 4.25 m, when hazel dominated the pollen rain. At 3.75 m the pollen suggests that a mixed woodland of birch, pine, elm, and hazel occurred on the drier ground; some fern spores and low values of herb pollen were recorded indicating small openings in the canopy. The stratigraphy suggests that terrestialisation occurred above 4.00 m and the macrofossil record of *Polytrichum* indicates drier conditions at the site. At 2.75 m, when the deposit is a wood peat, the pollen record suggests that pine may have been growing on the bog surface with *Sphagnum*, *Thelypteris* and undifferentiated fern spores. Pennington (1970, 41–79) considered that Cumbrian bogs became drier late in zone VI (Flandrian Id) and as a result pine was able to invade the surfaces. Oak and hazel were probably growing on the higher ground. The Flandrian I / II transition occurred at some level between 2.75 m and 1.75 m, given that at 1.75 m alder pollen has risen from 9 % to 35 %. Because pollen was scarce, the spectrum from 0.75 m has been excluded from the diagram (Fig 64), but the limited data suggest a mixed woodland on the surrounding dry land with sedges and ferns growing at the site.

Potential

It appears probable that the deposits have been truncated above 0.75 m. Nevertheless, they are capable of providing a record of vegetation changes during Flandrian I, II and part of Flandrian III, *ie* up to the Neolithic or Bronze Age. Pollen preservation was good below 4.00 m, but above the preservation was mixed and the concentrations low. This would make further work more difficult but would not necessarily preclude more detailed analysis. This site may be of considerable value due to the richness of the archaeological record and further study to confirm the age of the upper deposits would be beneficial.

Calvert Trust Land, Bassenthwaite (Fig 65)

NGR NY 2360 2720

Height OD 90 m

The site is located near to the inflow of the River Derwent into Bassenthwaite Lake from Derwent Water. The gyttja and silty clay deposits from 7.00 m to 3.00 m indicate that

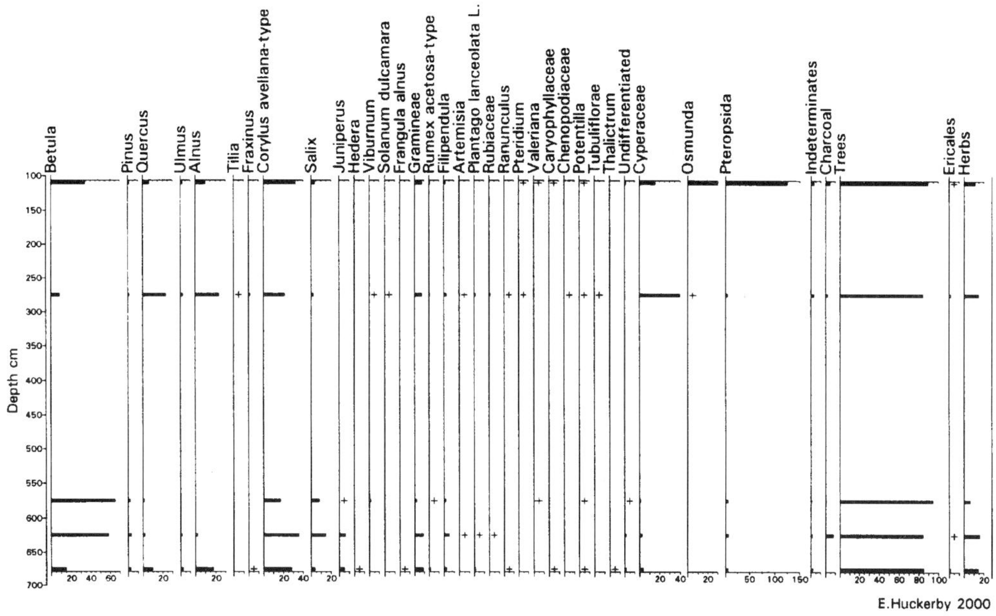

Fig 65 Calvert Trust Land, Bassenthwaite: simplified pollen diagram of palynological assessment

Bassenthwaite was once more extensive than it is today. The development of a poor fen carr with *Eriophorum* and *Calluna* is recorded above this and there is evidence of possible flooding episodes, with the deposition of silty clay. Today the site is rough pasture.

Depth (m)	**stratigraphy**
0–0.50	unsampled
0.50–1.00	light grey silty clay
1.00–2.00	H7 wood peat (Hazel nuts) (*1.10 m)
2.00–2.50	H7 *Eriophorum / Calluna* + wood fragments +monocot (*2.25 m)
2.50–3.00	H8 ditto (*2.75 m)
3.00–3.20	gyttja (*3.20 m)
3.20–3.50	light grey silty clay
3.50–6.00	gyttja (*4.50 m, 5.75 m)
6.00–7.00	grey silty gyttja (*6.25 m)
7.00+	light grey silty clay

Palynological assessment

The sample at 6.75 m has been disregarded in the light of the data from 6.25 m and 5.75 m because it was thought that 18 % alder pollen was unlikely, therefore suggesting contamination. The pollen spectra at 6.25 m and 5.75 m suggest that deposits started forming early in Flandrian I. At 6.25 m tree and shrub pollen is 84 % (birch, hazel, willow and juniper) and herb 15.5 %, together with 8 % charcoal. Above this, at 5.75 m, tree pollen is 94 %, mainly birch. No samples were examined between 5.75 m and 2.75 m. The sample at 2.75 m is probably post-Elm Decline with 14 % herb pollen, suggesting grassland; this sample was silty and required hydrofluoric acid treatment. At 1.10 m the pollen spectrum suggests secondary birch and hazel woods. Very high percentages of royal fern (*Osmunda regalis*) and undifferentiated fern spores indicate that there was a damp disturbed woodland. The low values of alder and oak pollen suggest that the deposits are post-Bronze Age at this level.

Potential

This site is situated adjacent to Bassenthwaite Lake and would provide a valuable record of vegetational changes for the major part of the Holocene, and possibly of changing lake levels through a study of the diatoms, given that at 2.75 m pollen of aquatic taxa is recorded. The stratigraphy suggests two levels at which additional pollen work might be rewarding. The first of these is the silty clay band at 3.20–3.50 m, which might indicate inwash into the deposits in Flandrian II or III, caused by woodland clearance. The second is the transition from wood peat to silty clay at 1.00 m. This stratigraphic change is of particular interest in the light of the Seathwaite, Borrowdale, site (Wild *et al* forthcoming), where there are indications of increased alluviation in the medieval period. Some caution must be exercised in interpreting the results of pollen analysis from this site, however, because of its location, since it is situated near the Derwent as it flows between the two lakes. During flooding episodes this could seriously affect the pollen rain entering the sediments as some could be of fluvial origin from the uplands and may represent contemporary or reworked pollen from older deposits eroded by the rivers. Pollen preservation was mixed in the samples analysed.

Castle How, Bassenthwaite (Fig 66)

NGR NY 202 308

Height OD 90 m

The site is situated close to the shore of the northern part of Bassenthwaite Lake, close to the Castle How hillfort (SMR 886). Only 1.70 m of peat was recorded and this provided evidence of the formation of a fen carr during Flandrian I.

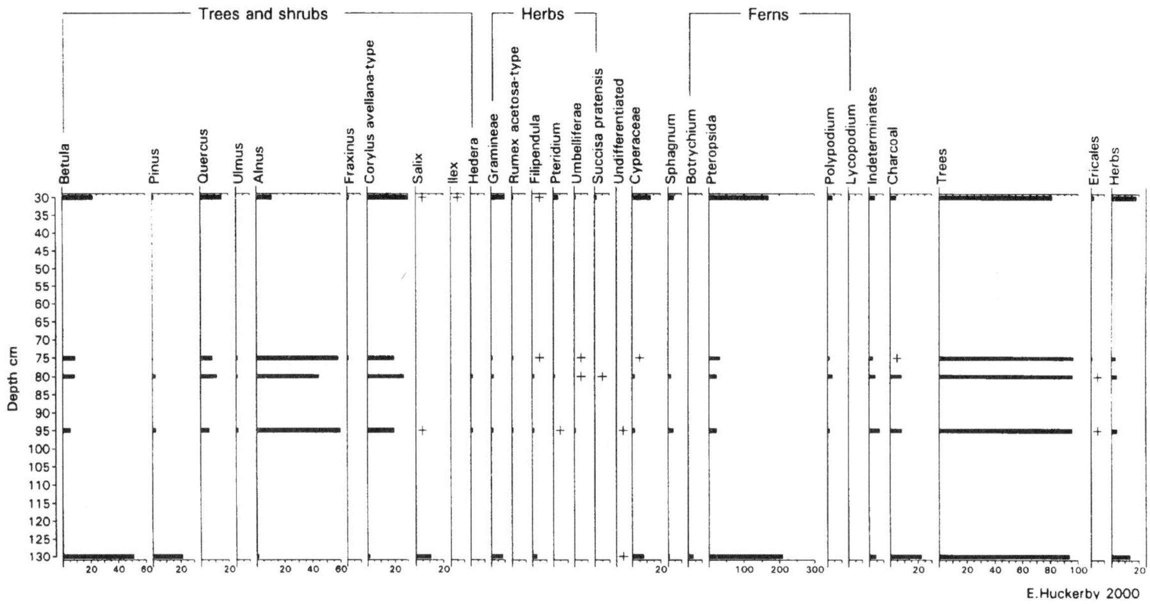

Fig 66 Castle How, Bassenthwaite: simplified pollen diagram of palynological assessment

This then developed into a fen above 0.90 m when a band of silty peat suggests possible evidence of inwash or a flooding horizon. It is likely that the deposits are heavily truncated. In addition to the hillfort, there is a deserted medieval village nearby (SMR 855).

Depth (m)	**stratigraphy**
0–0.75	H9 monocot (*0.30 m, 0.75 m)
0.75–0.90	silty peat (*0.80 m)
0.90–1.30	H8 wood peat (*0.95 m)
1.30+	light grey silty clay + stones

Palynological assessment

The data from this short core are inconclusive. It appears that the lowest sample represents some part of Flandrian I but hazel pollen is unusually low. The high percentages of pine, birch, willow and undifferentiated fern spores may represent the formation of birch, pine and willow woodland at the site, which is filtering out other pollen types. Alternatively, there is some contamination of this sample. At 0.95 m the sediments are most certainly post-Flandrian I/II boundary as 60 % alder pollen is recorded. The value of elm and other tree taxa suggest that the deposit is post-Elm Decline at this level, although this may be an artefact caused by local alder filtering out the other taxa. There is greater evidence of woodland clearance in the top sample but secondary woodland is still the major component of the vegetation, with some grassland. The very high values of fern spores suggest that the woodland is disturbed. Evidence of burning is low in all but the basal sample, where charcoal is present at over 20 %.

Potential

The site would be of only limited value for archaeological purposes as it appears to be heavily truncated and possibly only started forming in the later Mesolithic. Pollen preservation was mixed but at no levels was it good.

Coniston (Fig 67)
NGR SD 288 905
Height OD 49 m

The site is located on the western shore at the southern end of Coniston Water below the Blawith Fells. The 3.70 m of sediments suggest on stratigraphic evidence that a typical late-Glacial sequence (*ie* an organic silt between silty clay) was recorded. Above the upper clay layer, the gyttja indicates an aquatic environment, which became silted up allowing a fen carr to develop. The upper part of the sediments show the formation of poor-fen or acid mire with *Eriophorum* and *Calluna*. The site is now covered with rough pasture. The Blawith Fells are rich in prehistoric monuments and there is evidence also of Roman (rotary quern, SMR 5575) and medieval (house, SMR 2123) activity.

Depth (m)	**stratigraphy**
0–0.75	H8 *Eriophorum / Calluna* (*0.25 m)
0.75–1.50	H7 wood peat (*0.75 m)
1.50–2.00	H8 ditto + charcoal (*1.75 m)
2.00–2.50	H7 sedge + *Phragmites* + wood fragments (*2.25 m)
2.50–2.75	H8 ditto (*2.70 m)
2.75–2.90	H9 ditto (*2.90 m)
2.90–2.95	silt (*2.59 m)
2.95–3.00	organic gyttja (*3.00 m)
3.00–3.40	light grey silty clay
3.40–3.70	organic silt (*3.45 m, 3.70 m)
3.70+	very pure light grey silty clay (*3.75 m)

Palynological assessment

A sample from the basal organic silt at 3.70 m, which is not shown in the diagram (Fig 67) due to the scarcity of the pollen, suggests from the limited, but well-preserved, data available, that the vegetation was predominantly herbaceous. It is characteristic of the early stages of Late Devensian II. At 3.40 m the pollen indicates an open birch scrub and is possibly from the later stages of this zone. Although no samples were analysed between 3.40 m and 3.00 m it seems likely on stratigraphic criteria that the light grey silty clay at this depth represents Late Devensian III. If this is correct, the Flandrian I / Late Devensian III transition is represented at 3.00 m when the pollen spectrum suggests an open birch, juniper, and willow scrub. A silt band at 2.90–2.95 m, associated with an increase in herbaceous pollen, may indicate a period of soil instability in the area and / or a brief period of climatic deterioration in the early Holocene, a feature for which there is an increasing body of evidence (Lotter *et al* 1992; Johnsen *et al* 1992).

At 1.75 m, when the pollen is dominated by hazel, there is a peak of charcoal correlated with 11 % herbaceous pollen and very high values of fern spores, suggesting that fire caused some disturbance to the vegetation in the Mesolithic during the Holocene hazel maximum at this site. Hazel became a dominant member of the flora after *c* 9000–8000 cal BC (*c* 9500–9000 BP). The only other sample analysed is at 0.25 m and there the pollen spectrum suggests some clearances in the woodland, with grass and other herb pollen recorded. Heather was likely to be growing on the surface of the site. The sediment is probably post-Elm Decline in age.

Potential

The site appears to be truncated but would provide a record of the Late Devensian, and Flandrian I; it is more difficult to be certain whether Flandrian II or any part of Flandrian III is recorded. If the accumulation is very slow it is possible that both are represented but it appears more likely that the truncation of the deposits is very severe. The site might therefore provide a record of the early Mesolithic but, with no absolute dating, it remains uncertain if later periods are represented. However, a picture of the regional vegetational changes in the Holocene to the present day is provided from Coniston Water by Pennington (1997, 42–50). Pollen preservation was generally good but mixed at 1.75 m and 2.95 m.

Crook Hall (Fig 68)

NGR SD 448 944

Height OD 130 m

The site is situated north of the Lyth Valley and Underbarrow, in an area of low hills east of Windermere. It has well-preserved stratigraphy of 5.25 m, with basal sediments typical of a Late Devensian/early Holocene sequence, with an organic band of shell marl between two layers of silty clay. A lake formed in a hollow as the ice melted and continued there up to 3.25 m, when a fen carr developed for the major part of the sequence to 0.50 m. The upper 0.50 m is an amorphous peat with a band of silt at 0.30–0.50 m. This may represent an hiatus in the stratigraphy caused by peat-cutting. The site is now covered by a carr woodland. Bronze Age activity is known from the vicinity, in the form of two flint scrapers from ploughed ground at Yew Tree Farm, *c* 1 km north of the mire (Ellwood 1981) and a bronze palstave from near Crook (North 1942, 233).

Depth (m)	**stratigraphy**
0–0.30	H10 amorphous peat
0.30–0.50	ditto + silt (*0.30 m)
0.50–1.00	H9 wood peat (*0.80 m)
1.00–1.25	H8 ditto

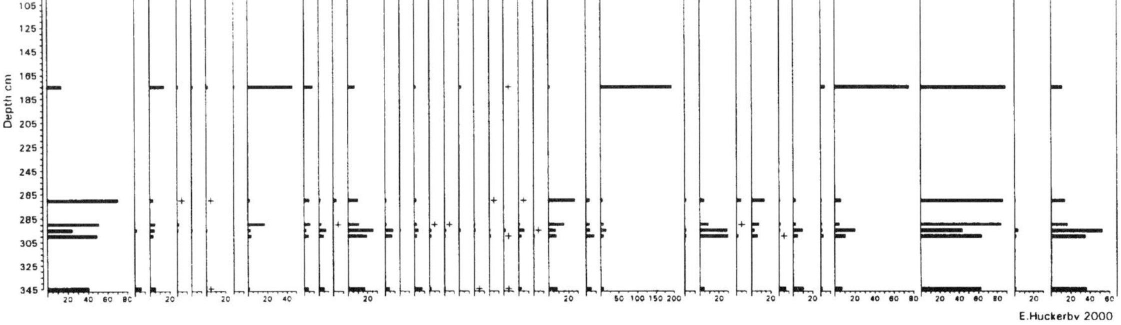

Fig 67 Coniston: simplified pollen diagram of palynological assessment

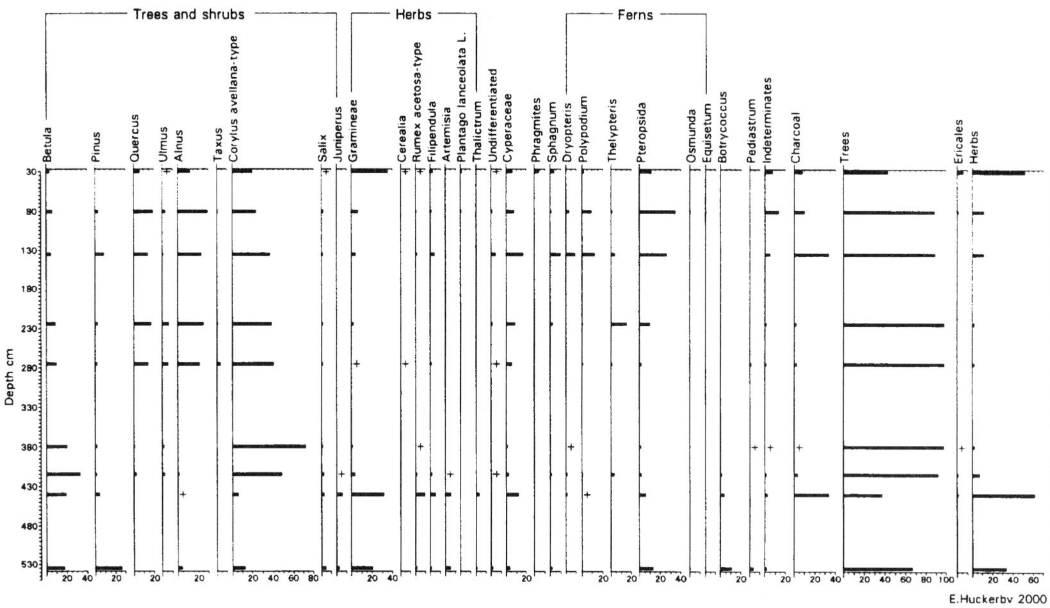

Fig 68 Crook Hall: simplified pollen diagram of palynological assessment

1.25–1.50	H7 ditto (*1.35 m)
1.50–2.00	ditto + *Phragmites* (*1.75 m)
2.00–2.50	ditto + charcoal (*2.25 m)
2.50–3.25	H7 ditto + *Potamogeton* (*2.75 m)
3.25–3.50	gyttja + abundant monocot
3.50–3.60	gyttja + *Menyaynthes trifoliata*
3.60–4.00	gyttja (*3.80 m, 4.15 m)
4.15–4.50	light grey silty clay (*4.40 m)
4.50–5.25	shell marl (*4.75 m)
5.25+	light grey silty clay + stones (*5.35 m)

Palynological assessment

Both the pollen and the stratigraphy suggest that the sediments started forming during the Late Devensian; the very base of the diagram probably represents Late Devensian I. The pine and hazel pollen may be the result of contamination or long-distance transport, a problem highlighted by Pennington (1970). No samples were analysed from the shell marl at 4.50–5.25 m but statigraphically this may represent Late Devensian II. At 4.40 m, when the deposits are a light grey silty clay, the landscape was predominantly open grassland with a little juniper and birch. It is perhaps worth commenting that this open vegetation is associated with quite high values of charcoal: this may be Late Devensian III. The pollen spectra from the samples at 4.15 m and 3.80 m are characteristic of Flandrian I, with the rapid development of a hazel–dominated community. The Flandrian I/II transition occurred prior to 2.75 m when there was 21% alder pollen with some oak, elm, hazel and pine. There was also 4 % yew (*Taxus*) pollen recorded at this level. By 1.35 m elm has declined and herbaceous pollen has risen to 11 % with 33 % charcoal, indicative of some local clearance. Pollen from the sample at 0.30 m suggests an extensively cleared landscape with a wide variety of herb taxa and cereal-type pollen recorded.

Potential

It appears likely that the site has been truncated but it may provide a record of the Late Devensian, Flandrian I, II and part of III, although the pollen evidence suggests that the later part of the prehistoric period and subsequent levels have been lost. Pollen was mainly good but more variable in the upper part of the core, which may be a factor of the peat type.

Derwent Water (Cannon Dub) (Fig 69)

NGR NY 262 189

Height OD 85 m

The site is situated at the southern end of Derwent Water near to the point where the Derwent enters the lake, in what is today a damp marshy area. The 3.90 m of well-preserved stratigraphy has alternating bands of peat, indicating mire or fen communities, and silty clays resulting from varying lake levels or hillwash from the surrounding fells deposited during flooding horizons. For a short period a fen carr (2.55–2.90 m) developed and at this level charcoal was recorded. There are records of a Neolithic axe (SMR 1234) and a cairn (SMR 12135) within 1 km of the site.

Depth (m)	stratigraphy
0–0.40	silty clay
0.40–1.10	silty peat
1.10–1.25	silty clay (*1.25 m)
1.25–1.35	H10 silty peat (*1.30 m)
1.35–1.50	silty clay
1.50–1.80	H9 amorphous peat
1.80–1.90	H10 amorphous peat (*1.90 m)
1.90–2.10	H10 amorphous peat (*2.10 m)
2.10–2.40	ditto + clay
2.40–2.55	silty clay (*2.40 m)
2.55–2.90	H8 wood peat + charcoal? (*2.70 m)
2.90–3.00	silty clay

3.00–3.20	H10 amorphous peat (*3.20 m)
3.20–3.90	silty clay + organics
3.90+	clay + stones

Palynological assessment

Because alder pollen was 12 % at 3.25 m it was thought that the organic deposits were initiated in the early part of Flandrian II. A mixed woodland of hazel, birch, pine, oak, and elm was the major plant community on the drier land. There is some evidence of disturbance, with moderate values of charcoal, fern spores and herbaceous pollen recorded. Alder dominated the local vegetation at 2.70 m and possibly filtered out other pollen types, causing a skewed pollen record. Values of herb and other tree pollen were very low, as was charcoal. From 2.40 m to 1.90 m there is quite marked evidence of a decline in the alder carr, with increasing herbaceous pollen from a number of taxa. Herb pollen rises to 19 % and there is a slight increase in charcoal at 2.10 m. Clay was recorded in the stratigraphy between 2.40 m and 2.10 m. This, together with the pollen spectra, suggests disturbance of the woodland. The increasing values of birch pollen suggest the formation of secondary woodland. There is even stronger evidence of woodland clearance at 1.30 m and 1.25 m, with herbaceous pollen reaching 35 % and charcoal 25 %. Cereal-type pollen was recorded at 1.20 m, 1.90 m, and 2.40 m, which might suggest cereal cultivation in the area; however, cereal-type pollen may also originate from some wild grasses (Andersen 1979).

Potential

The site would provide a record of the vegetation for Flandrian II and III but care must be exercised in interpreting the results because of its location, since it is situated adjacent to streams flowing into Derwent Water from the Borrowdale Valley. During flooding episodes, this could seriously affect the pollen rain entering the deposits as some could be of fluvial origin from the uplands of the Lake District and may also represent contemporary pollen or reworked pollen from older deposits. Pollen was exceptionally abundant and at some levels preservation was good, but at others it was mixed. This was particularly marked at 2.10 m when it was noted that there was the possibility of reworked sediments.

Dubbs Moss (Fig 70)

NGR	NY 103 293
Height OD	90 m

The site is located south–west of Cockermouth, near tributaries of the Cocker. The moss is approximately 17 ha and is a relict raised mire with the surrounding hills rising to *c* 110 m. The 3.70 m of well-preserved stratigraphy indicate that the moss developed from a lake as the ice melted after the last glaciation; bands of silty clay suggest that there was mineral inwash from the unstable soils surrounding the site in the early Holocene. The lake became terrestrialised with the formation of a reedswamp (*Phragmites*) then fen carr and finally an acid mire with *Eriophorum* and *Calluna* at 1.50 m. There is evidence for burning of the mire surface in the peats from 1.0 m to 1.50 m, when charcoal is recorded. Archaeological evidence for prehistoric activity exists in the area, with a Late Neolithic stone axe-hammer having been found at Moorland Close, 1.25 km south–west of Cockermouth (Jackson 1906; SMR 1076), and two unpolished axes found west of Eaglesfield (Dickinson 1878, 344; SMR 1060). Today the site is pasture.

Depth (m)	**stratigraphy**
0–0.70	H9 *Eriophorum/Calluna* (*0.25 m, 0.70 m)
0.70–1.00	H7 ditto
1.00–1.50	H6 ditto + charcoal (*1.25 m)

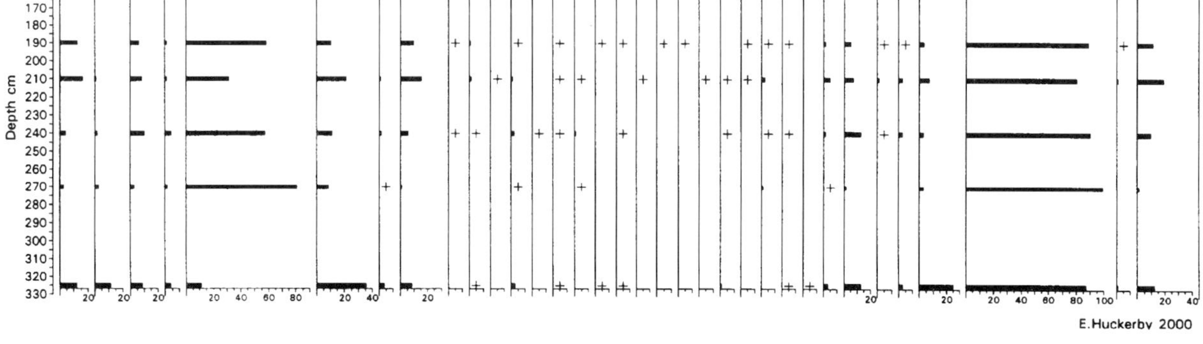

Fig 69 Derwent Water: simplified pollen diagram of palynological assessment

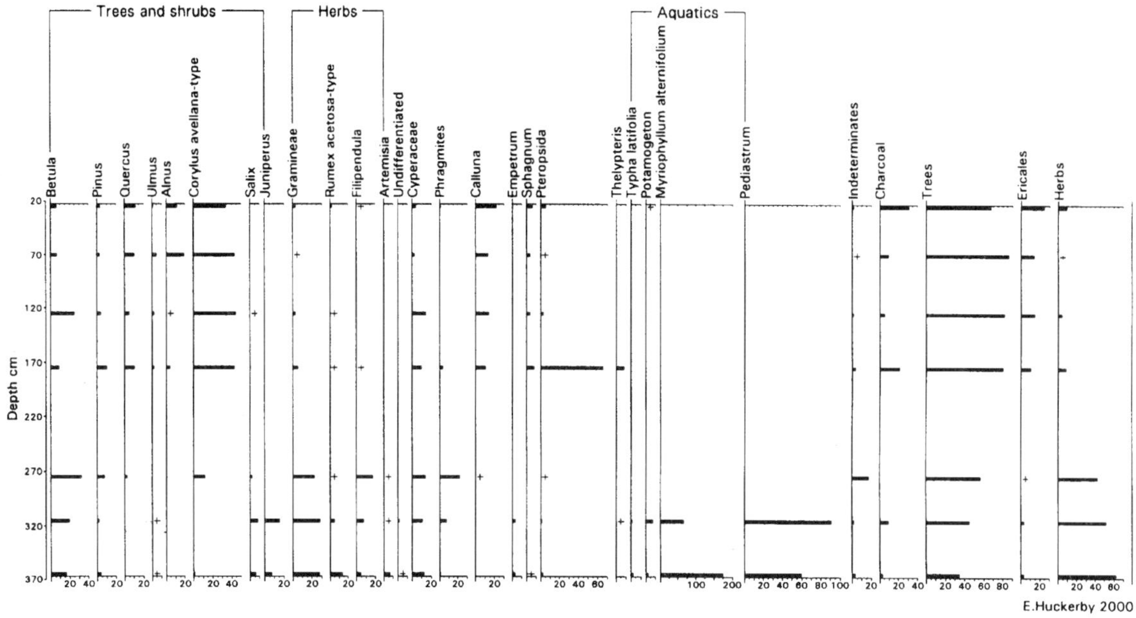

Fig 70 Dubbs Moss: simplified pollen diagram of palynological assessment

1.50–2.50	H6 wood peat + *Phragmites* + sedge (*1.75 m)
2.50–3.00	H5 *Phragmites* (*2.75 m)
3.00–3.10	gyttja
3.10–3.15	silty gyttja (*3.15 m)
3.15–3.65	silty clay (*3.65 m)
3.65–3.70	silty gyttja (*3.70 m) probably contaminated
3.70+	light grey silty clay + stones (*3.75 m) probably contaminated

Palynological assessment

Organic deposition was probably initiated early in Flandrian I when the sediments suggest an unstable landscape with periods of increased mineral inwash. However, this inwash may point to the Late Devensian being represented, although palynologically it is more likely that the Late Devensian/Flandrian I transition is recorded. The pollen spectrum suggests a community of willow and juniper, with some birch and crowberry developing, and herbaceous taxa becoming increasingly less important. This community is in turn replaced by an open birch scrub. Between 2.75 m and 1.75 m mixed deciduous woods developed with evidence of fire, in the form of charcoal, causing disturbance at 1.75 m in Flandrian Ic/Id. The high values of fern spores and some herb pollen supports the possibility of disturbance at 1.75 m. The Flandrian I/II transition occurs between 0.75 m and 1.25 m when the curve for alder pollen rises from <5 % to 19 %. The landscape was relatively wooded during this period, with heather on the site itself. The top sample analysed suggests that it is post-Elm Decline in age with high values of Ericaceous pollen, increased herb pollen, and 30 % charcoal recorded.

Potential

This site would provide a record of vegetational change for Flandrian I, II, and part of III. In archaeological terms this would encompass the Mesolithic and the transition to the Neolithic in the intermediate zone between the central uplands of the Lake District and the West Cumbrian Plain. It might therefore hold the potential to elucidate the problem of the highland/lowland divide in the Mesolithic. Pollen preservation was good, except at 1.75 m and 2.75 m, making detailed palynological work possible.

Eskrigg (Fig 71)

NGR SD 576 889

Height OD 190 m

The site is situated close to the M6 motorway south–east of Kendal. The 5.50 m of well-preserved stratigraphy suggest that the tarn was present in Late Devensian II, with a typical late-Glacial stratigraphic sequence of clay/organic/clay, although palynologically this is not obvious and organic deposition may have been initiated early in Flandrian I. Open water was replaced by a fen carr and then by fen in the upper stratigraphy. There are no SMR records within 1 km of the tarn.

Depth (m)	**stratigraphy**
0–0.05	H10 amorphous peat
0.05–0.50	H5 sedge (*0.15 m)
0.50–1.00	H6 sedge (*0.75 m)
1.00–2.50	H7 ditto (*2.25 m)
2.50–3.50	H8 wood peat + sedge
3.50–4.00	gyttja + wood fragments (*3.75 m)
4.00–5.00	gyttja (*4.40 m, 4.90 m)
5.00–5.10	light grey silty clay
5.10–5.40	organic clay (*5.10 m, 5.35 m, 5.50 m)

5.40–5.50	light grey silty clay
5.50+	light grey silty clay + stones

Palynological assessment

The results of the palynological assessment suggest that at 5.35 m and 5.50 m the vegetation was a relatively open birch scrub with some juniper and willow. Probably the sample at 5.50 m is contaminated, or hazel pollen was deposited as the result of long distance transport. It is not possible to ascertain whether these two spectra formed during Late Devensian II, or early Flandrian I. The stratigraphy suggests that it may be Late Devensian II, with the typical inorganic/organic/inorganic fluctuation, but alternatively the clay band at 5.00–5.10 m may represent unstable conditions at the start of the Holocene. At 4.40 m and 4.90 m hazel dominated the vegetation, with pine, oak and elm increasing in importance at 4.40 m. Herbaceous pollen is at very low values. The sample at 2.25 m suggests a post-Elm Decline date, with 8 % herb pollen and 4 % ash pollen recorded, suggesting some opening of the canopy. The upper sample at 0.15 m indicates a declining woodland cover with some heathland and grassland. There is a little evidence of arable agriculture with cereal-type, Chenopodiaceae and Compositae pollen recorded, but secondary woodland still dominated the landscape. Charcoal is recorded at values of 1.5–9 %, the higher values at the base and top of the core.

Potential

It is considered that the site has been truncated at some level above 0.15 m, possibly at 0.05 m, where the stratigraphy changes to an amorphous peat which may represent regeneration of the surface vegetation. The site will provide a record of the Flandrian I, II, and part of III, perhaps encompassing the prehistoric periods. It is possible that the Late Devensian late-Glacial is also represented. Pollen preservation was predominantly good, except at 5.50 m, where it was suspected that some contamination had occurred.

Gatebeck (Fig 72)
NGR SD 549 859
Height OD 70 m

The site is situated in the valley of Peasey Beck, north–east of Endmoor, near Kendal. The stratigraphy is well-preserved and is 3.0 m in depth, but is probably severely truncated. The shell marl recorded at 2.70 m to 3.00 m suggests that a calcareous lake developed in Late Devensian II or early in Flandrian I. The presence of a band of silty gyttja from 2.70 m to 2.60 m suggests either unstable soil conditions in Late Devensian III or in Flandrian I but, with no pollen sample from this band, it is not possible to be certain when this formed. The upper 2.0 m of peats indicate that terrestrialisation occurred with the formation of a reedswamp and fen carr, when charcoal is recorded. No acid-mire peats are recorded and it seems likely that the peats are heavily truncated, although today the site is under pasture. The Cumbria SMR has no records of any archaeology within 1 km of the site.

Depth (m)	**stratigraphy**
0–0.50	H8 wood peat (*0.20 m)
0.50–1.00	ditto + *Phragmites* (*0.50 m, 0.70 m)
1.00–1.70	ditto + charcoal (*1.25 m)
1.70–2.00	H8 *Phragmites* + *Polytrichum* (*1.80 m)
2.00–2.60	gyttja (*2.50 m)
2.60–2.70	silty gyttja
2.70–3.00	shell marl (*2.70 m)

Palynological assessment

The lower two samples at 2.70 m and 2.50 m suggest that the landscape was relatively open with some birch. The

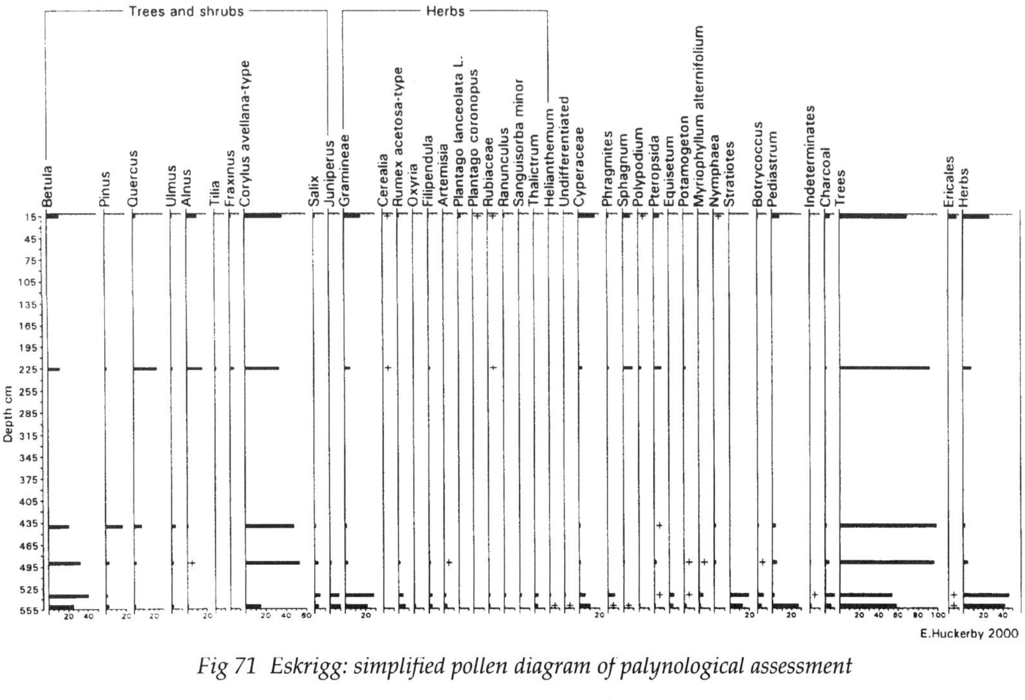

Fig 71 Eskrigg: simplified pollen diagram of palynological assessment

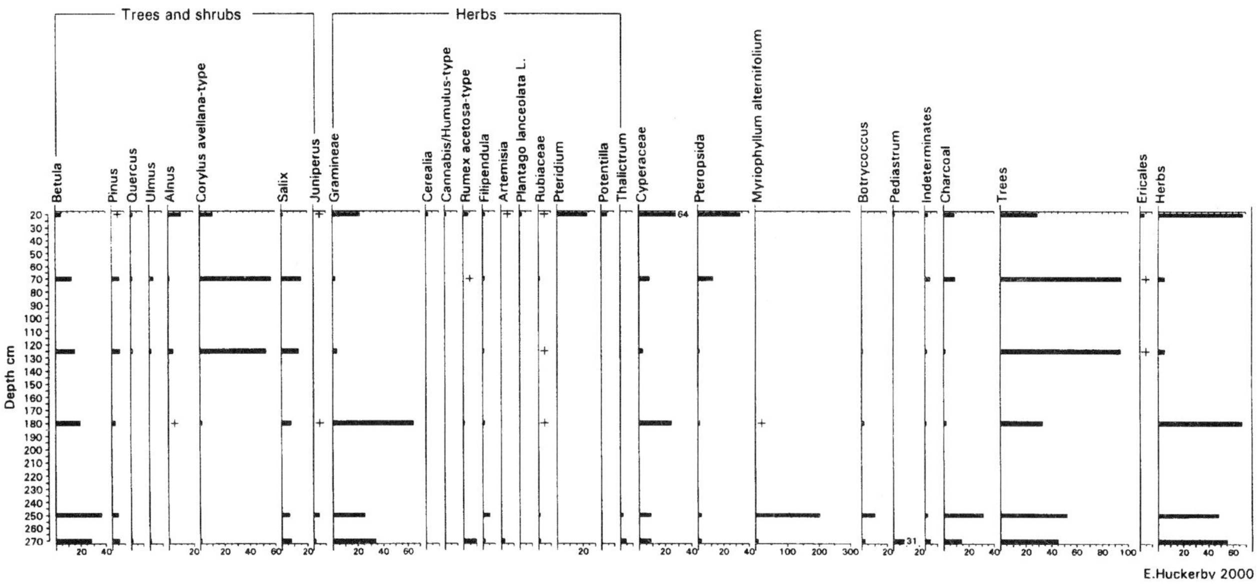

Fig 72 Gatebeck: simplified pollen diagram of palynological assessment

apparent increase in grassland at 1.80 m is probably attributable to *Phragmites* pollen and is therefore associated with the local vegetation rather than the landscape in general. By 1.25 m and 0.70 m hazel dominated the vegetation with some birch, willow and a little pine, oak, elm, and alder. Herb pollen is recorded at low values, suggesting a closed canopy. However, at 0.20 m, although the peat type is wood, the pollen spectrum suggests a predominantly treeless landscape with some evidence of arable and pastoral agriculture. This suggests that there may have been an hiatus in the deposition of the peat at some level between 0.70 m and 0.20 m.

Potential

The deposit would provide a record of the vegetation of the early Mesolithic but probably with a gap until relatively recent times. The upper part of the stratigraphy may represent the recent past, which is perhaps of importance, given the rapidly diminishing resource of later peat generally. Pollen preservation was good on the whole but very poor at 0.70 m. This may be associated with the peat type; alternatively, if peat-cutting occurred during the Middle Ages, it may have damaged the upper part of the peat stratigraphy by alteration of the water-table and oxygen content, thus causing damage to the pollen grains.

Glebe Farm (Fig 73)

NGR NY185 346

Height OD 130–140 m

The site lies in the north–west of the county just inside the boundary of the Lake District National Park; it is a basin mire and covers *c* 20 ha. Today the peat is used for pasture, but well-preserved sediments reach a depth of >7 m. A lake was present in the early Holocene, as indicated by 2 m of freshwater sediments and, as it became terrestrialised in the mid-Holocene, a reedswamp developed followed by fen carr with the formation of an acid-mire community above this.

Depth (m)	stratigraphy
0–0.50	H4 bryophytes (*0.40 m)
0.50–1.00	ditto + *Eriophorum*
1.00–2.50	unsampled
2.50–3.00	H9 amorphous peat (*2.75 m)
3.00–3.50	H9 wood peat + *Polytrichum* (*3.25 m)
3.50–5.50	H9/10 ditto + *Phragmites* (*4.25 m)
5.50–7.00	gyttja (*5.75 m, 6.80 m)
7.00+	silty clay

Palynological assessment

Four palynological samples were analysed from this 7 m core. The lowest suggests that the gyttja started accumulating in Flandrian I during the hazel maximum. Herb pollen was low and the spectrum was dominated by hazel and birch. The central part of the core at 3.25 m and 2.75 m is probably post-Elm Decline in age and there are indications of woodland clearance with evidence for arable (cereal-type, *Artemisia* and Chenopodiaceae pollen) and pastoral agriculture (grass and *Plantago lanceolata* pollen).The site was still fairly wet at this stage, with aquatic taxa and diatoms recorded. At the top of the core, high percentages of birch, grass and *Calluna* pollen suggest regeneration of open birch woods with some heathland. Charcoal was low throughout the core except at 0.40 m where it reaches 20 %.

Potential

Pollen preservation was mainly good and the site would probably record the vegetational changes throughout the major part of Flandrian I, II, and III, but if the sediments are truncated potential is likely to be very limited. However,

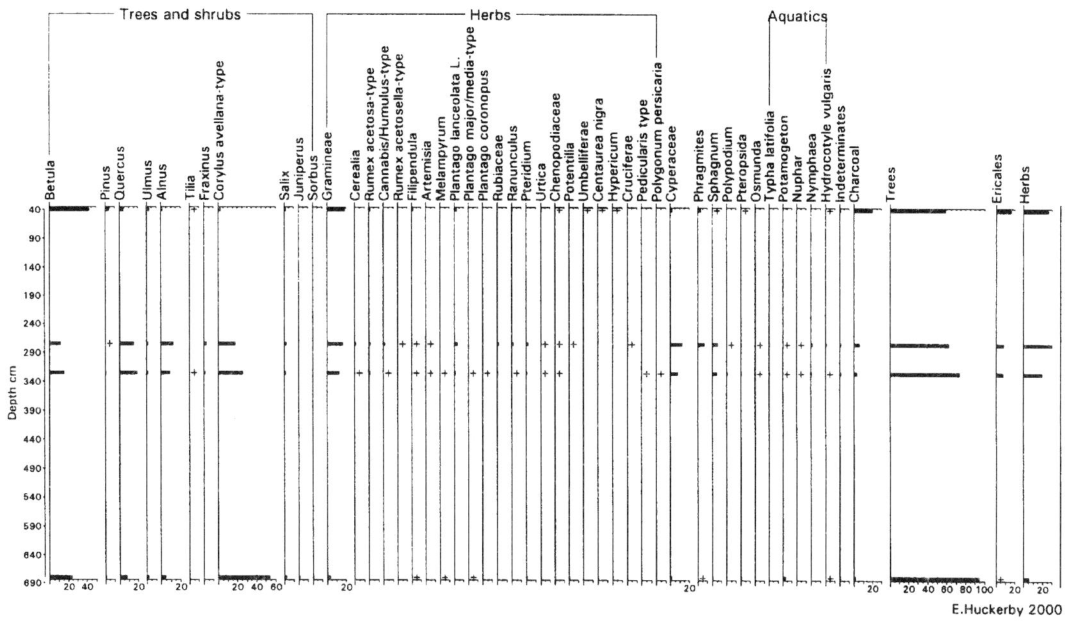

Fig 73 Glebe Farm: simplified pollen diagram of palynological assessment

this site might contribute significantly to our knowledge of the development of the landscape, which would be of importance because of its close proximity to Roman forts in the hinterland of Hadrian's Wall and their connecting road system.

Graythwaite (Fig 74)
NGR NY 1135 2300
Height OD 130 m

The site is situated 1 km north–west of Loweswater in a valley containing several small streams, all of which flow into the lake.The moss is *c* 23 ha and is a relict valley mire.

The 2.10 m of well-preserved stratigraphy suggests that organic deposition was initiated on silty clay in Flandrian I. The site became wetter at 1.70–1.80 m when a fen community formed and subsequently it became a fen carr. Today the site is under pasture.

Depth (m)	**stratigraphy**
0–0.50	H8 wood peat (*0.40 m)
0.50–1.70	H7 ditto (*0.80 m, 1.50 m)
1.70–1.80	H6 monocot (*1.70 m)
1.80–2.10	H10 amorphous peat
2.10+	light grey silty clay + stones

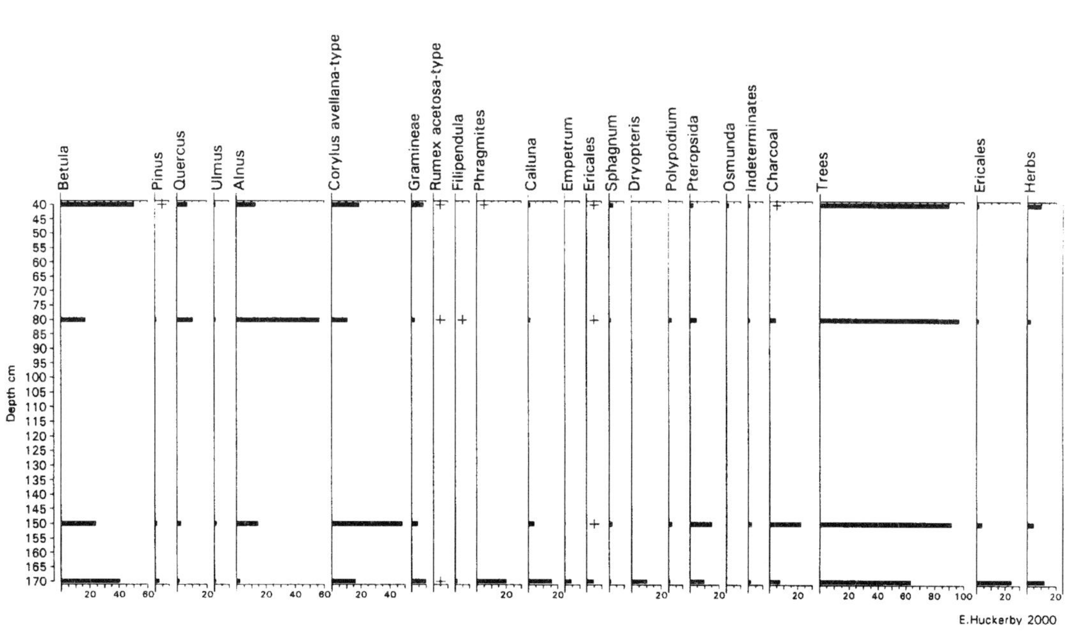

Fig 74 Graythwaite: simplified pollen diagram of palynological assessment

Palynological assessment

The data from the four samples analysed palynologically were inconclusive. The pollen spectrum from 1.75 m suggests that the deposit is early Flandrian I in age, with a birch and hazel community with significant heath and grassland present. At first glance the pollen at 1.50 m appears to be from Flandrian III, with low elm, oak and pine, 4.5 % herbs and 14 % alder pollen. However, stratigraphically it is unlikely that 0.20 m of peat represents both Flandrian Id and II. There are two possible explanations for this. The first is that organic accumulation was initiated after the Elm Decline and that the pollen spectrum represents a locally regenerated birch and hazel wood. Alternatively, sedimentation ceased for an unspecified period of time at the site, causing an hiatus.

Potential

Palynological assessment was inconclusive and therefore without the benefit of radiocarbon dates the site would be of very little value. Indeed, there is no indication to which chronozones deposits from this site belong. Pollen was abundant and preservation good. However, at 1.50 m, a level rich in charcoal, there were problems with the sample clumping after preparation.

Heathwaite (Fig 75)

NGR SD 255 869

Height OD 120 m

The site lies to the west of Subberthwaite Common at the head of the Furness Peninsula and south–west of a possible ring cairn (SMR 9407). The 3.95 m of stratigraphy indicate that a lake was once present and that this became terrestrialised with the formation of a fen carr, and subsequently a more acidic mire with *Eriophorum*, *Calluna* and *Sphagnum*. The surrounding upland margins are exceptionally rich in archaeological activity, consisting primarily of evidence for Bronze Age and medieval agriculture, comprising cairns, field systems, and settlements (*eg* SMR 9394–9423 and SMR 2160).

Depth (m)	stratigraphy
0–0.50	H8 *Eriophorum / Sphagnum / Calluna*
0.50–1.00	H8 wood peat (*0.60 m)
1.00–2.00	H8 / 9 ditto (*1.25 m, 1.75 m)
2.00–2.50	H9 ditto
2.50–3.50	H9 / 10 ditto
3.50–3.95	silty gyttja (*3.75 m)
3.95+	light grey silty clay + stones

Palynological assessment

The lower section of this core proved disappointing palynologically and the sample from 3.75 m yielded no data. The pollen analysis from the samples at 1.75 m, 1.25 m and 0.60 m suggests that peat at these levels formed after the Elm Decline, with an alder carr growing near the site and the formation of secondary woodland of hazel, but with some oak and birch. High levels of fern spores at 1.25 m indicate that the woodland was disturbed. The top sample (0.60 m) has high values of charcoal recorded, with 16 % herbaceous pollen.

Potential

The potential of this site is ambiguous and it is not possible to make any firm statements about its archaeological use without additional data, except to say that above 1.75 m the peats date from Flandrian III and are probably truncated. Pollen preservation was at best mixed but was more generally poor.

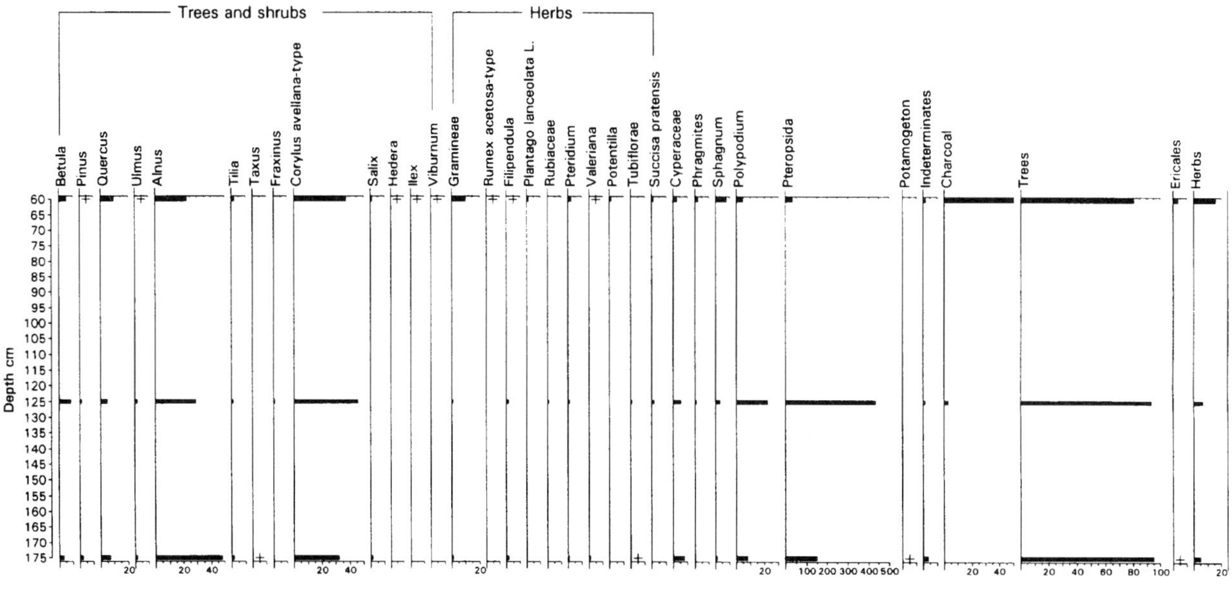

Fig 75 Heathwaite: simplified pollen diagram of palynological assessment

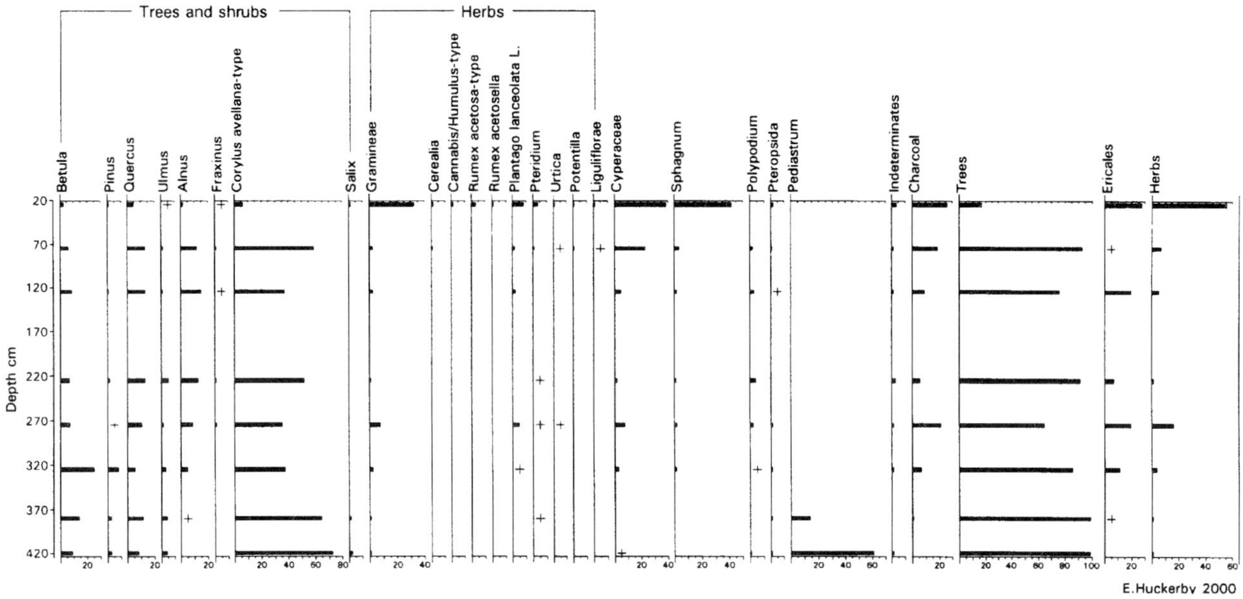

Fig 76 Heights: simplified pollen diagram of palynological assessment

Heights (Egholme Peat Moss) (Fig 76)
NGR SD 603 861
Height OD 220 m

The site is located on higher ground to the west of the River Lune, 8 km south–east of Kendal, in an area under pasture today. The 4.20 m of well-preserved stratigraphy indicates that a lake formed early in the Holocene, becoming terrestrialised with the formation of an acid mire for the remainder of the time-span encompassed by the deposits. There are few archaeological records within 1 km of the site except a small undated enclosure (SMR 4285) and a tomb-like object at Rigmarden Park (SMR 4368). However, to the east there are records of considerable Roman activity in the Lune Valley.

Depth (m)	stratigraphy
0–0.10	H10 amorphous peat
0.10–0.60	H4 *Sphagnum* (*0.20–0.30 m)
0.60–1.00	H7 *Eriophorum / Calluna* (*0.70–0.80 m)
1.00–2.50	ditto (*1.25 m, 1.75 m, 225 m)
2..50–3.00	H8 ditto (*2.75 m)
3.00–3.50	H7 ditto (*3.25 m)
3.50–3.75	H6 *Sphagnum* / sedge (*3.80 m)
3.75–4.00	gyttja (*3.80 m)
4.00–4.20	ditto (*4.20 m)
4.20+	light grey silty clay

Palynological assessment

The lower two samples indicate that the vegetation was dominated by hazel, with increasing numbers of thermophilous trees at 3.25 m where there is a decrease in the value of hazel pollen and a concomitant rise in birch, pine, oak, alder and heather, associated with a little charcoal. The Flandrian I / II boundary occurred around 2.75 m and, as at many other sites looked at during the course of the survey, for instance Thwaite House Moss (Middleton *et al* 1995), Danes Moss (Leah *et al* 1997), and Wolfshead Moss (Leah *et al* 1998), this was associated with a fall in tree pollen, a rise in *Calluna*, and also in the quantity and variety of herb pollen, and a peak in charcoal. By 2.25 m, tree pollen has recovered. The boundary between Flandrian II and III took place between 2.25 m and 1.25 m, at which depth tree pollen had declined again and herbaceous pollen was slightly higher. The values of herb pollen are at lower levels than at the I / II transition, which suggests less disturbance. The uppermost sample (0.25 m) is rich in herbaceous pollen and tree pollen has fallen to 17 %. Both cereal-type and *Humulus / Cannabis*-type pollen are recorded, indicating some arable agriculture.

Potential

This site would record the vegetational changes for the Mesolithic, Neolithic and possibly later periods. However, without the benefits of absolute dating, it is not possible to say what depth of peat has been removed. Pollen preservation was basically good.

Hood Tarn (Fig 77)
NGR SD 571 876
Height OD 160 m

The site is located east of the M6 motorway south–east of Old Hutton in the south–eastern extreme of the county. The stratigraphy of 5.50 m is from the margins of the present tarn and indicates that its extent in the early Flandrian I was greater than it is today, with gyttja recorded. Terrestrialisation of the margins took place with the formation of a fen carr. This became more acidic for a time with *Eriophorum* and *Calluna* recorded, before reverting again to fen carr. There are no archaeological records within 1 km of the coring site, but nevertheless the palaeoecological record could prove valuable in providing information about possible woodland clearance.

Depth (m)	stratigraphy
0–0.10	H10 amorphous peat
0.10–0.50	H8 wood peat (*0.50 m)
0.50–1.50	H6 wood peat + *Phragmites* (*0.75 m)
1.50–2.00	H9 ditto
2.00–2.50	H8 ditto
2.50–3.00	H8 *Eriophorum / Calluna* (*2.75 m)
3.00–3.50	H9 ditto + wood fragments
3.50–4.20	H8 sedge (*3.75 m)
4.20–4.50	gyttja
4.50–5.00	ditto (*4.90 m)
5.00–5.10	H4 *Hylocomium*? (*5.10 m)
5.10–5.50	gyttja (*5.40 m)
5.50+	light grey silty clay (*5.50 m)

Palynological assessment

Organic deposition was initiated in Flandrian I just prior to the rise in alder pollen, which is associated with a peak in charcoal at 5.10 m, a reduction in hazel, an increase in pine, oak, and elm, but only low values of herb pollen. The Elm Decline occurred possibly about 3.75 m, although the fall in elm pollen is only small; there were slightly reduced values of tree pollen and 6–8 % herb pollen. However, because of the very coarse sampling interval, this may only be a minor fluctuation and the actual decline is at some level prior to 2.75 m when elm pollen is only 0.5 %. At 0.75–0.50 m tree pollen falls to 60 % and herb reaches *c* 36 %, with a wide variety of taxa recorded. Cereal-type pollen is recorded in both these samples. It seems likely that the peat is truncated at some point above this and perhaps the stratigraphic change at 0.10 m indicates the level at which this occurred.

Potential

This site would possibly provide a record from the late Mesolithic to the late Iron Age / Romano-British period. From the limited data available it would appear that more recent deposits have been destroyed. The three lower samples had plentiful, well-preserved pollen. However, in the upper samples, in particular 0.50 m, preservation was at its best mixed but was generally poor.

Little Tarn (Fig 78)
NGR NY 2495 3385
Height OD 190 m

Little Tarn is a small expanse of water north of Skiddaw, less than 1 km from the Uldale Fells, situated on low-lying ground to the north–west of the central massif of the Lake District. The 6.70 m of well-preserved stratigraphy from the peats surrounding the present tarn suggest that it was more extensive in the early Holocene than today. A fen carr brought about its partial terrestrialisation sometime in Flandrian I or II. This carr was only replaced towards the top of the stratigraphy by a *Phragmites* peat, suggesting perhaps that conditions became wetter. Palynologically the sediments appear untruncated and therefore provide a unique opportunity to study pollen records from the transition from the lowlands of West Cumbria to the central uplands of Skiddaw and the Uldale Fells. Archaeological records from the immediate area are not abundant, but there is evidence for a moated medieval site (SMR 885), a non-defensive enclosure (SMR 879), and place-name evidence that suggests a castle (SMR 11034) in the vicinity.

Depth (m)	stratigraphy
0–0.20	silty peat
0.20–0.50	H7 *Phragmites* (*0.30 m)
0.50–1.00	H8 sedge + wood fragments

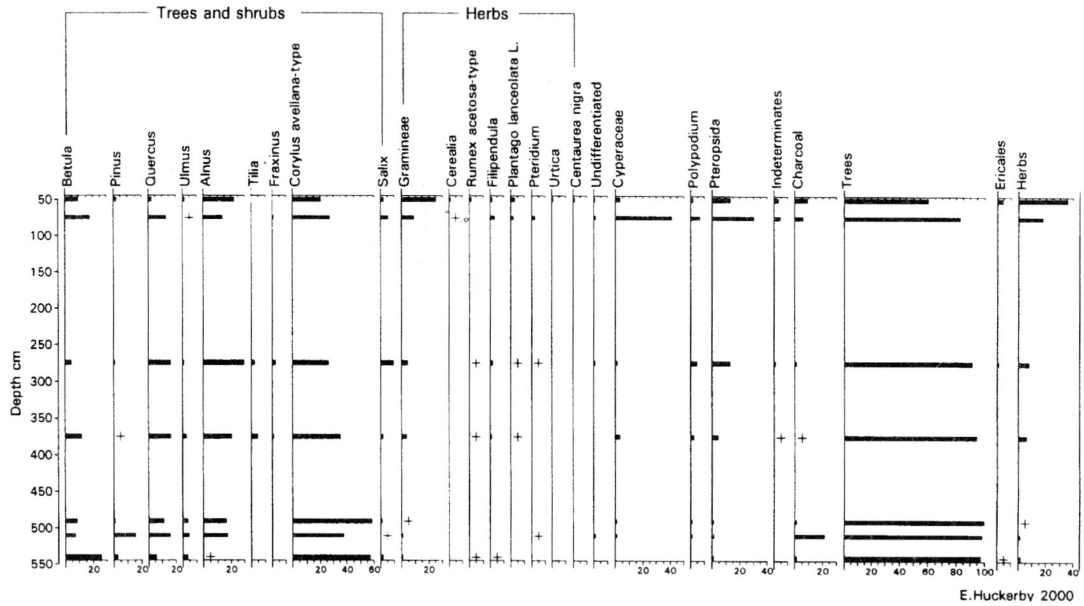

Fig 77 Hood Tarn: simplified pollen diagram of palynological assessment

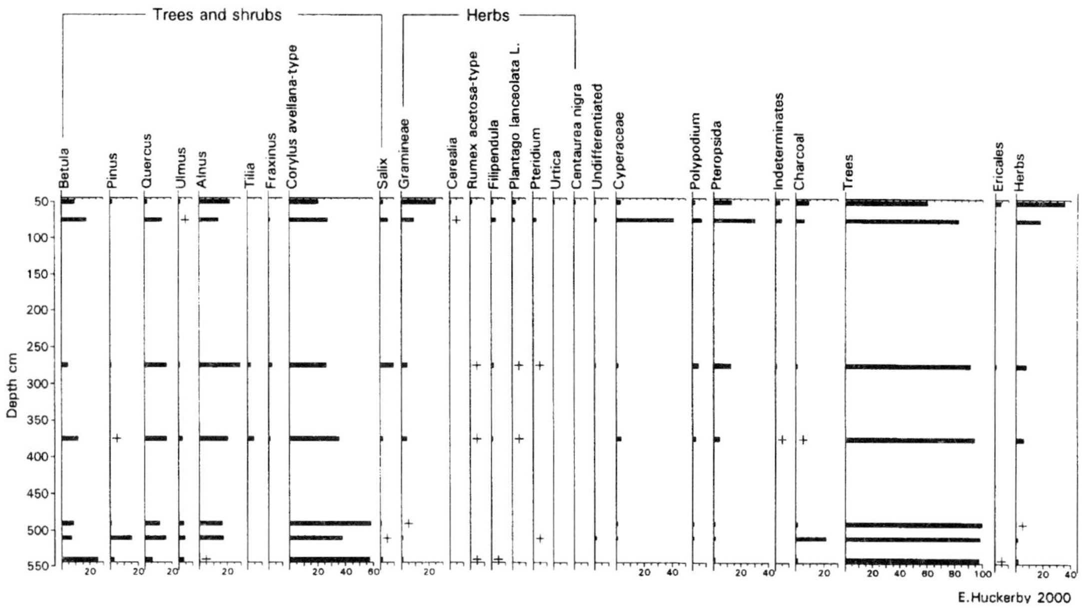

Fig 78 Little Tarn: simplified pollen diagram of palynological assessment

1.00–2.50	ditto + *Phragmites* (*1.25 m, 2.25 m)
2.50–3.00	H9 ditto
3.00–3.50	unsampled
3.50–6.00	H8 wood peat (*3.75 m, 4.75 m)
6.00–6.10	black gyttja
610–6.20	grey green gyttja (*6.20 m)
6.20–6.70	black gyttja (*6.70 m)
6.70+	light grey silty clay + stones

Palynological assessment

The two lower samples formed early in Flandrian I, as hazel, oak, elm, and pine replaced the birch scrub. The Flandrian I/II transition took place at some level between 6.25 m and 4.75 m as, by the upper sample, there is 24 % alder pollen. The pollen spectra at 4.75 m and 3.75 m suggest an elm decline had occurred but there is very little evidence of any clearance of woodland. From 2.25 m to 0.30 m the diagram suggests that a secondary woodland of birch and hazel replaced the oak and alder from the central part of the core: Pennington (1970) considered that oak and alder was extensively cleared in the Bronze Age. The high values of sedge, reed and aquatic pollen at 0.30 m suggest that the site became wetter in more recent times. This is associated with a pollen spectrum that indicates an extensively cleared landscape, with some cereal-type pollen and a wide variety of herb taxa recorded.

Potential

This site would provide a record of vegetation change from the low ground immediately to the west of the Uldale fells and north–west of Skiddaw in the central uplands of the Lake District. The assessment suggests that the record is complete from the Late Devensian to the present and would therefore be of very high value in assessing the land-use, throughout later prehistory and the historic periods, from a transitional landscape between the highland and lowland zones. Pollen was abundant and preservation was mainly good.

Littlewater (Fig 79)
NGR NY 5090 1705
Height OD 240 m

Littlewater is a small tarn north–east of the Haweswater reservoir, situated on raised ground to the west of Haweswater Beck below Bampton Common. Sampling was left incomplete as a result of the depth of the deposits, but at a depth of 6.50–8.50 m the tarn was shown to have been more extensive than it is at present. As the water level fell or terrestrialisation commenced, a reedswamp (2.0 m of monocot peat was recorded) developed, followed by a fen carr for the major part of the remaining stratigraphy, reverting to a *Phragmites* community for the upper 0.50 m. Today the ground is under pasture with some open woodland. The uplands around Haweswater are exceptionally rich in archaeological records (Quartermaine forthcoming) and within 1km of Littlewater there are, for example, finds of Neolithic axes (SMR 1542), Bronze Age cairns (SMR 1563), settlements (SMR 5281), cairnfields at Burn Banks (SMR 8325), Romano-British farmsteads (SMR 1518), amongst others.

Depth (m)	stratigraphy
0–0.25	H9 *Phragmites* (*0.25 m)
0.25–0.50	H6 ditto
0.50–1.00	H5 ditto + wood fragments (*0.75 m)
1.00–1.50	unsampled
1.50–2.00	H9 alder peat
2.00–2.50	H9/10 wood peat (*2.25 m)
2.50–3.00	unsampled
3.00–3.50	H8 bryophytes (*Hylocomium*?) + *Alnus* (*3.25 m)
3.50–4.00	H9/10 ditto

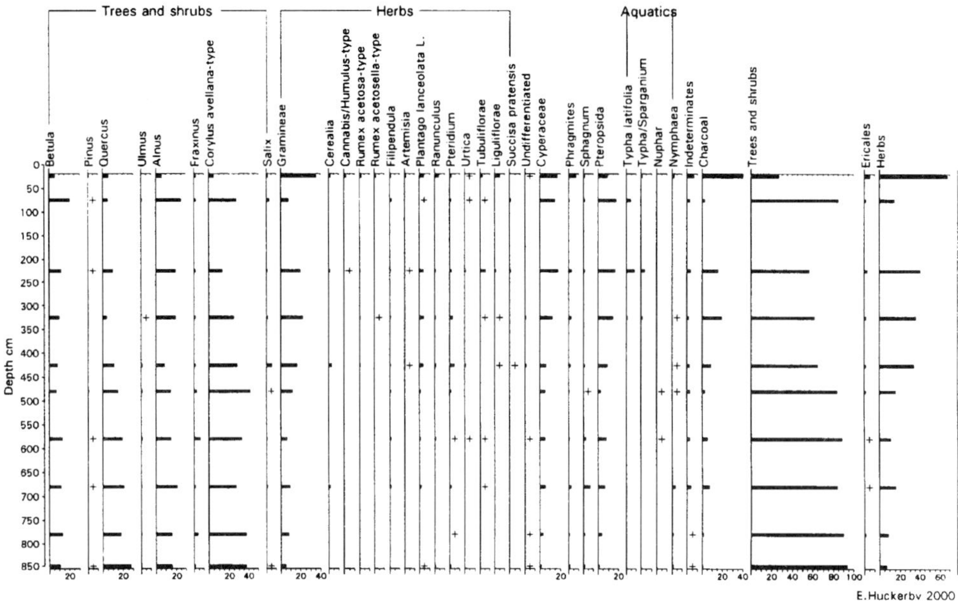

Fig 79 Littlewater: simplified pollen diagram of palynological assessment

4.00–4.50	H9 ditto + silt (*4.25 m)
4.50–5.00	H9 monocot + gyttja (*4.80 m)
5.00–5.50	ditto + *Polytrichum* (*5.80 m)
5.50–6.50	H8 monocot (*5.80 m)
6.50–8.50	organic gyttja (*6.80 m, 7.40 m, 7.80 m, 8.50 m)
8.50+	out of rods

Palynological assessment

The pollen analytical data from this 8.50 m+ core suggests that the accumulation rate may have been extremely rapid. This premise is based on the fact that elm pollen, even recalculating it as a percentage of tree species excluding hazel, herbs and heather, only reaches a maximum of 2.8 % whereas values of >10 % could be expected if it was prior to an elm decline. Tree pollen throughout the lower part of the core does not exceed 95 % and this, together with the consistent record of ash pollen, suggests some lessening of the canopy. Above 4.50 m evidence of clearance becomes greater and, at 4.25 m, 35 % herbaceous pollen is recorded, including 2.8 % cereal-type, indicating arable agriculture. At this level silt is recorded in the stratigraphy which may suggest inwash into the deposits as a result of the instability of the surrounding landscape. The date of this phase is a matter of conjecture as there are no radiocarbon dates for these samples, although recently peat from another core, from slightly deeper in the stratigraphy, has been dated to 900–790 cal BC (2645±50 BP, AA–33593 (GU–8249)). There is a single period of woodland regeneration recorded at 0.75 m, when birch, hazel, and alder rise steeply and herbaceous pollen falls. However, other regeneration phases may exist, since the site has only been assessed crudely for its potential. The uppermost sample (0.25 m) analysed shows obvious evidence of clearance, with 68 % herbaceous and 27 % tree pollen. Both cereal-type and *Humulus*/*Cannabis*-type pollen are recorded at values of 2 %, along with significant percentages of other herbaceous taxa.

Potential

This site has the potential to provide a comprehensive record of the vegetation in an area rich in archaeological remains (Quartermaine forthcoming). The extent of the record is unknown at present, as sampling was incomplete due to the depth of the deposits, and also because there is a possibility that the elm pollen record is not behaving in a characteristic manner. This possible abnormal record may be the result of local edaphic conditions, with only a few trees growing around the site in Flandrian II. The one indisputable fact, because of the values of alder pollen, is that at 8.50 m the deposits are post-Flandrian I/II transition in age. It does appear, however, that the entire historic period is represented at this site, which is unusual, and therefore makes it of great importance archaeologically. As at many sites, pollen preservation was good at the lower levels but more mixed in the upper half, particularly in the silt band at 4.25 m. The mixed nature of the pollen preservation, together with the silt at 4.25 m, may suggest that secondary pollen was being washed into the deposits along with the silt at this level.

Mosedale (possibly Bramely) (Fig 80)

NGR NY 109 234

Height OD 170 m

The site is situated in a valley created by a number of small streams flowing into Loweswater. The stratigraphy of 2.90 m of well-preserved deposits suggests that a sedge and reed community developed on the clay as the ice melted after Late Devensian III. It appears that a wetter, more acidic community, with *Scheuchzeria* and *Menyanthes trifoliata*, formed, becoming a fen carr and finally an acid mire.

Depth (m)	stratigraphy
0–0.20	unsampled
0.20–1.00	H9 amorphous peat (*0.50 m)
1.00–2.00	H8 wood peat + charcoal (*1.20 m)
2.00–2.60	H7 ditto + monocot (*2.30 m)
2.60–2.80	H6 bryophytes + *Scheuchzeria* + *Menyaynthes trifoliata*
2.80–2.90	H9 monocot (*2.90 m)
2.90+	clay + slate fragments

Palynological assessment

Four samples were analysed palynologically. The pollen spectrum from the sample at 2.90 m has 76 % tree and shrub, 21 % herb, and 3 % Ericaceous pollen. The major tree component is birch with some hazel, willow and juniper. The herbaceous taxa are varied and include grass, *Thalictrum*, *Helianthemum* and *Artemisia*, amongst others. Sedge pollen is abundant and there is 17 % *Sphagnum* spores; aquatic pollen from *Myriophyllum alterniflorum* and other taxa are recorded. This suggests a damp sedge and reed community with some open water at the site, surrounded by an open birch scrub. At 2.30 m tree and shrub pollen is 89 % and is dominated by hazel and birch. Ericaceous pollen is similar to the previous sample and herb pollen is less varied, only reaching 8 %. The site was drier at this stage with possibly some birch and heather growing on the surface and hazel dominating the wider landscape. At 1.20 m the spectrum suggested a Flandrian II or III date, with 78 % trees and shrub, 13 % Ericales and 9 % herbaceous pollen. Charcoal is quite high; the major tree taxa are birch, hazel and alder with very little elm, the local vegetation probably being an alder carr community with some heather and herbs. The low level of elm pollen suggests at least a post-3000 cal BC (*c* 5000 BP) date at this level. The sample at 0.50 m is a conundrum: the pollen suggests a wooded landscape with evidence of burning and a high percentage of pine pollen, together with 15 % alder pollen. Either this represents possibly a re-invasion of the bog surface by pine during the late Neolithic, a feature of sites in Cheshire, for instance White Moss (Lageard 1992) and Lindow Moss (Leah *et al* 1997, 61) which has not been recorded previously from Cumbrian bogs, or alternatively the stratigraphy is very disturbed.

Potential

This site may provide an insight into the changes in the landscape from the Mesolithic to the Late Neolithic / Bronze Age. Palaeoecologically it would be of great interest to verify the status of the pine at Mosedale. The peats at 0.50 m need to be dated by radiometric analysis to check whether the sequence is disturbed or if it is a previously undetected feature in West Cumbrian bogs. Pollen preservation was either good or mixed.

Oxenholme (The Helm) (Fig 81)

NGR SD 5290 8910

Height OD 100 m

The site is a small basin mire situated *c* 1 km north–east of Castlesteads hillfort, east of Kendal (SMR 2641). Shell marl is recorded in the well-preserved stratigraphy at 3.60–4.40 m and indicates that a calcareous lake was present, probably in Late Devensian II, with a narrow silty clay band indicating cooler and/or unstable soil conditions. An aquatic environment continued for a brief time before an acid mire, or poor fen, formed becoming more base-rich and eventually producing a fen carr and possibly a fen. Today the area is under pasture. Potentially the site is important

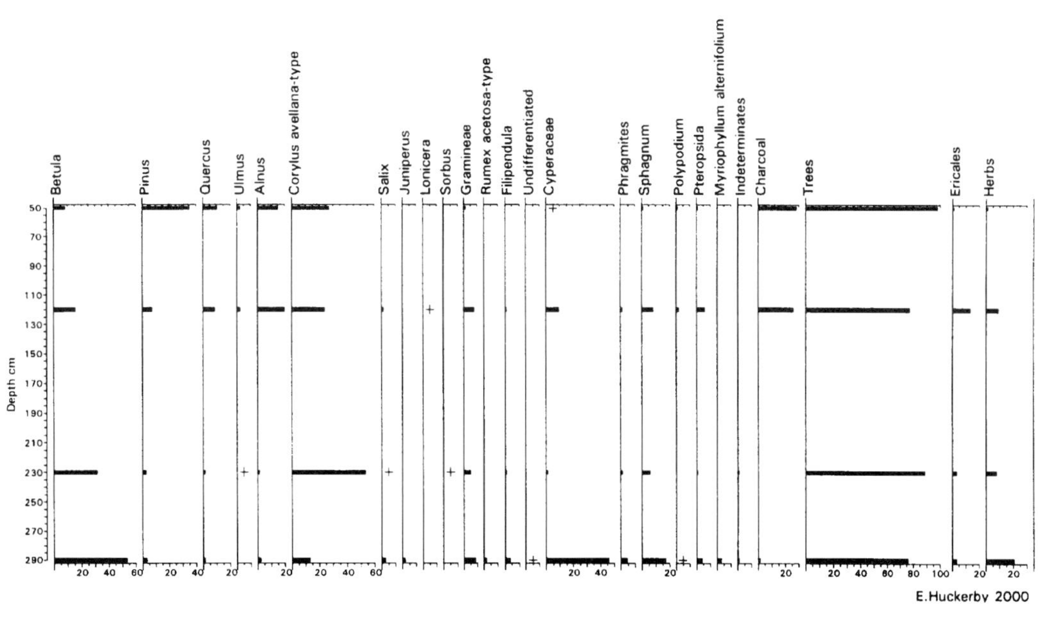

Fig 80 Mosedale: simplified pollen diagram of palynological assessment

because of its proximity to the hillfort, a cairnfield (SMR 14846) and field system (SMR 14743), both shown on aerial photographs.

Depth (m)	**stratigraphy**
0–0.50	unsampled (sample retrieved from location 20 m away)
0.50–1.00	ditto (*0.75 m)
1.00–1.50	H7 monocot (*1.10 m)
1.50–2.00	H8/9 monocot + wood fragments
2.00–2.50	H6 bryophyte
2.50–3.10	H7 *Eriophorum / Calluna* (*2.75 m)
3.10–3.50	gyttja (*3.25 m)
3.50–3.60	light grey silty clay (*3.55 m)
3.60–4.40	shell marl (*3.65 m)
4.40–4.50	light grey silty clay (*4.35 m, 4.50 m)

Palynological assessment

The pollen data indicates that the deposits contain a record of the Late Devensian Late-Glacial. The pollen spectrum from the base of this core at 4.50 m suggests a predominantly herbaceous vegetation characteristic of Late Devensian I. It is being replaced by an open birch scrub at 4.35 m and 3.65 m but this still has a rich component of herbs. This is succeeded at 3.55 m by a plant community, which is richer in herbs, such as grass, *Rumex acetosa*-type, *Artemisia*, and *Thalictrum*: this pollen spectrum is characteristic of Late Devensian III. At 3.25 m this open community has been replaced by an open birch scrub as is normally recorded early in Flandrian I. However, the light grey silty clay band at 3.50 m to 3.60 m may indicate a temporary climatic deterioration in the early Holocene. No samples were analysed between 3.25 m and 1.10 m when the pollen suggests a Flandrian III date with 62 % tree and 38 % herb pollen. Because of the apparent difference in age between 3.25 m and 1.10 m, it is possible that there is an hiatus in the deposition. However, if the accumulation rate is very slow, as at Wet Sleddale (Chinn and Innes 1995), the entire post-Glacial sequence may be represented, but an hiatus appears to be the more likely scenario. Pollen from both 0.75 m and 0.40 m suggests these samples are from the upper part of Flandrian III, because tree pollen is at very low values and the herbaceous component, including cereal-type and *Humulus / Cannabis*-type pollen, is increasing. It is, however, possible that this extensive clearance might indicate very early manipulation of the landscape if there has been no break in deposition and the accumulation rate is constant.

Potential

Palaeoecologically this site would provide a record of the Late Devensian and as such is important because of its location. There is greater ambiguity about the archaeological value of the site, without further work and radiocarbon dating. However, it is likely that even if the deposits are truncated they could provide a record of the historic period, following regeneration of the site at some point in time. Pollen preservation was good in all samples.

Robbed Cairn, Subberthwaite Common (Fig 82)
NGR SD 257 881
Height OD 150 m

This site is close to the Giant's Grave, a cairn on Subberthwaite Common (SMR 9407). The well-preserved stratigraphy of 5.10 m suggests that a lake developed after the Late Devensian, which gradually became infilled with reedswamp at 4.70 m before developing into a fen carr, with charcoal recorded in these peats. This in turn became an acid mire or poor fen, with abundant *Eriophorum* and *Calluna* in the upper 1.5 m of peats. Today the site is in an area of hill pasture. There is a dense concentration of archaeological sites in the area, for instance, SMR 9394–9423 and SMR 2160.

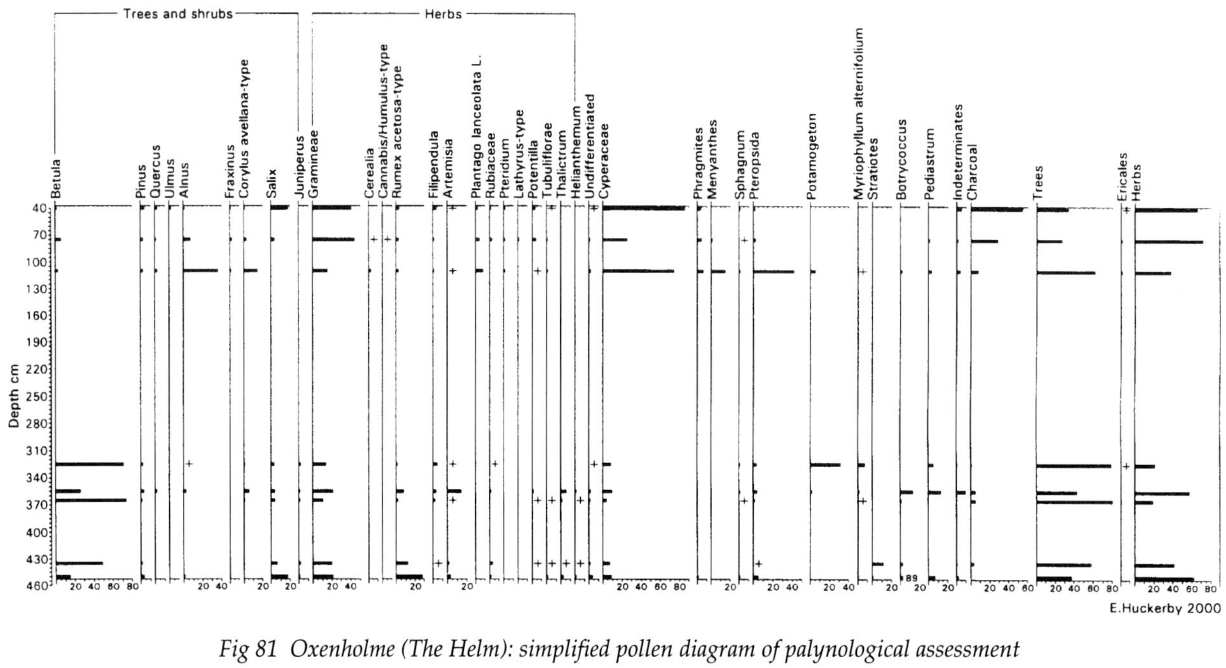

Fig 81 Oxenholme (The Helm): simplified pollen diagram of palynological assessment

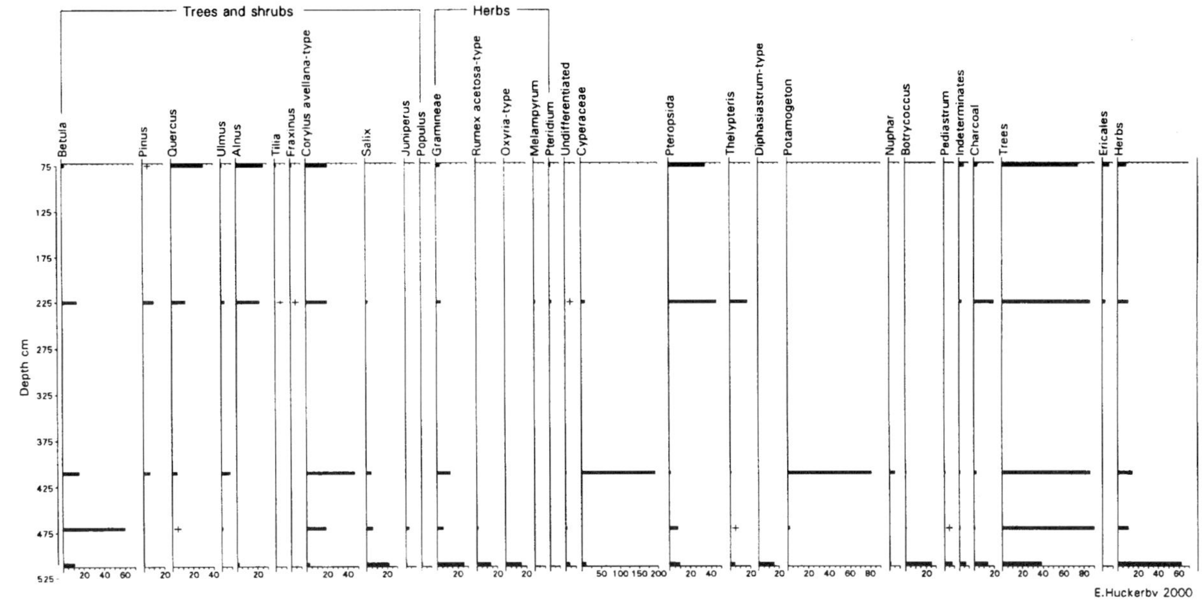

Fig 82 Robbed Cairn, Subberthwaite Common: simplified pollen diagram of palynological assessment

Depth (m)	**stratigraphy**
0–0.50	H7 *Sphagnum / Polytrichum / Calluna / Eriophorum*
0.50–1.00	H9 ditto (*0.75 m)
1.00–1.50	H8 *Eriophorum / Calluna* (*1.25 m)
1.50–2.50	H8 wood peat + charcoal (*1.75 m, 2.25 m)
2.50–3.25	H6 *Phragmites* (*2.75 m)
3.25–3.50	H5 wood peat + monocot (*3.30 m)
3.50–4.00	H6 ditto (*3.75 m)
4.00–4.20	H5 *Phragmites* + charcoal (*4.10 m)
4.20–4.70	H8 ditto
4.70–5.10	silty gyttja (*4.70 m)
5.10+	light grey silty clay (*5.10 m)

Palynological assessment

Organic deposition was initiated in the early Holocene, when an open herbaceous community with some willow and birch scrub surrounded an open water community. This herbaceous vegetation was replaced by an open scrub of birch, juniper, and willow, with some hazel; subsequently hazel dominated the landscape. This sequence of events is characteristic of Flandrian I. The Flandrian I/II transition occurred at some level between 4.10 m and 2.25 m as alder is absent at 4.10 m but by 2.25 m it has risen to 22 %: the criteria for defining the Flandrian I/II boundary is the increase in alder pollen to values >10 % (Hibbert *et al* 1971). At 2.25 m elm pollen is only present at 2 % and therefore deposits at this depth may be post-3000 cal BC (*c* 5000 BP) in age. The low value of elm pollen is associated with higher values of charcoal and increased values of herbaceous pollen and fern spores, providing evidence of woodland disturbance. This sample and that from 0.75 m are likely therefore to be of Flandrian III age. At 0.75 m there is 80 % tree and shrub pollen, with 25 % oak and 35 % alder recorded. This suggests that the peat at this level is Neolithic in age as Pennington (1970) considered that oak and alder were extensively cleared during the Bronze Age. The deposits may be truncated after the Bronze Age.

Potential

This site would provide a local record from the Mesolithic, possibly up to and including the Bronze Age, in an area rich in archaeological records. Pollen preservation was mixed but this is characteristic of fen and fen carr peats, which form the major part of the deposits.

St Sunday's Beck Bog (Fig 83)
NGR SD 553 895
Height OD 100 m

This bog is located in a hollow above St Sunday's Beck, south–east of Oxenholme. The 5.0 m of well-preserved stratigraphy indicate that a calcareous lake was present, either in Late Devensian II or early Flandrian I, becoming less calcareous from 4.70 m to 4.00 m when the sediment is a gyttja. Laminations were recorded in the lower part of these deposits; these laminations may have been caused by annual fluctuations in sedimentation. This lake became terrestrialised by a community of *Polytrichum* moss and *Phragmites*, prior to the formation of acid mire conditions initially with *Aulocomnium palustre* and *Calluna*, and then various communities with *Eriophorum* in which there are phases with a variety of taxa. Above 0.50 m the sediment is an amorphous peat: palynologically evidence suggests that the upper stratigraphy is severely truncated. Today the area is under pasture. Archaeologically there is evidence of a prehistoric (stone axe, SMR 4085) and medieval (SMR 4018, SMR 4019) presence around the bog.

Depth (m)	**stratigraphy**
0–0.25	H10 amorphous peat
0.25–0.50	H9 ditto

0.50–0.80	H8 *Eriophorum/Calluna* (*0. 50 m)
0.80–1.00	H7 *Eriophorum*
1.00–1.50	H6 ditto + *Sphagnum* (*1.25 m)
1.50–2.00	ditto + *Calluna/Hylocomium*
2.00–2.50	H5 *Aulocomnium palustre* + *Calluna* (*2.25 m)
2.50–3.00	H5 *Polytrichum* +*Phragmites* (*2.75 m)
3.00–4.00	H8 ditto
400–450	gyttja (*4.25 m)
450–470	laminated gyttja (*4.70 m)
470–490	shell marl
4.90–5.00	light grey silty clay (*5.00 m)

Palynological assessment

The records of hazel, oak and elm, an assemblage typical of late Flandrian I, at 5.00 m suggest that this sample may be contaminated, since those from 4.70 m and 4.25 m appear to be older. At 4.70 m there was a relatively open scrub community of birch, juniper and willow with a variety of herb taxa; this type of community is characteristic of Late Devensian II or early Flandrian I. By 4.25 m the pollen spectrum is unambiguously Flandrian I with rapidly rising values of birch and hazel, a feature characteristic of North West Britain. At 2.75 m the pollen spectrum is typical of late Flandrian I (Hibbert *et al* 1971) prior to the transition to Flandrian II, when a mixed woodland of pine, elm, oak, hazel, and birch dominated the landscape. The three samples from 2.25 m to 0.50 m are all characteristic of Flandrian II earlier than 3000 cal BC (*c* 5000 BP), with a woodland in which alder has replaced the pine. Some lime is also recorded.

Potential

This site would provide a record of the vegetation during the Mesolithic but the upper peats have been truncated and therefore it would be of limited use. The pollen preservation was good throughout the core.

Stable Harvey Moss (Fig 84)
NGR SD 278 918
Height OD 200 m

Stable Harvey Moss lies west of Coniston Water, on Torver Low Common below the Blawith Fells. The stratigraphy suggests that a lake developed in a kettlehole in the early Holocene. This gradually became terrestrialised with a succession of plant communities, initially of *Phragmites*, sedges, fen carr and finally a *Sphagnum* acid mire. The peat is probably truncated but, in view of the high concentration of prehistoric and later archaeological sites in the surrounding landscape (*eg* SMR 3235, SMR 1636, SMR 15929 and SMR 16433), the site is of considerable value.

Depth (m)	**stratigraphy**
0–0.50	H5 *Sphagnum*
0.50–1.00	H7 ditto (*0.75 m)
1.00–2.00	H7 alder peat + *Phragmites* (*1.25 m)
2.00–2.50	H6 monocot (*1.75 m)
2.50–3.00	H5 sedge
3.00–3.50	H4 *Phragmites* + sedge (*3.25 m)
3.50–4.15	gyttja + organics (*3.75 m)
4.15–4.70	silty gyttja (*4.15 m)
4.70+	light grey silty clay (*4.70 m)

Palynological assessment

The lower part of the deposits formed early in Flandrian I. Significant quantities of dwarf Arctic birch, *Betula nana*, pollen was recorded at 4.70 m, with juniper, willows and a variety of herbaceous taxa, suggesting a pioneer community as the climatic conditions ameliorated. This was succeeded by tree birches and increasing quantities of hazel. Charcoal

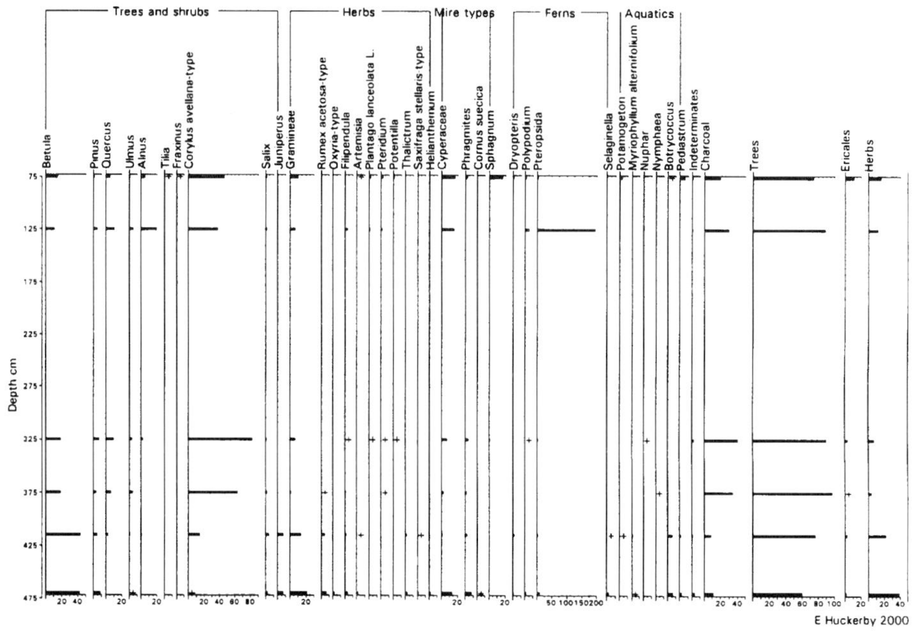

Fig 83 St Sunday's Beck Bog: simplified pollen diagram of palynological assessment

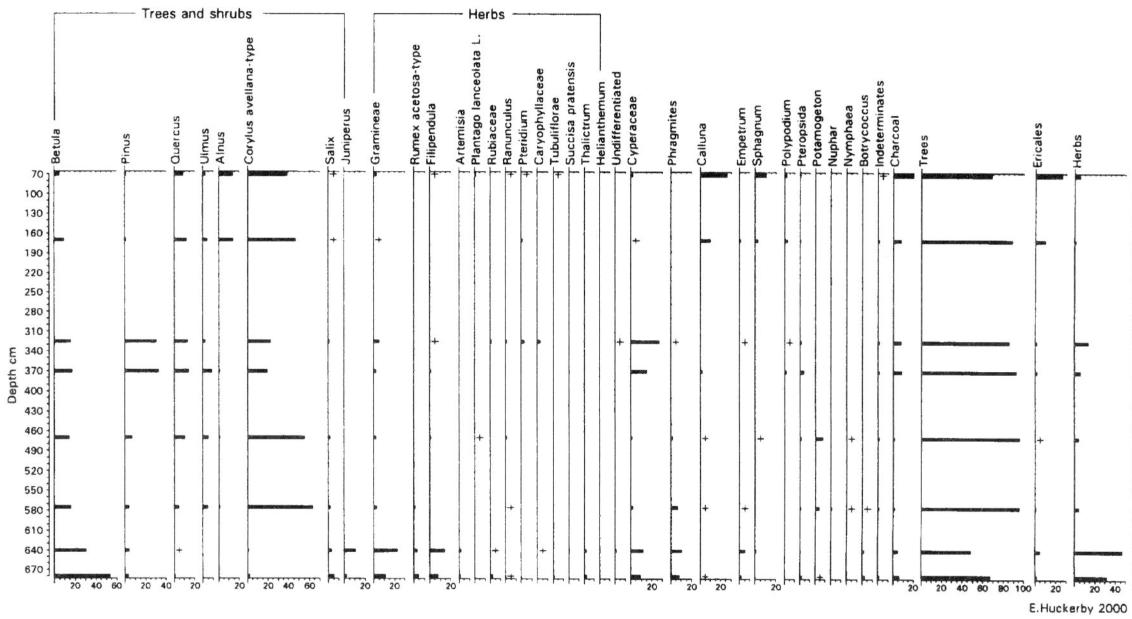

Fig 84 Stable Harvey Moss: simplified pollen diagram of palynological assessment

was recorded at high levels as hazel increased, together with pine, oak and elm pollen. At 3.75 m and 3.25 m, 34–41 % charcoal was recorded, together with 3–7 % herbaceous pollen, suggesting that fire was disturbing the vegetation during the Mesolithic prior to the Flandrian I/II transition. This boundary occurred at some depth between 3.25 m and 1.25 m. At 1.25 m there is evidence of a damp alder carr, when the peat type is wood. Oak, elm, pine and birch woods were growing on the surrounding drier land. Values of nearly 30 % charcoal, a very high representation of undifferentiated fern spores, and 12 % herb pollen, suggest considerable disturbance to the woodland vegetation.

Potential

It seems likely that although the deposits are truncated this site would yield a record of Flandrian I, II and part of III. Therefore it would provide a record of the Mesolithic, Neolithic and possibly up to or including the Bronze Age. The high frequencies of charcoal in most of the samples, together with the high density of archaeological remains in the surrounding area, suggest that further study of this site would provide valuable additional information about the possible modification of the landscape throughout prehistory. Pollen preservation was good below 3.50m but more variable above, a feature of fen and fen carr peats.

Three Mile Bog (Tarneybank) (Fig 85)

NGR SD 594 875

Height OD 180 m

The site is situated west of the River Lune south-west of Sedbergh, in the valley of a tributary of the Lune. It would appear that the site is a valley mire, but 7 m of well-preserved stratigraphy indicates that a lake was present in the early Holocene. This was terrestrialised with the formation of a reedswamp which developed into an acid-mire with *Sphagnum* and *Eriophorum*, succeeded by a fen carr for a brief period before an *Eriophorum* and *Calluna* community formed. The uppermost peats are amorphous and this, in conjunction with the pollen evidence, suggests that the site is truncated. Today the area is under damp pasture. There are several known archaeological sites around the bog, with records of prehistoric (SMR 4085), Roman (SMR 16530) and medieval (SMR 4018, SMR 4019) activity.

Depth (m)	**stratigraphy**
0–0.50	H10 amorphous peat
0.50–1.00	H8 *Eriophorum / Calluna* (*0.70 m)
1.00–2.10	H8 *Eriophorum / Calluna* + charcoal (*1.20 m, 1.70 m)
2.10–2.20	H8 *Alnus*
2.20–2.50	H7 *Sphagnum / Eriophorum* (*2.30–2.40 m)
2.50–3.00	H8/9 ditto
3.00–3.50	H7 *S sect Cuspidata* (*3.20–3.30 m)
3.50–4.00	H5 *Phragmites* / bryophytes (*3.90 m)
4.00–4.50	H4 *Phragmites*
4.50–5.00	H8 ditto (*4.90 m)
5.50–6.00	gyttja (*5.75 m)
6.00–6.50	gyttja (*6.40m)
6.50–7.00	ditto (*6.80 m)

Palynological assessment

The pollen evidence from 6.80 m suggests an open birch scrub community with some willow. However, at 6.40 m, the vegetation appears to become more open with a significant component of juniper and *Empetrum* pollen. The lower sample (6.80 m) may be contaminated as, characteristically, juniper and *Empetrum* precede the

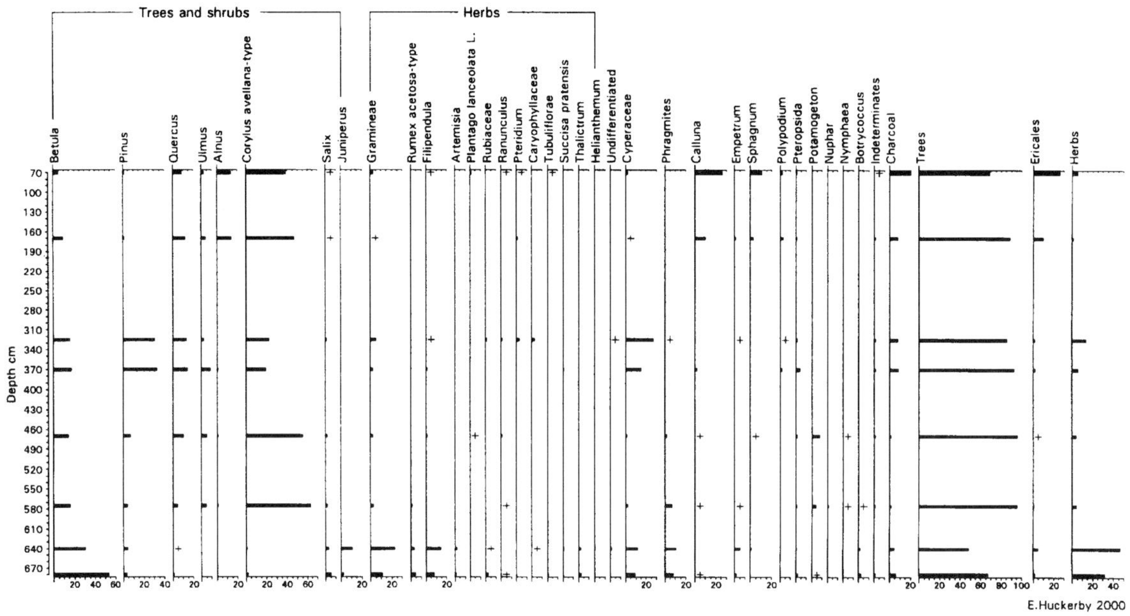

Fig 85 Three Mile Bog: simplified pollen diagram of palynological assessment

domination by birch in the early Holocene as climatic conditions ameliorated. If there is no contamination at 6.80 m the pollen spectrum recorded is unusual. The stratigraphy above this sample provides no evidence, in the form of silt bands, of soil instability around the site, which could suggest climatic fluctuations. The pollen spectra from the samples at 4.70 m and 5.75 m are characteristic of the hazel maximum of Flandrian Ic (Hibbert *et al* 1971) in North West Britain. At 3.70 m and 3.25 m, the major tree components are pine, oak, and elm, along with hazel and birch, and the deposits appear to be Flandrian Id in date. There is some evidence of fire disturbance in both these samples, with 8 % charcoal and 6–14 % herb pollen recorded and, at 3.25 m, a fall in elm pollen. The sample from 1.70m suggests a Flandrian II date, with 13.5% alder, and pine pollen no longer recorded. Again there is some evidence of clearings with increasing heather pollen. The elm decline occurred between 1.70 m and 0.70 m as, at 0.70 m, elm pollen has fallen to <2 %. Some disturbance to the local vegetation is suggested, with heather pollen rising to 26 %, herbs 5 %, and charcoal to 20 %.

Potential

This site would provide a record of vegetational change during the Mesolithic period. The possible forest disturbance at 3.25 m and 3.70 m prior to the Flandrian I/II boundary might be of great significance and further work could be most rewarding. In general, pollen preservation was good although it was poor at 3.70 m and mixed at 3.20–3.30 m.

White Moss (Fig 86)

NGR NY 3510 0650

Height OD <70 m

The site is situated on the lake shore at the western tip of Rydal Water near to the watercourse flowing between Rydal Water and Grasmere. Coring was left incomplete due to the great depth of the sediments. The well-preserved stratigraphy demonstrated that Rydal Water was once more extensive than at present, with gyttja recorded from 2.00 m to 8.50 m. Palynological records of plants characteristic of shallow water, and a band of gyttja plus clay at 2.0–2.10 m, suggest a fall in water level prior to the formation of a fen carr before the development of reedswamp. Today the site is covered by carr woodland.

Depth (m)	stratigraphy
0–0.50	unsampled
0.50–1.00	H8 *Phragmites* (*0.75 m)
1.00–1.50	unsampled
1.50–200	H9 wood peat+monocot (*1.75 m)
2.00–2.10	gyttja + clay (*2.05 m, 2.10 m)
2.10–8.50	gyttja (*2.15 m, 3.80 m, 4.80 m, 5.40 m, 6.50 m, 8.00 m, and 8.50 m)

Palynological assessment

From 8.50 m to 3.75 m the pollen data suggest that the landscape was densely wooded with very minor clearings, although the records of ash pollen, a light-demanding taxon, from above 6.50 m suggest that some openings probably existed. However, above 2.50 m there is strong evidence of clearance with reductions in tree pollen, higher values of herb pollen, and fern spores. Cereal-type pollen was recorded in all but one sample above 2.50 m. It is possible that water levels at the site fell above 2.50 m, as values of sedge and reed pollen rise, indicating that terrestrialization was occurring, and at the same time pollen of aquatic plants characteristic of shallow water (*eg Isoetes* and *Potamogeton*) are recorded.

Potential

The site could provide a record of the vegetation adjacent to Rydal Water and Grasmere for Flandrian II and III. However, it proved impossible to core to the bottom during the assessment and it is possible that earlier sediments exist. The depositional sequence of the site appears to be undisturbed and consequently the environmental archive has the potential to span from the Mesolithic to the present day. The site therefore represents a scarce resource and is of great importance. Pollen preservation was good at 2.40 m and below 3.80 m but mixed to poor above.

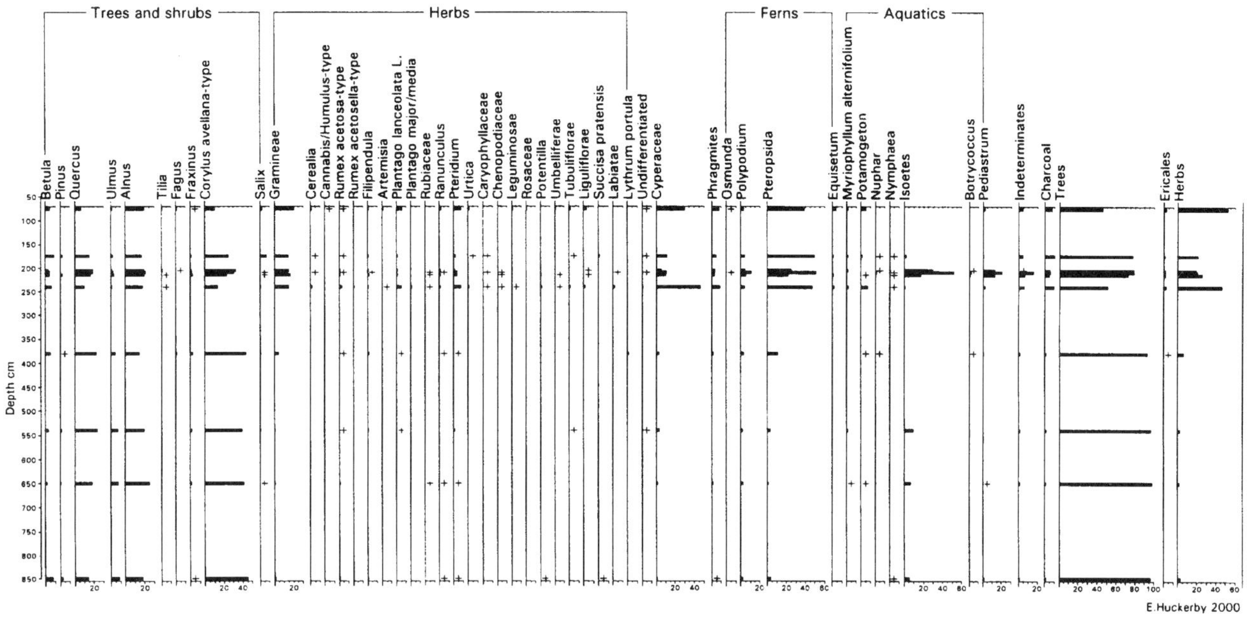

Fig 86 White Moss: simplified pollen diagram of palynological assessment

Appendix 6
Palaeoecological studies at Solway Moss

The finding of animal remains on Solway Moss (*Ch 4*) prompted the undertaking of detailed palaeoecological studies in an attempt to provide a contemporary environmental context for the finds. Detailed pollen and macrofossil analyses were undertaken from the site of the sheep discovery which was set in a wider context by a study of the general stratigraphy of the whole mire.

Pollen and macrofossil analysis

A large diameter (90 mm) Russian corer (Jowsey 1966; Barber 1984) was used to obtain a peat core immediately adjacent to the location of the sheep. A core 1.75 m in length was obtained which was analysed for pollen, plant macrofossils, and mineral content. Five radiocarbon dates have also been obtained, enabling a chronology to be constructed for the vegetation changes.

Methods
The methods used in preparation and analysis of samples were as discussed in Chapter 1.

Results

Site stratigraphy
Macrofossil analysis (Fig 87) indicates that peat accumulation began shortly after *c* 6500–6000 cal BC (*c* 7700–7200 BP) when fen carr became established. Fire disturbance appears to have been continuous from the beginning of this phase and reached a peak at *c* 5000 cal BC (*c* 6100 BP) when there is a sudden drop in the number of wood fragments and *Eriophorum* (Cotton grass) appears for the first time, indicating ombrotrophic or poor-fen inception. Wood representation recovers briefly after this episode before petering out completely, allowing the development of an *Eriophorum* and *Calluna*-dominated mire vegetation. The frequency of burnt plant fragments also declines to a lower level, but remains an almost constant feature of the profile, only disappearing near the top of the core. After *c* 4500–4000 cal BC (*c* 5700–5200 BP), *Sphagnum sect Acutifolia* becomes a prominent component of the *Eriophorum*-dominated mire. This vegetation continues almost to the top of the core, except at *c* 0.20–0.30 m where a pool dominated by *Sphagnum sect Cuspidata*, and dated to *c* 3300–2700 cal BC (*c* 4500–4150 BP), intervenes briefly. The abrupt change from *Eriophorum*–dominated mire to one characterized by an abundance of *S imbricatum* was noted in the last 0.05 m of the profile.

Gross stratigraphy
A short stratigraphic section from the edge of the present mire surface across the findspot of the sheep was extended northwards to cross the entire moss (Fig 51). Two other transects (crossing the site in north–west to south–east and west to east directions) have also been completed (Figs 48–50). The results suggest the widespread applicability of the succession outlined above to much of the site. Charcoal is a frequent feature of the deposits and there is evidence for a widespread and distinct phase change to wetter conditions at some point in the prehistoric period, characterised by a band of *Sphagnum sect Cuspidata*-dominated peats.

Pollen analysis
The pollen diagram (Fig 88) can be divided into three phases:

SM–a 1.70–1.30 m

This period is dominated by *Betula* (birch), *Alnus* (alder), *Quercus* (oak), and Coryloid (hazel and bog myrtle) pollen. The upper boundary is defined by the fall of birch and rise of *Calluna* pollen. Peat at 1.70–1.60 m was dated to 6080–5790 cal BC (7080±50 BP, GU–5277) and at 1.50–1.42 m to 5240–4940 cal BC (6180±60 BP, GU–5315).

SM–b 1.30–0.50 m

This phase is characterised by *Alnus, Quercus, Calluna* (heather), and Coryloid pollen. *Ulmus* (elm) is a consistent component during these two earlier phases, with a number of temporary fluctuations in its values. Peat at 0.93–0.87 m was dated to 4460–4165 cal BC (5470±60 BP, GU–52760).

SM–c 0.50–0.00 m

The lower boundary is defined by the decline in *Ulmus* (elm) and the start of a curve of *Plantago lanceolata* (ribwort plantain) pollen. Peat at 0.30–0.25 m was dated to 3340–2707 cal BC (4370±90 BP, GU–5315) and at 0.25–0.20 m to 3350–2890 cal BC (4400±80 BP, GU–5314).

Discussion

The peats examined from this core fall within Flandrian II. The initiation of peat development commenced immediately prior to the rapid expansion of *Alnus* pollen, and the relative decline of *Betula* pollen, although the relationship between the two taxa may be illusory, since the pollen concentration data indicate that *Betula* pollen remained constant, suggesting therefore that *Alnus* did not replace *Betula* in the community. The relative percentage diagram suggests a comparatively open *Betula* / *Alnus* wood, with *Hedera* (ivy), *Pteridium* (bracken), and herbs (with very little grass) to a depth of 1.34 m. The concentrations of *Betula* and *Alnus* decline sharply above 1.42 m while that of *Calluna* increases. This corresponds to abundant carbonised

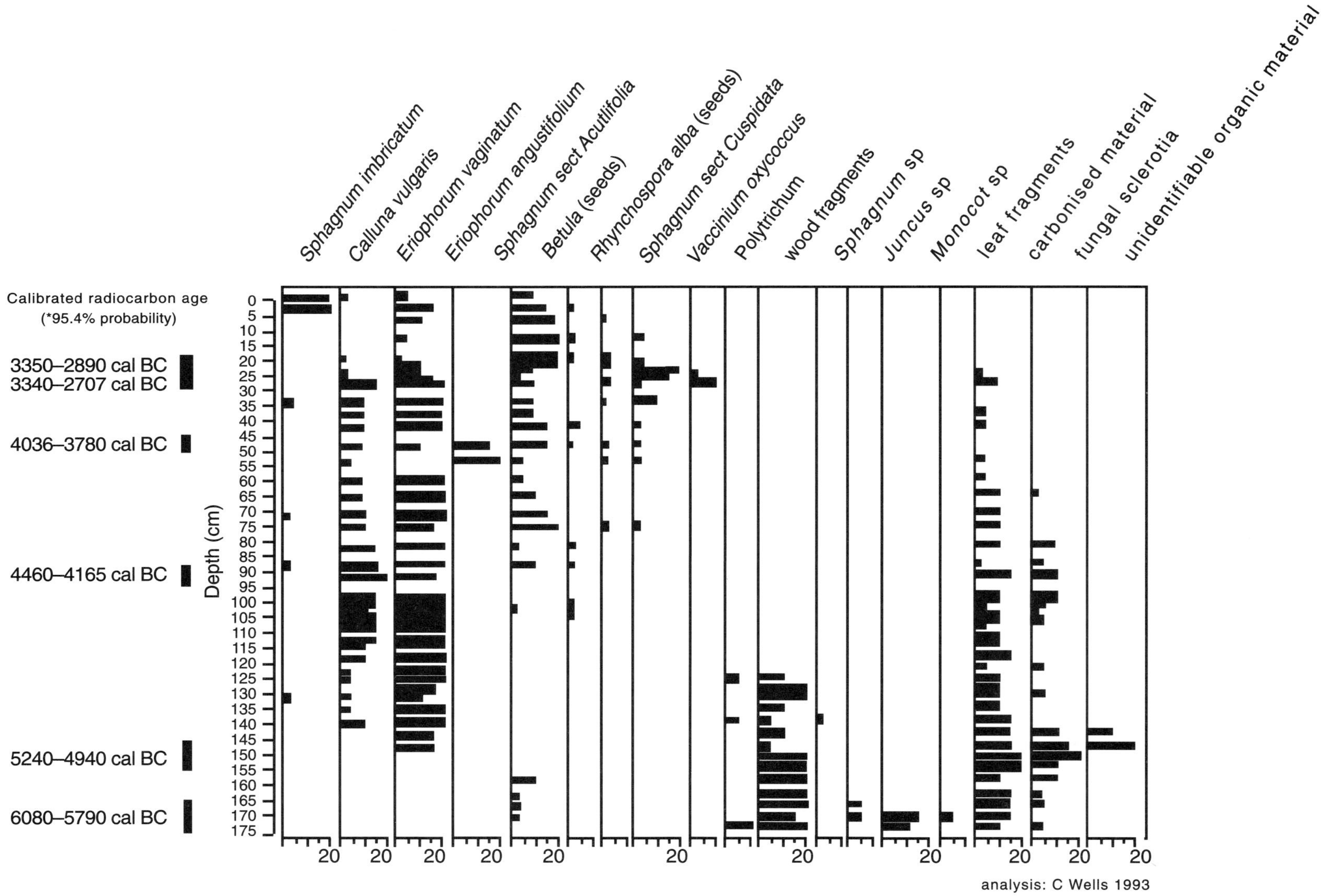

Fig 87 Macrofossil diagram from Solway Moss

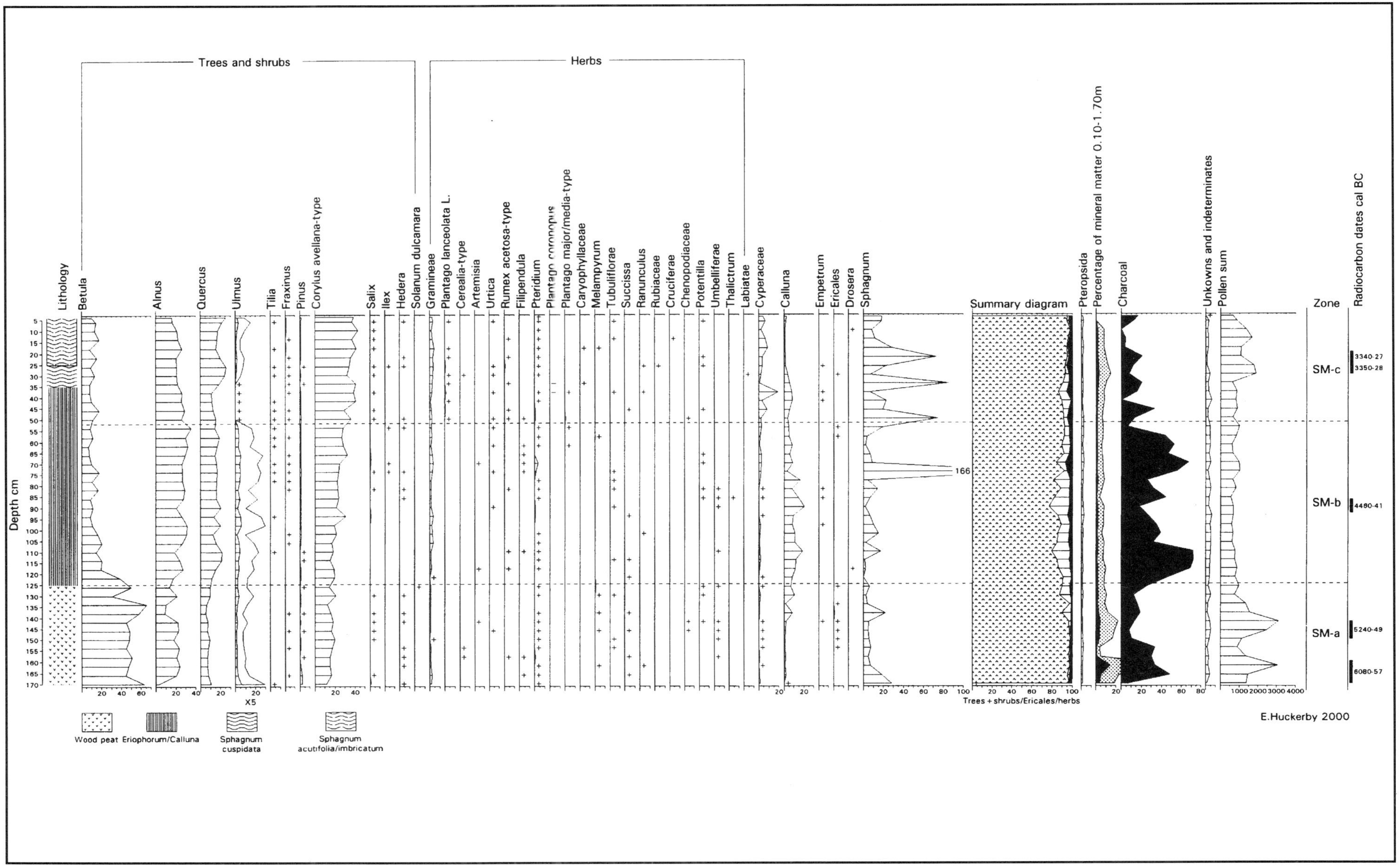

Fig 88 Percentage pollen diagram of selected taxa from Solway Moss

material in the macrofossil record dated to *c* 5240–4940 cal BC (6180±60 BP, GU–5315), but lower values of microscopic charcoal, suggesting the possibility that carr woodland was fired during the Mesolithic period, resulting in the establishment of an open *Calluna*-dominated community. A peak of mineral material and an increase in pollen concentrations occurs between 1.50 m and 1.36 m, either indicating clearance of the surrounding landscape or possibly the slower growth of peat due to drier climatic conditions. The microscopic charcoal record from the central portion of the pollen diagram, together with low peaks in herbaceous pollen, *eg* Gramineae, *Pteridium* and *Succissa*, suggests that regionally and locally there may have been some manipulation of the vegetation in the Mesolithic period prior to 3000 cal BC (*c* 5000 BP).

An elm decline is clearly shown in the diagram at 0.54 m. Above this level, *Plantago lanceolata* (ribwort plantain) and *Rumex* (sorrel) pollen is recorded, together with cereal-type pollen (*ie* grass pollen with diameter >40 µm) at 0.30 m (*c* 3350–2890 cal BC, 4400±80 BP, GU–5313). A possible cereal-type grain has also been tentatively identified at 0.46 m from a monolith near to the findspot of the sheep, dated to 4036–3780 cal BC (5110±60 BP, GU–5275). If these identifications are proved accurate, they would suggest Neolithic cereal cultivation in North Cumbria. The earliest previous record in this area for arable agriculture is that of Walker (1966), who recorded cereal pollen at *c* 2000 cal BC (*c* 3500 BP). It is perhaps significant that an increase in mineral matter from 1.5 % to 2.9 % occurred between 0.34 m and 0.22 m — *ie* at the level of cereal-type pollen. This is a considerable increase in an ombrotrophic peat as it reflects mineral matter derived from wind-blown sources, suggesting possible erosion from some nearby woodland clearance (Vuorela 1983).

Summary

The complementary study of all major aspects of peat stratigraphy (macrofossil, gross stratigraphic survey, and palynology) at the findspot of the sheep remains (*Ch 4*) has enabled the identification of possible clearance episodes where more conventional palynological indicators of anthropogenic disturbance are absent. Macrofossil analysis suggests the possibility that clearance may have been taking place *in situ* during the Mesolithic, with the disturbance of carr woodland by fire and the development of a *Calluna*-dominated community. The stratigraphic survey has confirmed the widespread occurrence of charcoal across the site, whilst palynology has indicated the possibility of arable agriculture in the area commencing some 1500–2000 years earlier than previously recorded.

Appendix 7
Deforestation and erosion episodes in the Howgill Fells

The landscape of the modern Howgills is characterised by smooth-sided hills, scarred by networks of deeply-eroded gullies, and covered with a vegetation dominated by an acid rushy grassland. How this distinctive scene evolved has begun to be revealed by a series of palaeoenvironmental studies. Shallow blanket mire covers much of the ground above *c* 400 m, with occasional deeper pockets of peat where subsurface hollows have enabled basin mires to develop. One of these mires, Archer Moss (NY 6330 0062), has provided the only extended vegetational history relevant to the hills to date (Cundill 1976; Fig 89). The base of the peat has been dated to 2124–1530 cal BC (3480±100 BP, GAK–4533) and the lower part of the pollen record hints at deforestation episodes, probably relating to the late Bronze Age or early Iron Age. The probable post-Roman segment of the diagram shows an initial regeneration of trees and shrubs, followed by a gradual decline and a large expansion of heather. The upper part of the record clearly shows that the predominantly matt-grass-covered hillsides have succeeded heather-rich communities very recently, probably as a result of increased stocking of sheep in the last 2–300 years, which is known to have a deleterious effect on heather regeneration above certain threshold levels (Halliday 1997, 76–7). Although limited, the data which has been generated from this site have acted as a 'control' with which to place in context palaeoenvironmental studies of short organic profiles buried beneath alluvium in the stream channels beneath the hillslopes (Cundill 1976; Harvey *et al* 1981). Because it is extremely unusual to find peat beds located in such dynamic depositional environments as alluvial fans, their existence implies sudden burial and possibly a major disturbance to the catchment hydrology of the streams after a prolonged period of stability. The studies to date strongly suggest human agency to have been involved in these changes. Unfortunately, the unavoidable dearth of radiocarbon dates from the studies means that much ambiguity has surrounded the precise timing of periods of apparent land-use change.

The Carlingill organic section

Original studies (Cundill 1976) were centred on Carlingill, the nexus of streams which are seen to incise the western flank of the Howgills so prominently when viewed from the Lune Gorge. Here, a short (1.2 m) organic profile was identified, stratified beneath an alluvial fan at the base of the gill (NY 6275 9980), the organic layers being intercalated by mineral bands. Radiocarbon dating showed that the lower part of the profile dated to 752–180 cal BC (2290±80 BP, GAK–4532) and pollen analysis suggested that phases of forest clearance preceded the deposition of clay bands and the final burial of the profile by alluvium. It seems likely from this that the lower clay bands signify evidence of erosion induced by the effects of Iron Age deforestation. The age of the uppermost disturbance episodes remained ambiguous due to the lack of independent dating and uncertainty about the rate of peat accumulation; it could have signified intensified forest clearance and erosion dating any time from the late Iron Age to the early medieval period or, possibly, it has been suggested, even the monastic period (Lambert 1996, 50).

The case for a late-Iron Age phase of disturbance, however, was strengthened by the increasing evidence for a widespread increase in deforestation across the north–west of England at this time. Oldfield and Statham (1965), Wimble (1986), Wells *et al* (1997), and Dumayne and Barber (1994) have all shown sharp falls in arboreal pollen relating to this period from sites in North Lancashire, and South and North Cumbria, while Mackay and Tallis (1994) recorded the same trend from the topographically very similar Bowland Fells, a western outlier of the Pennines situated *c* 40 km to the south of the Howgills.

It seemed plausible, therefore, to view major deforestation-related erosion events at Carlingill as fitting into this pattern of late Iron Age disturbance. However, Roman period activity was an equal possibility, given the abundance of archaeological evidence from the adjacent Lune Gorge (Lambert 1996, 46, 48). This was certainly the provenance preferred by Cundill (1976) for the episode. However, Harvey *et al* (1981) have made a strong argument for attributing the major phase of gully erosion to early medieval agricultural developments. In a detailed study at Middle Langdale (NY 665 005), another gullied stream-bed site with a buried organic layer situated 5 km north–east of Carlingill, they showed that geomorphological conditions were relatively stable between *c* 834–450 cal BC (2580±55 BP, UB–2212) and cal AD 890–1270 (940±95 BP, UB–2213). Some evidence for low intensity erosion was discernible, however, in the palaeomagnetic record from the middle of the profile, which coincided with a switch in the palynological signal indicating a reduction in arboreal pollen and an increase in ericaceous pollen, suggestive of an open heather moorland. Harvey *et al* (1981) considered that this could represent late Iron Age or Romano-British catchment disturbance, although it could possibly relate to a later date. After cal AD 890–1270 (940±95 BP, UB–2213), however, intense erosion and gully development was triggered. Harvey *et al* (*op cit*) suggest that the main cause of this change could have been the expansion of sheep farming in a period of Scandinavian settlement, leading to overgrazing of the vegetation cover and the crossing of a threshold of geomorphological equilibrium. This event is clearly much younger than the Roman period erosion event which Cundill supposed to have ended organic deposition at Carlingill but it is by no means certain that it can be attributed to any Norse settlement, as it could easily be later medieval in date. Caution is clearly required in interpreting imprecisely dated late Holocene deforestation signals.

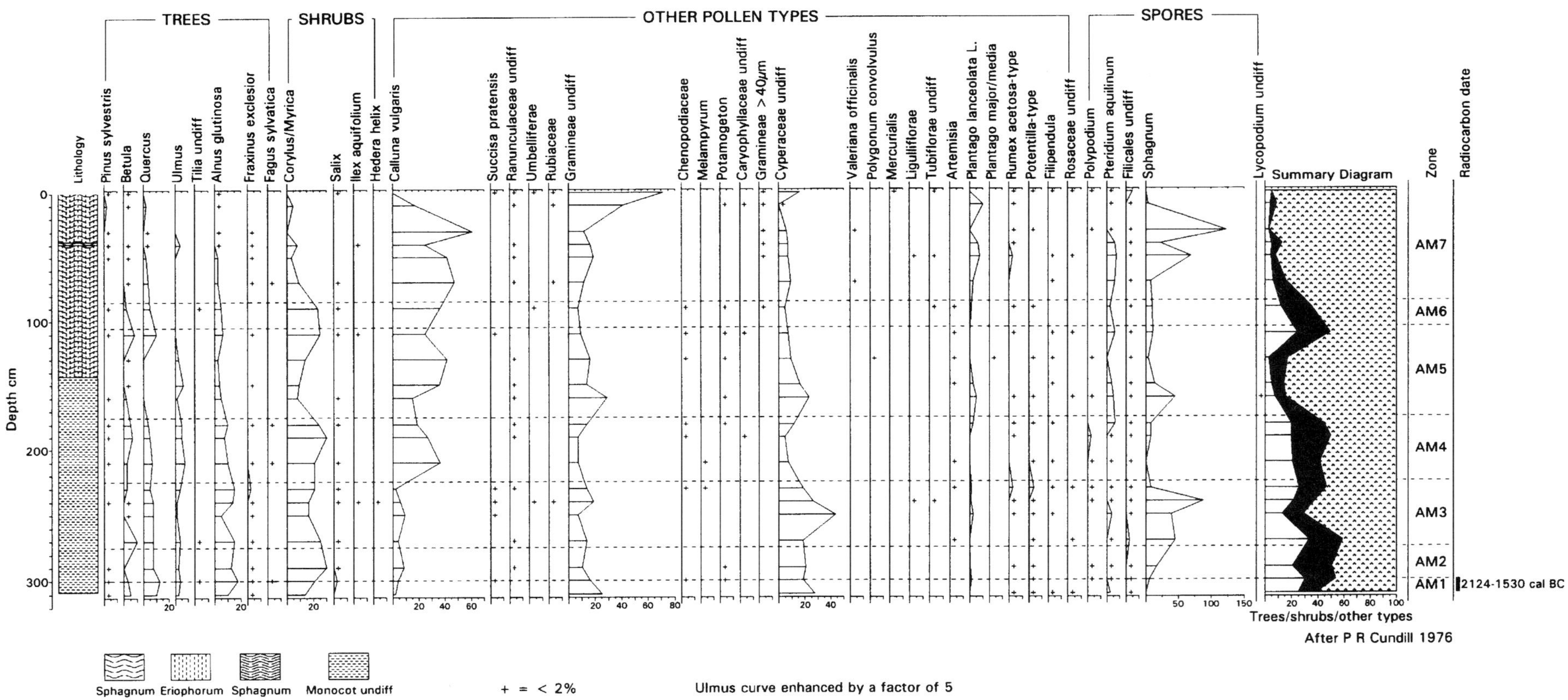

Fig 89 Percentage pollen diagram from Archer Moss (after Cundill 1976). Pollen sum includes all pollen types excluding spores

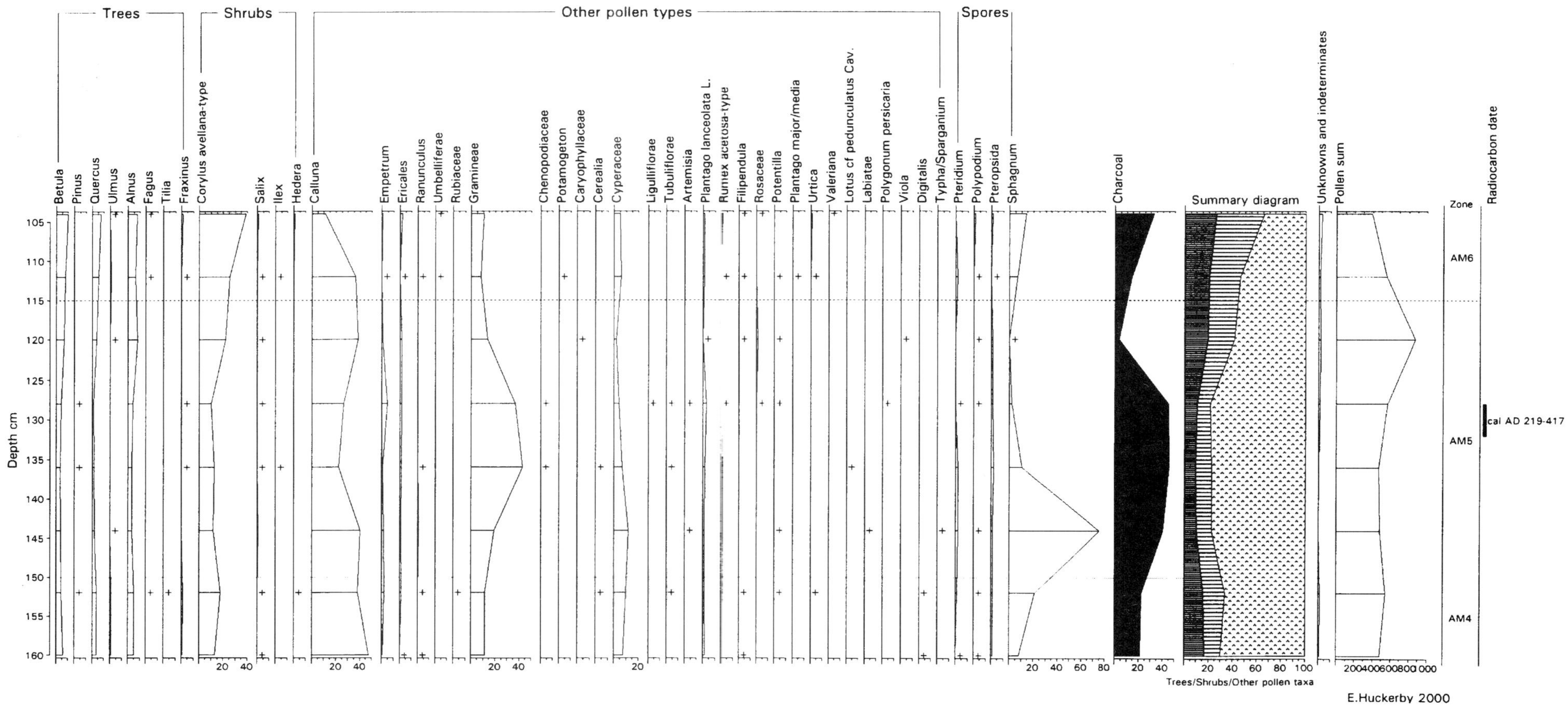

Fig 90 Percentage pollen diagram from a segment of peat at Archer Moss. Pollen sum includes all pollen types excluding spores

Harvey *et al* (*op cit*) sought to explain the possible discrepancy in dating between the Langdale and Carlingill sites by suggesting that the latter section may have given an anomolously old radiocarbon date on account of it having been been redeposited from above. However, there is absolutely no evidence for this assertion. Instead, the control site of Archer Moss, situated *c* 200 m above on the plateau, shows a strong similarity in pollen curves between 1.15 m and 2.40 m with the Carlingill section and lends support to the interpretation of the buried profile representing an *in situ* feature.

With this in mind, the Archer Moss stratigraphy was re-sampled and examined palynologically. This enabled a segment of pollen stratigraphy equating to the main erosion phase (upper alluvial contact) in the Carlingill section to be identified (Fig 90) and dated using AMS techniques. The results show that the decline in arboreal pollen and the rise in open ground indicators, which immediately preceded the erosive event, dated to the Roman period, *c* cal AD 219–417 (1720±45 BP, AA–28398), thereby confirming Cundill's (1976) original suggestion.

Conclusions

The study has confirmed that at least two major phases of erosion may have affected parts of the Howgills, one perhaps beginning in the late Iron Age but definitely intensifying in the Roman period, and one in the medieval period.

Appendix 8
Details of radiocarbon dates quoted in text

(Radiocarbon dates upto 8100 BP are calibrated using Calib v2.0. Dates beyond 8100 BP are calibrated using v3.0.3c)

Radiocarbon Age (BP)	Location	Calibrated date range (95% confidence)	Lab Code	Authority
11500±120	St Bees	11806–11196 cal BC	Birm–647	Coope 1994
11180±120	St Bees	11419–10888 cal BC	Birm–649	Coope 1994
10160±207	Scaleby Moss	11092–10155 cal BC	Q–147+Q–148	Godwin and Willis 1959
10700±200	Kirkhead Cave	11075–10169 cal BC	HAR–1059	Gale and Hunt 1985
10160±193	Scaleby Moss	10542–9028 cal BC	Q–152	Godwin and Willis 1959
9305±45	Bolton Fell Moss	8432–8099 cal BC	SRR–4945	Hughes 1997
9270±200	Heysham Harbour	8980–7943 cal BC	Birm–141	Shotton *et al* 1970
9009±194	Scaleby Moss	8425–7576 cal BC	Q–161	Godwin and Willis 1959
8470±50	Bolton Fell Moss	7561–7427 cal BC	SRR–4942	Hughes 1997
8135±150	Redkirk Point	7490–6604 cal BC	Q–637	Jardine 1980
8000±65	Redkirk Point	7045–6614 cal BC	UB–2445	Masters 1981
7995±80	Levens	7060–6707 cal BC	Hv–3362	Tooley 1978
7935±110	Redkirk Point	7193–6470 cal BC	UB–2470	Masters 1981
7361±146	Scaleby Moss	6457–5888 cal BC	Q–167	Godwin and Willis 1959
7160±60	Bart's Shelter	6110–5883 cal BC	OxA–8069	D Stables pers comm
7080±50	Solway Moss	6080–5790 cal BC	GU–5277	Huckerby and Wells 1993; this volume
6870±95	Wedholme Flow	5970–5550 cal BC	Hv–5228	Huddart *et al* 1977
6850±60	Bowness Common	5820–5630 cal BC	Hv–6208	Huddart *et al* 1977
6810±130	Crosscanonby	5970–5480 cal BC	Hv–6209	Huddart *et al* 1977
6752±156	Monk Moors	5970–5360 cal BC	BM–1216	Burleigh and Matthews 1982
6495±95	Crosscanonby	5613–5240 cal BC	Hv–5210	Tooley 1978
6320±85	Williamson's Moss	5473–5074 cal BC	Hv–5227	Tooley 1982
6180±60	Solway Moss	5240–4940 cal BC	GU–5315	Huckerby and Wells 1993; this volume
5875±220	Bowness Common	5298–4340 cal BC	Hv–6207	Huddart *et al* 1977
5865±115	Silverdale Moss	5033–4470 cal BC	Q–261	Huddart *et al* 1977
5760±60	Williamson's Moss	4780–4470 cal BC	SRR–2658	Tipping 1994
5734±129	Silverdale Moss	4932–4350 cal BC	Q–256	Huddart *et al* 1977
5650±50	Williamson's Moss	4667–4360 cal BC	UB–2546	Bonsall *et al* 1994
5555±40	Williamson's Moss	4756–4246 cal BC	UB–2545	Bonsall *et al* 1994
5520±85	Williamson's Moss	4530–4166 cal BC	UB–2712	Bonsall *et al* 1994
5500±70	Williamson's Moss	4490–4169 cal BC	GU–1664	Bonsall *et al* 1994
5480±90	Williamson's Moss	4510–4047 cal BC	UB–2713	Bonsall *et al* 1994
5470±60	Solway Moss	4460–4165 cal BC	GU–5276	this volume
5440±70	Williamson's Moss	4458–4047 cal BC	SRR–3065	Tipping 1994
5435±105	Levens	4490–4003 cal BC	Hv–3844	Huddart *et al* 1977
5421±130	Scaleby Moss	4520–3990 cal BC	Q–120	Walker 1966
5385±280	Bowness Common	4893–3640 cal BC	Hv–4714	Huddart *et al* 1977
5360±116	Newbie Cottages	4460–3970 cal BC	Birm–200	Jardine 1980
5340±120	Barfield Tarn	4457–3825 cal BC	K–1057	Pennington 1970
5315±35	Solway Moss	4314–4006 cal BC	SRR–4631	Barber *et al* 1994

Radiocarbon Age (BP)	Location	Calibrated date range (95% confidence)	Lab Code	Authority
5277±120	Heslington Moss	4360–3813 cal BC	Q–85	Huddart *et al* 1977
5110±60	Solway Moss	4036–3780 cal BC	GU–5275	Huckerby and Wells 1997; this volume
5090±60	Plasketlands	4032–3720 cal BC	GU–2572	Bewley 1993
5060±80	Red Moss	4034–3700 cal BC	Q–913	Hibbert *et al* 1971
5015±100	Arnside Moss	4033–3550 cal BC	Hv–3460	Huddart *et al* 1977
5010±70	Williamson's Moss	3990–3650 cal BC	SRR–3067	Tipping 1994
4964±300	Ehenside Tarn	4500–3040 cal BC	C–462	Arnold and Libby 1951
4940±90	Plasketlands	3970–3525 cal BC	GU–2573	Bewley 1993
4895±95	Heyhouses Lane	3950–3384 cal BC	Hv–4344	Tooley 1978
4850±80	Williamson's Moss	3893–3381 cal BC	SRR–3068	Tipping 1994
4845±100	Pelutho	3934–3370 cal BC	Hv–4418	Huddart *et al* 1977
4830±140	Moss Farm	3970–3340 cal BC	Hv–4347	Tooley 1978
4810±60	Plasketlands	3775–3380 cal BC	GU–2571	Bewley 1993
4725±65	Rhyl Beach	3680–3350cal BC	Hv–4348	Tooley 1978
4595±90	Solway Moss	3622–3040 cal BC	OxA–4214	Stallibrass 1993; this volume
4575±45	Bolton Fell Moss	3496–3104 cal BC	SRR–4937	Hughes 1997
4520±50	Solway Moss	3370–3040 cal BC	GU–5274	Huckerby and Wells 1993; this volume
4420±90	Solway Moss	3360–2890 cal BC	GU–5312	Huckerby and Wells 1993; this volume
4400±80	Solway Moss	3350–2890 cal BC	GU–5313	Huckerby and Wells 1993;this volume
4370±90	Solway Moss	3340–2707 cal BC	GU–5314	Huckerby and Wells 1993; this volume
4320±50	Bolton Fell Moss	3080–2780 cal BC	SRR–4811	K Barber pers comm
4135±55	Drigg	2900–2507 cal BC	UB–906	Cherry 1982
4125±115	Ehenside Tarn	3016–2407 cal BC	Q–303	Godwin and Willis 1960
4051±115	Ehenside Tarn	2910–2290 cal BC	Q–303	Godwin and Willis 1960
4050±55	Monk Moors	2883–2403 cal BC	BM–1385	Bonsall *et al* 1994
3890±90	Red Tarn Moss	2600–2049 cal BC	NPL–122	Pennington 1975
3840±85	Solway Moss	2569–2039 cal BC	OxA–4213	Stallibrass 1993; this volume
3870±70	Foulshaw Moss	2570–2140 cal BC	CAR–544	Dresser 1985
3780±55	Drigg	2456–2039 cal BC	UB–905	Cherry 1982
3755±105	Williamson's Moss	2480–1890 cal BC	BM–1396	Bonsall *et al* 1994
3700±120	Ewanrigg	2470–1760 cal BC	HAR–7077	Bewley *et al* 1992
3700±90	Ewanrigg	2454–1880 cal BC	HAR–5961	Bewley *et al* 1992
3665±40	Williamson's Moss	2179–1926 cal BC	UB–2568	Bonsall *et al* 1994
3690±70	Foulshaw Moss	2300–1890 cal BC	CAR–554	Dresser 1985
3650±120	Monk Moors	2456–1705 cal BC	BM–1395	Bonsall *et al* 1994
3640±90	Ewanrigg	2290–1760 cal BC	HAR–5959	Bewley *et al* 1992
3570±80	Ewanrigg	2170–1704 cal BC	HAR–7072	Bewley *et al* 1992
3530±150	Ehenside Tarn	2300–1520 cal BC	BM–68	Barker and Mackey 1961
3480±100	Archer Moss	2124–1530 cal BC	GAK–4533	Cundill 1976

Radiocarbon Age (BP)	Location	Calibrated date range (95% confidence)	Lab Code	Authority
3470±70	Ewanrigg	2012–1630 cal BC	HAR–5960	Bewley *et al* 1992
3400±70	Ewanrigg	1890–1520 cal BC	HAR–7071	Bewley *et al* 1992
3350±55	Wet Sleddale	1856–1520 cal BC	AA–28375	this volume
3270±80	Manor Farm	1740–1410 cal BC	HAR–5658	Olivier 1987
3140±70	Foulshaw Moss	1592–1260 cal BC	CAR–553	Dresser 1985
3040±140	Seathwaite Tarn	1620–910 cal BC	NPL–124	Pennington 1970
2859±49	Monk Moors	1252–910 cal BC	BM–1386	Burleigh and Matthews 1982
2645±50	Littlewater	900–790 cal BC	AA–33593	this volume
2580±55	Middle Langdale	834–450 cal BC	UB–2212	Harvey *et al* 1981
2420±80	Roudsea	800–380 cal BC	Beta–88434	Zong 1997
2430±60	Drumburgh Moss	730–338 cal BC	B–103261	Lloyd *et al* 1999
2410±40	Bolton Fell Moss	764–396 cal BC	SRR–4555	Dumayne and Barber 1994
2355±65	Archer Moss	761–260 cal BC	AA–28399	this volume
2290±80	Carlingill	752–180 cal BC	GAK–4532	Cundill 1976
2040±80	Midgeholme Moss	355 cal BC–cal AD 125	OxA–2325	Innes 1988
2015±45	Burnfoothill Moss	157 cal BC–cal AD 76	SRR–3752	Tipping 1995a
2000±45	Walton Moss	111 cal BC–cal AD 83	SRR–4531	Dumayne and Barber 1994
1925±40	Walton Moss	8 cal BC–cal AD 139	SRR–4530	Dumayne and Barber 1994
1980±50	Coniston Water	110 cal BC–cal AD 120	SRR–1872	Pennington 1997
1860±60	Bolton Fell Moss	cal AD 4–322	Hv–3085	Barber *et al* 1994
1785±50	Wet Sleddale	cal AD 110–380	AA–28374	this volume
1750±130	Devoke Water	20 cal BC–cal AD 592	NPL–117	Pennington 1970
1720±45	Archer Moss	cal AD 219–417	AA–28398	this volume
1560±130	Burnmoor Tarn	cal AD 210–680	NPL–116	Pennington 1970
1514±100	Heslington Moss	cal AD 266–680	Q–83	Godwin and Willis 1960
1425±40	Bolton Fell Moss	cal AD 555–666	SRR–4554	Dumayne and Barber 1994
1220±75	Solway Cow	cal AD 660–980	OxA–3642	Middleton 1992; this volume
1220±60	Glasson Moss	cal AD 670–960	WK–5479	Cox *et al* 1999
1210±80	Solway Cow	cal AD 660–1056	OxA–3641	Middleton 1992; this volume
1200±70	Fenton Cottage	cal AD 670–990	GU–5156	Wells *et al* 1997
1185±70	Solway Cow	cal AD 670–1056	OxA–3640	Middleton 1992; this volume
1170±50	Bolton Fell Moss	cal AD 714–980	Hv–3084	Barber 1981
940±95	Middle Langdale	cal AD 890–1270	UB–2213	Harvey *et al* 1981
820±50	Fenton Cottage	cal AD 1047–1280	GU–5142	Middleton *et al* 1997
790±60	Glasson Moss	cal AD 1060–1290	WK–5478	Cox *et al* 1999
680±50	Bolton Fell Moss	cal AD 1258–1395	Hv–3078	Barber *et al* 1981
560±35	Bolton Fell Moss	cal AD 1298–1429	Hv–3081	Barber *et al* 1981
265±95	Solway Sheep	cal AD 1440–1955	OxA–4215	Stallibrass 1993; this volume
230±75	Solway Sheep	cal AD 1470–1955	OxA–4216	Stallibrass 1993; this volume

Bibliography

Primary map sources

CRO(C) D/Lons/L/Kendal Barony/3/3 – Lyth 1590
CRO(K) WD/D/He/3/31 – Lyth 1636/1755
CRO(K) WD/D/He/3/31/7 – Lyth 1711
CRO(W) YSPC/16/7 – Drigg Enclosure Award 1828
CRO(C) QRE/1/30 – Egremont Enclosure Award 1783
CRO(C) QRE/1/120 – Eskmeals, Bootle Enclosure Award 1870
CRO(C) D/Lons/L5/2/41/13 – K34 Burgh by Sands Enclosure Award 1848
CRO(C) D/Lons/W4/16 – 1750 Maxwell report Wedholme
CRO(C) C/D/Lons/W.26 HR – 1755 enclosure
CRO(C) QRE/1/148 Holme Cultram Enclosure Award 1814
CRO(C) D&C/Machell Mss
CRO(C) LRO DDMc30/4 The Backbarrow Company Journal
LRO AE/Ulverston Enclosure Award 1913
LRO AE/ Broughton-in-Furness Enclosure Award 1847
Greenwood, C, 1823 Map of Cumberland
Greenwood, C, 1824 Map of Westmorland
Jefferys, T, 1770 Map of Westmorland

Secondary sources

Andersen, S Th, 1979 Identification of wild grass and cereal pollen, *Danm Geol Unders*, 1978, 69–92

Annable, R, 1987 *The later prehistory of Northern England*, BAR, Brit Ser, **160**, Oxford

Anon, 1903 Stone implements, in Addenda Antiquaria, *Trans Cumberland Westmorland Antiq Archaeol Soc*, n ser, **3**, 410–20

Anon, 1915 A bronze spearhead from Caldbeck, *Trans Cumberland Westmorland Antiq Archaeol Soc*, n ser, **15**, 191

Anon, 1917 Axe-hammer found near Silloth, *Trans Cumberland Westmorland Antiq Archaeol Soc*, n ser, **17**, 254

Anon, 1921 Bronze spearhead from Whitbarrow, *Trans Cumberland Westmorland Antiq Archaeol Soc*, n ser, **21**, 273

Anon, 1944 Addenda Antiquaria, *Trans Cumberland Westmorland Antiq Archaeol Soc*, n ser, **44**, 160–2

Armitage, G E, *c* 1980 *Agrarian change in Bowness on Solway 16th to 19th centuries*, CRO(C) Lib, unpubl doc

Armstrong, A M, Mawer, A, Stenton, F M, and Dickens, B, 1950 *The place-names of Cumberland*, 3 vols, Engl Place-name Soc, **20–22**, Cambridge

Arnold, J J, and Libby, W F, 1951 Radiocarbon dates, *Science*, **113**, 111

Atkinson, J C, 1887a *The Coucher Book of Furness Abbey, part II*, Chetham Soc, n ser, **11**, Manchester

Atkinson, J C, 1887b *The Coucher Book of Furness Abbey, part III*, Chetham Soc, n ser, **14**, Manchester

Bagot, A, and Munby, J, 1988 *All things is well here: Letters from Hugh James of Leven to James Grahme, 1692–95*, Cumberland Westmorland Antiq Archaeol Soc, Rec ser, **10**, Kendal

Bailey, R N, 1980 *Viking age sculpture in northern England*, London

Bailey, R N, and Cramp, R J, 1988 *Corpus of Anglo-Saxon stone scuplture, 2: Cumberland, Westmorland and Lancashire North-of-the-Sands*, Oxford

Baines, E, 1836 *The history of the County Palatine and Duchy of Lancaster*, London

Baines, E, 1870, *The history of the County Palatine and Duchy of Lancaster* (ed J Harland), 2 vols, London

Baines, E, 1893 *The history of the County Palatine and Duchy of Lancaster* (rev J Croston), 5 vols, Manchester

Barber, J W, and Crone, B A, 1993 Crannogs; a diminishing resource? A Survey of the crannogs of South West Scotland and excavations at Buiston Crannog, *Antiquity*, **67**, 520–33

Barber, K E, 1981 *Peat stratigraphy and climatic change: a palaeoecological test of the theory of cyclic peat bog regeneration*, Rotterdam

Barber, K E, 1984 A large-capacity Russian-pattern sediment sampler, *Quat Newsletter*, **44**, 29–31

Barber, K E, Dumayne, L, and Stoneman, R, 1993 Climate change and human impact during the late Holocene in northern Britain, in *Climate change and human impact on the landscape* (ed F M Chambers), London, 225–36

Barber, K E, Chambers, F M, Maddy, D, Stoneman, R, and Brew, J S, 1994 A sensitive high-resolution record of late-Holocene climatic change from a raised bog in Northern England, *Holocene*, **4(2)**, 198–205

Barber, K E, Dumayne-Peaty, L, Hughes, P D M, Mauquoy, D, and Scaife, R, 1998 Replicability and variability of the recent macrofossil and proxy-climate record from raised bogs: field stratigraphy and macrofossil data from Bolton Fell Moss and Walton Moss, Cumbria, England, *J Quat Sci*, **13**, 515–28

Barker, H, and Mackey, J, 1961 British Museum natural radiocarbon measurements, III, *Radiocarbon*, **3**, 39

Barnes, F, 1956 Pottery from prehistoric sites, North End, Walney Island, *Trans Cumberland Westmorland Antiq Archaeol Soc*, n ser, **56**, 1–16

Barnes, F, 1963 Discovery of four roughed-out stone axes at Skelmore Heads, July 1959, in Powell, 1963, 27–30

Barnes, F, 1968 *Barrow and District: an illustrated history*, 2nd edn, Barrow-in-Furness

Barnes, F, 1970 Microlithic sites in Walney Island, *Trans Cumberland Westmorland Antiq Archaeol Soc*, n ser, **70**, 277–80

Barnes, J A, 1904 Ancient corduroy roads near Gilpin Bridge, *Trans Cumberland Westmorland Antiq Archaeol Soc*, n ser, **4**, 207–10

Bartley, D D, Chambers, C, and Hart-Jones, B, 1976 The vegetational history of parts of south and east Durham, *New Phytol*, **77**, 437–68

Bartosiewicz, L, Neer, W van, and Lentacker, A, 1993 Metapodial symmetry in draft cattle, *Int J Osteoarchaeol*, **3**, 69–75

Baxter, W, 1914 The Granges of Holm Cultram, *Trans Cumberland Westmorland Antiq Archaeol Soc*, n ser, **14**, 274–80

Beijerenck, W, 1974 *Zadenatlas der Nederlandsche flora*, Wageningen

Bell, M, and Walker, M J C, 1992 *Late Quaternary environmental change: physical and human perspectives*, Harlow

Bellhouse, R L, 1984 Roman sites on the Cumberland coast: the new tower on Rise How, *Trans Cumberland Westmorland Antiq Archaeol Soc*, n ser, **84**, 41–59

Bellhouse, R L, 1989 *Roman Sites on the Cumberland Coast: a new schedule of coastal sites*, Cumberland Westmorland Antiq Archaeol Soc, Res ser, **3**, Kendal

Bellhouse, R L, and Richardson, G G S, 1982 The Trajanic fort at Kirkbride, the terminus of the Stanegate frontier, *Trans Cumberland Westmorland Antiq Archaeol Soc*, n ser, **82**, 80–99

Benson, D, and Bland, K, 1963 The Dog Hole, Haverbrack, *Trans Cumberland Westmorland Antiq Archaeol Soc*, n ser, **63**, 61–76

Berggren, G, 1969 *Atlas of seeds*, part 2 Cyperaceae, Stockholm

Berggren, G, 1981 *Atlas of seeds*, part 3 Salicaceae, Stockholm

Berry, A Q, Gale, F, Daniels, J L, and Allmark, B (eds), 1996 *Fenn's and Whixall Mosses*, Mold

Bewley, R H, 1986, Survey and excavation in the Solway Plain, Cumbria (1982–4), *Trans Cumberland Westmorland Antiq Archaeol Soc*, n ser, **86**, 19–38

Bewley, R H, 1992 Excavation of two cropmark sites in the Solway Plain, Cumbria: Ewanrigg Settlement and Swarthy Hill 1986–1988, *Trans Cumberland Westmorland Antiq Archaeol Soc*, n ser, **92**, 23–47

Bewley, R H, 1993 Survey and excavation at a crop-mark enclosure, Plasketlands, Cumbria, *Trans Cumberland Westmorland Antiq Archaeol Soc*, n ser, **93**, 1–18

Bewley, R H, 1994 *Prehistoric and Romano-British settlement in the Solway Plain, Cumbria*, Oxbow Monog, **36**, Oxford

Bewley, R H, 1998 Survey and excavations of a cropmark enclosure at Edderside, Cumbria, 1989–90, *Trans Cumberland Westmorland Antiq Archaeol Soc*, n ser, **98**, 107–18

Bewley, R H, Longworth, I H, Browne, S, Huntley, J P, and Varnell, G, 1992 Excavation of a Bronze Age cemetery at Ewanrigg, Maryport, Cumbria, *Proc Prehist Soc*, **58**, 325–54

Bidwell, P (ed), 1999 *Hadrian's Wall 1989–99*, Newcastle upon Tyne

Bingham, R K, 1987 *The Chronicles of Milnthorpe*, Milnthorpe

Bingham, R K, 1995 *Kendal: a social history*, Milnthorpe

Birks, H J B, 1965 Pollen analytical investigation at Holgate Moss, Lancashire and Lindow Moss, Cheshire, *J Ecol*, **53**, 299–314

Birks, H J B, 1973 *Past and present vegetation of the Isle of Skye*, London

Birks, H J B, 1982 Mid-Flandrian forest history of Roudsea Wood National Nature Reserve, Cumbria, *New Phytol*, **90**, 339–54

Bland, K P, 1994 Cause of death of the humans recovered from Dog Hole, Haverbrack, *Trans Cumberland Westmorland Antiq Archaeol Soc*, n ser, **94**, 285–86

Bonney, A P, 1972 A method for determining absolute pollen frequencies in lake sediments, *New Phytol*, **71**, 393–405

Bonsall, J C, 1981 The coastal factor in the Mesolithic settlement of North West England, in *Mesolithikum in Europa* (ed B Gramsch), Veröffentlichungen des Museums für Ur – und Frühgeschite, Potsdam, 451–72

Bonsall, C, 1983 Appendix: Trial excavations at St Bees Site VIII (Golf-course field), in Cherry and Cherry 1983, 13–14

Bonsall, C, Sutherland, D, Tipping, R, and Cherry, J, 1986 The Eskmeals project 1981–5: an interim report, *Northern Archaeol*, **7/1**, 3–30

Bonsall, C, Sutherland, D G, Tipping, R M, and Cherry, J, 1991 The Eskmeals Project: late Mesolithic settlement and environment in north–west England, in *The Mesolithic in Europe* (ed C Bonsall), Edinburgh, 175–205

Bonsall, C, Sutherland, D G, and Payton R W, 1994 The Eskmeals coastal foreland: archaeology and shoreline development, in *Cumbria Field Guide* (eds J Boardman and J Walden), Quat Res Assoc, Oxford, 90–103

Boyer, R, 1991 Sacrifice in Germano-Norse paganism, in *Mythologies* (ed Y Bonnefoy), 2 vols, Chicago, 289–90

Bradley, R, and Edmonds, M, 1993 *Interpreting the axe trade: production and exchange in Neolithic Britain*, Cambridge

Brady, J, 1971 An ancient metalled causeway by Goldmire Bridge, Thwaite Flat, near Dalton-in-Furness, *Proc Barrow Nat Field Club*, **10**, 32–4

Breeze, D J, and Dobson, B, 1976 *Hadrian's Wall*, London

Brooks, D, and Thomas, K W, 1967 The distribution of pollen grains on microscopic slides. I. The non-randomness of the distribution, *Pollen et Spores*, **9**, 621–29

Brownbill, J, 1916 *The Coucher Book of Furness Abbey*, **2/2**, Chetham Soc, n ser, **76**, Manchester

Brownbill, J, 1919 *The Coucher Book of Furness Abbey*, **2/3**, Chetham Soc, n ser, **78**, Manchester

Bruce, J Collingwood, 1867 *The Roman Wall*, 3rd edn, Newcastle upon Tyne

Brunning, R, 1999 The *in situ* preservation of the Sweet Track, in *Bog bodies, sacred sites, and wetland archaeology* (eds B Coles, J Coles, and M S Jørgensen), Exeter, 33–7

Burl, H A W, 1976 *The stone circles of the British Isles*, London

Burleigh, R, and Matthews, K, 1982 British Museum natural radiocarbon measurements XIII, *Radiocarbon*, **24**, 198–205

Burton, R G O, and Hodgson, J M (eds), 1987 *Lowland peat in England*, Soil Surv Spec Surv, **15**, Harpenden

Butler, S, 1992 Vegetational history and land use, in *Hillforts of Southern Scotland* (ed J S Ridout), Edinburgh, 151–69

Canivez, J M (ed), 1933 *Statuta Ordinis Cisterciensis*, Louvain

Catt, J A, 1977 Loess and Coversands, in *British Quaternary studies: recent advances* (ed F W Shotton), Oxford, 221–30

Caruana, I, 1989 Fieldwalking in the Solway Plain, 1983–4, *Trans Cumberland Westmorland Antiq Archaeol Soc*, n ser, **89**, 51–68

Caruana, I, and Cherry, J, 1994 A microlith from Carlisle, *Trans Cumberland Westmorland Antiq Archaeol Soc*, n ser, **94**, 281–282

Chambers, F, 1991 Peat hummification: proxy climatic record or indicator of land-use history, in *Soil and Human Settlement* (eds S Limbrey), Welsh Soils Discussion Gp Rep, **26**, Bangor, 27–44

Chapman, W, 1832 Account of some Antiquities presented to the Society by William Chapman Esq, Civil Engineer, Communicated by him to the Secretaries, *Archaeol Aeliana*, ser 1, **2**, 115–19

Cherry, J, 1963 Eskmeals sand-dunes occupation sites, Phase I, flint workings, *Trans Cumberland Westmorland Antiq Archaeol Soc*, n ser, **63**, 31–52

Cherry, J, 1965 Flint chipping sites at Drigg, *Trans Cumberland Westmorland Antiq Archaeol Soc*, n ser, **65**, 66–85

Cherry, J, 1966 Eskmeals sand-dunes occupation sites — Phase II, iron manufacture, *Trans Cumberland Westmorland Antiq Archaeol Soc*, n ser, **66**, 46–56

Cherry, J, 1967 Prehistoric habitation sites at Seascale, *Trans Cumberland Westmorland Antiq Archaeol Soc*, n ser, **67**, 1–16

Cherry, J, 1969 Early Neolithic sites at Eskmeals, *Trans Cumberland Westmorland Antiq Archaeol Soc*, n ser, **69**, 40–54

Cherry, J, 1982 Sea cliff erosion at Drigg, Cumbria: evidence of prehistoric habitation, *Trans Cumberland Westmorland Antiq Archaeol Soc*, n ser, **82**, 1–6

Cherry, J, and Cherry, P J, 1973 Mesolithic habitation sites at St Bees, *Trans Cumberland Westmorland Antiq Archaeol Soc*, n ser, **73**, 47–66

Cherry, J, and Cherry, P J, 1983 Prehistoric habitation sites in West Cumbria: Part I, the St Bees area and north to the Solway, *Trans Cumberland Westmorland Antiq Archaeol Soc*, n ser, **83**, 1–14

Cherry, J, and Cherry, P J, 1984 Prehistoric habitation sites in West Cumbria Part II, the Nethertown and Seascale areas, *Trans Cumberland Westmorland Antiq Archaeol Soc*, n ser, **84**, 1–17

Cherry, J, and Cherry, P J, 1985 Prehistoric habitation sites in West Cumbria: Part III, the Drigg and Ravenglass areas, *Trans Cumberland Westmorland Antiq Archaeol Soc*, n ser, **85**, 1–10

Cherry, J, and Cherry, P J, 1987a Prehistoric habitation sites in West Cumbria, Part V: Eskmeals to Haverigg, *Trans Cumberland Westmorland Antiq Archaeol Soc*, n ser, **87**, 1–10, Kendal

Cherry, J, and Cherry, P J, 1987b *Prehistoric habitation sites on the limestone uplands of Eastern Cumbria*, Cumberland Westmorland Antiq Archaeol Soc, Res ser, **2**, Kendal

Cherry, J, and Cherry, P J, 1996 Coastline and upland in the Cumbrian Neolithic, *Northern Archaeol*, **13/14**, 61–6

Chevalier, R, 1976 *Roman Roads*, London

Chinn, S J, and Innes, J B, 1995 Pollen analysis from Wet Sleddale, in Prehistoric habitation sites of the Cumbrian limestone uplands: occupation sites found between 1986 and 1993 (J Cherry and P J Cherry), *Trans Cumberland Westmorland Antiq Archaeol Soc*, n ser, **95**, 19–22

Claris, P, and Quartermaine, J, 1989 The Neolithic quarries and axe factory sites of Great Langdale and Scafell Pike: a new field survey, *Proc Prehist Soc*, **55**, 1–25

CLAU, 1983 *Kent Estuary Mosses Project Interim Report 1982–3*, unpubl rep, Lancaster

Clough, T McK, 1968 The Beaker period in Cumbria, *Trans Cumberland Westmorland Antiq Archaeol Soc*, n ser, **68**, 1–21

Clough, T McK, 1969 Bronze Age metalwork from Cumbria, *Trans Cumberland Westmorland Antiq Archaeol Soc*, n ser, **69**, 1–39

Clough, T McK, and Cummins, W A (eds), 1988 *Stone Axe Studies 2*, CBA Res Rep, **67**, London

Coles, B, 1995 *Wetland management*, Exeter

Coles, B, and Coles, J M, 1986 *Sweet Track to Glastonbury*, London

Coles, B, and Coles, J, 1989 *People of the wetlands*, London

Coles, J, 1961 The Salta Moss Rapier, *Trans Cumberland Westmorland Antiq Archaeol Soc*, n ser, **61**, 16–24

Coles, J M, 1989 The Somerset Levels project, 1973–1988, in *Somerset Levels Papers* (ed J M Coles), **15**, 5–14

Colgrave, B, and Mynors, R A B (eds), 1940 *Bede's ecclesiastical history of the English people*, Oxford

Collingwood, R G, 1933a An introduction to the prehistory of Cumberland, Westmorland and Lancashire north of the sands, *Trans Cumberland Westmorland Antiq Archaeol Soc*, n ser, **33**, 163–200

Collingwood, R G, 1933b Prehistoric settlements near Crosby Ravensworth, *Trans Cumberland Westmorland Antiq Archaeol Soc*, n ser, **33**, 201–26

Collingwood, W G, 1900 The so-called cockpit at Monk Foss, in the Parish of Whitbeck, Cumberland, *Trans Cumberland Westmorland Antiq Archaol Soc*, ser 1, **16**, 117–9

Collingwood, W G, 1923 An inventory of the ancient monuments fo Cumberland, *Trans Cumberland Westmorland Antiq Archaeol Soc*, n ser, **23**, 206–76

Collingwood, W G, 1926 An inventory of the ancient monuments of Westmorland and Lancashire North-of-the-Sands, *Trans Cumberland Westmorland Antiq Archaeol Soc*, n ser, **26**, 1–62

Coope, G R, 1994 The late glacial coleoptera from St Bees, Cumbria, in *Cumbria Field Guide* (eds J Boardman and J Walden), Quat Res Assoc, Oxford, 86–9

Coope, G R, and Joachim, M J, 1980 Late glacial environmental changes interpreted from fossil Coleoptera from St Bees, Cumbria, North West England, in *Studies in the Late glacial of North–West Europe* (eds J J Lowe, J M Gray, and J E Robinson), Oxford, 55–68

Countryside Commission, 1995 *The Solway coast landscape*, Cheltenham

Countryside Commission, 1998 *Countryside character*, **2**: *North West*, Cheltenham

Cowell, R, and Innes, J, 1994 *The wetlands of Merseyside*, Lancaster Imprints **2**, Lancaster

Cox, C, 1992 Satellite imagery, aerial photography, and wetland archaeology, *World Archaeol*, **24(2)**, 249–67

Cox, M, Chandler, J, Cox, C, Jones, J, and Tinsley, H, in press, The archaeological significance of patterns of anomalous vegetation on a raised mire in the Solway Estuary and the processes involved in their formation, *J Archaeol Sci*

Cressey, M, and Toolis, R, 1996 *Solway coastal assessment 1996*, Centre Field Archaeol Rep, **312**, unpubl rep

Crofton, H T, 1902 How Chat Moss broke out in 1526, *Trans Lancashire Cheshire Antiq Soc*, **20**, 139–45

Cross, Hon M, 1950 Two stone axe-hammers from Silecroft, in Anon, Addenda Antiqua 1: Prehistoric sites and discoveries, *Trans Cumberland Westmorland Antiq Archaeol Soc*, n ser, **49**, 213–4

Cundill, P R, 1976 Late Flandrian vegetation and soils in Carlingill valley, Howgill Fells, *Trans Inst Brit Geogr*, n ser, **1**, 301–9

Cunliffe, B, 1991 *Iron Age communities in Britain* , 3rd edn, London

Daniels, C M, 1978 *Handbook to the Roman wall*, 13th edn, Newcastle upon Tyne

Daniels, C M, 1979 Fact and theory on Hadrian's Wall, *Britannia*, **10**, 357–64

Darbishire, R, 1873 Notes on discoveries at Ehenside Tarn, Cumberland, *Archaeologia*, **44**, 273–92

Davenport, J, 1996 The Bewcastle cauldron, *Trans Cumberland Westmorland Antiq Archaeol Soc*, n ser, **96**, 228–30

Davidson, H R E, 1988 *Myths and symbols in pagan Europe. Early Scandinavian and Celtic religions*, New York

Davies, C, and Turner, J, 1979 Pollen diagrams from Northumberland, *New Phytol*, **82**, 783–804

Denton, J 1887 *An account of the most considerable estates and families in the County of Cumberland since the Conquest till the year 1610*, Cumberland Westmorland Antiq Archaeol Soc, Tract Ser, **2**, Kendal

DETR, 1999 *Monitoring and assessment of peat and alternative products for growing media and soil improvers in the UK*, London

DOE 1990 Archaeology and planning, *Planning Policy Guidance Note*, **16**, London

DOE 1995 Environment Act 1995: review of mineral planning permissions, *Minerals Planning Policy Guidance Note*, **14**, London

Dickinson, W, 1975 Recurrence surfaces in Rusland Moss, Cumbria, *J Ecol*, **63**, 913–93

Dilley, R S, 1972 *Common land in Cumberland 1500–1750*, unpubl thesis, Univ Cambridge

Dilley, R S, 1991 *Agricultural change and common land in Cumberland 1500–1750*, unpubl thesis, Univ McMaster

Dobson, J, 1907 Urswick Stone Walls, *Trans Cumberland Westmorland Antiq Archaeol Soc*, n ser, **7**, 72–94

Donaldson, A M, and Turner, J, 1977 A pollen diagram from Hallowell Moss, near Durham City, UK, *J Biogeog*, **4**, 25–33

Donkin, R A, 1978 *The Cistercians: studies in the geography of medieval England and Wales*, Toronto

Dresser, Q, 1985 University College Cardiff radiocarbon dates 1, *Radiocarbon*, **27(2B)**, 338–85

Dumayne, L, 1992 *The Late Holocene palaeoecology and human impact on the environment of northern Britain*, unpubl thesis, Univ Southampton

Dumayne, L, 1993a Iron Age and Roman vegetation clearance in northern Britain: further evidence, *Bot J Scotland*, **46**, 385–92

Dumayne, L, 1993b Invader or native? — vegetation clearance in northern Britain during Romano-British time, *Vegetation Hist Archaeobot*, **2**, 29–36

Dumayne, L, 1994 The effect of the Roman occupation on the environment of Hadrian's Wall: a pollen diagram from Fozy Moss, Northumbria, *Britannia*, **25**, 217–24

Dumayne, L, and Barber, K E, 1994 The effect of the Romans on the environment of northern Britain: pollen data from three sites close to Hadrian's Wall, *Holocene*, **4(2)**, 165–73

Dumayne, L, Stoneman, R, Barber, K E, and Harkness, D, 1995 Problems associated with correlating radiocarbon-dated pollen diagrams with historical events, *Holocene*, **5**, 118–23

Eastwood, T, Dixon, E E L, Hollingworth, S E, and Smith, B, 1931 *Geology of the Whitehaven and Workington district*, Mem Geol Surv England, UK Sheet 28, London

Edmonds, M, 1995 *Stone tools and society: Working stone in Neolithic and Bronze Age Britain*, London

Ellwood, A E, 1981 Finds from Crosby Ravensworth, Orton Scar and Crook, *Trans Cumberland Westmorland Antiq Archaeol Soc*, n ser, **81**, 158–9

English Nature, 1985 *Black Moss*, unpubl doc

English Nature, 1991 *Position statement on lowland peatland*, Peterborough

English Nature, 1997 *Cumbria fells and dales Natural Area profile*, Peterborough

Esmonde-Cleary, A S, 1997 England, in Roman Britain for 1996 (eds B C Burnham, L J F Keppie, A S Esmonde-Cleary, and R S O Tomlin), *Britannia*, **28**, 414–53

Evans, J, 1897 *Ancient stone implements of Great Britain*, London

Faegri, K, Kaland, P E, and Krzywinski, K, 1989 *Textbook of pollen analysis*, Chichester

Fair, M C, 1932 A reconsideration of the lakeside site at Ehenside Tarn, *Trans Cumberland Westmorland Antiq Archaeol Soc*, n ser, **32**, 57–62

Fair, M C, 1936 A sandhill site at Eskmeals, *Trans Cumberland Westmorland Antiq Archaeol Soc*, n ser, **36**, 20–3

Fair, M C, 1937 West Cumberland sandhill occupation, flint-working in the sandhills, *Trans Cumberland Westmorland Antiq Archaeol Soc*, n ser, **37**, 214–15

Fair, M C, 1943 The Gosforth area in pre-history, *Trans Cumberland Westmorland Antiq Archaeol Soc*, n ser, **43**, 50–54

Fair, M C, 1946a Bronze Age swords and daggers of Cumberland, Westmorland and Lancashire North-of-the-Sands, *Trans Cumberland Westmorland Antiq Archaeol Soc*, n ser, **45**, 34–38

Fair, M C, 1946b An interim review of types of bronze spear-heads and axes of Cumberland, Westmorland and Lancashire North-of-the-Sands, *Trans Cumberland Westmorland Antiq Archaeol Soc*, n ser, **45**, 172–8

Farrer, W, 1923–4 *Records relating to the Barony of Kendale*, Cumberland Westmorland Antiq Archaeol Soc, Record Ser (ed J Curwen), **4–5**, Kendal

Farrer, W, and Brownbill, J (eds), 1914 *Victoria history of the counties of England: Lancashire*, **8**, London

Fell, C I, 1940 Bronze Age connections between the Lake Counties and Ireland, *Trans Cumberland Westmorland Antiq Archaeol Soc*, n ser, **40**, 118–30

Fell, C I, 1953 A Beaker burial on Sizergh Fell, near Kendal, *Trans Cumberland Westmorland Antiq Archaeol Soc*, n ser, **53**, 1–5

Fell, C I, 1967 Tanged and barbed flint arrow- or lance-head, possibly from Ehenside Tarn, *Trans Cumberland Westmorland Antiq Archaeol Soc*, n ser, **67**, 229–30

Fell, C I, 1973 Prehistoric archaeology, in *The Lake District* (eds W H Pearsall, and W Pennington), London, 196–211

Fell, C I, 1985 Flints, *Trans Cumberland Westmorland Antiq Archaeol Soc*, n ser, **85**, 255

Fell, C I, 1991 Reworked axe from Little Strickland Hill, Witherslack, *Trans Cumberland Westmorland Antiq Archaeol Soc*, n ser, **91**, 265–6

Fell, C I, and Coles, J M, 1965 Reconsideration of the Ambleside Hoard and the burial at Butts Beck quarry, Dalton-in-Furness, *Trans Cumberland Westmorland Antiq Archaeol Soc*, n ser, **65**, 38–53

Fell, C I, and Davis, R V, 1988 The petrological identification of stone implements from Cumbria, in *Stone Axe Studies 2* (eds T, McK Clough and W A, Cummins), CBA Res Rep, **67**, London, 71–7

Fleming, A, 1988 *The Dartmoor Reaves: Investigating prehistoric land divisions*, London

Fletcher, M, and Miller, I, 1997 Recent archaeological work on the Holm Cultram sea dyke, Skinburness, Cumbria, *Trans Cumberland Westmorland Antiq Archaeol Soc*, n ser, **97**, 203–15

Fletcher, W, 1957 Greycroft stone circle, Seascale, *Trans Cumberland Westmorland Antiq Archaeol Soc*, n ser, **57**, 1–8

Forestry Commission, 1999 *A new focus for England's woodlands: strategic priorities and programmes*, Cambridge

Fulford, M, Champion, T, and Land, N, 1997 *England's coastal heritage*, Engish Heritage Archaeol Rep **15**, London

Gale, S J, and Hunt, C O, 1985 The stratigraphy of Kirkhead Cave, an Upper Palaeolithic site in northern England, *Proc Prehist Soc*, **51**, 283–304

Gale, S J, and Hunt, C O, 1990 The stratigraphy of Kirkhead Cave, an Upper Palaeolithic site in northern England: discussion, *Proc Prehist Soc*, **56**, 51–6

Gaythorpe, H, 1897 Prehistoric implements in Furness, *Trans Cumberland Westmorland Antiq Archaeol Soc*, ser 1, **14**, 442–7

Gelderd, C, Randall, J, and Dobson, J, 1914 Some Birkrigg Barrows, *Trans Cumberland Westmorland Antiq Archaeol Soc*, n ser, **14**, 466–79

Gibson, A M, 1986 *Late Neolithic and Bronze Age pottery*, Princes Risborough

Gidney, L P, and Huntley, J P, 1996 A *Horse skull from Winster*, Durham Environ Archaeol Rep, **33**, unpubl rep

Gilks, J A, 1973 The Neolithic and early Bronze Age pottery from Elbolton Cave, Wharfedale, *Yorkshire Archaeol J*, **45**, 41–54

Gilks, J A, 1987 Later Bronze Age pottery from Kirkhead Cave, Cumbria, *Trans Cumberland Westmorland Antiq Archaeol Soc*, n ser, **87**, 37–42

Godwin, H, 1975 *History of the British flora*, Cambridge

Godwin, H, and Willis, E H, 1959 Cambridge University natural radiocarbon measurements I, *Radiocarbon*, **1**, 63–5

Godwin, H, and Willis, E H, 1960 Cambridge University natural radiocarbon measurements II, *American J Sci, Radiocarbon suppl*, **2**, 62

Graham, T I P (ed), 1934 *The Barony of Gilsland: Lord William Howard's survey taken in 1603*, Cumberland Westmorland Antiq Archaeol Soc, Extra ser, **16**, Kendal

Grainger, F, and Collingwood, W G, 1929 *The Register and Records of Holm Cultram*, Cumberland Westmorland Antiq Archaeol Soc, Rec ser, **7**, Kendal

Granlund, E, 1932 De Svenska Hogmossarnas geologi, *Sver Geol Unders*, **26**, 1–93

Grimm, E, 1991 *TILIA and TILIAGRAPH**, Illinois

Grosse-Brauckmann, G, 1972 Über Planzlithe Makrofossilien Mitteleuropaischer Torfe, 1: Gewebereste Krautiger Pflanzen und ihre Merkmale, *Telma*, **2**, 19–55

Grove, J, 1988 *The Little Ice Age*, Cambridge

Hadfield, C, and Biddle, G, 1970 *The canals of North West England*, Newton Abbot

Haggart, B 1989 Variations in the pattern and rate of isostatic uplift indicated by a comparison of Holocene sea-level curves from Scotland, *J Quat Sci*, **4(1)**, 67–76

Hall, B R, and Folland, C J, 1970 *Soils of Lancashire*, Harpenden

Hall, D, and Coles, J M, 1994 *Fenland survey*, English Heritage Rep, **1**, London

Hall, D, Wells, C E, and Huckerby, E, 1995 *The Wetlands of Greater Manchester*, Lancaster Imprints, **3**, Lancaster

Hallam, J S, Edwards, B J N, Barnes, B, and Stuart, A J, 1973 A Late Glacial elk with associated barbed points from High Furlong, Lancashire, *Proc Prehist Soc*, **39**, 100–28

Halliday, G, 1997 *A Flora of Cumbria*, Lancaster

Harris, H, 1992 *The industrial archaeology of Dartmoor*, 4th edn, Newton Abbot

Harvey, A M, Oldfield, F, and Baron, F, 1981 Dating of post-glacial landforms in the central Howgills, *Earth Surface Processes and Landforms*, **6**, 401–12

Haselgrove, C, 1996 The Iron Age, in *The archaeology of Lancashire* (ed R Newman), Lancaster, 61–74

Haverfield, F, 1895 Report of the Cumberland Excavation Committee for 1894, *Trans Cumberland Westmorland Antiq Archaeol Soc*, ser 1, **12**, 453–69

Haverfield, F, 1900 Report of the Cumberland Excavation Committee for 1899, *Trans Cumberland Westmorland Antiq Archaeol Soc*, ser 1, **16**, 80–99

Hay, T, 1936 Early settlements near the head of Ullswater, *Trans Cumberland Westmorland Antiq Archaeol Soc*, n ser, **36**, 71–5

Healy, F, 1983 Are first impressions only topsoil deep? The evidence from Tattershall Thorpe, Lincolnshire, *Lithics*, **4**, 28–33

Henderson, G, 1911 *Survivals in belief amongst the Celts*, Glasgow

Herne, A, 1988 A time and a place for the Grimston bowl, in *The archaeology of context in the Neolithic and Bronze Age* (eds J C Barrett and I A Kinnes), Sheffield, 9–29

Hibbert, F S, Switsur, V R, and West, R G, 1971 Radiocarbon dating of Flandrian pollen zones at Red Moss, Lancashire, *Proc Roy Soc London*, **B 177**, 161–76

Higham, N J, 1981 Two enclosures at Dobcross Hall, Dalston, *Trans Cumberland Westmorland Antiq Archaeol Soc*, n ser, **81**, 1–6

Higham, N J, 1986 *The Northern Counties to AD 1000*, London

Higham, N J, and Jones, G D B, 1975 Frontiers, forts, and farmers, Cumbrian aerial survey, 1974–75, *Archaeol J*, **132**, 16–53

Hodgson, J, 1840 *History of Northumberland*, **2/3**, London

Hodgson, T H, 1895 Bone spear or harpoon head from Terra del Fuego, found on peat near Crosby-on-Eden, *Trans Cumberland Westmorland Antiq Archaeol Soc*, ser 1, **13**, 402

Hodgson, T H, 1904 An ancient palisade on Bowness Common, *Trans Cumberland Westmorland Antiq Archaeol Soc*, n ser, **4**, 211–2

Horsley, J, 1732 *Britannia Romana*, London

Howard-Davis, C L E, Innes, J, and Stocks, C, 1988 *Peat and the past*, Lancaster

Howard-Davis, C L E, and Buxton, K, forthcoming Excavations at Church Moss, Davenham, Cheshire 1995–6, *J Chester Archaeol Soc*

Huckerby, E, and Wells, C, 1993 Recent work at Solway Moss, Cumbria, in Middleton 1993a, 36–42

Huddart, D, Tooley, M J, and Carter, P A, 1977 The coasts of North–West England in eds Kidson and Tooley, 1977, 119–54

Hughes, P D M, 1997 *The palaeoecology of the fen/bog Transition during the early- to mid-Holocene in Britain*, unpubl thesis, Univ Southampton

Hughes, P D M, Mauquoy, D, Barber, K E, and Langdon, P G, 2000 Mire development pathways and palaeoclimatic records from a full Holocene peat archive at Walton Moss, Cumbria, England, *Holocene*, **10(4)**, 465–79

Huntley, B, 1993 Rapid early-Holocene migration and high abundance of hazel (*Corylus avellana L*): alternative hypotheses, in *Climate change and human impact on the landscape* (ed F M Chambers), London, 205–15

Huntley, B, and Birks, H J B, 1983 *An atlas of past and present pollen maps for Europe 0–13,000 years ago*, Cambridge

Huntley, J P, 1988 Palaeobotanical investigations, in B Harbottle, R Fraser, and F C Burton, The Westgate Road Milecastle, Newcastle upon Tyne, *Britannia*, **19**, 160–2

Huntley, J P, 1999 Environmental evidence for Hadrian's Wall, in Bidwell 1999, 48–64

Hutchinson, W, 1794 *The history of the County of Cumberland and some places adjacent*, 2 vols, Carlisle

Innes, J B, 1988 Report on pollen analysis from Midgeholme Moss, Birdoswald, Cumbria, unpubl rep, *Inst Prehist Sci Archaeol*, Univ Liverpool

Innes, J B, and Shennan, I, 1991 Palynology of archaeological and mire sediments from Dod, Borders Region, Scotland, *Archaeol J*, **148**, 1–45

Iversen, J, 1949 The influence of prehistoric man on vegetation, *Danm Geol Unders*, ser 4, **3(6)**, 1–25

Iversen, J, 1954 The late-glacial flora of Denmark and its relation to climate and soil, *Danm Geol Unders*, ser 4, **4(3)**, 5–31

Jackson, E, 1906 On a stone celt found near Cockermouth, *Trans Cumberland Westmorland Antiq Archaeol Soc*, n ser, **6**, 149–150

Jackson, K H, 1953 *Language and history in early Britain*, Edinburgh

Jackson, S T, 1990 Pollen source area and representation in small lakes in north–eastern United States, *Rev Palaeobot Palynol*, **63**, 53–76

Jackson, S T, 1994 Pollen and spores in Quaternary lake sediments as sensors of vegetation composition: theoretical models and empirical evidence, in *Sedimentation of organic particles* (ed A Traverse), Cambridge, 253–86

Jacobson, G L, and Bradshaw, R H W, 1981 The selection of sites for paleovegetational [*sic*] studies, *Quat Res*, **16**, 80–96

Janssen, C R, 1966 Recent pollen spectra from deciduous and coniferous–deciduous forests of north–eastern Minnesota: study in pollen dispersal, *Ecology*, **47**, 804–25

Jardine, W G, 1980 Holocene raised coastal sediments and former shorelines of Dumfries and Galloway, *Trans Dumfries Galloway Nat Hist Antiq Soc*, **55**, 1–59

Jarvis, R A, Bendelow, V C, Bradley, R I, Carroll, D M, Furness, R R, Kilgour, I N L, and King, S J, 1984 *Soils and their use in northern England*, Harpenden

Jecock, M, 1998 *High Park, Lancashire and Cow Close, Cumbria, an archaeological survey report*, unpubl rep

Jefferson, S, 1842 *The history and antiquities of Allerdale Ward, above Derwent, in the County of Cumberland*, Carlisle

Jessen, K, 1955 Key to subfossil Potamogeton, *Botanisk Tidsskrift*, 1–7

Johnsen, S J, Clausen, H B, Dansgaard, W, Fuhrer, K, Gundestrup, N, Hammer, C U, Iversen, P, Jonzel, J, Stauffer, B, and Steffensen, J P, 1992 Irregular glacial interstadials recorded in a new Greenland ice core, *Nature*, **359**, 311–13

Johnson, B R, 1997 Monitoring peatland restoration, in *Conserving peatlands*, (eds I Parks, R E Stoneman, and H A P Ingram), Wallingford and New York, 323–31

Jones, G B D, 1976 The western extension of Hadrian's Wall from Bowness to Cardurnock, *Britannia*, **7**, 236–43

Jones, G B D, 1982 The Solway Frontier: Interim Report 1976–81, *Britannia*, **13**, 283–97

Jones, G B D 1989 Appendix 2: The Western Stanegate, in *The eleventh pilgrimage of Hadrian's Wall* (ed C M Daniels), Newcastle upon Tyne, 92–95

Jones, G P, 1969 King James I and the Western Border, *Trans CumberlandWestmorland Antiq Archaeol Soc*, **80**, 129–57

Jowsey, P C, 1966 An improved peat sampler, *New Phytol*, **65**, 245–9

Katz, N J, Katz, S V, and Kipani, M G, 1965 *Atlas and keys of fruits and seeds occurring in the Quaternary deposits of the USSR*, Moscow

Kidson, C, and Tooley, M J (eds), 1977 *The Quaternary history of the Irish Sea*, Geol J Spec Issue, **7**, Liverpool

King, A, 1974 A review of the archaeological work in the caves of north–west England, in *Limestones and caves of north–west England* (ed A C Waltham), Newton Abbot, 182–200

King, C A M, 1976 *The geomorphology of the British Isles: Northern England*, London

Kirby, D P, 1962 Strathclyde and Cumbria: a survey of historical development to 1092, *Trans Cumberland Westmorland Antiq Archaeol Soc*, n ser, **62**, 77–94

Kirby, P, 1992 *Habitat management for invertebrates: a practical handbook*, Sandy

Knowles, D, 1940 *The religious houses of medieval England*, London

Korhola, A A, 1994 Radiocarbon evidence for rates of lateral expansion in raised mires in southern Finland, *Quat Res*, **42**, 299–307

Korhola, A A, 1995 Holocene climatic variations in southern Finland, *Holocene*, **5**, 43–58

Kromer, B, Rhein, M, Bruns, M, Schlosh-Fisher, H, Münnich, K O, Stuiver, M, and Becker, B, 1986 Radiocarbon calibration data for the sixth to the eighth millennia BC, *Radiocarbon*, **28**, 954–60

Lamb, H H, 1977 *Climate past, present, and future, 2: climatic history and the modern world*, Cambridge

Lambert, J (ed), 1996 *Transect through time: the archaeological landscape of the Shell north–western ethylene pipeline*, Lancaster Imprints, **1**, Lancaster

LUAU, 1993 *North West Wetlands Survey: project design*, unpubl rep

LUAU, 1994 *Torver High Common, Cumbria, Archaeological Survey*, unpubl rep

LUAU, 1995 *Torver Low and Blawith Commons, Cumbria. Archaeological Survey*, unpubl rep

LUAU, 1996 *Kidburngill, Cumbria. Stratigraphic survey report*, unpubl rep

LUAU, 1998 *Kidburngill, Cumbria. Archaeological survey and evaluation*, unpubl rep

LUAU, 2000 *Preliminary palynological analysis of post-Bronze Age deposits at Littlewater, Cumbria*, unpubl rep

Larousse, 1959 *Encyclopaedia of mythologies*, Paris

Lawes Agricultural Trust, 1983 *Soils of Northern England, 1:250,000*, Harpenden

Leah, M D, Wells, C E, Huckerby, E, and Appleby, C, 1997 *The wetlands of Cheshire*, Lancaster Imprints, **5**, Lancaster

Leah, M D, Wells, C E, Huckerby, E, Stamper, P, and Welch, C, 1998 *The wetlands of Shropshire and Staffordshire*, Lancaster Imprints, **7**, Lancaster

Leech, R H, 1983 Settlements and groups of small cairns on Birkby and Birker Fells, Eskdale, Cumbria, *Trans Cumberland Westmorland Antiq Archaeol Soc*, n ser, **83**, 15–23

Lewis, H T, 1977 Maskuta: the ecology of Indian fire in Northern Alberta, *Western Canadian J Anthropol*, **1**, 15–52

Lewis, H T, and Ferguson, T A, 1988 Yards, corridors and mosaics: how to burn a Boreal forest, *Human Ecol*, **16**, 57–77

Lindsay, R A, and Immirzi, C P, 1996 *An inventory of lowland raised bogs in Great Britain*, Scot Nat Her Res, Surv Mon Rep, **78**, Edinburgh

Linick, T W, Suess, H E, and Becker, B, 1985 La Jolla measurements of radiocarbon in south German oak tree–ring chronologies, *Radiocarbon*, **27**, 954–60

Lloyd, J M, Shennan, I, Kirby, J R, and Rutherford, M M, 1999 Holocene relative sea-level changes in the inner Solway Firth, *Quat Internat*, **60**, 83–105

Lotter, A F, Eicher, U, Siegenthaler, U, and Birks, H J B, 1992 Late-glacial climatic oscillations as recorded in Swiss lake sediments, *J Quat Sci*, **7(3)**, 187–204

Lowndes, R A C, 1964 Excavation of a Romano-British farmstead at Eller Beck, *Trans Cumberland Westmorland Antiq Archaeol Soc*, n ser, **64**, 6–13

Lyman, R L, 1984 Bone density and differential survivorship of fossil classes, *J Anthrop Archaeol*, **3**, 259–99

McCarthy, M R, 1990 *A Roman, Anglian, and medieval site at Blackfriars Street*, Cumberland Westmorland Antiq Archaeol Soc, Res ser, **4**, Kendal

McCarthy, M R, 1995 Archaeological and environmental evidence for the Roman impact on vegetation near Carlisle, Cumbria, *Holocene*, **5(4)**, 491–5

McEwen, L J, and Withers, C W J, 1989 Historical records and geomorphological events: the 1771 'eruption' of Solway Moss, *Scottish Geol Mag*, **105(3)**, 149–57

McIntire, W T, 1941 Solway Moss, *Trans Cumberland Westmorland Antiq Archaeol Soc*, n ser, **41**, 1–13

McIntire, W T, 1942 The salt pans of the Solway, *Trans Cumberland Westmorland Antiq Archaeol Soc*, n ser, **42**, 1–12

MacKay, A W, and Tallis, J H, 1994 The recent vegetational history of the forest of Bowland, Lancashire, UK, *New Phytol*, **128**, 571–84

McKenny Hughes, T, 1904a Some notes on mound opening, with a description of one recently explored on Sizergh Fell, Westmorland, *Trans Cumberland Westmorland Antiq Archaeol Soc*, n ser, **4**, 71–9

McKenny Hughes, T, 1904b On another tumulus on Sizergh Fell, *Trans Cumberland Westmorland Antiq Archaeol Soc*, n ser, **4**, 201–4

McKenny Hughes, T, 1912a On an ancient enclosure and interment on Heaves Fell, *Trans Cumberland Westmorland Antiq Archaeol Soc*, n ser, **12**, 397–402

McKenny Hughes, T, 1912b On some interments near Hyning, Westmorland, *Trans Cumberland Westmorland Antiq Archaeol Soc*, n ser, **12**, 403–5

MacPherson, Rev H A, 1892 *A vertebrate fauna of Lakeland*, Edinburgh

MAFF, The Welsh Office, Association of Drainage Authorities, English Nature, and National Rivers Authority, 1994 *Water Level Management Plans. A procedural guide for operating authorities*, London

Manning, A, Birley, R, and Tipping, R, 1997 Roman impact on the environment at Hadrian's Wall: precisely dated pollen analysis from *Vindolanda*, northern England, *Holocene*, **7(2)**, 175–86

Marshall, J D, 1958 *Furness and the Industrial Revolution*, Barrow-in-Furness

Marshall, J D, and Davies–Sheil, M, 1977 *Industrial archaeology in the Lake Counties*, Manchester

Marshall, J D, Helme, J, Wignall, J, and Braithwaite, J C, 1996 The lineaments of Newland iron furnace, 1747–1903, *Trans Cumberland Westmorland Antiq Archaeol Soc*, n ser, **96**, 195–214

Masters, L, 1981 A mesolithic hearth at Redkirk Point, Gretna, Annandale and Eskdale District, *Trans Dumfries Galloway Nat Hist Antiq Soc*, **56**, 111–14

Mellars, P A, 1976 Fire ecology, animal populations and man: a study of some ecological relationships in prehistory, *Proc Prehist Soc*, **42**, 15–45

Melville, J, and Hobbs, J L, 1951 *Early Railway History in Furness*, Cumberland Westmorland Antiq Archaeol Soc, Tract ser, **13**, Kendal

Messenger, P, 1975 Lowther Farmstead plans: a preliminary survey, *Trans Cumberland Westmorland Antiq Archaeol Soc*, n ser, **75**, 327–51

Middleton, R (ed), 1990 *North West Wetlands Survey annual report 1990*, Lancaster

Middleton, R, 1991a *North West Wetlands Survey annual report 1991*, Lancaster

Middleton, R, 1991b A Neolithic polished axe from the Solway Plain, Cumbria, in Middleton 1991a, 44

Middleton, R, 1992a *North West Wetlands Survey annual report 1992*, Lancaster

Middleton, R, 1992b Three animal excavations in Cumbria, in Middleton 1992a, 19–28

Middleton, R, 1993a *North West Wetlands Survey annual report 1993*, Lancaster

Middleton, R, 1993b Landscape archaeology in North West England and the definition of surface lithic scatter sites, in Middleton 1993a, 1–8

Middleton, R, 1993c A new polished stone axe find from the Solway Plain, Cumbria, in Middleton 1993a, 57–9

Middleton, R, 1993d The dating of the sheep remains, in Huckerby and Wells 1993, 38–41

Middleton, R, 1996 The Neolithic and Bronze Age, in *The archaeology of Lancashire* (ed R Newman), Lancaster, 35–60

Middleton, R, 1999 Unleashing the potential? Wetland survey and management in North West England, in *Bog bodies, sacred sites, and wetland archaeology* (eds B Coles, J Coles, and M S Jørgensen), Exeter, 157–64

Middleton, R, and Newman, R M (eds), 1994 *North West Wetlands Survey annual report 1994*, Lancaster

Middleton, R, Tooley, M J, and Innes, J B, forthcoming *The wetlands of South West Lancashire*, Lancaster Imprints, Lancaster

Middleton, R, and Wells, C E, 1990 A research design for the North West Wetlands, in Middleton 1990, 1–12

Middleton, R, Wells, C E, and Huckerby, E, 1995 *The wetlands of North Lancashire*, Lancaster Imprints, **4**, Lancaster

Moore, P D, Webb, J A, and Collinson, M E, 1991 *Pollen Analysis*, 2nd edn, Oxford

Morrison, A, 1981 The coastal Mesolithic in south–west Scotland, in *Mesolithikum in Europa*, Veröffentlichungen des Museums fur Ur- und Frühgeschite (ed B Gramsch), Potsdam, 441–50

Munn Rankin, W, 1910 The peat mosses of Lonsdale, *Naturalist*, **119**, 153

Nanson, W, 1886 Solway Moss, *Trans Cumberland Westmorland Antiq Archaeol Soc*, ser 1, **8**, 257–62

Newman, R, 1999 Iron working and mining enterprises, *Keer to Kent*, **38**, 14–15

Nicholson, F, and Nixon, E, 1915 *The older Noncomformists in Kendal: a history of the Unitarian Chapel in Market Place*, Kendal

Nickson, D, and Macdonald, J H, 1955 A preliminary report on a microlithic site at Drigg, Cumberland, *Trans Cumberland Westmorland Antiq Archaeol Soc*, n ser, **55**, 17–29

North, O H, 1934 Some recent local finds of stone implements, *Trans Cumberland Westmorland Antiq Archaeol Soc*, n ser, **34**, 113–15

North, O H, 1942 A bronze axe and other local finds, *Trans Cumberland Westmorland Antiq Archaeol Soc*, n ser, **42**, 233

Oldfield, F, 1960 Studies in the post–glacial history of British vegetation: lowland Lonsdale, *New Phytol*, **59**, 192–217

Oldfield, F, 1963 Pollen analysis and man's role in the ecological history of the south–east Lake District, *Geografiska Annaler*, **45(1)**, 23–40

Oldfield, F, 1965 Problems of mid–post–glacial pollen zonation in parts of north–west England, *J Ecol*, **53**, 247–60

Oldfield, F, 1969 Pollen analysis and the history of land use, *Advancement Sci*, March 1969, 298–311

Oldfield, F, and Statham, D C, 1963 Pollen–analytical data from Urswick Tarn and Ellerside Moss, North Lancashire, *New Phytol*, **62**, 53–66

Oldfield, F, and Statham, D C, 1965 Stratigraphy and pollen analysis on Cockerham and Pilling Mosses, North Lancashire, *Mem Proc Manchester Lit Phil Soc*, **107**, 70–85

Olivier, A C H, 1987 Excavation of a Bronze Age funerary cairn at Manor Farm, near Borwick, North Lancashire, *Proc Prehist Soc*, **53**, 129–86

O'Sullivan, D, 1985 A review of some Dark Age problems in North West England, in *The Scandinavians in Cumbria* (eds J R Baldwin and I D Whyte), Edinburgh, 17–36

Palmer, W T, 1952 *Byways in Lakeland*, London

Parker, A G, Anderson, D E, and Boardman, J, 1994 Seathwaite Vally: buried organic deposit, in *Cumbria field guide*, (eds J Boardman, and J Walden), Quat Res Assoc, Oxford, 143–51

Parker, F H M, 1905 Inglewood Forest, *Trans Cumberland Westmorland Antiq Archaeol Soc*, n ser, **5**, 35–61

Pearsall, W H, and Pennington, W, 1973 *The Lake District*, London

Pearson, G W, 1987 How to cope with calibration, *Antiquity*, **61**, 98–103

Pearson, G W, Pilcher, J R, Baillie, M G L, Corbett, D M, and Qua, F, 1986 High-precision ^{14}C measurement of Irish oaks to show the natural ^{14}C variations from AD 1840–5210 BC, *Radiocarbon*, **28**, 911–34

Pearson, G W, and Stuiver, M, 1986 High–precision calibration of the radiocarbon time scale, *Radiocarbon*, **28**, 839–62

Pearson, R G, 1962 Coleoptera from a Late-glacial deposit at St Bees, West Cumberland, *J Animal Ecol*, **31**, 129–50

Peglar, S M, 1993 The mid-Holocene *Ulmus* decline at Diss Mere, Norfolk, UK: a year-by-year pollen stratigraphy from annual laminations, *Holocene*, **3(1)**, 1–13

Pennant, T, 1774 *A tour in Scotland and voyage to the Hebrides 1772*, Chester

Pennington, W, 1947 Studies of the post-glacial history of British vegetation 7. Lake sediments: Pollen diagrams from bottom depostis of the north basin of Windermere, *Phil Trans Royal Soc*, B **233**, 137–75

Pennington, W, 1964 Pollen analysis from the deposits of six upland tarns in the Lake District, *Phil Trans Royal Soc*, B **248**, 205–44

Pennington, W, 1965 Appendix, in Cherry 1965, 82–5

Pennington, W, 1970 Vegetation history in North West England, in *Studies in the vegetational history of the British Isles* (eds D Walker and R G West), Cambridge, 41–80

Pennington, W, 1975 The effect of Neolithic man on the environment in North West England: the use of absolute pollen diagrams, in *The effect of man on the landscape: the Highland Zone* (eds J G Evans, S Limbrey, and H Cleare), CBA Res Rep, **11**, London, 74–86

Pennington, W, 1997 Vegetational history, in Halliday, 1997, 42–50

Phillips, C B, 1977 The Cumbrian iron industry in the seventeenth century, in *Trade and Transport: essays in economic history in honour of T S Willan* (eds W H Chaloner and B M Ratcliffe), Manchester, 34–45

Pigott, C D, and Huntley, J P, 1981 Factors controlling the distribution of *Tilia cordata* at the northern limits of its geographical range III: nature and causes of seed sterility, *New Phytol*, **87**, 817–39

Pigott, C D, and Pigott, M E, 1963 Late-glacial and post-glacial deposits at Malham, Yorkshire, *New Phytol*, **62**, 317–34

Piggott, S, 1954 *Neolithic cultures of the British Isles*, Cambridge

Piggott, S, 1962 Heads and hooves, *Antiquity*, **36**, 110–18

Plantlife, 1992 *Commission of inquiry into peats and peatlands. Commissioner's report: conclusions and recommendations*, London

Potter, T W, 1979 *Romans in North West England*, Cumberland Westmorland Antiq Archaeol Soc, Res ser, **1**, Kendal

Powell, T G E, 1963 Excavations at Skelmore Heads near Ulverston, 1957 and 1959, *Trans Cumberland Westmorland Antiq Archaeol Soc*, n ser, **63**, 1–30

Pratt, K E, 1996 *Development of methods for investigating settlement and land use using pollen data: a case study from north-east England,* circa *8000 cal BC–AD 500*, unpubl thesis, Univ Durham

Pryor, F, 1982 *Fengate*, Aylesbury

Quartermaine, J, 1989 Interim results of survey work on Stockdale Moor and Town Bank, West Cumbria, *Trans Cumberland Westmorland Antiq Archaeol Soc*, n ser, **89**, 25–31

Quartermaine, J, forthcoming *The archaeological landscapes of Haweswater and Thirlmere: a study of two reservoir catchments*

Quartermaine, J, and Leech, R H, forthcoming *The later prehistory of the Lake District, results of a recent surveys*

Ratcliffe, D A, and Oswald, P H (eds), 1988 *The flow country*, Peterborough

Ratcliffe, D A, 1997 Vegetation, in Halliday, 1997, 51–63

Richmond, I A, and Gillam, J P, 1952 Milecastle 79 (Solway), in Report of the Cumberland Excavation Committee for 1947–1949, *Trans Cumberland Westmorland Antiq Archaeol Soc*, n ser, **52**, 17–40

Roberts, B K, Turner, J, and Ward, P F, 1973 Recent forest history an ald use in Weardale, Northern England, in *Quaternary Plant Ecology* (eds H J B Birks, and R G West), Oxford, 207–20

Robinson, H, 1985 Some recent finds of stone axes in Furness, *Trans Cumberland Westmorland Antiq Archaeol Soc*, n ser, **85**, 35–42

Rodwell, J S, 1991 *British plant communities:* **2**, *mires and heaths*, Cambridge

Roe, F E S, 1979 Typology of stone implements with shaft-holes, in *Stone Axe Studies* (eds T H McK Clough and W A Cummins), CBA Res Rep, **23**, London, 23–48

Rollinson, W, 1967 *A history of man in the Lake District*, Dent

Rollinson, W, 1974 *Life and tradition in the Lake District*, Clapham

Rollinson, W, 1981 *Life and tradition in the Lake District*, 2nd edn, Clapham

Ross, A, 1974 *Pagan Celtic Britain*, London

Rowell, T K, and Turner, J, 1985 Litho-, humic and pollen stratigraphy at Quick Moss, Northumberland, *J Ecol*, **73**, 11–25

RSNC, 1990 *The peat report*, Lincoln

RSPB, 1993 *Out of the mire*, Sandy

RSPB, 1997 *The wet grassland guide*, Sandy

Ryder, M L, 1983 *Sheep and man*, London

Ryder, M L, 1990 The Kendal Pattern book, *Textile Hist*, **21**, 223–43

Ryder, M L, 1994a *Report on hair on ox skull from Solway Moss*, Engl Her AML Rep, **23/94**, unpubl rep

Ryder, M L, 1994b *The fleece of the Solway sheep*, unpubl rep

Salisbury, C R, 1986 Comments on Kirkhead Cave, an Upper Palaeolithic site in northern England, *Proc Prehist Soc*, **52**, 321–3

Salisbury, C R, 1988 Late Upper Palaeolithic artefacts from Lindale Low Caves, Cumbria, *Antiquity*, **62**, 510–3

Salisbury, C R, 1991 Late Pleistocene human occupation of the Morecambe Bay area, *Contrebis*, **15**, 15–23

Salisbury, C R, and Coupe, J, 1995 The Mesolithic occupation of Mill Dam Meadow, Gleaston, Cumbria, *Contrebis*, **20**, 4–8

Salisbury, C R, and Sheppard, D, 1994 The Mesolithic occupation of Heysham Head, Lancashire, *Trans Lancashire Cheshire Antiq Soc*, **87**, 141–9

Satchell, J E, 1984 A history of Meathop Woods: part II – the Middle Ages to the present, *Trans Cumberland Westmorland Antiq Archaeol Soc*, n ser, **84**, 84–98

Schubert, H R, 1957 *A history of the British iron and steel industry* c *450 BC –AD 1775*, London

Shotter, D A, 1990 *Roman coins from North–West England*, Lancaster

Shotter, D A, 1993 *Romans and Britons in North–West England*, Lancaster

Shotter, D A, 1994 Rome and the Brigantes: early hostilities, *Trans Cumberland Westmorland Antiq Archaeol Soc*, n ser, **94**, 21–34

Shotter, D A, 1997 *Romans and Britons in North–West England*, Lancaster

Shotton, F W, and Williams, R E G, 1971 Birmingham University radiocarbon dates, *Radiocarbon*, **12**, 385–99

Simmons, I G, 1996 *The environmental impact of later Mesolithic cultures*, Edinburgh

Simmons, I G, and Innes, J B, 1987 Mid-Holocene adaptations and later Mesolithic disturbance in northern England, *J Archaeol Sci*, **14**, 1–20

Smith, A G, 1958 Two lacustrine deposits in the south of the English Lake District, *New Phytol*, **57**, 363–86

Smith, A G, 1959 The mires of south–western Westmorland, *New Phytol*, **58**, 105–27

Smith, B, 1912 The glaciation of the Black Combe district, *Quart J Geol Soc London*, **68**, 402–48

Smith, R A, 1907 A bronze fragment of late-Keltic [*sic*] engraving, *Trans Cumberland Westmorland Antiq Archaeol Soc*, n ser, **7**, 95–9

Sobee, F J, 1953 *The history of Pilling*, Exeter

Spence, J E, 1937 Report of the committee for prehistoric studies, 1936, *Trans Cumberland Westmorland Antiq Archaeol Soc*, n ser, **37**, 100–3

Stallibrass, S M, 1986 *Some taphonomic effects of scavenging canids on the bones of ungulate species. Some actualistic research and a Romano-British case study*, unpubl thesis, Univ Sheffield

Stallibrass, S M, 1993 *Animal bones from excavations in the southern area of The Lanes, Carlisle, Cumbria, 1981–1982*, Engl Her AML Rep, **96/93**, unpubl rep

Stallibrass, S M, 1996 Sacred cows from Solway: bog–bodies on the Anglo–Scottish border, in *Ritual treatment of human and animal remains* (eds S Anderson and K Boyle), Oxford, 35–43

Stead, I M, Bourke, J B, and Brothwell, D (eds), 1986 *Lindow Man*, London

Steer, T P, 1991 *Landscape change in the Lyth Valley, 1770 to the twentieth century*, unpubl thesis, Univ Lancaster

Stockdale, J, 1872 *Annales Caermoelenses*, rev edn 1978, Ulverston

Stockmarr, J, 1971 Tablets with spores used in absolute pollen analysis, *Pollen et Spores*, **13**, 615–21

Stoneman, R, 1993 *Holocene palaeoclimates from peat stratigraphy: extending and refining the model*, unpubl thesis, Univ Southampton

Stuiver, M, Kromer, B, Becker, B, and Ferguson, C W, 1986 Radiocarbon age calibration back to 13,300 years BP and the ^{14}C age-matching of the German oak and US bristlecone pine chronologies, *Radiocarbon*, **28**, 969–79

Stuiver, M, and Pearson, G W, 1986 High precision of the radiocarbon time scale, AD 1950–500 BC, *Radiocarbon*, **28**, 805–38

Stuiver, M, and Reimer, P J, 1986 A computer program for radiocarbon age calibration, *Radiocarbon*, **28**, 1022–30

Stuiver, M, and Reimer, P J, 1993 Extended ^{14}C data base and revised Calib 3.0 ^{14}C calibration program, *Radiocarbon*, **35**, 1, 215–30

Switzur, V R, 1986 884BP and all that! *Antiquity*, **61**, 241–6

Symeonis Dunelmensis, 1868 *Opera et collectanea*, **1**, Surtees Soc, **51**, Durham

Taylor, J J, Innes, J B, and Jones, M D H, 1994 Locating prehistoric wetlands sites by an integrated palaeo-environmental / geophysical survey strategy at Little Hawes Water, Lancs, in *Whither environmental archaeology* (ed R Luff and P Rowley-Conwy), Oxbow Monog, **38**, Oxford, 13–23

Taylor, J J, Jones, M D H, Innes, J B, and Oldfield, F, 1987 *The Little Hawes Water Project, Preliminary Report*, Univ Liverpool, unpubl doc

Tipping, R, 1986 The stratigraphy of Kirkhead Cave: an Upper Palaeolithic site in northern England: a comment, *Proc Prehist Soc*, **52**, 323–6

Tipping, R, 1994 Williamson's Moss: Palynological evidence for the Mesolithic–Neolithic *Trans*ition in Cumbria, in *Cumbria field guide* (eds J Boardman and J Walden), Quat Res Assoc, Oxford, 104–27

Tipping, R, 1995a Holocene evolution of a lowland Scottish landscape: Kirkpatrick Fleming. Part I, peat- and pollen-stratigraphic evidence for raised moss development and climatic change, *Holocene*, **5(1)**, 69–81

Tipping, R, 1995b Holocene evolution of a lowland Scottish landscape: Kirkpatrick Fleming. Part II, regional vegetation and land-use change, *Holocene*, **5(1)**, 83–96

Tipping, R, 1995c Holocene evolution of a lowland Scottish landscape: Kirkpatrick Fleming. Part III, fluvial history, *Holocene*, **5(2)**, 184–95

Todd, J M (ed), 1997 *The Lanercost Cartulary*, Cumberland Westmorland Antiq Archaeol Soc, Rec ser, **11**; Surtees Soc, **203**, Gateshead

Tooley, M J, 1978 *Sea-level changes: the coast of north–west England during the Flandrian Stage*, Oxford

Troels–Smith, J, 1955 Characterisation of unconsolidated deposits, *Danm Geol Unders*, **IVR**, 3(101), 1–73

Trotter, F M, and Hollingworth, S E, 1932 The glacial sequence in the north of England, *Geol Mag*, **64**, 374

Turnbull, P, 1995 A lost bronze bucket from Westmorland, *Trans Cumberland Westmorland Arch Antiq Soc*, n ser, **95**, 55–9

Turnbull, P, and Walsh, D, 1996 A Beaker burial in Levens Park, *Trans Cumberland Westmorland Antiq Archaeol Soc*, n ser, **96**, 13–26

Turner, J, 1965 A contribution to the history of forest clearance, *Proc Royal Soc*, B **161**, 343–54

Turner, J, 1979 The environment of northwest England during Roman times as shown by pollen analysis, *J Archaeol Sci*, **6**, 285–90

Turner, R C, 1988 A Cumbrian bog body from Scaleby, *Trans Cumberland Westmorland Antiq Archaeol Soc*, n ser, **88**, 1–7

Turner, R C, 1989 Another Cumbrian bog body, found in Seascale Moss in 1834, *Trans Cumberland Westmorland Antiq Archaeol Soc*, n ser, **89**, 21–3

Turner, R C, and Scaife, R G, 1995 *Bog bodies, new discoveries and new perspectives*, London

UK Biodiversity Steering Group, 1995 *Biodiversity: the UK Steering Group Report, 2: action plans*, London

University of Washington Quaternary Isotope Laboratory, 1987 *CALIB Manual*, Washington

Van de Noort, R, and Etté, J, 1995 The Humber Wetlands Survey: background, objectives, and methodology, in *Wetland heritage of Holderness: an archaeological survey* (eds R Van de Noort and S Ellis), Hull, 357–60

Van der Sanden, W, 1996 *Through nature to eternity*, Amsterdam

Vincent, P J, and Lee, M P, 1981 Some observations on the loess around Morecambe Bay, North–west England, *Proc Yorkshire Geol Soc*, **43**, 281–94

Vuorela, I, 1973 Relative pollen rain around cultivated fields, *Acta Bot Fennica*, **102**, 1–27

Vuorela, I, 1983 Field erosion by wind as indicated by fluctuations in the ash content of *Sphagnum* peat, *Geol Soc Finland*, **55(1)**, 25–33

Walker, D, 1956 A Late-glacial deposit at St. Bees, Cumberland, *Quart J Geol Soc London*, **112**, 93–101

Walker, D, 1966 The late Quaternary history of the Cumberland lowland, *Phil Trans Royal Soc*, B **251**, 1–120

Walker, M J C, Coope, G R, and Lowe, J J, 1993 The Devensian (Weichselian) Late glacial palaeoenvironmental record from Gransmoor, East Yorkshire, England, *Quat Sci Rev*, **12**, 659–80

Walton Rogers, P, 1994 *Tests for colorant in samples of fibre from a cattle skin from Solway Moss, Cumbria*, Textile Research in Archaeology, York, unpubl rep

Walton Rogers, P, and Taylor, G, 1991 The characterisation of dyes in archaeological textiles, *Chromatography and analysis*, **17**, 5–7

Ward, G K, and Wilson, S R, 1978 Procedures for comparing and combining radiocarbon age determinations: a critique, *Archaeolmetry*, **20**, 19–31

Waterhouse, J, 1985 *The stone circles of Cumbria*, Chichester

Watson, W H, 1903 Stone implements found at Braystones, Cumberland, with remarks on probable Neolithic settlements in the neighbourhood, *Trans Cumberland Westmorland Antiq Archaeol soc*, n ser, **3**, 91–3

Watts, W A, and Winter, T C, 1966 Plant macrofossils from Kirchner Marsh, Minnesota: a palaeoecological study, *Geol Soc Amer Bull*, **77**, 1339–60

Webster, R A, 1969 *The Romano-British settlement in Westmorland — a study in cultural ecology*, unpubl thesis, Univ Reading

Wells, C, 1991 The environmental history of part of South cumbria (High Furness, Duddon and Lyth Valleys) from the Early Bronze Age to Historic times, in Middleton 1991a, 38–9

Wells, C, 1992 Preliminary stratigraphic investigations at Solway Moss, in Middleton 1992a, 26–7

Wells, C, Huckerby E, and Hall, V, 1997 Mid- and late-Holocene vegetation history and tephra studies at Fenton Cottage, Lancashire, UK, *Vegetation History and Archaeobotany*, **6(3)**, 153–66

Wells, J M, Mighall, T M, Smith, D E, and Dawson, A G, 1999 Brighouse Bay, south-west Scotland: Holocene vegetational history and human impact at a small coastal valley mire, in *Late Cenozoic environments and Hominid evolution: a tribute to Bill Bishop* (eds P Andrews and P Barker), Geol Soc, London, 217–233

West, T, 1805 *The antiquities of Furness* (rev edn W Close), Ulverston

Whellan, W, 1850 *The history and topography of the counties of Cumberland and Westmorland*, Pontefract

Wild, C, Wells, C, Anderson, D, Boardman, J, Parker, A, and Parker, I forthcoming Recent work at Seathwaite, Borrowdale, *Trans Cumberland Westmorland Arch Antiq Soc*

Williams, D J, Richardson, J A, and Richardson, R S, 1987 Mesolithic sites at Malham Tarn and Great Close Mire, North Yorkshire, *Proc Prehist Soc*, **53**, 362–83

Willis, K H, Sümegi, P, Braun, M, Bennett, K D and Tóth, A, 1998 Prehistoric land degradation in Hungary; Who, how and why? *Antiquity*, **72,** 101–13

Wilmott, T 1997 *Birdoswald; Excavations of a Roman fort on Hadrian's Wall and its successor settlements: 1987–92*, Eng Her Archaeol Rep, **14**, London

Wilmott, T, in prep Hadrian's Wall from Burgh-by-Sands to Moricambe: a view, *Trans Cumberland Westmorland Antiq Archaeol Soc*

Wilson, J (ed), 1915 *The Register of the Priory of St Bees*, Surtees Soc, **194**, Durham and London

Wiltshire, P E J, 1992 *Interim palynological report from Birdoswald Roman fort and Appletree turf wall*, Engl Her AML Rep, **3/92**, unpubl rep

Wimble, G A, 1986 *The palaeoecology of Lowland Coastal Raised Mires of South Cumbria*, unpubl thesis, Univ Wales

Winchester, A J L, 1987 *Landscape and society in medieval Cumbria*, Edinburgh

Winchester, A J L (ed), 1994 *The diary of Isaac Fletcher of Underwood, Cumberland 1756–1781*, Cumberland Westmorland Antiq Archaeol Soc, Extra Ser, **27**, Kendal

Wood, M G, 1968 *Notes on the reclamation of the Lyth Mosses and the social and economic effects*, CRO(K), unpubl doc

Wooliscroft, D S, 1994 Signalling and the design of the Cumberland Coast System, *Trans Cumberland Westmorland Antiq Archaeol Soc*, n ser, **94**, 55–64

Wymer, J J (ed), 1977 *Gazetteer of Mesolithic sites in England and Wales*, CBA Res Rep, **20**, Norwich

Wymer, J J, 1981 The Palaeolithic, in *The Environment in British Prehistory* (eds I Simmons and M J Tooley), London, 49–81

Zong, Y, 1997 Mid-Late Holocene sea level changes in Roudsea Marsh, north west England, *Holocene*, **7**, 309–21

Zvelebil, M, and Rowley-Conwy, P, 1986 Foragers and farmers in Atlantic Europe, in *Hunters in Transition* (ed M Zvelebil), Cambridge, 67–94

Index

Note: References in *italic* denote Illustrations and Plates. There may also be textual references on these pages.